W9-DFK-001

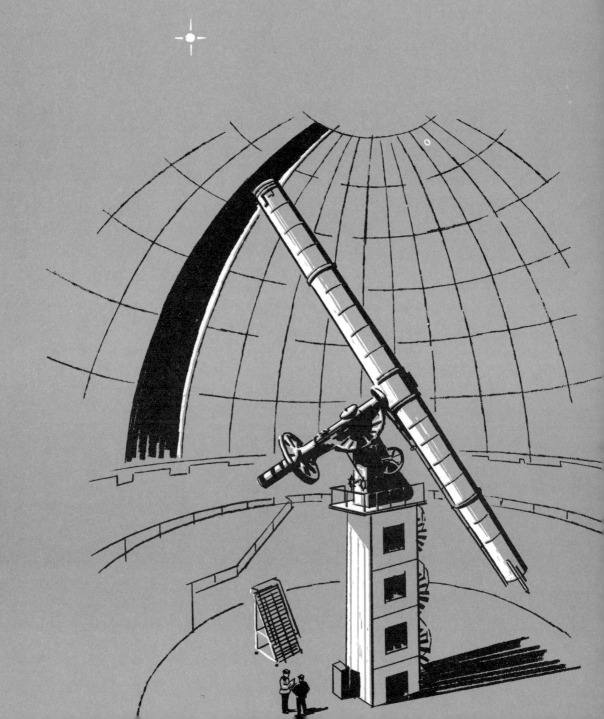

Bethlehem Steel Co.

Working the final piece of fabricated structural steel into place atop a new bank building in New York City.

THE BOOK OF
POPULAR
SCIENCE

Volume 8

Distributed in the United States by
THE GROLIER SOCIETY INC.

Distributed in Canada by
GROLIER LIMITED

Cover photograph: night view
of the large oil refinery at
Lake Charles, in Louisiana.

Anthony Linck—Cities Service

COPYRIGHT © 1971 BY  Grolier
INCORPORATED

COPYRIGHT © 1971 BY GROLIER LIMITED

ISBN 0-7172-1202-5
Library of Congress Catalog Card Number: 79-118653

Copyright © 1971, 1970, 1969, 1968, 1967, 1966, 1965, 1964, 1963, 1962, 1961, 1960,
1959, 1958, 1957, 1956, 1955, 1954, 1953, 1952, 1951, 1950, 1947, 1946,
1945, 1943, 1941, 1939 by GROLIER INCORPORATED

Copyright © 1931, 1930, 1929, 1928, 1926, 1924 by THE GROLIER SOCIETY

Copyright © in Canada 1971, 1970, 1969, 1968 by GROLIER LIMITED; 1967, 1966, 1965,
1964, 1963, 1962, 1961, 1960, 1959, 1958, 1957, 1956, 1955, 1954, 1953,
1952, 1951, 1950 by GROLIER OF CANADA LIMITED

No part of THE BOOK OF POPULAR SCIENCE may be reproduced without special permission in writing from the publishers

PRINTED IN U.S.A.

CONTENTS OF VOLUME VIII

EXPERIMENTS WITH WATER

The Most Familiar of Chemical Compounds

BY M. F. VESSEL

LIKE air, water is found almost everywhere. It is familiar to us in different forms — as drinking water, rain, water vapor, ice and snow; it is found in vast quantities in oceans, lakes and rivers. Living creatures, including man, consist largely of water. We would suffer seriously from lack of this precious liquid before we suffered from lack of food.

Experiments with frozen water

Water has some surprising properties. For example, when most liquids freeze, they become denser; water becomes lighter. If you fill a pan full of cracked ice, you will note that as the melting process goes on, the unmelted ice particles will remain at the surface of the liquid that is forming. This is because frozen water is lighter than water in liquid form.

Water expands as it becomes ice. As it expands it exerts pressure, as the following experiment will show. Provide yourself with a glass container with a screw cap. Fill the container full of water and then screw the cap on as tightly as you can. Wrap the container in a cloth and place it in the freezing compartment of your refrigerator. Examine it after a few hours. You will find that the water has changed to ice and that the expanding ice has broken the bottle. Water can crack water pipes and automobile engine jackets when it freezes. The damage that has been done becomes

All photos, Kellner Associates—Dick Krueger

The little girl is being raised by the hydraulic lift pictured in Figure 3 and described in the text on that page.

evident only when the ice melts, and water begins to make its way through the cracks. When plants freeze, the water they contain forms large crystals; these rupture the walls of the plant cells.

Set an ice cube on top of a soda bottle. Place a piece of fine wire on top of the cube; attach several large nails to either end of the wire, as shown in Figure 1. Grasp the nails in your fingers and press down on the ice cube. The pressure will make the ice melt, since it will cause the freezing point to be lowered. The wire will begin cutting its way through the melting ice; in time it will go completely through the ice cube. But you will not have two pieces of ice; the cube will be intact.

The reason is that as the wire forces its way through the ice cube, the melted

1

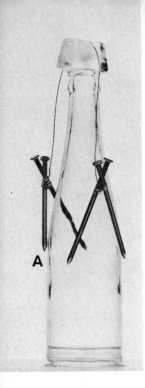

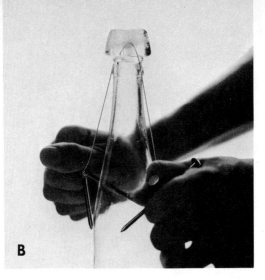

A

B

1. An ice cube is set on top of a soda bottle. A piece of fine wire is then placed on top of the cube, and several large nails are attached to either end of the wire, as shown in A. If you grasp the nails in your fingers and press down on the ice cube, the wire will keep cutting its way through the cube, which will remain in one piece (B).

water that gathers at the top of the wire freezes again. That is why the cube will be solid after the wire has passed through it. Usually enough air bubbles will be trapped at the place where the cube has been cut to leave a distinct line. If you tap the cube at this line, you will see that the two halves have frozen solidly together. The cube breaks no more easily along the cut (where the wire has passed through it) than it does in any other section.

You are undoubtedly familiar with other pressure effects upon frozen water. You press frozen water in the form of snow when you make a snowball. The pressure melts the snow, and the newly formed liquid freezes again; if you squeeze long enough you will have a ball of ice. When you skate on ice, the pressure of the ice-skate blades on the ice melts the part of it directly under your skates; hence you really skate on a thin film of water.

The incompressibility of water: Pascal's law

An important property of water in liquid form is that, like other liquids, it is very nearly incompressible. Because of this fact, pressure applied to any part of a liquid in a closed vessel is transmitted throughout the vessel with equal force. This is called Pascal's law, because it was first discovered by the seventeenth-century French scientist and mathematician Blaise Pascal. The operation of Pascal's law and its applications are described in detail in the article Forces within Liquids and Gases, in Volume 2.

Devices based on Pascal's law

Pascal's law is applied in the toy called the Cartesian diver (Figure 2), which will seem very mysterious unless one knows what makes it work. Here is how you can make your own Cartesian diver. Get a gallon glass jug, a medicine dropper and a tall drinking glass. Fill the jug and the glass with water of the same temperature. Draw up water into the medicine dropper from the glass until it holds just enough to allow the dropper to float in the glass with its tip barely touching the top. Carefully transfer the dropper to the gallon jug without losing any water from the dropper. The dropper will now just float in the water at the top of the jug.

Press the fleshy part of your hand on the open mouth of the jug. The dropper will go down. When you ease up on the pressure, the dropper will rise. It will go down again if you press down with your palm as before. Thus you will be able to make the dropper rise or fall at will.

How does the Cartesian diver work? Remember that pressure on the water at the top of the jug will be felt equally throughout the liquid. Water will be pushed up into the open end of the medicine dropper even though the pressure you apply is in a downward direction. The water entering

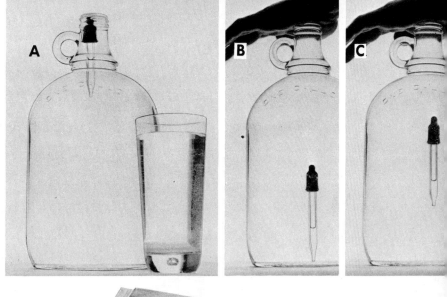

2. A Cartesian diver. First the diver floats in the water in the jug with its tip at the surface (A). If you press your palm on the open mouth of the jug, the diver will go down (B). But as the pressure of your palm is eased, the diver will ascend (C).

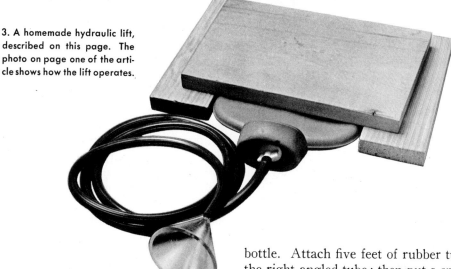

3. A homemade hydraulic lift, described on this page. The photo on page one of the article shows how the lift operates.

the dropper will compress the air within it. Since the dropper just barely floats at the start, the added water will make it heavier and it will sink. If the pressure is lessened, the compressed air in the dropper will push some of the water out. The dropper will therefore become lighter and it will rise. If the dropper is carefully adjusted, it will sink when only a very small quantity of water is forced into it.

The force transmitted by a thin column of water will be strong enough to lift you up, as the next experiment will show (Figure 3). First lay a hot-water bottle on the floor. Place two thin boards to the left and right of it, as shown in Figure 3. Set a third board over these so that it rests on the hot-water bottle. Put a right-angled glass tube through a one-hole stopper set in the

bottle. Attach five feet of rubber tubing to the right-angled tube; then put a small funnel in place at the other end of the tubing. Be sure that the stopper is firmly in place in the hot-water bottle and then stand on the board. Have somebody raise the rubber tubing as high as it will go and then have him pour water through the funnel into the hot-water bottle. Soon you will find that the water pouring into the bottle will lift you up. A column of water five feet high, such as you would have with the five-foot tubing, will support a hundred-pound person. To support persons weighing more than this, a greater length of tubing should be used.

The hydraulic lift that raises a car in a garage works in much the same way as the apparatus we have just described. Air pressure supplied by a compressor is used to push oil under a large cylinder. The cylinder then rises and lifts the car.

3

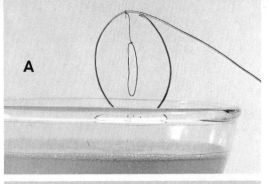

A

B

C

4. Above: a striking effect of surface tension. A loop of thread is attached to a wire loop (A). The wire loop is dipped into a soapy solution. When removed from the solution, it will be filled with a soapy film, to which the thread loop will adhere (B). Note what happens when you touch the film within the thread loop with a slightly heated needle point (C).

Attracting forces — cohesion and adhesion

Watch small insects skating over the surface of a pond; you will note that their feet make depressions in the water. The surface of the water is tough because water molecules have a strong attraction for each other. This force of attraction is called cohesion. The tendency of the surface to act as a tough elastic membrane is called surface tension.

To illustrate an interesting effect of surface tension, make a loop about two inches in diameter out of the end of a piece of stiff wire, as shown in Figure 4. Tie a silk thread to the top part of the wire loop and arrange the bottom part of the thread to form another loop, as indicated in the drawing. Prepare a soap solution and dip the wire loop in it. When you remove the loop, it will be filled with a soapy film, and the loop of thread will be seen adhering to the film, forming an irregular outline. Now touch the film at any place within the loop of thread with a needle point that you have passed over a gas flame or other heat source. The film within the thread loop will suddenly disappear, though the rest of the film will remain intact. The loop will now be perfectly circular in form. When the film within the loop of thread disappeared, surface tension caused the rest of the film to be drawn together, causing the loop to form a circle.

Fill a glass brimful of water, being careful not to spill any. Slowly drop some

5. Some paper clips are dropped into a glass brimful of water (A). The surface of the water will rise, forming a heaped-up mass (B).

A

B

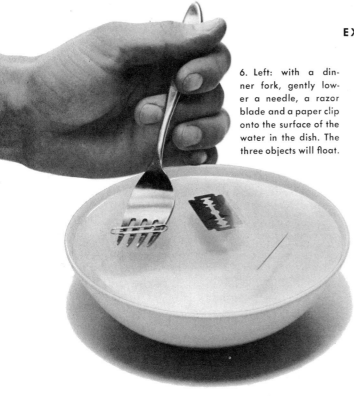

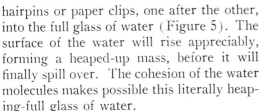

6. Left: with a dinner fork, gently lower a needle, a razor blade and a paper clip onto the surface of the water in the dish. The three objects will float.

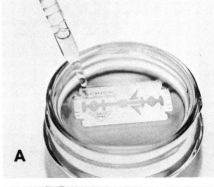

A

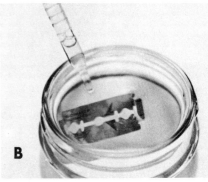

B

7. The razor blade shown in A, above, floats on the surface of the water. But if detergent is added to the water, the blade will sink (B).

hairpins or paper clips, one after the other, into the full glass of water (Figure 5). The surface of the water will rise appreciably, forming a heaped-up mass, before it will finally spill over. The cohesion of the water molecules makes possible this literally heaping-full glass of water.

Provide yourself with a needle, paper clip and razor blade. Rub these objects with your fingers to give them a slight coating of oil from your hands. Fill a dish with cold water. Using a dinner fork, gently lower the needle, clip and razor blade onto the surface of the water (Figure 6). If you are careful, you can make them float, because of the cohesive force of the water molecules. Next, fill a dish with boiling-hot water and try to float the three articles on it. You will find it difficult, since heat weakens the cohesive force of the molecules. You will have difficulty if you try to float your three objects in a soapy solution. In this case, too, the surface film of the liquid has been weakened.

The attraction of molecules of different substances for one another is called adhesion. Since soaps and synthetic detergents increase the adhesive force between water and other objects, they are called wetting agents. These agents also decrease the cohesive force of water, as an interesting experiment will show. Put a little soap or synthetic detergent in a jar of water on which an insect such as a water strider (see Index) is skating about. Almost immediately the little creature will break through the water film and it will drown unless it is rescued. If water striders or similar insects are not available, you can carry out this experiment with a razor blade. It will float on the surface of the water until detergent is added to the water. Then the blade will sink (Figure 7).

Try to float a needle, razor blade and paper clip on the surface of rubbing alcohol, carbon tetrachloride and kerosene. You will find that the skin of these liquids is not nearly so tough as that of water. In other words, the cohesive force of their molecules is not so great.

Moisten two pocket mirrors about three inches by two inches. Put the mirrors together and try to pull them apart without sliding them sideways (Figure 8). It will

8. If you moisten two pocket mirrors and pla e them together as shown, it will be almost impossible to pull them apart without sliding them sideways.

be almost impossible. But if you dry the mirrors, it will not be at all difficult to separate them. Two forces acting upon the wet glass plates tend to keep them together: the cohesive action of the water particles attracted to one another, and the adhesive action between the water particles and the glass.

When capillary action is at work

Sometimes the forces of cohesion and adhesion oppose each other; this brings about the effect known as capillarity. Fill a container partly full of water and place a very thin open glass tube in it (Figure 9). You will note that the water rises in the tube. What has happened? You set two forces in play when you put the glass tube in water: adhesion and cohesion. The force of adhesion between the water and the tube attracts the water to the sides of the tube and causes the surface of the water to become concave as shown. However, surface tension, caused by cohesion, causes the surface to contract, becoming flat. Again water molecules will adhere to the glass, producing a concave surface; again cohesion will flatten out the surface. As this process continues, water will rise in the tube until the weight of the column of liquid will counteract the forces that tend to make the liquid rise. The thinner the tube, the higher the column of water will go, as you will see if you put glass tubes of different diameters in the water.

Put some mercury in a test tube and then attach the tube by means of a clamp to a ring stand, as shown in Figure 10. If you put a thin tube in the mercury, pressing it against the wall of the larger tube, you will note that the level of mercury in the thin tube will be lower than the level in the bigger tube. In this case, the force of cohesion between the molecules of the liquid is greater than the force of adhesion between the liquid and the glass. The

9. Place a thin open glass tube in a container partly full of water. Water will rise in the tube.

10. Put some mercury in a test tube; attach the tube by a clamp to a ring stand (A). Set a thin open tube against the side of the upper tube. The mercury level in the thin tube will be lower than the level in the bigger tube (B).

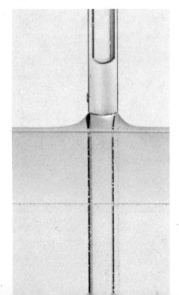

A

B

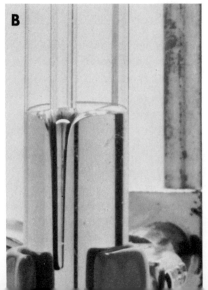

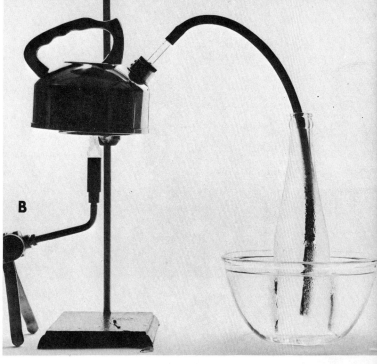

A B

surface therefore will become convex, as shown. Adhesion will tend to flatten out the surface; cohesion will make it convex again, and so on. In this case the liquid in the tube will fall until upward pressure upon the liquid in the tube counteracts the forces that tend to make the mercury fall in the tube.

Capillary action is at work when water rises in the soil, kerosene in a lamp wick and ink in blotting paper. If there is excessive capillary action in the soil, water will be drawn up to the surface, where it will continually evaporate. If the soil near the surface is loosened by cultivation, the pores become too big for capillary action to take place; as a result, evaporation at the surface is reduced, though water will still be available to the roots. This is called dry farming.

Distilled water and hard water

The pure water that is poured into the storage battery of a car or into a steam iron is called distilled water; it has been purified through the process of distillation. You can prepare a simple distilling device, as shown in Figure 11. Fill a teakettle with muddy water. Fit a one-hole stopper into the spout of the kettle and attach a rubber tube to it. Lead the tube into a milk bottle that has been previously set

11. A homemade distilling device. A teakettle is filled with muddy water (A). A one-hole stopper is fitted into the spout of the kettle and one end of a rubber tube is attached to the stopper. The other end of the tube is led into a milk bottle set in a pan containing cold water (B). The water in the kettle is heated to the boiling point. Steam passes from the rubber tube into the bottle and condenses there. The water that collects at the bottom of the bottle is pure water (C).

C

in a saucepan containing cold water. Heat the water in the kettle to the boiling point; the steam will pass out of the rubber tube into the milk bottle. Here it will condense as it comes in contact with the chilled sides of the bottle. The water that collects in the bottom of the bottle will be pure water. The impurities the water formerly contained will be left behind in the teakettle.

Rainwater is almost pure water, for it has passed through a somewhat similar process. In this case surface water (also water derived from other sources, such as plants) evaporates, passes into the at-

12. The glass tumbler in photograph A, above, contains distilled water; the tumblers in B, C and D contain hard tap water. Note what happens when powdered soap is added to the distilled water in A and to the hard water in B; when borax is added to the hard water in C; when synthetic detergent is added to the hard water in D.

iron; such water is called hard. Drinking water that is pumped into homes from underground sources is usually hard, unless it has been specially treated. This becomes evident when one tries to wash one's hands and discovers that it takes a good deal of soap to make a lather.

To tell whether the water you use is hard, examine your teakettle. If the water is hard, the kettle will be encrusted with a grayish-white, chalklike substance, sometimes called fur. This is the same material that accumulates in pipes and hot-water boilers, where it is called boiler scale; it is made up largely of the salts of calcium and magnesium.

To examine the effects of hard water, set up four drinking glasses (Figure 12). Put rainwater or distilled water in one glass and hard tap water in the rest; each glass is to be about half full. If your tap water is soft, you can make it hard by adding about a teaspoonful of plaster of Paris and some Epsom salts to a quart of water.

Pour a little powdered soap into the glass containing the rainwater or distilled water. Note how little of the soap is necessary to form suds. Put some powdered soap into one of the other glasses and stir it. A curd will be formed as the soap combines with the minerals in the water. If you were washing clothes, these curds would remain in your clothing, giving them a gray appearance even though they had been thoroughly washed.

Place about half a teaspoonful of borax or washing soda in one of the other glasses and then add the powdered soap. You will need less soap now because the borax or washing soda got rid of most of the mineral salts. Test the water in the last glass with one of the synthetic detergents that are so familiar nowadays. Note how quickly the water lathers whether the water is hard or not. The synthetic detergents do not combine with the mineral salts in the water; therefore no curd is formed. These detergents are very effective wetting agents; they make it possible for water to cling to the dirt particles, which are then washed out of a material.

See also Vol. 10, p. 289: "Physics, Experiments in."

mosphere and becomes the gas called water vapor. As water vapor comes in contact with colder layers of air, it condenses in the form of clouds. The rain that is produced when clouds yield up their moisture is almost pure water; the dust particles in the air and various gases are the only contaminating substances.

When water seeps through the soil, it dissolves mineral salts such as common salt and compounds of calcium, magnesium and

THE WORK OF RUNNING WATER

How Streams and Rivers Carve the Land

BY NORMAN E. A. HINDS

RUNNING water is one of the most important agents eroding the surface of the earth. During and after rainstorms and the melting of snow and ice, water may run downhill in a sheet over the land, forming what is known as a sheet flood. Great quantities of water also circulate either temporarily or permanently in the more concentrated flow of streams and rivers. Sheet floods wash immense quantities of rock fragments down slopes; streams transport even larger quantities as they carve out their valleys. As a result of all this activity, a great variety of landscapes are created.

Other articles of THE BOOK OF POPULAR SCIENCE discuss the conditions that bring about precipitation and show how clouds release their load of moisture in the form of rain, snow, hail, sleet and mist. Part of the water that falls on the earth's surface goes into streams, rivers, lakes and oceans; this is called runoff. Part (flyoff) is evaporated back into the atmosphere or is transpired by plants. (See Index, under Transpiration.) Part (sinkin) is added to the underground supply of water by sinking into the myriads of open spaces in rocks and in rock mantle — the rock debris that covers bedrock. Part of the precipitation is locked up in fields of ice and glaciers — in some cases, for many thousands of years.

As far as we know, rain or snow falls on all parts of the earth's surface, but the amount and frequency vary widely. At

Both photos on this page, USDA

When plant cover is inadequate or entirely absent, sheet floods form gulleys, which are usually V-shaped trenches and which may be several hundred feet deep. Above is shown the Rutledge Gulley, Stewart County, Georgia.

This butte, in Colorado, represents a block which is made up of resistant strata of rock. The butte remained standing after much of the surrounding area had been worn away by the combined agencies of running water and wind.

Cairo, Egypt, the annual rainfall averages 1.2 inches and at Yuma, Arizona, 3.4 inches. Over parts of the Atacama Desert, in Chile, there may not be any rain for a period of from 5 to 10 years. On the other hand, in the more populous sections of the earth, annual precipitation ranges from less than 20 to more than 60 inches. Within and near the tropical zones, annual rainfall of over 900 inches has been measured.

The rise of sheet floods

Much damage to the surface of the earth may be wrought by catastrophic flooding: that is, by unusually heavy falls of rain causing extensive sheet floods. If little or no water is evaporated or absorbed by the ground, immense quantities of water pour down along the surfaces of slopes.* Sheet floods may also occur when sudden rises in temperature in winter and spring quickly melt snow and ice. Such floods are particularly devastating when quick melting is accompanied by heavy rains.

The erosive effects of sheet floods are determined in large part by the amount of water flowing down a slope, by the steepness of the slope and by the velocity of the flood. The resistance of the materials over which the flood travels also affects its erosive action; this action is particularly great if the sheet of water attacks weak rocks or rock mantle. Sheet floods are not nearly so erosive if the plant cover is reasonably dense. The impact of falling rain is received chiefly by the plants, through which and down which the water drips or trickles to the ground. The living plants and the litter of dead plants on the ground divide the sheet flood as it moves down slopes, thus lessening its force. Roots, especially the intricate root systems of the smaller plants, bind the soil particles together and make them resistant to the onrush of the flood.

How sheet floods sculpture the landscape

In all regions, sheet floods are sculptors, shaping the land by washing loose rock fragments down slopes and eventually into streams that carry them away. In arid lands, especially in deserts, the erosive effect of falling rain and sheet floods is most conspicuous because vegetation is scanty or entirely absent and because rainstorms are frequently violent. The floods also cause great damage in humid regions where the plant cover has been partially or completely removed by natural causes or by man. Wind erosion often works together with sheet floods in producing landscape changes.

Where plant cover is inadequate or absent, sheet floods form gullies, which are generally narrow V-shaped trenches. As gullies are eroded, sediment is swept into

* Such sheet floods may also lead to stream flooding if they fill watercourses beyond normal capacity. Sudden disastrous flows of water are generally called flash floods. We discuss stream flooding later in the article.

Union Pacific Railroad

Pyramidlike mass in the Grand Canyon. This magnificent canyon was carved out in the course of the centuries by the Colorado River, flowing through the vast Colorado Plateau. Certain strata resisted the force of the stream more than others; this accounts for the wide variety of rock masses in the above photograph.

streams, dry valley bottoms or low areas beside slopes. Starting as shallow furrows, the gullies may become tens, scores or even hundreds of feet deep.

The chief type of landscape developed primarily by sheet floods is known as the badlands; it is a labyrinth of deep, mostly narrow gullies, separated by ridges with rounded or narrow crests. Badlands are most likely to evolve in arid or semiarid regions, such as the western Dakotas and central Wyoming. Those that develop in more humid areas often represent a serious problem, since valuable or potentially valuable farm or pasture land becomes entirely unproductive.

If sheet-flood erosion takes place along closely spaced fractures in weak rock or rock mantle, isolated pillars of various sizes are formed. Pillars are also developed in areas where weak rock or rock mantle is covered at intervals by boulders or flat slabs. These protect the weak materials beneath them from erosion, while the intervening areas are etched out. At first the pillars retain their capstones, but eventually these fall, leaving more or less sharp-pointed pinnacles. Continued erosion causes these pillars to collapse.

In most rock masses, some sections are less resistant to erosion than others. On exposed faces, the weaker or more weathered parts are eroded by rainwash; this erosion gives rise to etched or honeycombed cliffs or slopes and to cavernous openings in rock walls. Occasionally natural arches are formed.

If masses of resistant rock are surrounded by weaker rock formations, the latter may be so washed away that the resistant materials will stand out as eminences above the general surface. Ship Rock Peak, in northern New Mexico, is a formation of this kind; it is the neck of an ancient volcano and is surrounded chiefly by soft clays. Another good example is the Devil's Tower, in northern Wyoming.

Falling rain and sheet floods play an important part in modifying the surface of arid regions where there are horizontal or nearly horizontal strata. If these strata vary in their resistance to erosion, torrential downpours slowly excavate or weaken the materials below the more resistant layers. Eventually the mass above is insufficiently supported and joint blocks break off. Frost wedging and other agents also aid in this process. If such erosion is long continued and the weak zones are fairly thick, cliffs may retreat many miles, and gently sloping terraces are developed.

As cliffs retreat, blocks of resistant strata are isolated from the main mass and form mesas, buttes, monuments and pillars, showing a great variety of forms. Magnificent examples of all these features can be seen in the Grand Canyon of the Colorado River in Arizona and in other parts of the Colorado Plateau, through which this river flows.

11

The flow of
streams and rivers

An immense quantity of runoff water is collected, as we have seen, in streams and rivers. (A stream is any natural watercourse, large or small; a river is a large stream.) These drainage lines are developed in natural depressions, some of which are due in part to sheet floods. If a depression has a free, or open, side, water will flow through it and a stream will develop; otherwise a lake or swamp will be formed. Streams and lakes may also be fed by underground water seeping into their beds.

The depression, or valley, through which the stream starts to flow is gradually modified in form. It may be widened and deepened as a result of erosion until it becomes a huge excavation like the Grand Canyon of the Colorado. On the other hand, the main activity of many streams is the deposition of sediment, which gradually fills the original depressions. This condition is found along great stretches of the Mississippi and Nile rivers.

Streams vary greatly in length and width. The principal rivers of the earth are trunk streams which receive the direct or indirect flow of countless tributaries; the whole is called a river system. The area that is drained by the river is known as the drainage basin. In the case of large rivers, this basin may cover a considerable part of a continent; for example, the Amazon and its numberless tributaries drain about 40 per cent of South America's land area.

In the humid parts of the tropical and temperate zones, large streams and most of their tributaries are permanent. In more arid regions, some main streams and most tributaries are intermittent — that is, their flow ceases for a certain period of time. Few streams are permanent in desert regions; they flow principally during and after heavy rains or the melting of snow and ice in nearby mountains. Where winters are long, the smaller streams are deeply frozen during the colder months so that little or no water flows through their channels; large streams have a cover of ice several feet thick, which reduces their volume.

The channel, or bed, of a stream is a trench having a roughly U-shaped cross section; it is the space normally covered with water. In rivers that spread over adjacent flats or bottoms, the channel normally containing the water is called the minor bed; the area covered by flood waters is the major bed.

The velocity of a stream varies along its length and also in different parts of a given cross section. The controlling factors are the gradient, or rate of slope, of the bed, the volume of water, the friction between water and bed, the shape of the bed and the load of rock fragments that the stream carries.

The flow of a stream is called its discharge. In most cases, discharge increases downstream because tributaries, seeps and springs all contribute water to the stream. In arid regions, however, there is generally a lessening of discharge downstream because water is lost as it sinks into the ground or evaporates. Practically all streams pass through a volume cycle between an extreme low to a flood peak; there are also minor cycles between these limits.

Flood waters never overtop the walls of deep, narrow canyons; hence the record of their passing is revealed only by erosion and deposition along the canyon walls. In the case of lowland streams, the waters of the ordinary flow are contained within the normal or minor bed. When this is completely filled, the stream has reached the bankfull stage; it is at the overbank stage when its waters flow over its banks and flood the adjacent lowlands.

In small streams, there is generally quick flooding; the flood crest rushes swiftly downstream. In large rivers the water rise is more gradual, and the flood crest proceeds in more leisurely fashion. It takes several weeks for the floods starting in the upper part of the Ohio or the Mississippi to reach the mouths of these rivers.

Streams transport varying quantities of weathered rock fragments and a great deal of animal and plant debris; they also carry, dissolved in their waters, many kinds of mineral and organic substances. The

NYSPIX—Commerce

Ausable Chasm, a deep gorge in the Adirondack Mountains of New York. This striking gorge was created as the Ausable River cut through formations of sandstone.

sum total of this material is called the load of the stream.

The undissolved load of a stream may range from minute particles of suspended clay to large boulders, which are rolled, bumped or pushed along the bed. The distances that rock fragments travel in water before being dropped depend on their weights, sizes and shapes, as well as on the volume and flow of the stream itself. The heavier, larger and rounder bodies usually require fast-moving or even turbulent waters in order to be transported. They are often picked up last (if at all) and then released first when speed or volume of flow diminishes somewhat. The finer materials are frequently carried the farthest and are deposited only when the current slackens still more or ceases entirely. However, if the drop in water volume and speed is sudden and extreme, practically the entire load (coarse and fine) may be deposited at once. Dissolved materials may be precipitated out when the stream loses much water through evaporation or when it enters the sea, where the current slackens.

Conditions may be such that fine sediments are caught up at the same time as the coarser matter, or after it or perhaps not at all. Dense clays, adhering firmly to the bed, may resist being moved along by the stream. The latter, if turbulent enough, may transport sand and pebbles instead and perhaps drop them a considerable distance away. However, once the fine sediments are in suspension, they are usually moved farthest even by gentle currents. The net results of all the activities described above are stream deposits of highly varied character, ranging from beds of fine clay to piles of large stones and often including mixtures of clay and stones.

Some streams carry chiefly sand and coarse fragments except during times of flood, when much fine debris is washed in by tributaries and sheet floods; such streams are clear when the water is low. Other streams are muddy most of the time since the greater part of their load consists of fine material kept in suspension. Large boulders are moved only by major floods and are found principally along mountain and high-plateau streams, which have steep gradients and consequently high velocities. Small boulders, pebbles and gravel are most abundant along such streams; however, they may be swept onto lowlands by great floods. Sand, especially the finer grades, and mud are found particularly in low-gradient stretches.

The material in solution comes principally from spring waters which have dissolved mineral and organic materials underground. The amount of dissolved substances transported by streams is very great. Each year the Mississippi carries an average of 136,000,000 tons of material in solution and, at the same time, an average of 380,000,000 tons of rock fragments.

As fragments are carried downstream, they abrade the beds of streams: that is, they wear them away by friction, thus widening and deepening them. This is one of the principal factors in the erosive action of streams. Flowing water, using its load of sediment as a gouging tool, can carve out canyons thousands of feet deep even in the most resistant rock. Another important factor is hydraulic action; the impact of flowing water on weak rock or rock mantle loosens and quarries out fragments or even masses. Flowing water also dissolves small amounts of soluble materials in rock, especially in such formations as limestone or gypsum.

By removing weathered material and abrading the bottom and sides of its bed, a stream gradually excavates its valley. In doing so, it works in two directions — downward and from side to side. For a time after a stream begins its flow over the land, downward erosion is rapid if both the elevation of the land and the slope are great enough. The resulting stream velocity is high; therefore the abrasive action of particles coming in contact with the bedrock is very powerful. As the slope of the bed decreases, downcutting falls off.

Sideward erosion goes on because all streams have irregular courses and the water, with its load of sediment, is thrown first against one bank and then against the other. This type of erosion is due to the removal of weak materials by the force of water and to the more gradual undercutting of resistant rock by abrasion.

In the course of time, cavernous openings are excavated in the valley sides. Eventually great masses of unsupported rock above the cut fall, and as a result the valley is widened. Sideward erosion is most active along concave banks against which

Three stages in the formation of an oxbow. A. A meander (winding) develops. B. A new channel is carved out. C. The old channel is partly filled up and an oxbow develops. The oxbow itself may be filled in time.

NEW CHANNEL   OXBOW

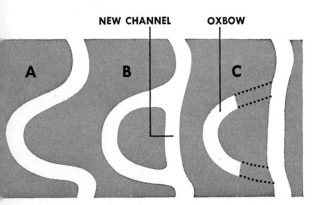

A   B   C

the water strikes with greatest force. The channel gradually becomes more crooked and the windings called meanders are developed. (The word is derived from *Maiandros*, an Asiatic river famous in antiquity for its windings.) Later, channels are cut through the necks of meanders, forming islands; the main current then follows the shorter route. Gradually the entrances to the bypassed channels are filled with sediment, leaving crescent-shaped lakes, called oxbows. The oxbows formed in this way may likewise be filled in the course of time.

Valleys are lengthened both downstream and upstream; upstream lengthening comes about because the waters of the stream erode backward into the valley head. The downstream extension of the valley comes to an end when the stream joins another stream, flows into a lake or ocean or disappears by evaporating or sinking into the soil. Upstream lengthening continues until the power of the backwaters is insufficient to erode back the valley head.

Tributary valleys are formed at the same time as a main valley because water flows down the depressions that are always present in valley walls. Most tributaries enter a main stream at the level of the latter; this is called accordance of level. In some places, however, the mouth of a tributary hangs above the main stream and the water plunges over a precipice to join the master channel.

Streams erode their channel downward until they reach the basin into which they empty. The level of this basin is called the erosion base; once this level is reached, the erosive action of the stream comes to an end. Most drainage systems flow into the sea. The erosion base in such cases is theoretically sea level; actually, however, large rivers scour their channels a little below sea level.

If a stream empties into a lake or inland sea or disappears in an arid basin, the surface of the lake, sea or basin is the erosion base. Such bases are generally above sea level; in some instances, however, they are considerably below it. For example, the Jordan River flows into the Dead Sea at a point 1,292 feet below sea level.

The formation
of falls and rapids

Falls and rapids are developed under a variety of conditions. When a stream passes from more resistant to less resistant rock, erosion is more rapid in the less resistant formation; consequently, a cliff or steep slope marks the boundary between the two sections. The Great Falls of the Yellowstone River in Wyoming was formed in this way. In other cases, undermining or sapping along weak layers causes unsupported rock to break off along fractures and to develop steep slopes or cliffs. This structure is found at Niagara Falls. As we have already pointed out, certain tributaries join their main streams by plunging over steep precipices.

Waterfalls are common in areas where there has been uplifting of fault blocks. They have also developed in regions once covered by glaciers. As a result of glacial action, the main valleys have been deepened more than their tributaries. After the ice disappears, the water plunges down the steep slopes marking the boundaries between tributary valleys and main valleys. Great numbers of waterfalls have originated in this way, as in Switzerland and western North America, where mountain canyons were occupied by immense glaciers in comparatively recent geological time.

How natural
bridges develop

Most large natural bridges result from the cutting through of narrow promontories that form the inner curves of many meanders in deep gorges or valleys. If the stream erodes caverns on both sides of the rock barrier, the intervening wall may eventually be worn through, allowing part of the stream to pass through the aperture. As the erosion continues, the opening is enlarged until all of the stream flows through it; the former channel is filled up in the course of time. The opening continues to be carved out while the thickness of the natural bridge decreases as its sides are weathered. The famous Rainbow Bridge, in the southern part of Utah, was formed in

USDA

Above: Koosah Falls, in the MacKenzie River, Oregon. This is one of several picturesque falls along the course of the river. Below: a majestic natural bridge in Utah's Bryce Canyon, as seen from the canyon's brim.

Union Pacific Railroad

this way. So was the Pont d'Arc ("Arched Bridge"), in southern France.

Natural bridges develop in various other ways. Sometimes the narrow divide between two closely spaced tributaries is undercut. In some places a section of the roof of a cave may collapse, leaving part of the roof as a bridge. Wave erosion along shorelines accounts for many natural bridges, too. In the case of waterfalls, part of the water may work through a joint in the bed and come out below the brink of the fall. If the passageway formed in this fashion accommodates all the water of the stream, the brink becomes a natural bridge.

U. S. Geological Survey

Alluvial fan in the Mojave Desert. Such fans develop when the velocity of a stream quickly decreases as it flows from a mountain canyon onto an adjacent lowland. It gives up most of the sediment it was transporting.

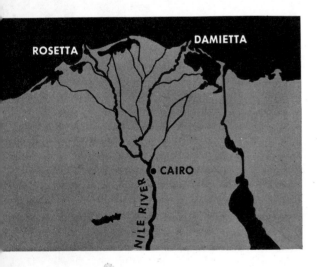

ROSETTA

DAMIETTA

CAIRO

NILE RIVER

Above: the delta of the Nile River, in Egypt. Note the complicated system of channels into which the Nile has divided. Below: a cross section of a typical delta.

DELTA PLAIN

SEA LEVEL

UNDERLYING ROCK

When streams deposit their loads

As the volume or velocity of a stream decreases, its transporting power is diminished and it is forced to deposit part or all of its load. Deposition occurs at various places along the course of a stream and always at its mouth.

Bars of sand and gravel are formed along the channels of a good many streams. They may be found along convex banks where the velocity of the stream is least. They also develop at the place where a fast-flowing tributary joins a main stream if the tributary is bringing in more sediment than the main stream can carry away. In some cases, bars develop at low water and are swept away in time of flood; sometimes they are a permanent feature of a stream, though changing both their form and their position from one season to the next.

As a stream widens its valley, a plain or terrace is formed. During floods, the stream spreads over part or all of this area; as the flood waters recede, a layer of sediment is deposited upon it. The area containing the sedimentary fill, which becomes more extensive with every flood, is called the flood plain, or deposition terrace.

When a swift stream flows from a mountain canyon onto an adjacent lowland, its velocity quickly decreases and it gives up most of its load, forming a deposit known as an alluvial fan. The slope of young fans ranges up to fifteen degrees; that of the older ones is much gentler. Alluvial fans are particularly well developed along the fronts of high, steep mountains in arid regions, where streams not only decrease in velocity as they leave the canyons, but shrink greatly in volume through evaporation and because of the sinking of water into the ground. As the deposit grows, it is projected farther and farther over the lowland and headward into the canyon. The union of many fans along a mountain front forms what is known as an alluvial apron.

As a stream enters a lake or ocean, the abrupt decrease in velocity causes it to deposit the debris that it carries. The deposition of sediment goes on more or less continuously; it is particularly great during floods. If the action of waves or currents is not too powerful at the mouth of the stream, a deposit called a delta is built up. The delta is a fan-shaped, layered formation sloping from the river mouth outward under the surface of the ocean or lake. The name comes from the Greek capital letter "delta," written Δ; the formation has been compared to a Δ, with the tip of the letter pointing upstream.

Certain parts of the delta may be built up above water level; in time embankments called natural levees are formed and they are projected seaward. The river spreads over this new land during every flood, and upon retreating adds sediment to it. Thus this deposition gradually increases the area of the newly emerged formation, which becomes known as a delta plain. In many rivers the delta plain is continuous with the flood plain.

During floods, water breaks through weak points in the natural levees and a branching system of distributary channels is formed. Between the levees of these channels there are depressions, which are covered by arms of the ocean or by lakes (if the depressions are isolated). The depressions are gradually filled as more and more sediment is deposited upon them, until in time they become part of the delta plain. Some delta plains are very large; that of the Nile starts a few miles below Cairo and is about 125 miles long.

Stream activities
interrupted by accidents

The activities of streams may be temporarily interrupted by various accidents. A portion of a valley may be flooded by a lava flow or by a succession of such flows. The lava may dam the valley and impound the stream waters, thus forming a lake. Eventually the water overflows and erodes a channel in the lava; this channel lowers the level of the lake and may cause it to disappear. The products of volcanic explosions — ashes, pumice stone and so on — produce the same results as lava flows in interrupting stream activities; they settle into river channels and fill them.

Landslides often dam up streams and form lakes. Some of the newly formed barriers are resistant; overflow drainage damages them very little. Others are easily cut into and may be partly or wholly demolished. In some instances, log rafts carried by main streams have dammed up the mouths of tributaries. The lakes that have been formed in this way have existed for many years in certain cases.

Young, mature and
old streams and valleys

Geologists classify streams as young, mature and old. The age of the stream is not the important factor; the classification is based on the sort of work the stream does and the type of valley it erodes. These factors in turn depend on the elevation and stability of the land, the initial slopes and the composition and structure of the rocks. Even within a single given region, different streams of roughly the same chronological age may be youthful or old in development.

Young streams and valleys. Young streams carry coarse debris as they head swiftly down more or less steep slopes; they are rarely loaded to capacity with sediment. They downcut vigorously, eroding V-shaped gorges and valleys. The widening of the valleys of young streams is accomplished partly by the attack of the stream upon its banks; however, it is chiefly due to other erosive agents, which cause loose fragments or masses of rock mantle to slide down the slope. The chief deposits laid down by young streams are bars of gravel and sand.

Mature streams and valleys. As the slope of a stream bed is reduced by continued erosion, downcutting power diminishes. The stream's energy then serves principally to transport sediment and to bring about sideward erosion. Meanders become more prominent and the erosion terrace is widened. A mature stream is loaded with sediment to about its capacity at all times. When this sediment spreads over the terrace, flood plains are formed.

Old streams and valleys. When a stream reaches old age, its downcutting power is negligible; its activities are limited almost exclusively to transportation of load and sideward erosion. The erosion terrace becomes very wide and is continuous along both sides of the stream. Water spreads over parts·of this terrace, causing major floods and leaving layer upon layer of sediment. Flood-plain deposits and erosion terraces are far more extensive than those of mature streams.

In its old age, a stream flows through a low, gently undulating land surface, which is called a peneplain (from Latin *paene planus*: "almost flat"). The peneplain consists of erosion terraces and flood plains of

U. S. Dept. of the Interior

Above: a part of the canyon of the Yellowstone River. The young stream shown here has downcut vigorously, eroding a deep, V-shaped gorge. Below: a mature river in Aroostook County, Maine. This type of river carries a good deal of sediment and brings about sideward erosion. Note the prominent meander that has developed.

USDA

old streams and the inconspicuous divides separating the various drainage lines. Here and there, bold eminences appear above the general surface. These heights are called monadnocks, after Mount Monadnock, in New Hampshire, which rises from the plain in just this way.

The land areas between valleys undergo their own cycle of erosive development, in association with that of the valleys. The interstream profiles are often the reverse of those of valleys — broad and flat or rounded in youth, narrow and possibly jagged in maturity. This may be explained by the fact that the divides between the streams are usually undergoing more or less constant reduction by erosion as the valleys are extended and broadened at their expense. As the slopes of the divides continue to develop, however, they tend to become more gentle, with the remnants of the old uplands surviving, perhaps, on the heights. Should the general cycle go into extreme old age, which practically never happens, the valley floors and divide slopes may combine to form a gently rolling landscape, with isolated rounded rises here and there representing the old divides. This landscape may resemble the peneplain with monadnocks described above.

Generally speaking, the evolution of streams and valleys is not an orderly progression from youth to old age. For one thing, the titanic forces at work under the crust of the earth may cause the land to be uplifted and tilted; the resulting steep gradients will cause streams to flow with greatly increased velocity and will add to their erosive power. Such streams are said to be rejuvenated, or revived. Rejuvenation may also be brought through changes in climate or destruction of plant cover.

When streams are thus invigorated, they cut youthful V-shaped valleys and in general dissect the surface of the earth like any young stream. Sometimes landscapes representing several rejuvenations are found in the same area. In the Appalachian Mountains, for example, there are at least seven distinct landscapes; there are three in California's Sierra Nevada.

See also Vol. 10, p. 270: "Sculpture of Land."

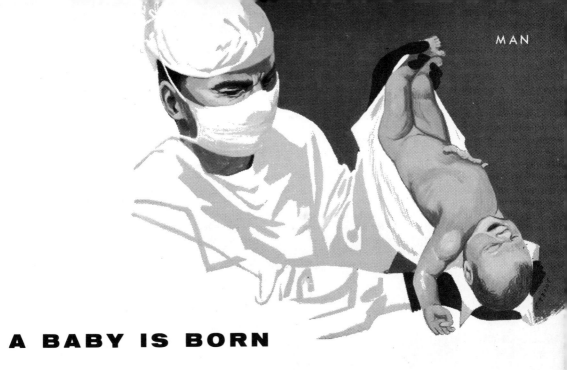

A BABY IS BORN

The Process of Reproduction in Man

BY CHARLES C. MACKLIN

MAN is mortal and his life span is comparatively brief, even under ideal conditions. Fortunately, like all other living things, human beings can reproduce their kind and thus assure survival of the species.

One-celled forms of life, such as amebas and diatoms, reproduce by fission: a cell divides into two, and each of these represents a new individual. Fission also occurs among certain multicelled organisms — that is, plants and animals consisting of more than one cell. Some of these forms of life reproduce vegetatively. A part of the original organism, removed from the main body, grows into another organism of the same type. Other plants and animals bring new representatives of their species into being by means of single cells called spores. In the higher forms of life (including man) and in many of the lower forms, too, sexual reproduction takes place. Two kinds of sex cells, or gametes, are involved in the process: the male gamete, or sperm, and the female gamete, which is called the ovum, or egg.* Only when an egg is fertilized by a sperm can a new individual arise under normal conditions.

The sex cells of man represent only a part of the trillions and trillions of cells found in the human body. (See the Basic Units of the Body, in Volume 1.) Each cell contains a core, or nucleus, in which there are a number of filaments, called chromosomes. At certain phases of cell division, the chromosomes are elongated and intertwined; at other phases, they become shorter and thicker, forming distinct units. Chromosomes carrying hereditary factors called genes (see Index). Human cells (except for mature sex cells) generally have forty-six chromosomes. Mature sex cells have only twenty-three.

The sex cells of man are produced in the gonads, or sex glands — the ovaries in the female and the testes in the male. A number of accessory sex organs also serve the reproductive process. These organs include the penis in man and the uterus, or womb, in woman. Certain secondary sex characters also develop. Among these are the growth of hair on the face in males and the development of breasts in females.

* The word "sperm" has the same form in the singular and plural. The plural of "ovum" is "ova."

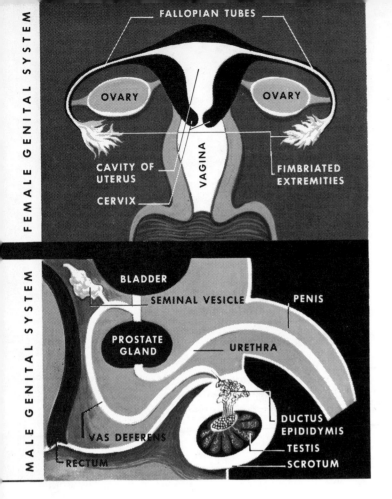

FEMALE GENITAL SYSTEM

FALLOPIAN TUBES

OVARY — OVARY

VAGINA

CAVITY OF UTERUS — FIMBRIATED EXTREMITIES

CERVIX

MALE GENITAL SYSTEM

BLADDER

SEMINAL VESICLE — PENIS

PROSTATE GLAND — URETHRA

VAS DEFERENS

DUCTUS EPIDIDYMIS

TESTIS

RECTUM — SCROTUM

LEFT:
the female and male genital systems.

RIGHT:
the travel of an egg that will develop into an embryo.

1. A Graafian follicle bursts and discharges the egg.

2. The egg makes its way toward the Fallopian tube.

3. A sperm (male cell) enters the egg.

4. The male and female pronuclei are still distinct.

5. The pronuclei fuse.

6. The fertilized egg divides again and again.

7. The egg has become a morula— a little ball of cells.

8. The ball of cells enters the uterus.

9. The product of fertilization—the conceptus— is now fully implanted. Pregnancy has taken place.

The ovaries, in which female sex cells — the ova — are produced, are almond-shaped structures, which lie in the pelvis. The ova develop from cube-shaped cells covering each of the two ovaries. Columns of these cubical cells penetrate deep within the ovary; tiny cell groups then break off from the penetrating columns. These groups are called primitive Graafian follicles. Each consists of a comparatively large cell, surrounded by a single layer of smaller cells. The central cell is the primitive ovum, which may later develop into a mature ovum.

Within the ovary of a sexually mature female, one of the primitive Graafian follicles begins to mature from time to time. The cells surrounding the primitive ovum multiply until they form several layers. Finally the follicle ruptures and the ovum is discharged from it. This process is called ovulation. The ovum, still immature, makes its way from the ovary into a slender tube — the Fallopian tube — where it becomes mature. There are two Fallopian tubes,

one leading from each ovary. It is while the ovum is in one of the Fallopian tubes that it may be fertilized by a sperm. The ovum finally makes its way into the broader end of the uterus, a hollow, pear-shaped organ. If the egg has been fertilized, it will develop into an embryo (if conditions are favorable) in the uterus. If it has not been fertilized, it will degenerate and will be passed out in the menstrual flow.

The small, narrow end of the uterus is called the cervix. The cervix opens into the tubular vagina, which receives the male organ in the act of copulation, or sexual union. In the case of virgins, the opening to the vagina is partly closed by a membrane called the hymen.

We noted that the breasts, or mammary glands, of women are secondary sex organs. They are made up of a mass of glandular tissue, which can secrete milk when properly stimulated. The milk is transported along ducts to the nipple, from which it is withdrawn by the suckling infant. The nipple is covered by dark skin;

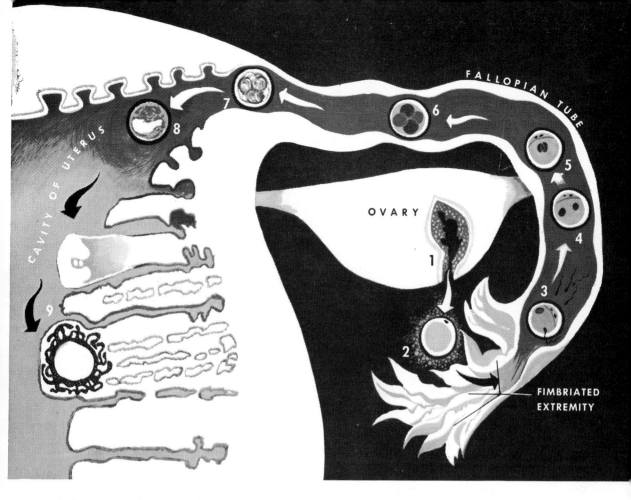

it is surrounded by a dark area called the areola.

The two testes of the male are egg-shaped bodies, which are enclosed in a pouch called the scrotum. The scrotum is outside the body cavity, at the bottom of the abdomen. The sperm produced in the testes are transported to the female sex organs by way of the external male organ called the penis. The penis also serves to discharge urine that has collected in the bladder.

After leaving the testis, the sperm travels along a slender twisted tube, over twenty feet in length — the ductus, or vas, epididymis. This elongated tube leads into the ductus, or vas, deferens. This is a muscular tube with a much larger caliber than that of the vas epididymis. The duct from the seminal vesicle empties into the vas deferens. Each of the two testes has the conducting system that we have just described.

In the course of the act of copulation, the seminal vesicles emit a secretion in which the sperm coming from the testes are suspended. These products are transported to the upper end of the urethra, the canal running through the penis. There they mix with the secretions from a gland called the prostate gland and from other glands of the urethra. The mixture of sperm and secretions is called semen.

The sperm have now become very active; under the microscope they are seen to be in violent motion. Each has a somewhat spadelike head, attached to a long filament, the flagellum, which propels the sperm by means of lashing movements. After the semen has been deposited in the vagina as the result of copulation, sperm make their way up the uterus by their own movements and by contractions of the uterus. Then they enter the Fallopian tubes. If there is a mature egg in one of the Fallopian tubes at this time, a sperm may penetrate the egg. When this happens, the male and female pronuclei (primitive nuclei) merge. This constitutes the act of fertilization. The egg now contains its full quota of chromosomes, half from the egg, half from the sperm. The materials from two lines of descent have thus been blended and the mechanism of heredity is set in motion. (This mechanism is discussed in detail in the article Heredity in Man, in Volume 8.) The cell divides again and again, as the growth of the organism continues.

Two or more offspring at a single birth

Fertilization in man generally results in the development of a single individual. However, there may be two or more offspring at a single birth. Identical twins arise from the same egg; they are always of the same sex. Twins may also arise from two separate ova that have been liberated and fertilized within a relatively short period of time. Such twins are known as fraternal or double-egg twins; they may be of the same sex or of different sexes. Usually they do not resemble one another any more closely than ordinary brothers and sisters.

The same general principles apply to the formation of triplets, quadruplets and quintuplets. Mixtures of single-egg and multiple-egg offspring may occur.

The early stages of development

During the earlier stages, the developing egg — the product of fertilization — bears no resemblance at all to the baby that is to be. It is a mere blob of microscopic cells, each with a nucleus that is capable of dividing and dividing again until ultimate growth is reached. Yet from the outset this little blob of matter is a self-determining and independent unit, which cannot be affected by the mother's mental processes or emotional experiences. The baby is the result of hereditary properties of the mother's and father's germ plasm. The only valid "prenatal influences" are those of the environment in which the unborn child develops. For example, it will be influenced by certain diseases that attack the mother and by extreme changes in the mother's diet.

At first the embryo consists of a little ball of cells, called a morula. This soon develops an outer and inner layer; the interior of the ball fills with fluid. Numerous fingerlike "sprouts" form on the exterior, which begins to look quite fuzzy. The outer covering of the tiny ball is called the chorion (the Greek word for "skin"); the fingerlike sprouts that form on it are known as villi. Meanwhile the embryo has arrived in the uterus.

Changes in the mucous membrane of the uterus

In the meantime the mucous membrane of the uterus has undergone certain vital changes in preparation for the possible arrival of a fertilized egg. Under the ac-

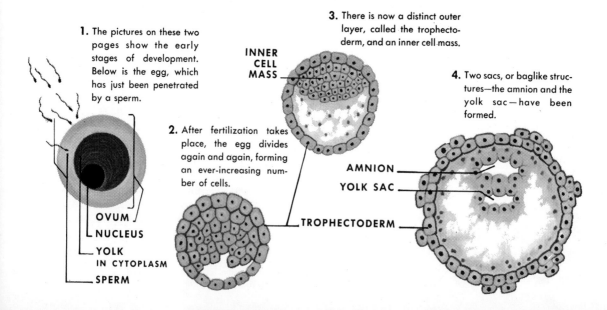

1. The pictures on these two pages show the early stages of development. Below is the egg, which has just been penetrated by a sperm.

2. After fertilization takes place, the egg divides again and again, forming an ever-increasing number of cells.

3. There is now a distinct outer layer, called the trophecto-derm, and an inner cell mass.

4. Two sacs, or baglike structures—the amnion and the yolk sac—have been formed.

INNER CELL MASS

AMNION
YOLK SAC
TROPHECTODERM

OVUM
NUCLEUS
YOLK
IN CYTOPLASM
SPERM

tion of the hormone called progesterone, the mucous membrane has come to form a soft tissue, which is abundantly supplied with blood vessels. This newly formed tissue is known as the decidua. The embryo embeds itself in the decidua in the process of implantation.

The process
of menstruation

The ovum makes its way to the uterus from the Fallopian tube whether or not it has been fertilized. If fertilization has not taken place, the ovum degenerates. The inner lining of the uterus, which had been prepared for the embryo, is shed, together with a certain amount of blood in the process of menstruation. The menstrual cycle lasts about twenty-eight days.

Ovulation — that is, the discharge of an ovum from a mature Graafian follicle — occurs about midway between two menstrual periods. The menstrual cycle is repeated as long as a female is capable of bearing young, except during pregnancy and also for a certain length of time after the period of pregnancy.

Suppose now that the embryo has become embedded in the mucous membrane of the uterus. Its inner cells come to form the walls of two sacs, or baglike structures, called the amnion and the yolk sac. A group of cells known as the embryonic shield develops at the place where the amnion comes in contact with the yolk sac; the body of the embryo will be formed from this shield. As time goes on the cavity of the amnion becomes larger and larger, while the yolk sac shrinks and ultimately disappears. At last the amnion completely surrounds the embryo. The amnion has come to be filled with liquid. Because of this development the name "bag of waters" is sometimes given to the amnion.

The embryo
develops steadily

In the meantime, the embryo has been developing steadily from the embryonic shield. Its cells soon form two layers — an outer one, called the ectoderm, and an inner one — the endoderm. A third layer, known as the mesoderm, soon develops between these two. All the structures of the body arise from these three layers. The ectoderm gives rise to the skin, the nervous system, the lining of the nose, the outer parts of the sense organs, the lens of the eye, the pituitary body, the glands of the mouth and the enamel of the teeth. From the mesoderm are developed the skeleton and the muscles, the heart and the blood and lymph vessels, the kidneys and ureters, the urinary bladder (but not including its lining) and the gonads and their ducts. The endoderm develops into the linings of the respiratory and alimentary tracts, various glands and the secreting cells of the two important organs known as the pancreas and the liver.

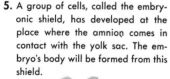

5. A group of cells, called the embryonic shield, has developed at the place where the amnion comes in contact with the yolk sac. The embryo's body will be formed from this shield.

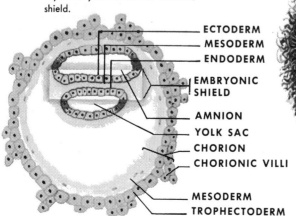

ECTODERM
MESODERM
ENDODERM
EMBRYONIC SHIELD
AMNION
YOLK SAC
CHORION
CHORIONIC VILLI
MESODERM
TROPHECTODERM

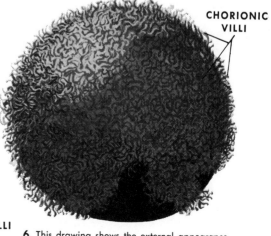

CHORIONIC VILLI

6. This drawing shows the external appearance of the product of fertilization. Its outer covering is called the chorion; it is covered with numerous "sprouts," called villi.

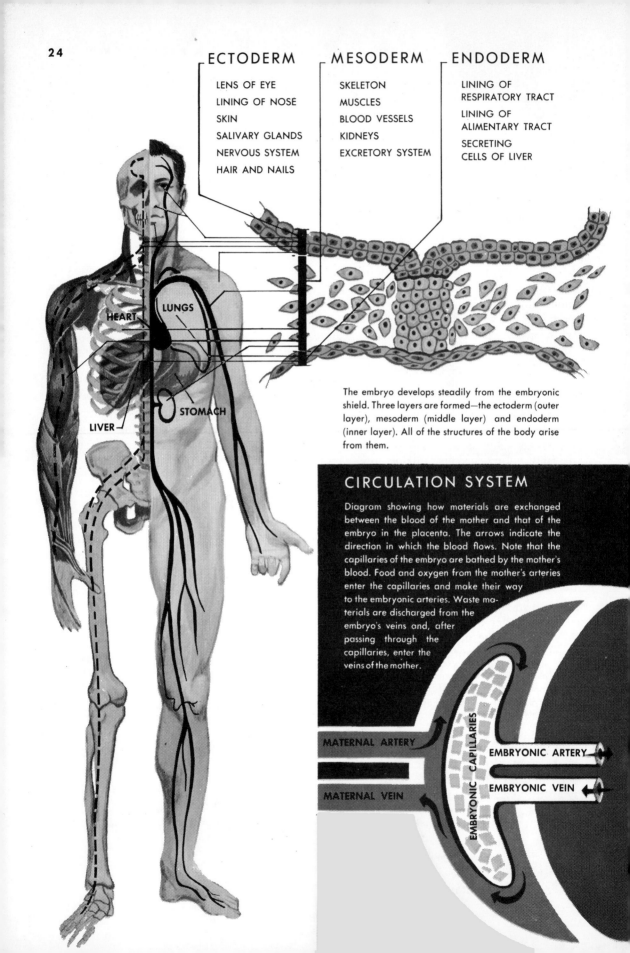

ECTODERM

LENS OF EYE
LINING OF NOSE
SKIN
SALIVARY GLANDS
NERVOUS SYSTEM
HAIR AND NAILS

MESODERM

SKELETON
MUSCLES
BLOOD VESSELS
KIDNEYS
EXCRETORY SYSTEM

ENDODERM

LINING OF
RESPIRATORY TRACT
LINING OF
ALIMENTARY TRACT
SECRETING
CELLS OF LIVER

LUNGS
HEART
STOMACH
LIVER

The embryo develops steadily from the embryonic
shield. Three layers are formed—the ectoderm (outer
layer), mesoderm (middle layer) and endoderm
(inner layer). All of the structures of the body arise
from them.

CIRCULATION SYSTEM

Diagram showing how materials are exchanged
between the blood of the mother and that of the
embryo in the placenta. The arrows indicate the
direction in which the blood flows. Note that the
capillaries of the embryo are bathed by the mother's
blood. Food and oxygen from the mother's arteries
enter the capillaries and make their way
to the embryonic arteries. Waste materials are discharged from the
embryo's veins and, after
passing through the
capillaries, enter the
veins of the mother.

MATERNAL ARTERY

MATERNAL VEIN

EMBRYONIC CAPILLARIES

EMBRYONIC ARTERY

EMBRYONIC VEIN

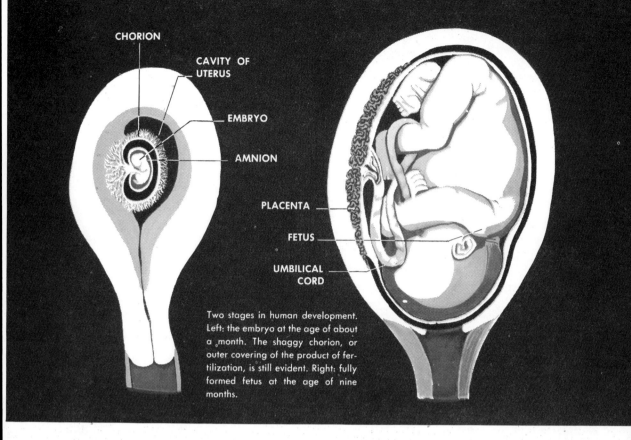

CHORION

CAVITY OF UTERUS

EMBRYO

AMNION

PLACENTA

FETUS

UMBILICAL CORD

Two stages in human development. Left: the embryo at the age of about a month. The shaggy chorion, or outer covering of the product of fertilization, is still evident. Right: fully formed fetus at the age of nine months.

For the first eight weeks or so of its growth, the unborn young is called the embryo. After this period of time it is known as the fetus.

The development of the placenta

When the embryo first becomes implanted in the uterus, the villi on the embryo's outer surface are surrounded by the lining of the uterus. The area where the villi project into the membranes of the uterus is called the placenta. This grows until it covers almost a third of the uterine lining. The fetus develops in the amniotic cavity; it is connected to the placenta by means of the umbilical cord. Two umbilical arteries passing through the cord carry blood from the fetus to the placenta; an umbilical vein returns blood from the placenta to the fetus.

The circulation of the mother and of the fetus come into close contact in the placenta. Yet the blood in the mother's circulation system does not mix with that in the vessels of the fetus. Oxygen and food elements contained in the capillaries

(see Index) of the mother pass to the capillaries of the fetus. Carbon dioxide and other wastes pass from the capillaries of the fetus to those of the mother.

The formation of the different body systems

Let us now see how the various parts of the unborn human young develop. We shall take up the different systems of the body in turn, bearing in mind that they may be developing at much the same time.

The heart and blood vessels. In the early stages nutrition and wastes can pass quite effectively by diffusion between the cells of the embryo and the mucous membrane of the uterus. As growth advances, however, this system becomes inadequate. Specialized blood vessels are formed to transport nutrition and wastes; together with the heart and blood, these vessels make up the extensive cardio-vascular system.* Without it, development could not proceed beyond the early stages.

* *Cardia* means "heart" in Greek; *vasculum* means "little vessel" in Latin. The cardio-vascular system, therefore, has to do with the circulation of the blood through the heart and blood vessels.

LENGTH
¼ INCH

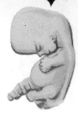 LENGTH
1 INCH

WEIGHT
1/30
OUNCE

We can trace the beginnings of the heart in the future neck region of the embryo. At first the heart (or rather the structure that will develop into the heart) is in the form of two tiny cavities lying side by side in the long axis of the embryo. Similar cavities develop in other places. They grow longer and send out branches, which join with other branches. In this way the larger arteries and veins develop. These channels soon unite with the primitive heart, whose cavities enlarge, elongate, fuse, twist and later subdivide into four chambers — the left and right ventricles and the left and right auricles. Valves arise at the entrances to the ventricles. Later, the heart descends to the thorax, or chest.

In the meantime groups of cells called blood islands have begun to form. The outer cells of these blood islands flatten and become the capillaries — the smallest blood vessels in the circulatory system. The inner cells develop into the different types of blood cells. The capillaries come to form a network, which is connected to the arteries and the veins. Fluid — the plasma of the blood — gathers around the blood cells.

As the lungs are formed, they develop their own blood vessels, which are also connected to the heart. This lesser, or pulmonary,* circulation, as it is called, does not operate fully until after birth, when the baby draws its first breath and the lungs begin to function.

Long before this happens, blood begins to circulate through the arteries and veins and capillaries of the greater, or systemic,

circulation, which passes through the rest of the body. Circulation is brought about by rhythmic contractions of the heart.* The heart keeps pumping blood; the arteries distribute it to the capillaries in different parts of the body. The capillaries deliver materials contained in the blood to the cells and collect waste materials from them. The veins return the waste-laden blood to the heart.

It does not suffice for blood to pass from the human embryo's heart to its tissues and back again. Some blood must make its way to the placenta to get food materials and oxygen and to be relieved of waste products. We remarked that blood is transported to the placenta through the umbilical arteries and from the placenta through the umbilical vein. It is returned to the heart through the liver.

The nervous system. At the same time that the cardio-vascular system is developing, the embryo's nervous system is being formed. First, parallel ridges arise along the body, flanking a central groove. The ridges form conspicuous bulges at the head end. In time they fuse along the midline, forming the neural tube. The larger end of the neural tube develops into the brain; the rest, into the spinal cord. The openings at the ends of the neural tube ultimately close.

The central nervous system develops from this primitive neural tube. Growth is most rapid of all in the brain. There are really two brains: a large brain, the cerebrum, and a small one — the cerebellum. In both there is ultimately an outer covering of gray matter, called the cortex,

---

* "Pulmonary" means "having to do with the lungs." (The Latin word *pulmo* means "lung.")

* It is possible to watch the first contraction of the heart muscle in the living chick through a window opened in the shell of the incubated egg.

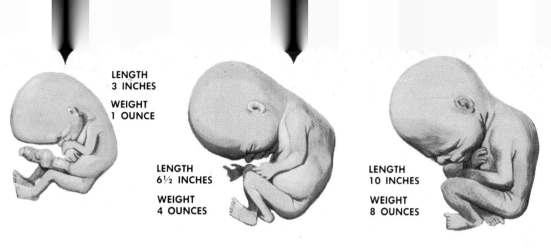

LENGTH
3 INCHES

WEIGHT
1 OUNCE

LENGTH
6½ INCHES

WEIGHT
4 OUNCES

LENGTH
10 INCHES

WEIGHT
8 OUNCES

and an inner mass, consisting of white matter. In the midline is the brain stem. This structure connects the brain with the spinal cord.

The cavity of the neural tube continues to grow larger, though not at the same rate as the tissues that surround it. As the brain develops, the cavity within it forms the ventricles — spaces filled with fluid. In the spinal column, the cavity of the neural tube makes up the inconspicuous central canal.

In the course of the development of the neural tube, two lines of tissue appear between the tube and the overlying skin. They become segmented and form a paired series of dorsal root ganglia — the cell stations of the sensory nerves leading to and from the brain and spinal column.

The skeletal and muscular systems. Underneath the impressive mass of the central nervous system, a rod of cells — the notochord — has grown forward from the tail end of the embryo. It forms a central axis, around which the vertebral column develops. The notochord extends forward into the base of the skull; in time it disappears almost entirely. However, the central part of each disk, between the vertebrae, is retained.

Hardly has the notochord made its appearance when strips of mesoderm form alongside it; these soon become a series of paired blocks. The first to appear are in the lower part of the head. They are then added in pairs farther down the notochord, until in the fourth week their number reaches forty. They are called mesodermal somites and are found in all vertebrates. Parts of the mesodermal somites will give rise to the vertebrae. Other parts of the somites will develop into muscles, supplied with nerves at an early stage. Later, when the muscles shift their positions to more distant parts of the body, they retain their original nerve connections. This is also true of various organs. Thus the heart and stomach, which develop in the lower region of the head, are supplied by nerves arising in that region.

The digestive system. The primitive digestive system originates at a very early stage in the development of the embryo. Its lining layer of endoderm is continuous with that of the yolk sac. From the cavity of the yolk sac a pouch called the foregut continues forward over the primitive heart and under the head region. Behind, extending into the tail region is a similar pouch — the hindgut. An intermediary region, known as the midgut (really part of the yolk sac) connects the two guts.

At first the foregut is separated from the primitive mouth, or stomadaeum, by a membrane; another membrane separates the hindgut from the primitive anus. In normal development, both these membranes ultimately disappear. The primitive digestive system then forms a continuous tube from mouth to anus.

The foregut, which at first looks like the finger of a tiny glove, grows rapidly and becomes differentiated into various structures. From the foregut arise the mouth, pharynx, stomach and much of the small intestine, including innumerable small glands. A lung bud forms on the foregut; it branches repeatedly and gives rise to the bronchial system and the lungs. The hindgut makes up part of the small intestine as well as the large intestine and the embryonic structure that is called the cloaca. We

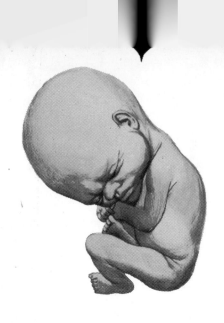

LENGTH
12 INCHES

WEIGHT
1½ POUNDS

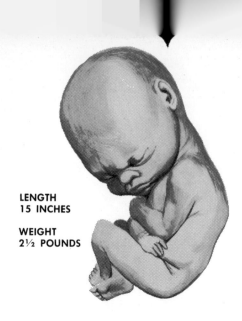

LENGTH
15 INCHES

WEIGHT
2½ POUNDS

shall have more to say about the cloaca when we discuss the urogenital system on this page (second column).

The body wall, coelom and mesentery. When the body wall develops, we can recognize the parts that are to become the thorax (chest), the abdomen and the pelvis. The body wall is made up of supporting connective tissue, with special sheets of muscle to serve for movement. Tissue called mesenchyme forms in parallel lines at right angles to the axis of the body. These parallel structures change into cartilage, and later mainly into bone. They are then recognizable as the ribs. The upper members of these elements are joined in front by two parallel bars of cartilage. These fuse to form the sternum, or breast bone.

As the body becomes walled in, the future chest and abdomen form a single hollow space called the coelom. It is separated lengthwise into two parts by a primitive membrane, called the mesentery. Soon a septum, or wall, which becomes the diaphragm, separates the chest from the abdomen. In front of the heart and intestine, the mesentery disappears; the rest of the mesentery in this part of the body undergoes great changes. The hind part of the mesentery lengthens as the intestine to which it is attached becomes longer. Part of the posterior (hind) mesentery lies against and is attached to the posterior abdominal wall.

*The urogenital system.** In the early stages of the human embryo, there is a primitive kidney, the pronephros, which gives way to a more advanced kind of kidney — the mesonephros. The mesonephros contributes important parts of the male reproductive system. Later the permanent kidney, the metanephros, is formed. This organ removes waste materials from the blood. In the unborn young these materials first pass through ducts, called ureters, into a chamber called the cloaca. Later the front part of the cloaca becomes the bladder; this discharges urine from time to time through the urethra.

The sex glands, or gonads, are alike in the early male and female embryo. Soon however, they become differentiated and can be distinguished as the ovaries and the testes. The testes are at first contained within the body; later they descend into the scrotum. Occasionally they fail to make their way to the scrotum, but remain within the body. The physician who delivers the baby is expected to discover this abnormality and to make it known to the parents.

The external sex organs of the male and female embryo are at first very similar. Soon, however, differences develop, and by the time of birth they can be easily distinguished. In some cases, however, the male and female characters are more or less blended. Individuals in whom this happens

* The urogenital system includes the organs having to do with the production or removal of urine and with reproduction.

28

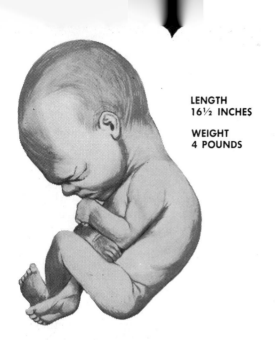

LENGTH
16½ INCHES

WEIGHT
4 POUNDS

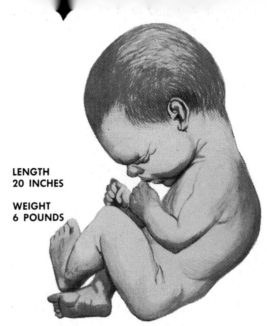

LENGTH
20 INCHES

WEIGHT
6 POUNDS

are called hermaphrodites — a name derived from the god Hermes and the goddess Aphrodite. The sex glands may be involved in this abnormality.

The arms and legs. In the lower neck region of the embryo paired buds appear; they will become the arms. In time the tips show five ridges, representing the future fingers. Cartilage is formed in the central axis of each arm; later this cartilage is transformed into bone. Joints appear at the end regions of the bones. Muscles developing around these parts become attached to the bones and bring about helpful movements. Blood vessels and nerves develop together with these structures. Far back in the trunk, paired buds, representing the legs-to-be, appear; they develop in much the same way as the arms.

In the limbs and throughout the body and head parts, a system of lymphatic vessels forms in the embryo. This will represent a tributary system for the veins, collecting fluids from various areas.

The head. Besides the skull and brain, the head includes the mouth, nose, eyes, ears and tongue. The skull is formed at first of soft connective tissue or membrane. Much of this is transformed later into cartilage. The cartilage never occurs over the roof of the brain, but only on the floor and sides. Eventually it develops into many of the bones of the skull.

Early in the third month of pregnancy, though the skull is only about as large as the tip of one's finger, it already shows in a rudimentary condition the different bones that form it. However, even at birth the brain is not completely enclosed by bone; in some areas, called fontanelles, it is covered only by membrane. The largest of these areas is on the top of the baby's head. It can be felt as a soft spot in which sensitive fingers can detect a pulse, transmitted from the arteries. The fontanelles are gradually closed up with bone.

In the early embryo the mouth and nose regions are merged in a single mouth-like part, called the stomadaeum. Shelves grow from its sides; normally they meet after a time to form the palate, which shuts off the mouth from the nose. If the process should fail, we have the serious defect called cleft palate. This is often combined with the condition called harelip, in which the upper lip is divided much like that of a hare.

The lower jaw is immovable and wide open in the early stages. It is first represented by an arch of cartilage. Later the mandible is formed by paired bones at the sides of the cartilage. The hyoid bone, under the tongue, and the larynx skeleton are also developed.

At birth there is still much cartilage in the skull as well as in other parts of the skeleton. Some of it persists throughout life in places where its properties make it an ideal structural material. It is found in the joints, the larynx, the nose, the juncture of the ribs and the breastbone and in various other parts of the body.

In the early stages the head is bent forward so that it almost meets the future chest, and the neck is practically nonexistent. In the course of time the neck develops and the head is raised somewhat. It is still bent forward, however, even up to the time of birth. The entire fetus is packed into the smallest space possible, curled up so that it is almost egg-shaped. This form is retained for some time after birth as the baby is held in the mother's arms.

In their earliest form the human eyes occur as two dimples on either side of the brain. Each dimple changes into a spoon-like structure, which becomes the optic cup. The inner membrane of the cup develops into a specialized layer of cells called rods and cones (see Index) and into neurons, or nerve cells, connected with them. The nerve cells establish connections with the optical centers in the brain by means of the optic nerve.

When the baby is born, therefore, it has a mechanism that can be stimulated by light and that can transmit the stimuli to the brain. The outer membrane of the optic cup forms a layer of densely pigmented cells, which keep out all light other than that coming in through the pupil. Generally the development of all these eye parts is almost perfect. Occasionally, however, defects occur and these may be of a serious nature.

The ears develop quite differently; they are related to the branchial arch region. While the embryo is still very small, a minute pit develops above the first branchial cleft. The pit becomes deeper, and is closed off in time to form a hollow ball. It soon becomes irregular as it grows; it is transformed into a set of three arched tubes — the semicircular canals — and a small seashell-like structure, the cochlea. The end organ of hearing — the organ of Corti — develops in the interior of the cochlea. Meanwhile the middle and the external ear are being formed. The outer part of the external ear develops a supporting shell-like elastic cartilage, which enables it to retain its shape. In the depths of the canal that leads from the opening of the external ear, a thin membrane — the eardrum — is formed; it closes up the canal. The eardrum vibrates when it is struck by sound waves. Three bones called ossicles arise in the middle ear; they transfer to the inner ear the vibrations that have been set up in the eardrum.

An arch-shaped structure develops on each jaw, and it divides into ten buds, which will form the milk, or primary, teeth. Thus there are twenty buds in all. Each bud forms an enamel organ, which is a minute ball of cells, almost hollow and bent in above or below. Other cells develop within the ball and form dentine. The enamel organ and the dentine work together; they secrete fibers that become impregnated with hard deposits rich in calcium. The soft inner core of the teeth develops blood vessels and nerves. The milk teeth are not visible until they are cut; the first is generally cut some months after birth. These primary teeth are shed during childhood and are replaced by the thirty-two regular teeth. We discuss the formation of the milk teeth and the development of the regular teeth elsewhere (see Index, under Teeth: starred article).

The skin and its derivatives. The skin proper, in the embryo, is continuous with the amnion, which is cast off at birth. An outer layer of cells, the epidermis, develops; underneath it there arises another layer, the derma, which contains nerves, blood vessels and sweat glands. In the embryo there is also a special layer of cells, called the periderm, on the outside of the epidermis. The periderm is shed at birth; most of it is washed off with the first thorough bath.

The hair and nails develop as special formations of the skin; so do the sweat and oil glands and the mammary glands of the female's breast. Certain other parts of the body are formed primarily from the skin; they include the lens of the eye, the cornea, the eyelids and the internal ear.

The control of
growth and differentiation

The myriads of cells engaged in forming a baby behave in an amazingly orderly manner. From a simple cell — the fertilized ovum — highly specialized cells occur in proper sequence. There are the gland cells of the stomach, the bone cells, the thin surface cells of the skin and nerve cells with fibers three feet long. There is nothing more extraordinary in nature than this systematic development.

Why do two buds for the future arms and legs make their appearance at the right time and place, with their various simple elements properly arranged? Why do they grow so perfectly in every part — elbow, hip, knee and ankle? How do the various muscles develop? How do they attach themselves to the right regions of the right bones? How do the muscles acquire nerves to move them and to register their state in the central nervous system? We could ask thousands of questions such as these.

The why and wherefore of the embryo's development is still largely a mystery. However, the science of heredity gives some help in understanding the processes involved. We assume, on the basis of our knowledge of heredity, that the fertilized ovum is packed with tendencies derived from all past generations and that development is simply the working out of these tendencies. We assume also that hereditary forces continue to act after birth throughout the life of the individual.

In the early stages of development there is a rapid sorting out of the different kinds of cells. Quickly, for example, the cells that are destined to make up the chorion are mustered out; from them are separated the cells that will form the villi and other parts. Once a cell is started on some special path, it tends to go on in that direction; its destiny is established.

The cells that will give rise to specialized tissues bear the suffix "blast" (from the Greek *blastos:* "bud"). There are neuroblasts, which will give rise to the different types of nerve cells; hemoblasts, which will develop into blood cells, and many other kinds of "blasts." We might compare the cells in the early stages of the embryo to young people of a community, some destined to become engineers, others chemists, still others doctors or lawyers. The comparison is not altogether apt, of course. Highly specialized cells of a given type cannot develop into other types of cells. On the other hand, a highly specialized pianist could conceivably change his profession and become, say, a highly specialized lawyer.

In many tissues differentiation goes on throughout the life of the individual in order to make good the losses of cells. For instance, the cells making up the surface layer of the skin are worn off in daily use and must constantly be replaced. The same is true of cells in other parts of the body, such as the intestinal tract and the blood. This means that such areas must be provided with nests of blast cells, capable of regenerating this or that tissue. In the deepest layers of the epidermis there are the chorioblasts, or Malpighian cells, that can give rise to the cells making up the surface of the skin. In the bone marrow there is a great mass of hemoblast cells that can differentiate into still more specialized types of blast cells. The latter will give rise to the red blood cells and the different kinds of white blood cells.

In the case of some tissues, such as those of the nervous system, there are no reserve blast cells; once a cell is destroyed, it cannot be replaced. The brain, for example, must get along throughout life with the cell quota it has at birth or shortly thereafter. That is why poliomyelitis, or infantile paralysis, is so serious. The cells that are destroyed because of its ravages cannot be replaced. Other groups of cells may be able to take over their functions, but unless this can be done, the loss is irreparable.

In the orderly development of the different parts of the body, often distant from one another, the hormones, which serve as chemical messengers, play a very important part. So does the nervous system. Doubtless, too, muscular movements have a more general effect than we realize. In fact,

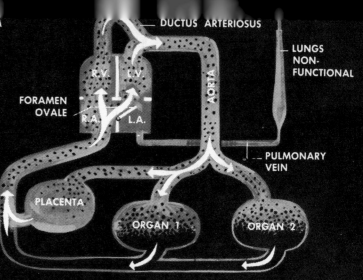

CIRCULATION SYSTEM BEFORE BIRTH

Aerated blood from the placenta mixes with blood from various bodily organs and flows to the right auricle. Some of this blood passes through the foramen ovale into the left heart, and from there to the aorta. Some of it enters the right ventricle and passes through the ductus arteriosus on its way to the aorta. Practically no blood flows through the lungs at this time.

DUCTUS ARTERIOSUS

LUNGS NON-FUNCTIONAL

FORAMEN OVALE

AORTA

R.V. L.V.

R.A. L.A.

PULMONARY VEIN

PLACENTA

ORGAN 1

ORGAN 2

---

everything taking place in the developing embryo or fetus probably reacts in some way with the organism as a whole.

The birth
of the baby

What we call birth — the emergence of the fetus from the body of the mother — is obviously not a beginning. It is rather the climax of a long sequence of events. Generally it occurs after the pregnancy has lasted from nine to nine and a half months. In all well-organized homes preparations have been made long in advance. The prospective mother has been under her physician's care; corrective measures have been taken at the appearance of any untoward symptom. Nature too has made its preparations for the coming event. The uterus has developed strong muscle fibers; the fetus will be expelled from the uterus as the result of strong muscular contractions. The outlets of both uterus and vagina have become softened and capable of great expansion.

As the time of birth approaches, the mother-to-be has labor pains, which are caused by contractions of the uterus. These contractions begin to force the fetus out of the uterus. They become more vigorous and the pains become more severe as time goes on. In the last stages of labor, the mother aids in forcing out the fetus by voluntarily contracting the abdominal muscles. The amnion, the closed liquid-filled sac surrounding the embryo, generally ruptures before birth. Sometimes it does not; in that case the baby must be rescued from its "prison" — an event called "birth in a caul."

Usually the baby comes from the mother's body head first. Sometimes it emerges feet first; this is called breech presentation. It may be necessary to take the baby from the uterus by cutting through both the abdominal wall and the uterus. This is known as a Caesarian section or operation, because it is believed that Julius Caesar was delivered in this way. All good medical schools train their students to be prepared for any eventuality. Modern methods of abolishing or dulling pain ease the mother's ordeal.

When the baby is born, it is still attached to the placenta by means of the umbilical cord. It draws nourishment and oxygen from the placenta and expels its wastes into it. The obstetrician ties up the cord close to the baby's body with tape; then he cuts the cord on the mother's side of the tape. About twenty minutes after the cord has been cut, the afterbirth (placenta and membranes) is passed from the mother's body.

When the placental circulation is cut off by the tying of the cord, there is an increase of the waste product carbon dioxide in the blood. This stimulates a center in the medulla oblongata, a part of the brain, and the baby usually begins to breathe

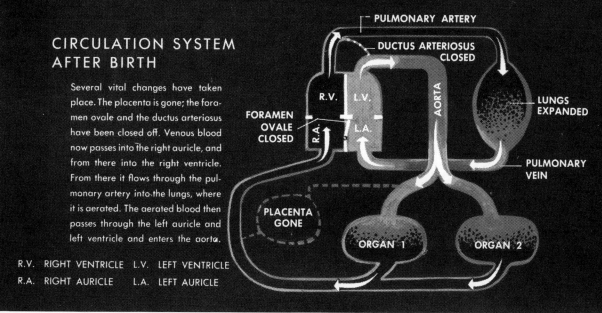

CIRCULATION SYSTEM AFTER BIRTH

Several vital changes have taken place. The placenta is gone; the foramen ovale and the ductus arteriosus have been closed off. Venous blood now passes into the right auricle, and from there into the right ventricle. From there it flows through the pulmonary artery into the lungs, where it is aerated. The aerated blood then passes through the left auricle and left ventricle and enters the aorta.

R.V. RIGHT VENTRICLE L.V. LEFT VENTRICLE
R.A. RIGHT AURICLE L.A. LEFT AURICLE

Adapted, with permission, from diagrams in THE MACHINERY OF THE BODY, by A. J. Carlson and V. Johnson, Univ. of Chicago Press © 1946

as a result. If respiration does not begin, the attending physician must give aid. He may stimulate the skin by slapping the buttocks or in other ways; he may administer artificial respiration; he may even introduce oxygen into the air tract by means of a tube.

Normally after birth the lungs get as much blood as the rest of the body put together. As we saw, they are on a special circulation circuit — the pulmonary circulation. In this, the blood travels from the right ventricle of the heart through the pulmonary arterial system to the pulmonary capillaries. Here it gives up its carbon dioxide and takes on oxygen. It then goes by way of the pulmonary veins to the left auricle of the heart and then to the left ventricle, to the aorta (an artery) and from there to all parts of the body.

Sometimes, because of faulty development of the blood vessels, too little blood is kept circulating through the lungs of the newborn baby. This often happens because the ductus arteriosus, a short tube connecting the pulmonary artery with the aorta in the fetus, fails to close. Much of the blood from the right ventricle passes into the aorta instead of into the pulmonary artery as it should after birth. There is weakness and coldness, due to the lack of oxygen. The skin becomes bluish; hence infants suffering from this condition are called "blue babies." Modern surgery can sometimes cure this disorder.

When a baby is born, it has the same bodily parts as the adult, but they are smaller and differently proportioned. The head is relatively large, the legs relatively small. The nose is tiny; but it serves well for the feeble breathing that is required of the infant.

A newborn baby is a rather unlovely sight. It is covered with blood and with a slippery, waxy substance called vernix caseosa ("cheeselike varnish," in Latin). This substance keeps the embryo from absorbing the liquid in which it is immersed. Of course it no longer serves, once the infant has been born. After the newborn child has been bathed and the blood and vernix caseosa have been removed, a startling transformation takes place. The infant has become sweet and clean and pink — a thoroughly delightful little creature.

After the afterbirth has been ejected from the uterus, the so-called puerperal period begins. It is during this period, which lasts from about six to ten weeks, that the uterus and the other structures involved in birth are restored to a normal condition. The uterus shrinks to almost its original size as the now useless tissue is absorbed. While this is taking place, there is discharge, consisting mostly of blood and debris, from the vagina. Gradually the discharge slackens and finally it ceases entirely. The puerperal period is now at an end.

See also Vol. 10, p. 276: "Physiology."

SOILLESS AGRICULTURE

A Weapon in the War against Hunger

The water-culture method of soilless agriculture developed by Julius von Sachs about 1860. Von Sachs allowed seeds to germinate in sawdust. Then he set each seedling in a perfo rated cork, with the roots in a nutrient solution.

IN THE last years of the eighteenth century, the English clergyman Thomas R. Malthus advanced the theory that the world's population was increasing so rapidly that it would in time outstrip the food supply.* His dire predictions have not yet been fulfilled. Yet there has been growing concern in many quarters about the fantastic increase in world population in recent years, particularly since the end of World War II. Will the food supply be able to keep pace with this "population explosion," as it has been aptly termed?

Various methods have been proposed to meet prospective food shortages in the years to come. A promising one is the application on a wide scale of soilless agriculture—the growing of plants in a medium other than natural soil. This is done by providing, in the form of a solution or a soluble substance, the essential chemical elements that the plant would normally draw from the ground. Soilless agriculture has also been called nutriculture (cultivating by means of nutrients, or essential food ele-

ments). Another name for it is hydroponics (literally "water labor": that is, performing the work of crop-raising by using water that contains nutrients).

Soilless agriculture is by no means a new development; it is over a hundred years old. The great German chemist Justus von Liebig (see Index) pointed out in 1840 in his CHEMISTRY IN ITS APPLICATIONS TO AGRICULTURE AND PHYSIOLOGY that plants require certain minerals and other elements that they obtain from the soil. By 1850, scientists had a fairly good general idea of the nutritive needs of plants. It was recognized that they require large amounts of carbon, hydrogen and oxygen, obtained from the atmosphere or from soil water, and also smaller quantities of such chemical elements as phosphorus, potassium and calcium, derived from the soil.

It now occurred to certain investigators that these essential chemicals could be supplied to the plant independent of the soil. The first experiments in this field were carried out by the French chemist Jean Boussingault. He set plants in insoluble artificial soils, made of substances such as

* The ideas of Malthus are discussed elsewhere; see Index, under Malthusian theory.

Left: packing cabbages in wooden crates at the Chofu, Tokyo, hydroponic farm. This big United States Army enterprise was discontinued in 1961.

U.S. Army Photograph

USIS

Above: a hydroponic establishment at Montebello, in California. Here seedlings are being grown in sand and subirrigated with a nutrient solution.

Right: Japanese women planting seeds in the gravel bed of an unusually large hydroponic farm operated by the United States Army at Chofu, Tokyo.

USIS

sand and quartz, and he watered these soils with nutrient solutions. Boussingault succeeded in growing various types of plants.

A German investigator, Julius von Sachs, decided to dispense with the solid medium altogether and to grow plants in water to which the required chemical elements had been added. This water-culture method was developed about 1860. Von Sachs allowed seeds to germinate in sawdust. When the seedlings were ready for transplanting, he set each one in a perforated cork, with the roots dipping into the nutrient solution. Various formulas for the solution were prepared; it was found that several different ones could be successfully applied.

For a long time, soilless agriculture was used in small-scale laboratory experiments in order to solve problems of plant nutrition and physiology. These experiments made it possible to complete the list of chemical elements normally drawn from the soil and necessary for plant life. They include potassium, calcium, magnesium, nitrogen, phosphorus and sulfur, and also such trace elements (required only in mi-

nute amounts) as iron, boron, manganese, copper, zinc and molybdenum.

It was not until the 1920's that investigators began to consider the possible application of soilless agriculture to the growing of crops on a commercial scale. In 1929, W. F. Gericke, of the University of California, created a sensation by growing tomato vines 25 feet high in a nutrient solution and obtaining quantities of fine tomatoes. The public at large was dazzled by the possibilities of soilless agriculture. Some American newspapers claimed that farmlands devoted to conventional farming would soon be replaced by hydroponicums, or hydroponic farms. In the cities, huge hydroponic establishments, atop skyscrapers, would supply all the fresh fruit and vegetables that the inhabitants would require for a balanced diet.

These glowing predictions were not fulfilled. Yet in the years that followed, commercial growing of vegetables and fruits by soilless-agriculture methods was introduced on a very modest scale. Hydroponic farms were established in Illinois, Ohio, California, Indiana and Florida. In most of them, so-called aggregate culture * was used. Plants were grown in watertight beds of sand, gravel or other inert materials, which were regularly flooded with water containing the necessary chemicals. These had to be replenished as they were used up by the plants.

Soilless agriculture acquired renewed importance for the United States during World War II, when American troops were sent to almost every quarter of the globe. The problem of supplying troops in remote areas with fresh fruits and vegetables was often a serious one. It was solved in several places by the setting up of hydroponic farms. These produced good results in Ascension Island (in the South Atlantic), British Guiana, Nanking (China) and the Japanese island of Iwo Jima after its capture by United States troops.

When American occupation forces moved into the Japanese islands after the end of the war, the Army Quartermaster Corps had the task of providing fresh vegetables not only for the troops but also for civilians taking part in the occupation. It was felt that available Japanese soil was unsuitable, for one thing, because of contamination, due to the age-old custom of utilizing "night soil" (human feces) as fertilizer. Likewise, the soil had been depleted as the result of centuries of intensive farming; and fertilizers were in short supply at the time. Hence several hydroponic farms were started in the islands in November 1945. Plants were set in beds of washed gravel and other solid materials, which were irrigated regularly with a nutrient solution supplied from tanks. Within six months, the yield was ample enough to make it unnecessary to ship most salad-type vegetables from the United States. The hydroponic farm at Chofu, Tokyo, occupied fifty-five acres and was the largest and most productive in the world.

In the course of time, the soil-contamination problem was solved in various Japanese areas. Soil-type farms, which cost less to operate than the hydroponic variety, were set up on an increasing scale to meet the

needs of the Americans. It was finally decided in 1961 to do away with the Army-operated hydroponic farms in Japan. It had been shown, however, that such farms could be established on a large scale and that they could produce abundant crops.

In the meantime, a new development in hydroponic farming had taken place in India. In 1946, Sholto Douglas, an agricultural expert at the Kalimpong experimental station in Bengal, had started to work with his associates on a simple and practical hydroponic technique that laymen could employ. Two years later, Douglas announced the development of what he called the Bengal system of hydroponics. Plants were grown in troughs, which could be built from any suitable material. They were filled with a mixture of two parts of sand and five parts of rock chips, gravel, cinders or broken bricks. The nutrients were supplied in dry form, either as a powder or as tablets. The first rain would cause them to go into solution and penetrate the mixture; or, if it did not rain, a light spray of water could be applied. The grower would have to follow certain simple rules for replenishing the nutrient supply, but he required no further knowledge.

The Bengal system of hydroponics has been applied in various parts of the world. It has become popular because it is so easily installed and operated and also because it can be adapted to a wide variety of climates.

There seems to be no likelihood that hydroponics will ever replace normal soil culture. Already, however, it is providing a supplement of food for the world's increasing population. As its techniques are improved, it will probably offer some exciting possibilities. It may be used in years to come to grow vast crops in deserts and other barren wastes and in the rocky terrain of various mountainous districts. It is possible that we may yet realize the old dreams of skyscraper hydroponicums, supplying fresh vegetables and fruits to great numbers of city dwellers. It is even quite conceivable that some day hydroponicums will be set up in manned space stations, orbiting around the earth, and in bases established on the moon.

---

* An aggregate is a clustered mass made up of small units..

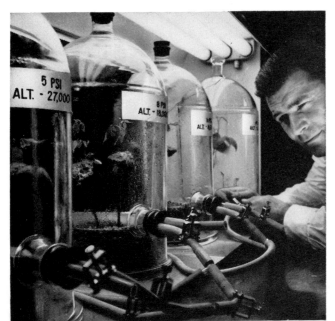

It is believed that hydroponic farms will be established on future moon bases. Some authorities hold that plants will be grown under low-pressure conditions. In a research project sponsored by the United States Air Force, scientists at a Republic Aviation laboratory in Farmingdale, Long Island, have been studying the effects of low pressure on plants. Left: one of the scientists examines beans growing in bell jars at pressures indicated on the outside of the jars. The pressures are given in pounds per square inch (psi). In each case, the pressure is that found at the altitude (on earth) that is indicated. Thus, at an altitude of 27,000 feet, the pressure would be about 5 pounds per square inch. Below: an artist's conception of a hydroponic farm on a moon base of the future. The "farmers" would enter through an airlock (extreme right). Light would be provided by lamps, and nutrients would be supplied to the different plants as required.

Photo and diagram, Republic Aviation Corp.

PROGRESS IN ACOUSTICS

Some Recent Advances in the Study of Sound

BY E. E. FREE

THE science of sound (or acoustics, as it is often called) has been made over radically within a comparatively short space of time. Not so long ago the lectures on sound in colleges and high schools dealt chiefly with the vibrations of such things as the air columns in organ pipes and the strings of pianos and violins, and with the tones that were produced in this way. Nowadays, however, thanks chiefly to a number of electronic instruments, engineers can study sounds and their behavior as effectively as they study mechanical forces or the phenomena of electricity. The result has been a new approach to research in sound; scientists have been able to make far-reaching discoveries in many fields of acoustics and they have been able to put many of these discoveries to practical use.

Foremost among the instruments that have revolutionized the study of acoustics are electronic sound-level meters (SLM), also known as sound meters and sound-intensity meters. These are effective devices that first convert sound waves into weak electric signals, then amplify the signals through electronic means and finally measure them.

The early sound meters were housed in boxes about the size of suitcases. Unfortunately for the engineers who were compelled to carry them, most of the meters were a good deal heavier than ordinary suitcases, unless these suitcases were loaded with bricks or books. Modern sound meters intended for use in the field are quite light even though, like the older models, they are battery-operated. The reason for the use of batteries is that when sound measurements are made in the field, the meters must often be operated at considerable distances from power lines.

Somewhere in the vicinity of the sound-meter case, usually behind a small screened window at one end of it, there is a microphone; it is very much like the type that is used in radio transmission. The sounds that enter this microphone are converted into tiny electric signals, just as they are by the microphone of a broadcasting station. The next step is also the same as in radio. The electric signals that are yielded by the microphone are altogether too feeble to be read on ordinary electric meters; consequently, they must be amplified in the same way as the signals from the radio microphone are amplified before they are broadcast. In the case of sound meters, this is brought about by a special vacuum-tube amplifier inside the case.

Finally, these amplified signals are fed into an electric meter, which records on a dial the precise amount of power that the signal contains. If the microphone always produces exactly the same quantity of electric signal for the same amount of sound and if the amount of amplification that is then given to this signal is always the same or else is definitely known, it is obvious that the reading of the electric meter will show the intensity of the original sound, whatever its source may have been. The engineers who tackled the problem of producing a satisfactory sound meter solved the problem once and for all when they made sure that the two things mentioned above did happen — that is, that the microphone always had the same sensitivity and that the amplifier always provided exactly the same degree of amplification.

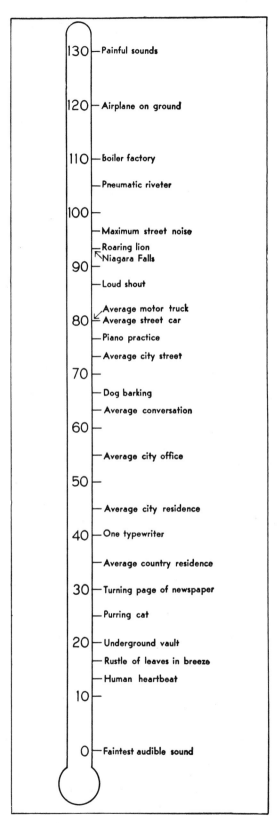

This "noise thermometer" is marked out in decibels.

Various types of sound-level meters are available to the sound engineer for the analysis of sound intensity. With them, one can measure sounds of any kind readily and accurately. Among other things, noise surveys of cities have been carried out and machinery has been tested and rated for its noise level. We shall discuss other uses of the meters in the following pages.

How sound intensity is measured

The sounds that reach our ears are due to disturbances — the beating of a drum, the clang of a bell, the crash of thunder — that cause vibrating pressure waves to travel out in all directions. The intensity of a sound wave is the rate of transfer of vibratory energy per unit of area of the wave. It is not the same thing as loudness, which is the actual sensation produced in the brain when a sound is heard. Loudness depends not only on the intensity of the sound that strikes our ears, but also on how well our hearing functions.

The intensity of a sound is measured in units called decibels.* A sound, taken arbitrarily as the standard in the noise-intensity scale, is rated "zero." It is the faintest sound that the unaided human ear can detect and is equivalent in power to one ten-millionth of a billionth of a watt per square centimeter. The decibel measures the ratio of the intensity of a given sound to the standard ("zero") sound. Practically, the decibel may be thought of as a measure of about the smallest difference in sound intensity that a normal human ear can distinguish. When the loudness ratio changes by one decibel, the ratio of intensity is changed by 26 per cent.

The decibel scale, or "noise thermometer"

The decibel scale, or "noise thermometer," ranges from 0 to 130. An intensity of 130 decibels is perceived not only as sound, but also as a painful sensation in the ear. It may be called the ear's "boiling point." The normal range of painlessly

* A decibel is the tenth part of a bel, a unit named after Alexander Graham Bell, inventor of the telephone.

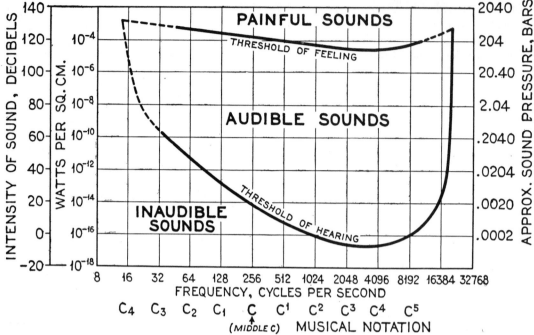

APPROXIMATE RESPONSE OF AVERAGE EAR TO SOUNDS OF DIFFERENT PITCHES AND INTENSITIES

This curve shows the sensitivity of the average ear to tones of different pitches and intensities.

audible sounds for the average human ear is about 120 decibels. For forms of life other than ourselves, the range may be quite different.

The ordinary sound meter, as we have seen, measures the intensity of a given sound, rather than its actual loudness. Under most conditions, however, it is a quite good indicator of loudness, since loudness is based to a considerable extent on the intensity of a sound.

Different noises and their intensities

Hundreds of thousands of noise measurements have been made with sound meters. Average conversation among three or four persons in a room registers at 65-70 decibels; piano practice, 75; a motor truck, 80-95; a roaring lion, 95; Niagara Falls, 95; thunder, 80-110; a railroad train at top speed, 105; and an airplane taking off, 115-120 decibels. Some of the softer sounds register as follows: chewing gum, 20 decibels; a human whisper, 25-30; turning a newspaper page, 30; and the purring of a cat, 35 decibels.

Probably the loudest known noise ever heard by human ears was that of the explosive eruption in August, 1883, of the volcano of Krakatoa, between the islands of Java and Sumatra in the East Indies. No electronic sound meters, of course, were in existence then; but since the sound was heard 3,000 miles away, physicists estimate that the sound at its source must have had an intensity of 190 decibels!

Among the softest sounds (about 15 decibels) detected by the unaided human ear are the beating of the heart, the rush of blood through arteries and veins and the

More than a thousand times as much energy is needed for conversation in a fairly noisy place as in a quiet home.

movement of air in the lungs. A living person with normal ears will hear these sounds — provided he is consciously aware of them — in the more quiet places.

The most silent spot attainable on earth is probably deep underground or up aloft, as in the gondola of a balloon about three miles or so above the earth (providing there is no wind). Where little or no atmosphere exists, as in outer space, silence reigns. Many persons are concerned about the harmful effects of noise in our mechanized civilization. The total or near-total absence of sound may be just as bad; it will undoubtedly be one of the many hazards faced by man should he ever succeed in traveling through space.

How loud can the human voice itself be? A shout, uttered in the open air, may hit 92 decibels. Lee Chrisman of Kentucky, the shouting champion of the United States, could be heard eight miles away; it is estimated that his voice may have registered as much as 97 decibels. According to the ancient poet Homer, the shouts of Stentor, a renowned Greek warrior, equaled the combined voices of fifty men. This would be about 100 decibels, equivalent to the roar of a speeding train.

The power output of even loud sounds is very small

Despite the loudness of such sounds in the ear, their power output is extremely small. Stentor's loudest shout could not have had an output of more than about one-hundredth of a watt, or about a five-thousandth of the power used by an ordinary 50-watt electric lamp. All the people in New York City, shouting together at full voice, could not produce enough power to move even the smallest automobile. It has been calculated that the faintest audible sound is so weak that it would have to go on for more than ten billion years before it could generate enough energy equal to that needed to light an ordinary electric lamp for a single second.

One might be led to believe that the production of many sounds all at once would result in the addition of their intensities and therefore a proportionate increase in loudness. Yet this may not always be the case.* In many large cities there are mixtures of noises from numerous sources, but they may not be excessively loud. City streets range in noisiness from about 50 to 80 decibels. The noise output may be higher — from 90 to 100 decibels — on certain corners of large cities.

Listening in on marine life

The development of the sound meter is only one of a number of important advances in the science of sound, or acoustics, in recent years. Experts have been able to analyze sounds whose very existence was unknown not so long ago. For one thing, they have discovered that fish and other kinds of marine life have "voices" and utter sounds that we normally cannot hear either when we are above the surface of the water or under it. This fact was not definitely known until the second World War. Observers on warships and submarines with sensitive sound-detecting devices picked up all kinds of strange noises that could not have been caused by enemy craft. It was finally learned that sea life was responsible for these sounds. Since that time, the Office of Naval Research has sponsored investigations of this phenomenon. Marie P. Fish, at the University of Rhode Island, has done notable work in recording these sounds in the aquaria of the Narragansett Marine Laboratory and in studying the noise-making organs of sea animals.

These noises have been described as croaking, moaning, pinging, bellowing, humming, whistling, beeping, grunting, clucking, squealing, drumming, whining, buzzing, squawking, grinding, snapping, crackling and popping sounds, depending on just what form of marine life is producing the noise and the particular situation in which it finds itself. Certain areas of the ocean have such characteristic sound patterns of their own that the Navy is developing new sonic instruments and training their operators to cope with this interfer-

---

* In fact, two sounds can be made to cancel each other so that complete silence follows, at least theoretically.

N. Y. World-Telegram & Sun

In the above picture we see an acoustical engineer using a sound meter to measure the amount of noise produced by a pneumatic drill on one of New York City's busy street corners in downtown Manhattan. Note the Empire State Building in the background.

ence. Maps have even been made showing the distribution of the animals making the noises.

Among the forms of sea life (besides fish) that emit sounds are porpoises, whales, shrimps and lobsters. Many marine animals lack vocal cords; they must use other parts of their bodies to produce noise. Fish vibrate their air bladders in different ways or rasp together their teeth, jaws or other bony structures. Shrimps and lobsters, like their distant cousins, the insects, "speak" by rubbing or snapping their legs or feelers against various parts of their bodies.

Some of these noises are no doubt by-products of other activities, but most are probably produced deliberately. As among terrestrial animals, the sounds in question are a form of communication — mating calls, warnings and notes of anger, fear and pleasure. Some fish may use sound to locate their whereabouts in different areas of the ocean. Schools of fish can be detected by their clamor. The device called sonar (see Index) has been used by fishing fleets in order to find their catch. Unfortunately, most food fish "shut up" when manmade craft approach them. The sounds made by whales, dolphins and porpoises are so complex, that some authorities believe they actually form an intelligent language.

The effects of noise on humans

Sound experts have also made detailed studies of familiar noises. Just what is a noise? The generally accepted definition is that it is any nonmusical sound. Not all noises are displeasing or harmful. A rhythmic sound — even the thumping of machinery or the pounding of drums — may inspire us to greater efforts. Soft, repeated noises — the patter of raindrops or the chirping of crickets — may have a hypnotic, lulling effect, which is desirable or undesirable, as the case may be.

Noises are often annoying, and they may have more serious consequences for the human organism. Loud night noises are known to cause dangerous blood-pressure rises in sleepers. Everyone is familiar with the deafening effect of a sudden loud sound, such as that of an explosion or a nearby thunderclap. In extreme cases, the eardrums may rupture and death may even ensue because of the blast pressure. Men who work in constantly noisy places, such as factories and electric-generator stations, tend to become somewhat deaf; their ears become insensitive to lesser sounds.

Aside from the physical harm it may cause, continuous noise, even if not very loud, makes it difficult for us to concentrate on mental work of any kind. More effort is actually required to keep one's attention fixed on the task in hand. The result is that we make mistakes and that we are unusually tired toward the end of the day's work. These observations have been confirmed by scientific tests.

Some of the ill effects in question may be due rather to psychological than to physiological causes. Loud, "frightening" noises, such as the din of a machine or a loud wind, may create a mental hazard. Try crossing a street when pneumatic drills are working on it, even though there may be no traffic to menace you. It takes some conditioning to ignore such a disturbance.

Our attitude toward a given noise may depend on our knowledge or ignorance of its source. A certain person may, for example, find a whistling sound pleasing because he is under the impression that it is a bird's. If he finds that it is being produced by a youngster blowing on a whistle, he may become decidedly annoyed. Sometimes a sound is irritating because it interferes with another one we were listening to or that we expected to hear. To persons attending an outdoor concert, the distracting drone of a plane passing overhead may be disagreeable. To a soldier in a trench the noise of an airplane may be terrifying, while the same sound is wonderful to someone lost in a wilderness.

How noise is abated

Noise may be abated, or reduced, in any number of ways. The most effective method is to eliminate the source of the sound altogether or, as an alternative, to muffle the noise-producer. Automotive and mechanical-appliance industries have spent large sums in research on noise and its prevention. The most common solution is to reduce vibration in the moving parts of machinery or to substitute machines with few or, in some cases, with no, moving parts. A machine can be effectively muffled by means of some special device.

One satisfactory solution of certain noise problems is to put the source of the noise far from centers of population. Unfortunately, this method is not always practicable. Stopping the ears with wax or some other soft material may provide a certain measure of relief.

Noise may be a by-product of certain activities or may be deliberately produced, for purposes of communication, warning or simply for its own sake. Whatever the reason for a noise, it may lead to certain abuses by careless or inconsiderate individuals. To combat excessive noise, particularly in large cities, anti-noise committees have been set up and various laws have been passed. In some cities, a municipal ordinance forbids unnecessary blowing of automobile horns. Strange as it may seem, this has reduced the number of road accidents. Generally speaking, anti-noise committees and laws are effective only insofar as they have the active support of the public at large.

Where noise cannot be eliminated at its source, soundproofing is advised. Rooms and interiors of vehicles, such as buses, trains and airplanes, are often lined with noise-absorbing materials that virtually silence all sounds coming from the outside, including the vibrations of motors. Noises originating from the inside of a room may also be muffled as well. In a room without soundproofing a small noise may be magnified by reverberation — that is, by bouncing off the walls and objects in the room. Further amplification results if the materials present should vibrate in sympathy with the original sound waves. Soundproofing is most effective in such cases. The only disadvantage is that even a slight noise, produced, say, by the fall of a coin, may be disturbing.

To soundproof a room and improve its acoustic properties, suitable materials and construction methods must be used. The conventional building materials — glass, wood, stone and plaster — are hard and reflect sound waves all too effectively. This is especially noticeable in an empty room. Even the presence of a few soft objects, including human bodies, cuts down the amount of reverberation and noise. Soft, porous substances, such as cork, spongy matter, drapes and thick rugs, have an appreciable silencing effect. Walls, ceilings and even floors may be covered with special acoustic linings, designed to absorb sound waves. Flat or convex walls with various partitions and projections prevent the bouncing around or concentration of undesirable noises.

Many of these ideas about sound and acoustics were worked out before the invention of adequate noise meters. But the use of these instruments has placed acoustics on a firm scientific basis and subjected it to exact mathematical analysis. As a result, much progress has been made.

The sound "microscope"

So far we have been concerned with the reduction of sound; but very interesting results have been obtained with its magnification. The ordinary microscope greatly enlarges our vision of things minute. Similarly, sound-intensity meters may be made to amplify noises that we normally cannot hear at all (or only very faintly) so that they become as loud as a pistol shot. So used, they are often called sound microscopes.* The sound of small insect larvae chewing away at wheat grains, prepared cereals or biscuits has been magnified by sound microscopes and has provided an efficient method of detecting the presence of these insect pests.

A sample of the infested or suspected grain is placed into a small cup attached to the sensitive microphone of a sound microscope. Every single move that one

---

* Not to be confused with a recently developed "microscope" that uses artificially produced ultrasonic waves (sound waves of very high frequency) to probe biological cells and tissues.

of the grubs makes as it eats is distinctly audible as a kind of click or thump. In a program that was sent out over the network of the Columbia Broadcasting System, the researchers of the E. E. Free Laboratories broadcast the sound of the grubs in a cupful of grain. On this occasion, the sound heard by radio listeners all over the United States was something like 10,000,-000,000,000 times louder than the actual sound made by the insects.

Why excessive sound magnification would be useless

Still greater magnifications than this are easily possible, so far as the instruments are concerned. They would be quite useless, however, because one would begin to hear sounds caused by atoms or subatomic particles. A few gas atoms are left in the vacuum tubes of the amplifier after the air has been exhausted; these atoms strike the sides of the tubes after the latter become heated. Free electrons are also flying back and forth through the tubes. In the tube filaments, billions of atoms of tungsten are set in agitation by the heat. Finally, the atoms of the outer air are in constant movement; they keep striking against the microphone, producing tiny sounds of their own.

It has been estimated that these atomic sounds would measure only a little below zero on the decibel scale. Tests indicate that we would hear them in the sound microscope if we pushed its magnification beyond 100,000,000,000,000 times, ten times the magnification used to detect the wheat insects. That seems to be the definitive limit of operation of the sound microphone. It also probably represents the maximum degree of sensitivity of the human ear. If the ear were much more sensitive than this, we would begin to hear the atoms striking against our eardrums.

Detecting the high-pitched sounds of insects

A special type of sound magnifier was developed by George Washington Pierce and his associates in the Cruft Laboratory of Harvard University. With this device it has been possible to detect and measure, although not actually to hear, the extremely high-pitched sounds made by certain insects. Experts had previously maintained, indeed, that these insects produce and also hear sounds that are too high in pitch to be perceptible to human ears. However, before the advent of the electronic sound magnifier, one could not prove that such sounds actually existed.

Sound microscopes have been used effectively in a number of industries. They make possible quick detection and location of dangerous vibrations or of imperfections in adjustment. Cracks and flaws in articles made of metal or of ceramic materials can be detected in the same way.

How physicians use the sound microscope

In the medical profession, sound microscopes are used for the easier and more complete study of familiar sounds such as those of breathing or of the heart beat, already familiar through the use of the old-fashioned stethoscope. With the sound meter, physicians can also listen to fainter sounds, never before heard or used in medical diagnosis. Among these sounds are the slight creaks in diseased joints and the tiny whispering noises produced in diseases involving the tubes that connect the throat and the ears.

Echoes and acoustical qualities

Certain instruments based on the same principles as the electronic sound meters serve in the study of the acoustical qualities of churches or public halls. These qualities are determined to a considerable extent by good or bad echoes, or by too much or too little of the general echoing called reverberation. For hundreds of years, it has been common knowledge that it is easy to hear in some auditoriums but very hard to hear in others. The pioneer work of Wallace C. Sabine of Harvard showed that acoustical qualities, good or bad, depended largely on echoes.

He demonstrated that in auditoriums with bad acoustical qualities, sounds kept reverberating to such an extent that a given musical sound or spoken syllable had to compete with the slowly dying reverberations of a number of preceding sounds. In halls where acoustical conditions were good, reverberations were not nearly so pronounced. By using various sound-absorbing devices to cut down the reverberation time, Sabine was able to improve greatly the acoustical qualities of auditoriums.

Acoustical conditions can be improved scientifically

Nowadays, electronic instruments measure auditorium echoes and reverberations precisely; they show what remedies are possible. Guesswork has given way to scientific knowledge. Gone are the days when so-called "experts" sought to improve the acoustics of an auditorium by stringing a few wires across it — a remedy that produced no effect, good or bad.

Electronic sound meters have added greatly to our understanding of musical sounds. It is well known that if the same tone is played on three such different instruments as a piano, a violin and a clarinet, the three sounds produced will not be the same, even though all have the same musical pitch. The pioneer German physicist Hermann von Helmholtz pointed out that these differences are due chiefly to differing overtones — accessory tones produced at the same time as the fundamental tone. Differences in overtones also account in large part for the striking contrast between a good singer and a bad singer.

When critics judge musical tones by ear

All this was necessarily a matter of personal judgment until the science of acoustics developed its electronic instruments. There was no appeal from the dictum of the experienced music critic; he stated arbitrarily that this or that musical tone was good or bad. As a matter of fact, the judgments of many critics are still as arbitrary as ever; they praise or condemn without hesitation, relying only on their sense of hearing. There is no longer any reason why this should be the case.

The unerring scientific analysis of musical tones

Sound meters now measure the exact intensities of musical tones; frequency meters can determine pitch more accurately than a musical ear can do. Other instruments, called frequency analyzers, separate the overtones or other sounds of a mixed tone so that each can be measured individually. These devices have been applied to the improvement of musical instruments as well as to musical training.

Most frequency analyzers work by means of sound filters. These are combinations of electric circuits or mechanical devices that allow certain sounds to pass but that hold back others. In addition to their use in studying musical quality, the instruments have been applied to the analysis of machine noises. Noise arising from a part of a machine, such as an automobile, is isolated from other noises, so that it can be traced and then measured.

The useful "sound sampler"

Certain sounds occur but seldom or else are not available when desired. For example, one may want to study a lion's roar, but may not be able to persuade a zoo lion to roar exactly on schedule. In such cases, a sound engineer often uses a "sound sampler," which is really no more than a portable phonographic recorder. The instrument can be taken to the place where the noise is to occur. It is started automatically as soon as the desired sound begins and thus makes available a sample for study with the sound microscope.

Bird songs have been studied in this way; so have heart sounds from patients with unusual kinds of heart disease. Just as the photographic plate freed astronomers from night-long vigils at their telescopes, automatic recorders have freed sound engineers from millions of hours of listening. This is certainly just as well, because these engineers are occupied at the present time with research projects that bid fair to add greatly to our knowledge of acoustics.

See also Vol. 10, p. 281: "Electronics."

THE GALAXIES

Island Universes in the Heavens

BY GÉRARD DE VAUCOULEURS

GALAXIES are vast systems of stars, similar to our own Galaxy (the Milky Way *) and of comparable dimensions. They populate the depths of space out to the limit of penetration of the largest telescopes. The total number that could be photographed with the 200-inch telescope on Palomar Mountain may come to a billion or so. It is reasonable to assume that many more could be brought into view with an even more powerful telescope.

Early notions
concerning galaxies

The idea that stellar systems existed outside our Milky Way Galaxy had been discussed by philosophers and astronomers as early as the eighteenth century. In the second half of the century, the great German philosopher Immanuel Kant advanced the theory that the celestial objects observed as faint, hazy patches of light and called nebulae were really systems of stars. Kant's theory was confirmed toward the end of the eighteenth century by the British astronomer William Herschel. With the aid

Editor's note: See the article The Milky Way, in this volume.

of large reflecting telescopes that he himself built, Herschel discovered thousands of new nebulae. In the early nineteenth century, the German scientist Alexander von Humboldt gave the nebulae the picturesque name of "island universes." Few astronomers of that period doubted that the nebulae were really made up of stars. It was believed that with the use of more powerful telescopes it would be possible to make out the individual stars in an island universe.

The problem of the nature of the nebulae was reopened in 1864. In that year, the British amateur astronomer William Huggins observed that many of the irregular, diffuse nebulae (such as the one in the constellation Orion) were composed, not of stars, but of extremely rarefied gas. The spectrum * of this gas consisted of isolated bright lines, of which a pair of green lines were the most conspicuous. On the other hand, other nebulae, like the one in the constellation Andromeda, were whitish in color, and their spectra appeared more or less continuous, like that of our own star, the sun. Apparently, then, there were two

Editor's note: For a discussion of the spectrum of light, see the Index, under the entry Spectrum.

THE HUBBLE CLASSIFICATION OF GALAXIES

Proposed by Edwin P. Hubble in 1925, this classification has won world-wide acceptance. In recent years several new types and subtypes have been added.

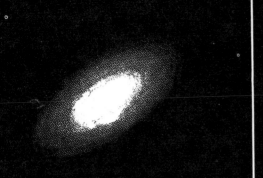

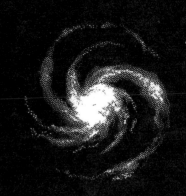

An elliptical galaxy. It has a smooth structure, from a bright center out to vaguely defined edges.

A normal spiral. It has a bright nucleus, from which emerge several distinct spiral arms, or whorls.

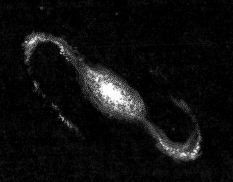

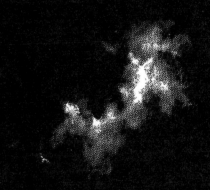

In the barred spiral, above, the spiral arms emerge at the extremities of a bar across the nucleus.

An irregular galaxy. It does not have the clearly defined features of the other types shown on this page.

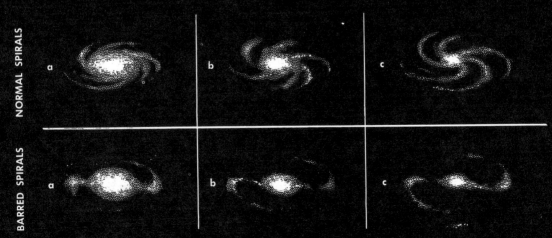

NORMAL SPIRALS

a   b   c

BARRED SPIRALS

a   b   c

In Hubble's system, there are three stages—a, b and c—among normal and barred spirals. The relative size of the nucleus decreases from a to c; the development of the arms increases from a to c.

kinds of nebulae — one kind made up of gases, the other of stars.

It was not until the first quarter of the twentieth century that the full meaning of the distinction between the gaseous nebulae and the stellar nebulae was firmly established. The existence of "extragalactic nebulae" (galaxies outside the Milky Way Galaxy) was then placed beyond doubt. This was due to the work of the American astronomers Heber D. Curtis and Edwin P. Hubble — particularly Hubble. In a decade — from 1924 to 1936 — Hubble pushed the exploration of extragalactic space from the nearer galaxies, at distances of the order of a million light-years,* to the limit of penetration of the 100-inch telescope at Mount Wilson Observatory, at distances a thousand times greater. The more distant galaxies that have been photographed with the 200-inch telescope on Palomar Mountain may be at distances of several billion light-years.

Catalogues and counts of galaxies

The first catalogue of nebulae and star clusters was published in 1782 by the French astronomer Charles Messier. The numbers in his catalogue are still in use for the brightest objects that he listed. For example, the Andromeda nebula (which we now know to be a galaxy) is often referred to as Messier 31, or M 31.

In the year 1783, William Herschel, in England, began a series of systematic surveys of the northern skies. His work was continued and extended to the Southern Hemisphere by his son, John Herschel, who published in 1864 a GENERAL CATALOGUE of 5,000 nebulae and clusters. In 1888, John L. E. Dreyer, of Armagh Observatory, Ireland, published a NEW GENERAL CATALOGUE (NGC), which included 7,814 objects. Two supplements, issued in 1895 and 1908, brought the total of catalogued nebulae and star clusters to over 13,000. The NGC numbers are in universal use; for instance, the Andromeda nebula is NGC 224, as well as M 31.

After the introduction of photography in astronomy, the number of recorded nebulae increased rapidly. For statistical purposes, homogeneous photographic surveys* complete to a certain magnitude limit (limit of brightness) are necessary. Such a survey was published in 1932 at the Harvard College Observatory by Harlow Shapley and Adelaide Ames; it included 1,249 galaxies brighter than the thirteenth magnitude.** A revision of this catalogue by the writer is now in preparation at Harvard College Observatory. A new and much more extensive survey of galaxies brighter than magnitude 15 and to the north of declination *** −30° is being carried out at the Mount Wilson-Palomar Observatories under the direction of Fritz Zwicky.

Exhaustive counts of galaxies brighter than the eighteenth magnitude have been made by the American astronomer Charles D. Shane and his collaborators at the Lick Observatory, in California. In 1936, E. P. Hubble published a series of sampling counts to the twentieth magnitude in small regions regularly distributed over the celestial sphere. From such counts as these, it has been estimated that there are about 1,000,000 galaxies brighter than magnitude 18 and over 50,000,000 brighter than magnitude 21.

The different catalogues and counts that have been prepared by astronomers from Messier's time to the present have provided the material for studies of the distribution of the galaxies.

* *Editor's note*: Homogeneous in this case means "conducted under the same conditions."
** *Editor's note*: The apparent brightness of a star is given in magnitudes. In the old catalogues, first magnitude was assigned to the brightest stars; sixth magnitude, to those just visible with the naked eye. Later, as powerful telescopes revealed stars that had never been seen before, the scale of magnitude was extended. The faintest stars that can now be observed are of about magnitude 23. It has been found that a few stars are brighter than first magnitude. They are given negative magnitudes; thus Sirius is of magnitude −1.52. The brightest star in the sky is the sun, with magnitude −26.72. All the magnitudes mentioned above are *apparent magnitudes*, which measure the brightness of stars as we see them. A star like our sun may seem to be particularly bright only because it is comparatively close to us. To find the intrinsic brightness, or *absolute magnitude*, of a star, we must know how far away it is, as well as how bright it appears to us.
*** *Editor's note*: A star's declination is its position, given in degrees (°), minutes (′) and seconds (″), north or south of the celestial equator (the extension of the earth's equator into space). North declination is marked plus; south declination, minus.

* *Editor's note*: The light-year is an astronomical unit of distance, equivalent to the distance that light travels in a year — about 6,000,000,000,000 miles.

Appearance and classification

As we have pointed out, the extra-galactic nebulae appeared in the small telescopes of the early observers as faint, diffuse patches of light; these seemed to be either circular or elliptical. Internal structure was first noted when larger telescopes became available. In 1845, Lord Rosse (William Parsons) and his assistants, at Parsonstown in Ireland, discovered that certain nebulae were spiral in form.

In 1925, E. P. Hubble proposed a galaxy classification that has been accepted, with certain modifications, by astronomers the world over. In its original form the classification divided galaxies into four main classes, as follows.

(1) *The ellipticals (E)*. They have a smooth structure, from a bright center out to vaguely defined edges.

(2) *The normal spirals (S)*. They show spiral arms or whorls emerging from a bright nucleus.

(3) *The barred spirals (SB)*. Their spiral arms emerge at the extremities of a bar across the nucleus.

(4) *The irregular galaxies (I)*. Some of these are of the same type as the two galaxies called the Magellanic Clouds and are classified as magellanic irregulars (Im). Others are so chaotic in appearance that they are simply listed as irregulars (I).

Hubble distinguished three stages among both normal and barred spirals, labeling them a, b, and c. The relative size of the nucleus decreases from a to c; the development of the arms increases from a to c.

The Hubble classification was recently revised in order to include certain new types and subtypes. The classification now includes the lenticular type (SO), having the form of a double-convex lens; it combines the smooth structure of the ellipticals with the luminosity distribution of the spirals. New stages d and m have been added to the spirals; these stages form the transition between stage c and the magellanic irregulars (Im).

Star populations in the galaxies

The German-American astronomer Walter Baade distinguished two basic types of star populations in galaxies.

Type I is found in irregular galaxies and along the arms of spirals. It includes the stars called blue giants, blue supergiants and red supergiants. The regions where these stars occur are characterized by the presence of interstellar gas and dust and by bright regions of glowing hydrogen gas, ionized by the ultraviolet radiation of hot stars.

Type II is found in elliptical and lenticular galaxies and in the nuclear regions of spirals. It includes red giant stars, subgiant stars intermediate between the giants and the dwarfs, and probably also subdwarf stars.

Types I and II are usually mixed in galaxies. At first, astronomers believed that the ellipticals are composed mostly of type II stars and the magellanic irregulars of type I stars. However, recent studies indicate that the diversity of stellar populations is probably greater than was first realized by astronomers.

The distances of the galaxies

In order to determine the distances of galaxies outside of our own, we must first set up an absolute distance scale for stars in our own galaxy. The direct method of measuring distances by parallax (see Index) would be of little use here, since it would apply only to stars within a radius of about 300 light-years — and our galaxy is about 80,000 light-years in diameter. Hence we have to adopt indirect methods, based on the proper motions of stars in the galaxy, their radial motions,* their apparent magnitudes and so on. By means of these methods, we can establish the distance and the intrinsic brightness of various types of stars in our galaxy. If similar star types can be recognized and observed in other galaxies, they can be used as indicators of distance.

* *Editor's note*: For an explanation of proper and radial motion, see the article The Motions of Stars, in Volume 7.

The extragalactic distance scale has been revised several times since 1952 and is still provisional. At the present time, the following values for the nearer galaxies are often adopted:

	Distance in millions of light-years
Magellanic Clouds	0.2
Andromeda group	2
Ursa Major group	8
Virgo cluster	32

For more distant galaxies, astronomers can make only rough estimates, ranging up to several hundred millions of light-years. The faintest galaxies that can be observed with the most powerful telescopes may be several billion light-years away.

The dimensions of the galaxies

Once the distance of a galaxy is known, its intrinsic (actual) dimensions can be derived from its apparent dimensions, measured on a photographic plate. However, since the galaxies do not have sharply defined boundaries, it is difficult to determine these dimensions exactly. To compare the dimensions of galaxies of various types, it is necessary to study a fairly large number of them; they must all be observed under the same conditions.

Generally speaking, galaxies vary in size from dwarf systems having diameters of 10,000 light-years or thereabouts, to giant systems with diameters ranging up to 100,-000 light-years. Dwarf galaxies are many times more numerous than the giants.

Spectral types and colors

The first spectrogram of a galaxy, the Andromeda nebula, was obtained in 1899 by the German astronomer J. Scheiner at Potsdam Observatory. It showed absorption lines similar to those found in the spectrum of the sun. In addition to absorption lines, the spectra of some galaxies show bright emission lines, like those of gaseous nebulosities in our own galaxy.

The system of spectral classification (that is, classification of spectra) adopted

for stars was recently extended to galaxies by W. W. Morgan at Yerkes Observatory. Of course the spectrum of a galaxy is a composite of the spectra of all the stars it contains. The following main spectral types can be recognized: B, A, F, G and K, corresponding respectively to blue, white, yellow, orange and red stars. These different types correspond to variations in the stellar populations of galaxies.

Rotation and masses of galaxies

The flatness of many galaxies, when seen edgewise, and the presence of spiral

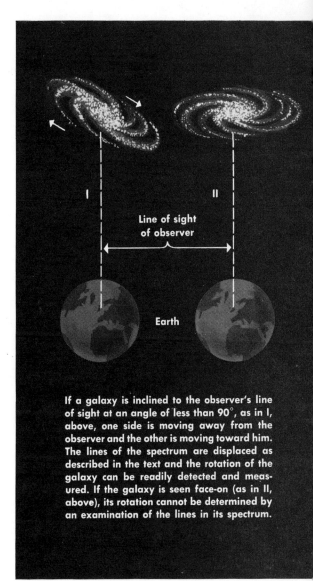

If a galaxy is inclined to the observer's line of sight at an angle of less than 90°, as in I, above, one side is moving away from the observer and the other is moving toward him. The lines of the spectrum are displaced as described in the text and the rotation of the galaxy can be readily detected and measured. If the galaxy is seen face-on (as in II, above), its rotation cannot be determined by an examination of the lines in its spectrum.

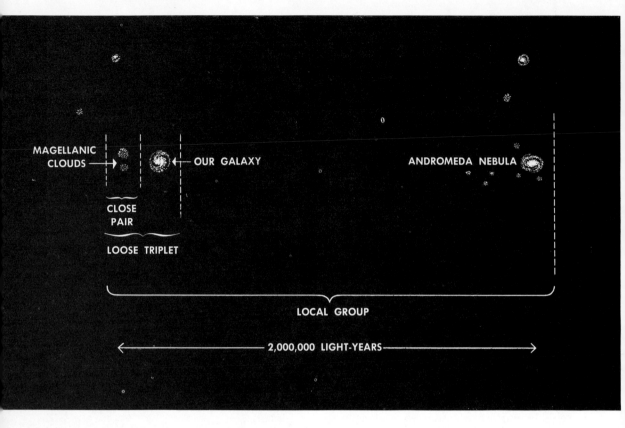

Galaxies often occur in pairs, which may form part of larger groups; the larger groups, in turn, may be parts of still larger ones. For example, the Large and Small Magellanic Clouds form a close pair of galaxies; this pair is associated with our own galaxy to form a loose triplet. The members of the triplet and various other galaxies, including M 31 (the Andromeda Nebula), make up the Local Group, which numbers a score of galaxies and is about 2,000,000 light-years across. There is evidence that this Local Group is part of a Local Supercluster, made up of several thousand galaxies and having an over-all diameter of about 100,000,000 light-years.

arms suggested very early that the stars that composed them were rotating around the nucleus or center of these systems. The rotation of a galaxy can be observed and measured by the displacement of the lines in its spectrum, compared with reference lines in the spectrum of a fixed terrestrial object. If the galaxy is inclined to the line of sight at an angle of less than 90°, one side is moving away from the observer, and the lines of this section are displaced toward the red part of the spectrum; the other side is approaching, and the lines are displaced toward the blue part of the spectrum. This is called the Doppler effect, after the name of its discoverer, the Austrian physicist Christian Doppler. The over-all effect of the rotation results in an inclination of the spectral lines except in the center of the galaxy. For a galaxy seen face-on, the ro-

tation cannot be detected by an examination of the lines in its spectrum.

The velocity of rotation of any part of a galaxy is proportional to the distance from the nucleus. The period of rotation increases outward from a few million years near the nucleus to a few hundred million years in the outer region. Stars, interstellar gas and dust all partake in the rotation, with about the same velocity.

If we can determine which side of a spiral galaxy is nearer to the observer, we can determine the direction of rotation of the spiral arms. In all the cases studied, the arms were found to be trailing in the rotation.

We have just pointed out that the velocity of rotation is proportional to the distance from the nucleus. If we analyze the variations in velocity, applying the law of universal gravitation (see Index), we have a method for determining the masses of galaxies. They range from a few billion times the mass of the sun, for dwarf systems, to several hundred billion times the solar mass, for giant systems.

Pairs, groups and clusters of galaxies

Galaxies occur frequently in pairs, triplets and larger groups. Thus the Large and Small Magellanic Clouds, in the southern skies, form a close pair, which is associated with our own galaxy to form a loose triplet. The Andromeda nebula (M 31) is the major member of a close triplet that includes M 32 and NGC 205. A score of nearby galaxies are members of the so-called Local Group of galaxies; our own galaxy, M 31 (the Andromeda nebula) and the Magellanic Clouds form part of this group, which is about 2,000,000 light-years across. More distant groups have also been observed. In many regions of the sky, galaxies form circular or elliptical clouds of roughly uniform density. The great ma-

jority of the members in these clouds are spiral galaxies, with very few ellipticals or lenticulars.

In some areas, there are huge clusters, consisting of hundreds and often thousands of galaxies. Such is the Virgo cluster which, at a distance of about 32,000,000 light-years, has a diameter of about 7,000,000 light-years. At still greater distances, even larger and denser clusters have been observed. The great majority of members of these big clusters are elliptical and lenticular galaxies. The nearest of these, the Coma cluster, is at an estimated distance of 120,000,000 light-years. It has a total population of more than 10,000 galaxies in all and a general diameter of from 20,000,000 to 25,000,000 light-years.

Evidence has been obtained that at least certain clusters form still larger groups, called superclusters. Since 1953, the writer has been presenting the evidence for the existence of a Local Supercluster. This flattened system seems to include about a thousand of the brighter galaxies and probably several thousand of the fainter

Neighboring galaxies sometimes interact with one another, producing spectacular effects. Ribbonlike filaments of matter may stream out from one galaxy to another.

54 THE GALAXIES

ones. The Virgo Cluster is at the center of the system; the Local Group, of which our own galaxy is a member, is apparently not far from the edge. The over-all diameter of the Local Supercluster is about 100,-000,000 light-years; its thickness, about 20,000,000 light-years. The flattening of this supercluster suggested that it might be rotating. The rotation was confirmed in 1958 through an analysis of the velocities of several hundred bright galaxies.

Interactions between galaxies

The close grouping of galaxies occasionally brings out spectacular interaction effects between neighboring systems. These effects have been specially investigated by F. Zwicky, at the Palomar Mountain Observatory. He found that they take a great variety of forms, depending on the distance between the galaxies, their sizes, their masses and probably various physical properties still little understood. Very often, ribbonlike filaments of matter stream out from one galaxy to another. A filament may also emerge in the opposite direction. In certain cases, the outer arm of a spiral joins with a corresponding arm of a neighboring galaxy. When two galaxies are in collision and intermingle, vast antennalike streamers emerge from a chaotic central mass. In a few instances, galaxies which seem isolated display various distortions for which no visible companion can be considered responsible.

The mechanism of these interactions is not yet clearly understood. The presence of bright emission lines in the spectra of interacting galaxies indicates an unusual state of excitation of the interstellar gas. But no emission lines have been observed in the extended filaments and appendages of weakly interacting galaxies. This suggests that the luminosity of these filaments is due to starlight and not to the excitation of interstellar gas.* It is difficult to understand how such long filaments of matter can remain stable for any length of time.

* Editor's note: For a discussion of how interstellar material reflects light or is excited by hot stars, see the article The Astronomer Explores Space, in Volume 9 of The Book of Popular Science.

Red shifts and receding galaxies

We have already pointed out that the Doppler effect provides a means for determining whether a star or group of stars is receding from an observer or whether it is approaching him. The first radial (that is, line-of-sight) velocities of galaxies were measured between 1914 and 1925 by the American astronomer Vesto M. Slipher at the Lowell Observatory in Arizona. Except for a few of the nearest galaxies, they were recession velocities; the galaxies were moving away from the earth. In 1929, E. P. Hubble discovered that the greater the distance of a galaxy, the greater its recession velocity.

Galaxies are speeding away from us at a truly awesome rate. Milton L. Humason, at Mount Wilson Observatory, measured velocities of up to 40,000 kilometers (25,000 miles) per second. Still greater ones have been observed with the 200-inch telescope on Palomar Mountain. They range up to 61,000 kilometers (38,000 miles) per second, or more than a fifth the velocity of light, for members of a distant cluster of galaxies in the constellation Hydra. The American astronomer William A. Baum has estimated that certain velocities are in excess of 100,000 kilometers (60,000 miles) per second, or about a third the speed of light. In all the cases we have mentioned above, it has been found that the recession velocity of a given galaxy is very nearly proportional to distance.

The theory of an expanding universe

Some astronomers have attempted to show that the red shift of the spectral lines is not due to the Doppler effect but· to other effects. Since they have failed in their efforts, we must accept the theory of an expanding universe. We may think of space itself as expanding and in this process carrying away the galaxies with it. As viewed from any galaxy, the effect would be the same; all other galaxies would appear to be receding.

See also Vol. 10, p. 268: "Galaxies."

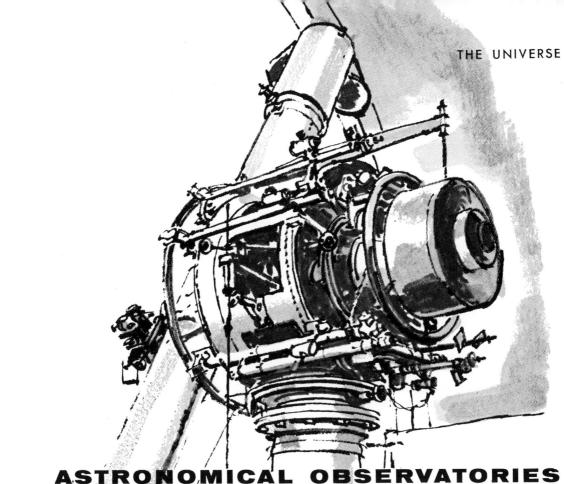

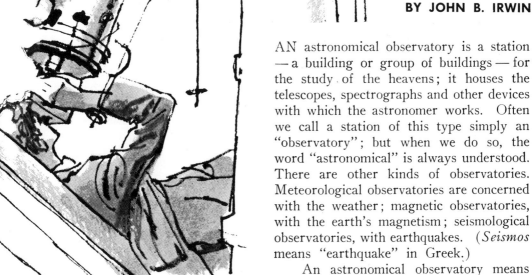

The vertical-circle telescope with which the positions of Southern Hemisphere stars have been catalogued at Australia's Mount Stromlo Observatory. Mount Stromlo is a 2,650-foot peak in the Australian Capital Territory.

ASTRONOMICAL OBSERVATORIES

An Account of the Astronomer's Headquarters

BY JOHN B. IRWIN

AN astronomical observatory is a station — a building or group of buildings — for the study of the heavens; it houses the telescopes, spectrographs and other devices with which the astronomer works. Often we call a station of this type simply an "observatory"; but when we do so, the word "astronomical" is always understood. There are other kinds of observatories. Meteorological observatories are concerned with the weather; magnetic observatories, with the earth's magnetism; seismological observatories, with earthquakes. (*Seismos* means "earthquake" in Greek.)

An astronomical observatory means different things to different people. To a letter carrier, it means the end of a winding mountain road; to an airplane pilot, a welcome and well-known landmark. A janitor may think of it as a place to keep clean; a place filled with complicated instruments that he must not touch. To an astronomer's wife, it may mean an unusual way of life — an isolated home with magnificent views; endless mountain trails over which one may ramble.

To an astronomer, the observatory may be his home; more often, it is a place where he goes for a few nights or weeks of intensive observing, returning to his home base to digest his observations and write his scientific papers. Everywhere astronomers are starting their night's observing. They set their telescopes, throw switches, push buttons and levers and start clocks; they glue their eyes to the eyepieces of their telescopes in order to hold a faint stellar image in place.

The site of an astronomical observatory must be as carefully selected as the instruments that will be housed in it. If the observatory is in a low-lying or comparatively low-lying area, the image that appears in the lens or the mirror of a telescope will be faint and distorted. The reason is that the lower layers of air are often filled with ground fog or low-lying stratus clouds, as well as with a haze of dust or smoke. Nearness to a city is as much of an obstacle to good viewing as low altitude. The lights of a great city can effectively blot out the fainter stars. Many an observatory originally built "out in the country" has had its effectiveness seriously reduced by the encroaching suburbs of a city.

For the best viewing, an observatory should be constructed on a height and in as isolated a place as possible. The great English scientist and mathematician Isaac Newton (1642-1727) pointed out the advantages of a high-altitude site. He wrote: "The Air through which we look upon the Stars is in a perpetual Tremor . . . The only Remedy is a most serene and quiet Air, such as may perhaps be found on the Tops of the highest Mountains above the grosser Clouds." It is true that even a mountain peak does not make a perfect site for an observatory. Its crest, as any experienced mountaineer will tell you, is sometimes anything but "serene and quiet" and may produce its own storms. Yet when conditions are right, mountain observatories offer un-

paralleled viewing. Visitors to such observatories often marvel at the brilliance and great numbers of visible stars and are awed by the magnificent appearance of the Milky Way.

The first thing that strikes the eye as one approaches a typical large observatory is the white, hemispherical structure called the dome. (There may be more than one.) The telescope is housed within this structure; the longer the telescope, the larger the dome. The dome protects the telescope from the sun, rain, wind, dust and heat. An opening, or slit, in it is kept covered when the telescope is not in use; it is opened up by means of sliding or folding panels when the astronomer is ready for his observing. At night, as the telescope follows a star in its apparent rising and setting, the dome is rotated. It is driven by electric motors and rolls on a smooth track.

The dome is painted white with a special paint, so that it will reflect as much of the sun's heat as possible and not absorb it. The temperature must be kept the same inside and outside the dome. If it is hotter inside, the air will well up and "bubble" through the slit, thus spoiling the image that strikes the telescope lens or mirror. Sometimes large fans are located on the inner walls in order to circulate the cool night air inside the dome. Often a canvas or metallic wind screen is provided; this can be raised or lowered in the slit opening in order to protect the long barrel of the telescope from buffeting by the wind.

Telescopes are the focal points of observatories. Some of them look like gigantic spy glasses, with a large lens at one end of a long tube and an eyepiece (or, more commonly, a plate-holder) at the other. Such telescopes are called refractors. Their size is indicated by the diameter of the large lens. The largest refractor in the world, at the Yerkes Observatory, has a lens with a diameter of 40 inches, and a tube that is approximately 60 feet long.

Other telescopes — the reflectors — are quite different in shape. They have a great reflecting mirror set at the bottom of the large tube, which is not usually a closed cylinder but a skeletal structure. The starlight that is reflected from the mirror converges upward and comes to a focus — called the prime focus — at or near the top of the tube. As we shall see, the converging rays are made available for observation in several ways. The largest telescopes in the world are of the reflector type; the very largest, the Hale telescope at the Palomar Observatory atop Palomar Mountain in California, has a mirror with a diameter of 200 inches, or 16⅔ feet.

A large telescope, whether of the refractor or reflector type, is mounted on one or two giant piers, solidly fixed in bedrock and built to a great height. The piers must be completely independent of the dome and the rest of the building. If the building were set to vibrating by the wind, by the motors or by people walking about, the vibration would be transmitted to the telescope, and the image would be hopelessly blurred.

The telescope rotates about two axes — polar and declination. The polar axis is parallel to the axis of the earth. As our planet rotates from west to east, the apparent motion of the heavens in the opposite direction must be compensated for in order to keep a given celestial body exactly centered. The telescope is made to turn about the polar axis, by means of a motor or clockwork, at just the right speed to keep up with the stars. The declination axis is perpendicular to the polar axis. Rotation about the declination axis sets the telescope in the north-south direction.

The astronomer looks through an eyepiece that may magnify 500 times or more. By means of small electric motors connected to gearings, he carefully and frequently adjusts the telescope pointing. The body that is being observed must be centered either on the magnified image of a pair of illuminated crosswires or on the narrow slit of a spectrograph,* through which the light of the star is to be directed.

* A spectrograph is a device that spreads out or disperses light into its component wave lengths or colors, which are then recorded photographically. (Each color corresponds to a definite wave length.) The spectrograph is a variety of spectroscope; for a detailed account of this important device, see the Index, under Spectroscopes.

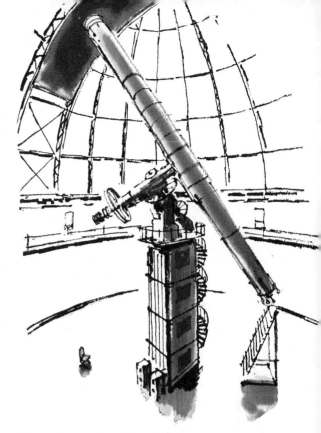

Above: the 40-inch Yerkes Observatory refractor, the largest refracting telescope in the world. Its tube is approximately 60 feet long. The observatory has a rising floor, which is shown at its lowest position here.

In the domes that house large telescopes, a platform or in some cases the entire floor is raised or lowered by means of electric motors or hydraulic pumps in order to make the eyepiece comfortably accessible. In some cases, an adjustable observing ladder is provided.

In the case of certain reflecting telescopes, the starlight reflected from the mirror at the bottom of the tube is made to strike a small mirror at or near the top. From there, it is directed at a 90° angle to an eyepiece set at the side of the tube. This is called the Newtonian focus. The observer is stationed here on an observing platform that may be raised or lowered. This is a most inconvenient place in which to work. One astronomer has suggested that the easiest way to get at the eyepiece in such a system would be to fill the dome with water and to paddle to the desired point in a canoe. My own thoughts, when I am stationed at the eyepiece near the top of a reflector tube, perhaps for ten or twelve hours at a stretch, may be summed up as follows: (1) Don't fall off the observation platform to the concrete floor of the observatory far below. (2) Don't drop your flashlight or anything else on the mirror at the other end of the tube.

The 200-inch Palomar reflector and the recently completed 120-inch reflector at the Lick Observatory are so large that it is possible for the observer to operate at the prime focus in a little cage centered in the top of the tube. The percentage of the star-light cut off by such a cage is not serious. As the telescope is focused on one star after another by the night assistant pushing buttons on the control panel far below, the astronomer in the cage may have a rather thrilling ride through the night.

There are more convenient systems for observing than those we have just described. In some telescopes, a small curved mirror is located near the top of the reflector tube. The mirror sends the reflected

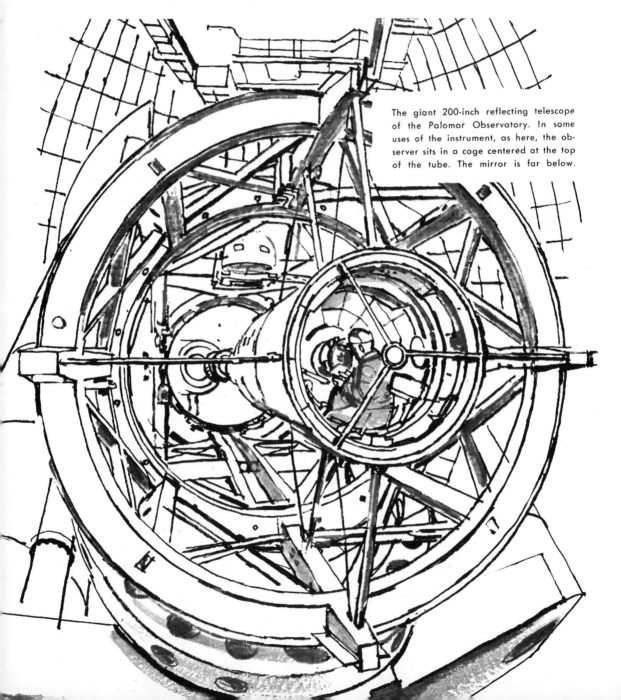

The giant 200-inch reflecting telescope of the Palomar Observatory. In some uses of the instrument, as here, the observer sits in a cage centered at the top of the tube. The mirror is far below.

light from the mirror back again down the tube and through a hole in the main mirror. The light then reaches what is called the Cassegrainian focus. This place — back of the big mirror and near the floor level — is much more convenient for observing than the Newtonian focus, near the top of the tube.

Even more convenient is the so-called coudé focus, at the lower end of the polar axis. It may take from two to four mirrors to get the light to this point, which is often in a temperature-controlled room. A very large, efficient and stationary spectrograph can be located here. At the Palomar Observatory, the coudé spectrograph room is large enough so that one can walk around in it; one of the spectrograph mirrors is 48 inches in diameter.

Photography with a coudé spectrograph may take hours; hence the astronomer can often relax with a book. He need only glance into the eyepiece now and then to make sure that the image is approximately bisected by the slit of the spectrograph. However, if the object is very faint, reading light and book must go. No stray light can be permitted to enter the slit; the observer's eye must be completely dark-adapted. The control panels and clocks can be dimly illuminated only with red light — if at all.

The very large reflectors, such as the 200-inch at the Palomar Observatory, have a small field of view. They can sound the great depths of space in this direction and that, but they cannot give the complete overall picture. To supplement the work of such instruments, smaller wide-angle telescopes, with an extended field of vision, are required. The smaller telescopes can do many jobs more quickly and effectively.

A telescope is a complex research device, which may serve in various ways. The earlier astronomers relied entirely on visual observation — that is, examining the desired object through the eyepiece of the telescope. This method is used now only to a limited extent by modern astronomers; it has been largely replaced by various other types of observation.

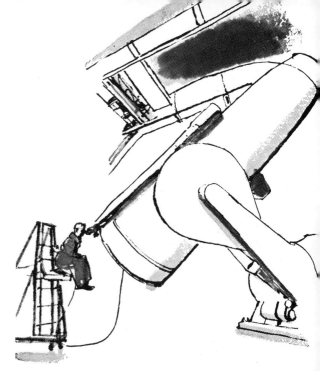

The Schmidt 49-inch telescope of the Palomar Observatory is really a big camera. The observer sights through a small visual telescope and operates the big telescope electrically by means of a control box that he holds.

The telescope serves very often nowadays as a giant camera, in which the telescope mirror or a mirror-lens combination (usually called a Schmidt telescope) is substituted for the camera lens. The light of a star or galaxy is directed to a photographic plate or film, where it is recorded. Much research astronomy is now conducted by this method. The entire sky has been photographed over and over again with a variety of focal lengths, emulsions and exposure times. A photographic plate can provide a permanent and duplicable record of a host of fine details in an object such as a galaxy — details much too faint to be seen by the observer stationed at the eyepiece.

Many persons who visit observatories have previously had their imaginations stirred by spectacular photographs of various celestial objects. Often they are sadly disappointed when they look at the same objects through a telescope. I remember the remark of a visitor who stepped up to the eyepiece of the large reflecting telescope of the McDonald Observatory, at Mt. Locke, Texas, after standing in line for a long time. The telescope was focused on a bright globular star cluster. The visitor looked in the instrument for a few seconds and then said feelingly: "Never again will I wait so long to see so little!"

Sometimes the light of a star or galaxy is not photographed directly, or even at all. A great telescope mirror may be used to gather up a large column of starlight and to focus the star — perhaps invisible at the eyepiece — onto a specially prepared alkaline surface. Through the magic of the photoelectric process (see Index, under Photoelectric cells), the light is converted into electric current, which can be readily measured. This current is often so weak that it may have to be amplified a million times in the phototube and a million times again in an electronic amplifier.

The colors of a star can be precisely analyzed by interposing a series of colored glass filters in the beam of light coming from the mirror of a reflector. The analysis of color helps us determine the temperature and brightness of stars. The reddening and absorption of light due to the dust between the stars and galaxies gives much valuable information about the composition of interstellar materials.

Every large reflector is used with a whole series of spectrographs. For the brightest objects, the observable range of color may cover many feet of photographic plate. For faint stars, it may be so small that the film on which it appears could be tucked under the astronomer's fingernail. By the analysis of spectrograms — photographic records made by spectrographs — the astronomer can obtain invaluable information about the composition of the stars, their motion in space and other matters.

It must not be thought that the astronomer dispenses entirely with visual observation when he does direct photography, or makes photoelectric measurements or uses color filters or spectrographs. In all these cases, he either looks at the object on which he is working in order to focus it accurately, or else he looks at a nearby guide or reference star.

Certain specially designed instruments are used at solar observatories. One of these is the coronagraph, which makes it possible to observe the sun's corona during those hours of sunlight when the atmosphere is immaculately clean. Another is the tower telescope, used to study solar spectra. These two instruments are described in the article The Sun, Our Star, in Volume 3 of THE BOOK OF POPULAR SCIENCE.

Large telescopes are usually operated through the night by a two-man team of astronomer and night assistant. These often alternate in their duties. The astronomer will frequently be developing one plate in a nearby darkroom, while another is being exposed by the night assistant. At Palomar, duties are alternated around midnight for a hot "lunch." After a long night's work, when the dawn light becomes so bright that the stars have begun to vanish, the astronomer walks down a path a few hundred yards to the "Monastery," a dormitory where a quiet, air-conditioned and completely darkened room awaits him. He is called to "breakfast" early in the afternoon.

It may be, of course, that no observational work is possible because of weather conditions. If this is the case, the astronomer may occupy himself with calculations based on the previous night's observations. Or he may while away the time in various ways. One of the rooms at the Palomar Observatory contains a pool table. During the long, cloudy California winter nights, frustrated astronomers often play endless games of pool; some of them have become really expert at the game.

Scene at Canada's David Dunlap Observatory, at Richmond Hill. Since the temperature inside the dome must be the same as the outside temperature, staff members wear heavy clothing when making observations in winter.

Observer at the United States Naval Observatory looking through the 26-inch refractor. A part of the mounting of the telescope is shown at the right.

A large modern observatory is a complex affair. Besides the domes and observing rooms, there are offices for the staff of observers and computers; a library; rooms for the reception of visitors; lecture rooms; laboratories of various kinds; dark-rooms for developing and copying photographs; and shops in which parts for various devices can be built or kept in repair. The larger observatories have residences and dormitories for the staff.

New devices are constantly being perfected. An observatory director may sometimes spend little time "directing" and a great deal of time obtaining money for new devices and helping to plan and build them. The importance of such devices cannot be exaggerated. A fifty-year-old telescope with brilliantly designed modern instrumentation can outperform a larger, newer telescope in an observatory where the newer developments in the field of astronomy have been ignored.

The work of the astronomical observatory

The astronomers who man the world's observatories carry out a variety of tasks. At some observatories, they determine the correct time by observing the passage of certain stars across the meridian.* Once the time has been obtained, it is sent out far and wide by telegraph and radio. The Naval Observatory of the United States has such a time service. Every clear night, trained astronomers at Naval Observatory stations in Washington, D.C., and in Richmond, Florida, photograph the stars as they pass nearly overhead. After much measurement and calculation, the results of these observations are used to correct master precision clocks. Radio Station WWV at the United States National Bureau of Standards broadcasts clock beats night and day on a number of wave lengths; these signals are controlled by master clocks, which in turn are corrected by the stars.

Astronomers also calculate exact latitude and longitude. International boundaries are fixed on paper by international treaties but are actually marked out on the earth through the observation of the stars. To determine latitude and longitude requires stellar observations of the most delicate nature, together with the knowledge of the positions and motions of thousands of stars.

Some astronomers calculate the changing positions of the stars in the heavens. These positions are given in degrees, minutes, seconds and hundredths of a second of arc.** A hundredth of a second of arc in the sky corresponds to a foot on the surface of the earth. If you walk a hundred feet, the sky will continue to look the same to you;

* The meridian is a great circle passing through the celestial poles and the zenith of a given place — the point of the heavens directly overhead.
** For an explanation of these units of measurement, see the Index, under Arc, degrees of.

but the proper telescope will reveal the resulting change in the positions of the stars as viewed from the earth.

The determination of time, latitude and longitude and the calculation of star positions form what is known as fundamental astronomy. At the present time, this type of work is done in a comparatively small number of observatories.

An important task of astronomers in observatories is to prepare photographic star maps showing the celestial bodies in the heavens. The National Geographic Society recently sponsored the preparation of such a series of sky maps, called the *National Geographic Society-Palomar Observatory Sky Atlas.* The *Sky Atlas* covers the more than three-fourths of the sky observable from Palomar Mountain; it includes stars a half-million times fainter than can be seen with the naked eye. The "camera" was the 49-inch Schmidt telescope at Mount Palomar. The Carnegie Institution hopes to establish a similar Schmidt telescope in north-central Chile, on an 8,000-foot peak not far from La Serena, in order to complete the history-making Palomar survey to the south celestial pole.

Under the auspices of the International Astronomical Union, about eighteen observatories all over the world began working in 1887 on a map of the sky (*Carte du Ciel,* in French). Telescopic cameras of the same focal length and scale were used in obtaining the necessary photographs.

This huge radio telescope of the United States Naval Research Laboratory at Maryland Point, on the Potomac River, detects radio noises that come from outer space. The "dish" of aluminum mesh has a diameter of 84 feet at the outer edge.

A big modern observatory is made up of a considerable number of units. At the right is shown the layout of the Mount Stromlo Observatory on Mount Stromlo, in the Australian Capital Territory.

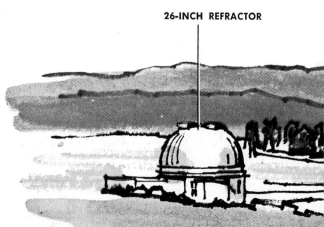

26-INCH REFRACTOR

Work on this gigantic co-operative venture is now being completed. About 4,500,000 different stars have been recorded on more than 20,000 different photographic plates. The exact positions of a great many of these stars have been measured and recorded in catalogues printed in English, French, Spanish, Italian and German.

Solar observatories study the surface features and the composition of the sun. They also conduct important researches on the nature and quality of solar radiation. This radiation affects the ionosphere, the atmospheric layer that plays an important part in radio transmission. (See Index, under Ionosphere.) The National Bureau of Standards now issues monthly forecasts of ionospheric conditions based on the study of solar radiation.

There are many other kinds of pure research carried on in observatories. Among the fields of investigation are the proper motions of the stars; their composition; their magnitudes (brightnesses) and the classification of their spectra; solar eclipses; the absorption of light by interstellar particles; and the general structure, chemical composition and age of the universe.

In research carried on in an observatory, a great deal depends not only on *how* the stars are observed but on *what* stars are observed. For example, an astronomer with a 36-inch telescope equipped with a good photoelectric photometer can measure — with some effort — the magnitude

(brightness) and color of any of a hundred million stars. He must make a choice as to what few stars he will observe. That choice is based partly on the astronomer's interests and training; it must also be based on what is already known and what other astronomers are doing. It may well be that halfway through his observing program, he may have to change it drastically because of some new fact that has turned up.

No article on astronomical observatories would be complete without some mention of the mushrooming number of radio observatories, featuring radio telescopes. A radio telescope is a device that detects radio waves emanating from various parts of the sky. (See the article Radio Astronomy in Volume 9.) The combination of radio and optical observations has been very fruitful. Powerful optical telescopes are needed to identify and photograph many heavenly radio sources.

The development of astronomical observatories

We have been dealing thus far with the modern astronomical observatory. We must bear in mind that it represents only the latest stage in a development that has been going on for thousands of years. Sites or posts for the observation of the heavens *

* Generally, the observatories of the ancients were simply places where portable astronomical instruments could be used. Sometimes structures such as obelisks served as instruments for observation.

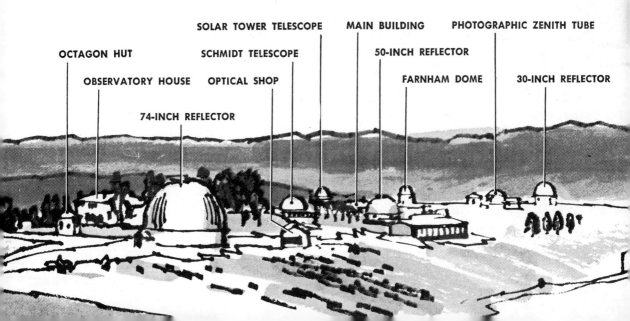

were employed by the ancient Babylonians, Egyptians and Greeks. The Greek astronomer Hipparchus, who lived in the second century B.C., prepared a catalogue of the stars from observations he made on an observatory on the island of Rhodes. In the early Middle Ages, the Arabs constructed noteworthy observatories at Damascus, Baghdad and elsewhere. Later, observatories were set up in Western Europe.

About 1420, Ulugh Beg, a Persian prince, established a famous observatory at Samarkand, in Central Asia. Here he erected large masonry instruments, one of them 180 feet high, in order to observe the sun, moon, planets and stars with the highest possible precision.

Another renowned observatory was the one built for the Danish astronomer Tycho Brahe (1546-1601), on the island of Ven (or Hven), now in southern Sweden. The main observatory building was called Uraniborg (Castle of the Heavens). Tycho and his students did much important work here, including a catalogue of the positions of more than a thousand stars.

Galileo Galilei (1564-1642) was the first to use the telescope in astronomical observation. In 1609, he began to scan the heavens with the newly invented device; his "observatory" was the balcony of a building in the Italian city of Padua. With the successful application of the telescope to the study of astronomy, astronomical observatories multiplied in the centuries that followed. Outstanding ones were founded in the Netherlands, France, Denmark, England, Germany, Russia and other countries.

In 1828, the first permanent observatory was set up in South Africa at Slang-Kop (Snake Hill), about three miles from Cape Town. Other observatories have been established in South Africa since that time; some of them represent branch stations of northern observatories.

The Southern Hemisphere has fine observing conditions. North-central Chile was selected by ten American universities as a site for a 158-inch reflecting telescope, being constructed near La Serena at the Cerro Tololo Inter-American Observatory. The European Southern Observatory is building a 144-inch reflector at Cerro La Silla, also near La Serena, Chile.

The first permanent astronomical observatory in the United States was founded at the University of North Carolina, at Chapel Hill, in 1831, though various temporary structures had been put up before that time. In 1875, James Lick, an eccentric San Francisco millionaire, left $700,-000 to the University of California for the establishment of the world's largest observatory. It was decided to erect it on the top of Mt. Hamilton, in California. A 36-inch refractor was housed in the observatory. The Lick venture was brilliantly successful. Recently, a 120-inch reflector was dedicated at the Lick Observatory.

In 1917, a 100-inch reflecting telescope — the largest in the world up to that time — was erected in an observatory on Mount Wilson, California, by the eminent American astronomer George Ellery Hale. An even larger telescope, with a mirror 200 inches in diameter, was installed in an observatory atop Palomar Mountain, California, in 1948. The story of this great telescope is told elsewhere (see Index, under Palomar Mountain telescope).

A large mountain observatory has been constructed on Kitt Peak, 6,875 feet high, some 40 miles southwest of Tucson, Arizona. This Kitt Peak National Observatory is financed by the National Science Foundation and is managed by a group of ten American universities.

Canada has several outsanding observatories. The Dominion Astrophysical Observatory, established in 1916 at Victoria, British Columbia, has a 73-inch reflecting telescope. The David Dunlap Observatory at Richmond Hill, Ontario, has an even larger reflector, with a 74-inch mirror.

The earth's atmosphere blots out many wavelengths of electromagnetic radiation before they reach the ground. To study these wavelengths, which include X rays, ultraviolet and infrared, the United States has launched an orbiting astronomical observatory into space around the earth.

Many other important observatories are listed in the table on page 66.

See also Vol. 10, p. 268: "Observatories."

Mount Wilson and Palomar Observatories

Mount Wilson Observatory, at Mount Wilson, California, houses the 100-inch Hooker telescope. The slit of the rotatable dome is open, indicating that the big telescope is being used.

IMPORTANT OBSERVATORIES OF THE WORLD

In this table, "O" stands for "Observatory"; "refl." for "reflector"; "refr." for "refractor"; "schm." for "Schmidt"; "int." for "intensity interferometer"; "sol." for "solar telescope." The figures given under "Telescopes" represent the diameters of reflector mirrors and refractor lenses. Only the largest instruments in a particular observatory are included here.

NAME OF OBSERVATORY	SITE	TELESCOPES
Academy of Sciences U.S.S.R. O.	near Zelenchukskaya (Caucasus Mtns.)	236-in. refl.*
Anglo-Australian O.	Siding Springs, Australia	150-in. refl.*
		40-in. refl.
Argentina, National O. of (Cordoba)	Bosque Alegre, Argentina	61-in. refl.
Boyden Station (Armagh O., Dunsink O., Hamburg O., Rep. of S. Africa, Stockholm O., Uccle O.)	Maaselspoort, Rep. of South Africa	60-in. refl.
		32-in. schm.
Cerro Tololo Inter-American O.	Cerro Tololo, Chile (near La Serena)	158-in. refl.*
		60-in. refl.
		24-in. schm.
Crimean Astrophysical O.	Central Crimea, U.S.S.R.	104-in. refl.
		50-in. refl.
David Dunlap O. (U. of Toronto)	Richmond Hill, Ontario, Canada	74-in. refl.
Dominion Astrophysical O.	Victoria, B.C., Canada	73-in. refl.
		48-in. refl.
European Southern O. (Belgium, Denmark, France, Holland, Sweden, West Germany)	Cerro La Silla, Chile (near La Serena)	144-in. refl.*
		60-in refl.*
		40-in. refl.
		40-in. schm.
Greenwich O., Royal	Herstmonceux Castle, Sussex, England	98-in. refl.
Harvard College O. (Oak Ridge Station)	Harvard, Mass.	61-in. refl.
Haute-Provence O.	Saint Michel—l'Observatoire, Basses-Alpes, France	77-in. refl.
		48-in. refl.
Helwan O.	Helwan, Egypt	74-in. refl.
Kitt Peak National O.	Kitt Peak, Arizona	158-in. refl.*
		84-in. refl.
		60-in. sol.
La Plata O.	Leoncito, Argentina	85-in. refl.*
Lick O. (U. of California)	Mt. Hamilton, Calif.	120-in. refl.
		36-in. refl.
		36-in. refr.
Lowell O.	Flagstaff, Arizona	72-in. refl.
		42-in. refl.
Lunar and Planetary Lab. (U. of Arizona)	Mt. Bigelow, Arizona	60-in. refl.
Mauna Kea O. (U. of Hawaii)	Mauna Kea, Hawaii	88-in. refl.
McDonald O. (U. of Texas and U. of Chicago)	Mt. Locke, Texas	107-in. refl.
		82-in. refl.
Mt. Stromlo O.	Canberra, Australia	74-in. refl.
		50-in. refl.
Mt. Wilson O. (Carnegie Inst. of Washington and Calif. Inst. of Technology)	Mt. Wilson, Calif.	100-in. refl.
		60-in. refl.
Narrabri O. (U. of Sydney)	Narrabri, Australia (2)	264-in. int.
National Astronomical O. Mexico (formerly Tonantzintla O.)	San Pedro Martir Mtns. (Baja Calif.)	60-in. refl.*
		26-in. schm.
Okayama Astrophysical O.	Kamogata, Japan	74-in. refl.
Ondrejov O.	Ondrejov, Czechoslovakia	79-in. refl.
Palomar O. (Carnegie Inst. of Washington and Calif. Inst. of Technology)	Palomar Mountain, California	200-in. refl.
		60-in. refl.
		49-in. refl.
Radcliffe O.	Pretoria, Rep. of South Africa	74-in. refl.
Republic O.	Johannesburg, Rep. of South Africa	74-in. refl.*
Royal Cape O.	Cape Town, Rep. of South Africa	40-in. refl.
Shemakha Astrophysical O.	Shemakha, Azerbaijan	79-in. refl.
Smithsonian Astrophysical O.	Mt. Hopkins, Arizona	60-in. refl.*
Steward O. (U. of Arizona)	Kitt Peak, Arizona	90-in. refl.
Tautenburg O.	Tautenburg, East Germany	53-in. schm.
U.S. Naval O.	Flagstaff, Arizona	61-in. refl.
		40-in refl.
Uppsala O. (Kvistaberg Station)	Bro, Sweden	39-in. schm.
Yerkes O.	Williams Bay, Wisconsin	40-in. refr.
		41-in. refl.

* Under construction.

COMPETITIONS FOR YOUNG SCIENTISTS

A Survey of Projects and Awards

BY THOMAS GORDON LAWRENCE

Westinghouse photo

John F. Kennedy, the late President of the United States, posing with winners of the 1962 Science Talent Search.

HOW does loud rock music affect your hearing ability? What is the population of various plants and animals in the community pond? How can we explain the formation of ammonia in outer space? These are a few of the problems studied in award-winning science projects.

There was a time, not so very long ago, when a youngster who completed a science project on his own could expect no more than a pat on the head or a word of praise from his parents or teachers. The vastly increased interest in science in recent years, however, has brought about a great change. Today there are literally thou-sands of awards for bright boys and girls who are proficient in science and who carry out acceptable projects. These awards include medallions, slide rules, sets of books, cash prizes and scholarships ranging up to $10,000.

You win in science competitions largely through your work in science projects — and you are old enough to start one if you are old enough to read this book. "But just how do I go about it?" you may ask. It may prove helpful if I tell you how I got started on my first science project, when I was thirteen years old. I had finished read-ing THE INSECT BOOK, by the American biologist Leland O. Howard. I was fascinated by his pictures of the praying mantis (see the illustrations on this page) and his account of how this fierce insect captures flies, grasshoppers and mosquitoes and then nibbles at them in much the same way that you might bite into a lamb chop.

I examined the bushes and grass in our yard, but I could find only some ants, a few crickets, several dozen spiders and some very busy flies. That night, though, a big green mantis flew in through the window and circled around the electric lamp. I finally managed to capture it without hurting it and put it into a tight wooden box with a flat piece of glass for a cover. My mother thought that the mantis might be thirsty. She suggested that I should moisten some crumpled tissue paper and put it in the box. (When she was a girl, she said, she had drowned a grasshopper that she wanted for a pet by pouring too much water into the aquarium that was to serve as its home.)

I was delighted when the mantis beat the air with its feelers, stalked over to the paper and began to "nibble at" and swallow drops of water. I had read somewhere that scientists always kept notes; I wrote down in a composition book all that I saw.

The next day I caught two flies, a small caterpillar and a cockroach and dropped them all in the mantis's box. The newcomers and the mantis paid no attention to each other at first. Then a fly came within two inches of the mantis. The latter swung its head around and stared at the fly; then it began creeping forward very slowly, trembling as if it were nervous. Suddenly, its front legs lunged forward; it seized the fly and began to take dainty bites. I wrote all of this down in my notebook.

What happened next day seemed almost too good to be true: I discovered that the mantis was laying eggs. She was depositing on the walls of her wooden box a mass of something that looked like egg white, beaten stiff and mixed with mucilage. From what I had read, I knew that she was placing her eggs in this mass as she stirred it with the tip of her abdomen, and that she was mixing in thousands of little air bubbles. By the time she walked away from her egg case, or "nest," it was hard and dry and could protect the eggs from enemies and bad weather.

The first science project of the author of this article involved the insects known as praying (or preying) mantises. At the left, below, we see a female praying mantis laying eggs; at the right, a group of young insects.

John H. Gerard, from Nat. Audubon Soc.

Robert C. Hermes, from Nat. Audubon Soc.

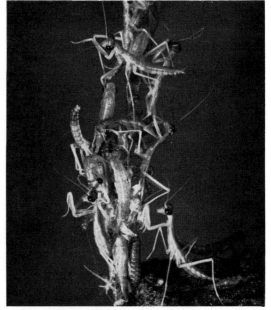

I never got to see the babies of this mantis hatch from the egg case. Later, though, I found a mantis egg case on a rosebush, cut off the branch and placed it under a glass jar. Early in the spring, a hundred infant mantises squirmed out of the egg case, looking like tiny mummies wrapped up in cellophane sheets. They dangled from the "nest," each suspended by a thin silk thread. Later, the mummy cases split to let the long-legged little mantises come out and walk around. I then made a very small discovery which I have never found in any book to this day.*

I never took my mantises to school, as I did not think that my teachers would be interested, and I never wrote a report on my observations. If I were a thirteen-year-old today and had carried out the project I have just described, I would be able to prepare an exhibit based on it. I would write up a report on my mantises and then submit this report, my specimens and any photographs or drawings I might have made to a science fair — and I think I would have a good chance of winning an award.

Projects suitable
for a science fair

Perhaps I should explain the term "science fair." It refers to a group of exhibits showing interesting specimens, collections, machines, models and the like. Each exhibit explains a scientific principle or tells an interesting story about some aspect of science.

You should design and build your own exhibit, but you may seek help from your friends, your teacher, your parents or anyone else who can help you. Suppose that you wished to enter an exhibit of rocks and minerals found in the district where you live. You might want to collect all the rocks in the exhibit yourself. There would be no objection, however, to your borrowing a few specimens from a friend or even from the school collection; but in that case you should not claim that you had collected all the specimens. You would look up the names

and characteristics of the rocks, and then you would arrange and label the collection. It would be all right for you to get help in naming your specimens from your teacher or from a scientist. It would be proper, also, for you to ask somebody else's opinion of what you had written and to have him make corrections in your spelling or facts.

You should remember that a science fair is not intended to appeal only to people who know about science. Your classmates should be able to understand your exhibit and they should enjoy looking at it. Don't try to explain things that you don't understand yourself and don't use words that you cannot explain. Arrange your specimens or models neatly. Print your title and your labels very plainly; give as much interesting information as you can. In this way, you can prepare a worthwhile exhibit even if your ideas are not original and your specimens are not uncommon. You may only have gathered rocks and minerals found along roadsides and in fields near your home; yet other boys and girls and their parents will enjoy learning the names of these rocks and minerals and finding out interesting facts about them.

You can make an attractive exhibit based on wild flowers and grasses and other plants along highways that are perhaps within a short distance of your own home. In order to prepare a worthwhile exhibit of this kind, you will have to learn how to collect and press flowers, leaves and small plants. If you do a good job, you will have beautiful specimens that will remain in good condition for years.

You can make many different kinds of exhibits for a science fair with the plants that you have collected. Here are a few possibilities:

(1) The commonest weeds and wild flowers along our highways.
(2) Flowers that bloom in the spring.
(3) Autumn leaves.
(4) Tree flowers.
(5) How to use leaves to tell one kind of tree from another.
(6) Various kinds of leaves from the same kind of plant; why they differ.

---

* I discovered that a baby mantis seems to require a good drink of water very soon after it starts to walk.

(7) Oak leaves of the United States — their leaves, twigs, bark and acorns.
(8) Legumes. (Beans, peas, sweet peas, clover, alfalfa, lespedeza, vetch and locust trees belong to this family.)
(9) Our enemies, the ragweeds, whose pollen causes hay fever. (If you show the flowers of these plants, be sure to enclose them in glass or plastic vials; otherwise you may be accused of causing hay-fever attacks.)

You might make your exhibit more interesting by also showing live plants, especially if you select those with flowers or fruits. Often you can cut off the tops of the plants and place these in containers of water. Or you might make a terrarium and show the plant life of a particular location (say, a swamp), or the plants that grow in shady woods or a miniature forest sprouting out of an old tree stump. Be sure that your specimens and container are clean and neatly arranged.

You can make a fine exhibit of an insect collection. In the article Collecting Insects, in Volume 3 of The Book of Popular Science, you are told how to collect, kill and display insects. You might arrange the insects according to the family to which they belong, or according to whether they are beneficial or injurious to man and his crops.

Live animals sometimes make interesting and attractive exhibits. The article Unusual Pets, in Volume 7 of The Book of Popular Science, tells you how to collect and take care of wild creatures. Always try to make your wild-animal exhibit as attractive as possible. For example, if you showed a few doleful-looking snakes in a wooden box with a glass cover, you would certainly not be likely to win an award.

However, if you placed the same snakes in a sparkling glass terrarium, containing a layer of clean white sand and a few healthy growing plants, the spectators would be more likely to admire your specimens. If you designed and drew handsome colored posters or charts giving the names of the snakes and telling what part they play in the balance of nature and where they are found, your exhibit might be outstanding.

Some very interesting exhibits are based on the explanation of scientific instruments. You can display a microscope, Geiger counter, telescope, X-ray tube, radio or other device and explain how it works by means of accurate diagrams. Remember that in all such exhibits you should *show* as much as possible. It is usually good to include written (or typed) information, but, as far as possible, your exhibit should explain itself.

Again, you might prepare an exhibit explaining some scientific principle or other. You could go all the way back to Newton and his great laws of motion; or you could explain Einstein's theory of relativity or the quantum theory — provided, of course, that you understand them well enough yourself. Geology and the other branches of earth science offer many possibilities for this type of exhibit. You could use models and drawings to show how lakes are formed and how they sooner or later disappear; how rivers "grow old" and then "grow young" again. You could explain humidity and then show how it is measured by means of the instrument called a psychrometer.

How to carry on scientific experiments

The projects that we have described thus far have consisted mostly of exhibits which might be interesting but which would not require much original work on your part. Other projects involve original observations or original experiments. In working on them, you should follow the basic approach used by scientific researchers and by inventors who use scientific principles to develop useful things. Each one of these men decides on the problem or problems that he wants to study. He reads about what has already been done in the field and examines the apparatus or machines that have already been constructed. Then he plans how to go beyond those who have already preceded him. He tries one avenue of approach after another. At every step, he makes careful observations and keeps a written record of all his results. In the end, his careful work may lead him exactly nowhere; or else he may come upon a scien-

Joe Covello, from Black Star

The author and a group of his students, all of whom won prizes in science competitions in 1962. Left to right: Miriam Herzfeld, Carl Feit, Peter Grafstein, Leslie Chess, Thomas G. Lawrence (the author), Ralph Zuckerman and Herbert Fried. Leslie Chess won a prize in the Future Scientists of America competition; the other five were among the forty Science Talent Search finalists.

tific principle that may revolutionize knowledge or a practical device that may benefit millions of people.

At the outset, let me give you a bit of advice that may surprise you: don't try too hard to be original. Above all, don't take chances. Don't play around with live wires; don't try to experiment with your television set while the current is on. Never, *never* mix chemicals together in haphazard fashion just to see what will happen. Nothing much may happen; but there is always the chance that a dangerous explosion will take place. Fortunately, relatively few young people do the foolish things I just warned you about.

A good many young would-be experimenters simply waste their time, begging their teachers or friends to suggest original experiments that they can work on or hoping desperately that some idea or other will occur to them. To such young people I say that they will never be original unless they know something. The more you know, the more likely you will be to think of a hypothesis — a scientific guess that can be tested. To know, you must read. You might start with your textbooks and THE BOOK OF POPULAR SCIENCE; read the chapters or articles that appeal to you most. Then study the reading lists given in these books and find out all that you can about the topics that particularly interest you.

If a hypothesis occurs to you, you must put it to the test; you must plan and perform experiments. Suppose that you had formed the hypothesis that the growth of the terminal bud of a stem (the bud that forms at the tip) checks the growth of most of the other buds on the stem. To test your hypothesis, you might grow coleus cuttings or bean seedlings. In each case, you would find that the terminal bud would develop rapidly, producing growth in the length of the stem. The other buds, producing new stems, would develop much more slowly or, in most cases, would not grow at all. You would note that some of the lower buds would promptly start to grow after you cut off the terminal bud. Thus you would show that your hypothesis was correct.

However, if you had the spirit of a true scientist, you would soon wonder: "*How* does the growth of the terminal bud check the growth of the buds lower down?" You would then consult your references and would probably learn that plant hormones greatly affect plant growth. This might lead you to repeat some of the plant-hormone experiments described in your references. The results of these experiments, in turn, might cause you to form new hypotheses, which would lead both to new experiments and new reading.

Science competitions
that you can enter

For boys and girls in elementary schools, the best chance for success in science competitions is usually in science fairs held in the school or in a central fair in the town where the school is located. Where there is such a central fair, the various schools first hold fairs of their own and then send only the best exhibits from each school to the central fair. If you were one of the top winners in such a fair, your exhibit would be shown at your main county fair or state fair. A number of awards are given for winners at local, county and state fairs.

Each spring, Science Service conducts the National Science Fair-International. The exhibits in this national fair are of top quality and no wonder; for they are the cream of the exhibits in 200 or more area fairs. An area fair may include exhibits from a large city, a county, a part of a state or an entire state. While boys and girls of any age can compete in many of the school fairs, you have to be in the last three years of senior high school to be eligible for acceptance in the National Science Fair-International. Your teacher can get information about science fairs, from the lowest to the highest level, from Science Clubs of America, 1719 N Street, N.W., Washington 6, D. C.

Those whose exhibits are selected for the National Science Fair-International are treated to a five-day, all-expense-paid trip by plane or fast train to Seattle, Los Angeles, Philadelphia or whatever city has been selected for the fair in that particular year. Here they meet other budding scientists from all over the United States and perhaps from other countries.

There are many awards, ranging up to $100, for both the biological and the physical sciences at the National Science Fair-International. Many organizations provide special prizes. For example, the American Society for Microbiology offers awards of up to $100 for studies of bacteria or other microorganisms. Two $100 awards are granted by the American Chemical Society for the best chemical exhibits. The National Pest Control Association gives $100 for the best work in any field of biology.

If you were sent to the National Science Fair-International as a local top winner, you might be invited to be a guest at the annual convention of one of the great scientific associations. At this convention, you would find exhibits showing the latest advances in biology, dentistry, medicine, pharmacy, veterinary medicine, aeronautics and space science. You might also have the opportunity to take an exciting trip. For example, the United States Army offers sixteen trips to see what is new in missiles, mathematics, electronics and chemistry. A boy exhibitor at the Fair might be selected as a Navy Science Cruiser; he would spend about a week on ships at sea and on tours of land-based research posts. Boy and girl exhibitors would have an equal chance for summer jobs in hospital laboratories.

Perhaps you might want to enter the Future Scientists of America competition for awards offered through the National Science Teachers Association. All students in grades 7 through 12 in any school in the United States and its territories and in Canada are eligible for this competition. So are students in schools conducted for the dependents of United States military personnel overseas. To enter the competition, you must hand in a report of a project in any field of science or mathematics. You should send only the *report* of your project and *not* the project itself. However, you may submit drawings, diagrams, charts, photographs and other illustrative material.

If you enter the Future Scientists of America competition, you may win one of 660 medallion awards or one of the 2,000 honorable-mention certificates. If you win an award, a special certificate will be given to your school. In addition, there are twenty-five scholarship awards of $250 each

Selections from
SCIENCE APTITUDE EXAMINATION
for
THE TWENTIETH ANNUAL
SCIENCE TALENT SEARCH
(Copyright 1960 by Science Service, Inc.)

DIRECTIONS: Four possible answers are given for each question. Choose that answer which is *most nearly correct.*

Bioluminescence is principally the result of
1. a chemical reaction
2. an electrical discharge
3. cosmic radiation
4. reflection of light rays

A reference to a "Moebius strip" would most likely be found in a text on
1. chromatography
2. cybernetics
3. physiology
4. topology

The Seebeck effect is associated with
1. contraction of cardiac muscle
2. design of nuclear reactors
3. meteorology
4. thermoelectricity

Cryogenics is the study of
1. computer logic
2. free radicals
3. the properties of matter under high pressure
4. the properties of matter at temperatures near absolute zero.

The stellerator is for
1. amplification of microwave signals
2. high-altitude observation of the surface of the sun
3. investigation of the spectra of very hot stars
4. research in control of fusion reactions

Charles Darwin, in his *Origin of Species,* formulated the theory of
1. mutations producing new species
2. natural selection producing new species
3. radiation producing mutations
4. use and disuse in forming new species

Reproduced by permission of Science Service, Inc.

Specimen questions from a science aptitude examination taken by contestants in a recent Science Talent Search. For the correct answers, consult the Index of our set.

for students in grades 11 and 12. Entry blanks are not sent directly to students; your teacher may obtain them by writing to Future Scientists of America, 1201 Sixteenth Street, N.W., Washington, D.C.

Local industry or professional groups, or both, often sponsor chapters of the Junior Engineering Technical Society (JETS) in high schools. The JETS program includes a National Project Exposition with awards and scholarships. Junior-high-school winners receive pieces of equipment, such as slide rules and drafting sets; high-school winners may get college scholarships. Your teacher can obtain information from JETS, P. O. Box 589, East Lansing, Michigan.

A new program that began in September 1970 is perhaps the largest and most extensive awards program for high-school projects in science and engineering. Called "Tomorrow's Scientists and Engineers," it is administered by the National Science Teachers Association, The Engineers Council for Professional Development and Scholastic Magazines, Inc. The top awards are ten scholarships of $6,000 each. A total of $87,000 is awarded each year. For information, write to the National Science Teachers Association, 1201 Sixteenth Street, N.W., Washington, D.C.

The oldest national competition is the Westinghouse Science Talent Search. The Search is conducted by the Science Clubs of America, an activity of Science Service, and is supported by the Westinghouse Educational Foundation. The competition is restricted to members of the graduating class of senior high schools. For information, write to Science Clubs of America, 1719 N Street, N.W., Washington, D.C.

Each contestant must (1) take a science aptitude examination, (2) submit his scholastic record and (3) send in a report on an original project or experiment.

You take the aptitude examination in December of your senior year. In preparing for the examination, it is important to study previous examinations (copies can be obtained for a modest fee from the Science Clubs of America). Of course, the more advanced your scientific reading and the wider its range, the more likely you will be able to do well in the examination. Generally speaking, contestants should be able to read and understand such periodicals as *Scientific American* and *Natural History* as well as college textbooks in at least one branch of science.

A transcript of your record — your courses and grades — will be sent to Washington, where the finals for the Talent Search are held. In addition to the transcript, you will have to list all of your activities in science or mathematics. You should also set down any other activities that may enable the judges to determine whether you have the qualities typical of a first-rate scientist — outstanding ability, perseverance, imagination and originality.

Your report should be about 1,000 words in length and not longer than 2,000 words. Give the details of your work, as far as possible, in charts, tables, diagrams, drawings, photographs or maps. You do not have to send in your equipment; some winning students, however, have included a few slides or short sections of films.

Out of the many thousands of boys and girls who enter the Science Talent Search, forty finalists are selected for the final competition, which takes place in Washington, D. C. Each finalist wins an award of $250 and receives a gold pin; he also is given a five-day trip to Washington, D. C., with all expenses paid. He discusses his favorite project with well-known scientists; he may meet the President of the United States. There are interviews, luncheons and dinners, as well as tours to such institutions as the National Bureau of Standards, the Smithsonian Institution and the National Gallery of Art. While the forty winners are in Washington, they compete for top scholarships, ranging up to $10,000. The total amount awarded is $67,500.

Besides the scholarship awards, which go only to finalists, Science Clubs of America presents about 10 per cent of the entrants in the Science Talent Search with Honors Certificates. You will find that colleges are very much interested in finalists or holders of Honors Certificates in the Science Talent Search, as well as in students who are successful in other science competitions. Such students will find it easier to enter the college of their choice and will have a good chance to get scholarships.

The report that you prepared for the Science Talent Search or one of the other competitions mentioned above may be published, if you are fortunate. It may appear in a magazine or yearbook gotten out by your own school. SCIENCE WORLD magazine runs a section called "Tomorrow's Scientists" in each issue; this section consists of accounts of projects by high-school students. Once each year, NATURAL HISTORY magazine publishes an article by a high-school scientist, telling about his successful biology project. Many original projects and researches by high-school students have been described by C. L. Stong in "The Amateur Scientist," a column appearing in the SCIENTIFIC AMERICAN. SCIENCE PROJECTS HANDBOOK, published by Science Service, includes about three dozen accounts of scientific research submitted in the Science Talent Search and the National Science Fair-International.

Suggestions for science projects

We have already indicated some possible projects for young scientists. Here are some more suggestions:

A. BACTERIOLOGY

Making a culture of bacteria from the skin, throat or nose.

What effects do different types of soap have on the growth of bacteria?

Effects of antibiotics on various cultures of bacteria.

Competition between different species of bacteria in sour milk.

What happens when *Penicillium* mold is grown in Petri dishes with various bacteria?

Effects of tetracycline on *Sarcina lutea*. (Tetracycline is an antibiotic; *Sarcina lutea* is a bacterium that grows in bright yellow colonies.)

Bacteria and molds found in a school classroom.

B. BOTANY

The parts of plants that we eat. (You can exhibit seeds, fruits, stems, roots and plant products, such as flour and sugar.)

How to tell a fruit from a vegetable.

Different types of seeds.

Experiments with plant hormones.

Using plant juices to make chemical indicators. (Red cabbage and some other plants will give beautiful results.)

Experiments with plants growing in different degrees of humidity.

Experiments with plants exposed to varying amounts of light.

Experiments with plants grown in light of different wave lengths — that is, different colors.

Repeat Mendel's experiments with pea plants. (Consult a biology book.)

Do all plants store starch in their leaves? (Use the iodine test for starch to find out.)

A study of root hairs.

A study of the vegetation of a swamp, forest, meadow, roadside, pond or stream. (Study, collect and classify the plants of the selected area; note the abundance, appearance and size of the plants.)

Changes in the growth of black mold caused by variations in heat and pH. (pH is a measure of the acidity or alkalinity of a substance.)

C. CHEMISTRY

How to make charcoal.

Experiments in growing crystals.

What you can do with paper chromatography.

A comparison of the chemical content of rain water from different areas, or of tap water from different pipes.

A study of dyes.

What are the uses of ion-exchange chemistry?

Experiments in the control of corrosion in iron or
other metals.
Catalytic effects of metallic oxides in the burning
of sugar.
Construction of a microbalance.
Chemical gardens.
Oil from the ground to you. (For an exhibit.)
Products made from petroleum or natural gas.
(For an exhibit.)
Removing stains from different fabrics.
The chemistry of rubber — natural and synthetic.
Tests for different kinds of cloth.

D. EARTH SCIENCE (GEOLOGY, METEOROLOGY)

Air currents.
A weather station.
Cross section of a volcano.
Cross section of the earth.
Cross section of an oil well and the surrounding
strata.
Measuring air pressure with barometers.
History of the earth and its inhabitants as revealed
in the rocks.
Rocks and minerals of your county.
Model of a geyser.
How we use radioisotopes to determine the age of
rocks.
The Ice Ages in North America.
Homemade rain — a study of cloud-seeding.
A study of caves.
Cities buried under volcanic ash.

Westinghouse photo

Westinghouse photo

BACTERIOLOGY (above). Display showing how
bacteria affect the properties of cheeses. Microbiology means "biology studied with the microscope."

BOTANY (below). Project based on a two-year
study of plant respiration and the effects of various inhibitors on this fundamental cell process.

Westinghouse photo

CHEMISTRY (above). Project and exhibit dealing with sugar-boron complexes: that is, chemical
combinations of boric acid with various sugars.

EARTH SCIENCE (right). Display of minerals
with unusual characteristics; demonstration of use
of chemicals in identifying rocks and minerals.

Science Service

Westinghouse photo

MATHEMATICS (above). Research project dealing with mathematical logic, as applied to a special field of mathematics known as topology.

Science Service

MEDICINE (above). The study of hypothermia, involving the application of extremely cold temperatures to the body, led to this interesting project.

PHYSICS (left). The unusual ion accelerator shown here was designed and built by the exhibitor; it won the top award in a Science Talent Search.

Westinghouse photo

PSYCHOLOGY (below). Prize-winning exhibit, based on a number of experiments which had to do with the singular courting behavior of the ring dove.

Westinghouse photo

Science Service

ZOOLOGY (above). This project involved determining the extent of fish hearing by implanting electrodes in the auditory areas of a fish's brain.

E. MATHEMATICS

Which is better — the decimal number system, which we now use, or the duodecimal number system, in which there are twelve symbols?

A study of the uses of the binary number system. (In this, there are only two symbols — 1 and 0.)

Design and construction of a computer.

What are the "imperfections" in traditional Euclidean geometry?

A study of flexagons.

What kinds of arguments are there in mathematics?

A new method for summing certain infinite series.

F. MEDICINE AND DENTISTRY

How to protect oneself against contamination by radioactive fallout.

Electronic devices in medicine.

Does the eating of apples help to check tooth decay and gum disorders?

Are detergents harmful when swallowed?

Has royal jelly, the food given to future queen bees, any beneficial effects on test animals?

How a tooth decays.

Plants valuable in medicine.

Effects of fluoridation on the teeth and the body in general.

G. PHYSICS

The flow of electricity.

How light is reflected and refracted.

Experiments with static electricity.

Simple machines and how they work for us.

What makes an airplane move?

Making crystal radio sets and experimenting with them.

Model airplanes that really fly.

Experiments with magnets.

Building an electromagnet.

Using heat to produce electricity.

A homemade cloud chamber.

Experiments with a homemade electric cell.

A study of sound waves and musical instruments.

The bubble chamber and how it is used to study subatomic particles.

H. PSYCHOLOGY

Conditioned reflexes in flatworms.

The perception of depth in creeping babies and young animals.

A study of the behavior of young children.

Investigation of various factors affecting the perception of vertical and horizontal lines.

Optical illusions.

Experiments on the perception of color.

Experiments on learning in canaries or other birds.

Training bees to visit artificial flowers.

I. ZOOLOGY

Wildlife in the backyard.

A collection of birds' nests.

Prehistoric animals.

The eyes of different animals.

The hearts of different animals.

A balanced aquarium.

Animal life contained in a cubic foot of soil.

Experiments with wrigglers (the larvae of mosquitoes).

Making a culture of protozoans.

An experimental study of ants of the family Formicidae.

Effect of X rays on *Paramecium* and *Blepharisma*.

Regeneration of the tail fin in goldfish.

A comparative study of electrocardiograms taken of various mammals.

Pamphlets and books on projects

You will find a great many projects listed in the following pamphlets and books:

A. PAMPHLETS AND PAPERBACK BOOKS

If You Want to Do a Science Project. National Science Teachers Association, Washington 6, D. C.

Science Projects Handbook. Science Service, Washington 6, D. C. (Actual projects of thirty-six students successful in Science Talent Search and in National Science Fair-International.

Sponsor Handbook. Science Service, Washington 6, D. C. (Gives much information on science projects, clubs and contests.)

Thousands of Science Projects. Science Service, Washington 6, D. C. (Lists several hundred experiments and projects that were shown at science fairs.)

B. HARD-COVER BOOKS

How to Do an Experiment, by Philip Goldstein. Harcourt, Brace and World Inc., N. Y. (Excellent introduction to subject.)

101 Simple Experiments with Insects, by H. Kalmus. Doubleday and Company, Inc., Garden City, N. Y.

The Living Laboratory, by J. D. and R. H. Witherspoon. Doubleday and Company, Inc., Garden City, N. Y. (Experiments in biology.)

Seven Hundred Science Experiments for Everyone, compiled by UNESCO. Doubleday and Company, Inc., Garden City, N. Y.

Science Teaching Today, in seven volumes. National Science Teachers Association, Washington 6, D. C. (Physical science experiments with water, air, heat, sound and so on.)

You will find a good many ideas for projects in THE BOOK OF POPULAR SCIENCE, particularly in the section called "Projects and Experiments." It would be a good idea, too, to look through your science textbooks to find facts and principles that can serve as bases for projects.

Margot Fuld, from Monkmeyer

Lawn bowling in New York's Central Park. Older people are benefited by amusements like this.

ON GROWING OLD

The Developing Science of Geriatrics

BY HELEN MERRICK

"GROW old along with me, the best is yet to be," sang the poet Robert Browning in a blithesome mood. The idea of growing old gracefully has been with us for a long time. Some few men and women achieve the feat, loved and honored by younger generations as well as their own. For all too many others, however, old age is dreary, useless and beset by infirmities. Yet we know today that the later years can be made rich and productive.

One tool in this effort is geriatrics, the branch of medical science concerned with human old age and its diseases. "Geriatrics" comes from two Greek words meaning "old age" and "healing." A broader term, also from the Greek — gerontology — is sometimes used. It means the study of aging, not only in human beings but also in other living creatures and in such nonliving substances as crystals.

No one would deny that to make old age brighter is a worth-while aim. There are other reasons, however, for the growing importance of geriatrics to society as a whole. In countries where the standard of living is high, more men and women than ever before are living to a ripe old age. In fact, in such countries as the United States, the elderly group is increasing more rapidly than any other. Two thousand years ago, in the Roman Empire, the average length of life was only 23 years, hardly past youth. In the United States around 1900, the average was 47. But consider how life expectancy has leaped within our own century. Today the United States average is about 70 years, a gain of nearly 25 years. For American men, life expectancy is almost 67 years; for women, 74, surpassing the Biblical three score years and ten. Since 1900 the total United States population has more than doubled. The age group 65 years and over has done more than keep pace, but has increased from about 4 per cent of the total population to about 10 per cent or somewhat over. In some states it may be higher. This proportion of oldsters will probably rise in future years.

The tremendous gain in life expectancy has come about largely because many of the hazards of childbirth and youth have been practically conquered. One-fifth of the babies born in the United States around 1900 died before they were old enough to go to school. Today, if a child reaches his first birthday, he will almost surely reach high-school age. Diseases that attack youth come for the most part from outside the body, carried by microorganisms. With the tremendous improvements in sanitation and medicine's victories over many infectious illnesses, a much greater number of people are surviving into middle and old age. Medicine has still to win a like conquest over the chronic diseases, which begin silently within the body and are a much greater threat in the later years. Some scientists believe that if human beings could enjoy complete freedom from disease, the normal life span would be 125 years.

The profound shift in the population structure has far-reaching implications. The economic aspect alone has tremendous importance because so many elderly persons have not been able to support themselves. Despite retirement, old-age-assistance, welfare and social-security payments, oldsters have a difficult time maintaining themselves in general. Few persons reach-

Black Star

Black Star

Advanced years hold no terrors for this skillful typist, who finds contentment in being useful.

ing 65 have enough savings to be financially independent. Nevertheless, benefits paid to old people have increased enormously over the past few decades, from about a half-billion dollars yearly in the 1930's to billions of dollars annually today. Governmental Medicare and Medicaid programs have solved some of the problems of sick older people, but much more remains to be done in this area.

We begin to age from birth and even before; it is all very gradual. Youth does not depart and age arrive at the stroke of midnight on a particular date. The organs of the same person age at different rates. Furthermore, the rate of biological aging varies widely among different individuals. Some persons are able to ward off amazingly the effects of old age. Thus, in one study of old people without disease, comparatively little impairment of bodily organs was found although the subjects were a hundred and more years old.

There is, however; a hazy dividing line on one side of which an organism is considered young and on the other side of which changes appear that indicate aging. Throughout life, there is a building-up (growth) and a breaking-down (atrophy) going on in most tissues. Youth is in the ascendant only while growth exceeds atrophy.

Some living things appear to live indefinitely as in the case of the one-celled organisms, which, when they have grown to their maximum size, simply divide to form two new young individuals. In one

experiment a single ameba went through two hundred successive divisions and was still multiplying merrily at the end of thirteen months. One of the strangest facts of all is that under ideal conditions, living material that ordinarily would age and die will remain young far past its usual span. The most famous experiment to prove this is the one begun by Dr. Alexis Carrel, of the Rockefeller Institute, in 1912. He took a bit of heart tissue from the embryo of a chick and placed it in a nourishing solution. Care was taken that the necessary kind of food (from other embryos) was provided regularly, waste was disposed of and no bacteria contaminated the solution; and from time to time the tissue was trimmed. It lived and grew until the experiment was deliberately ended in 1946, a period of thirty-four years, though the normal life span of a chicken is about fourteen years.

All the symptoms of aging can be summed up as a gradual decline in the body's powers of self-renewal. Gradually, bones lose their organic material, which is replaced by mineral matter. They become brittle and knit more slowly after a break. Tissues become drier as more fat seeps out of them. It becomes harder for the body to regulate its temperature and to maintain its chemical balance. The senses lose their keenness although, except for some loss of focusing power, the eye is good for 125 years. On the credit side of the ledger is the fact that the lungs and the digestive system alter comparatively little with age.

It is the heart and the blood vessels that first succumb to aging. In fact, breakdowns in the circulatory system account for almost two-thirds of the deaths after the age of sixty. The course of such breakdowns usually begins with deposits of calcium in the heart and arteries, followed by loss of elasticity in these organs, high blood pressure, heart strain, burst capillaries, blood clots and, finally, the failure of the circulation. The brain, nervous system and kidneys might hold out indefinitely were it not for the failure of the complicated circulation that supplies them. This seems to bear out the old saying that a man is as old as his arteries. Yet hardening of the arteries alone is too special a condition to use as a measure of physical age.

Dr. John H. Lawrence, of the University of California, discovered what seems to be a better measure—that a man is as old as his ability to get rid of nitrogen gas from the blood. At sea level about one thousand cubic centimeters of nitrogen gas are dissolved in the body fluids of an adult. The total amount stays the same but the fluids

Black Star

There is no substitute for the experience of years. This carpenter has the skill of hand that youth might envy.

constantly eliminate the nitrogen molecules they have and take in new ones during breathing. In his experiments, Dr. Lawrence had a group of people of all ages inhale small quantities of radioactive nitrogen as tracer material. He then determined how fast the subjects eliminated nitrogen by collecting the gases breathed out and counting the tagged nitrogen atoms with a Geiger counter. Youngsters of fifteen eliminated half the gas in a few minutes, while persons sixty-five or older took as long as five hours. Another indication that nitrogen-elimination rate is a measure of

physical fitness was shown by the fact that subjects in poor physical condition had abnormally slow turnover rates.

What aging does is fairly self-evident, but the reason (or reasons) for the decline still eludes us. As Dr. Edward J. Stieglitz puts it, we either wear out, from use, or rust out, from disuse, with abuse playing a part as it encourages degeneration. Dr. Stieglitz believes that the answer lies hidden in the matrix, the complex non-living material in which the body's cells are bathed. It holds the elements of the cells together, supports them, brings nourishment to them and carries waste away. The most familiar form of the matrix is the plasma, or liquid part, of the blood. It is essential to life that the chemical balance of the matrix be maintained within narrow limits; and with age, the mechanisms that control this balance grow slower and react with less vigor.

Reasoning along somewhat similar lines, many biologists suspect that enzymes have much to do with the aging process. Enzymes are exceedingly complex molecules. They seem to be protein compounds, with nitrogen as an important element. They are largely built up out of vitamins and minerals, and they act as catalysts in the various chemical processes of the body. That is, they must be present for the chemical reactions to take place, though the enzymes themselves remain little changed or not at all. Without enzymes, no vital process can continue. Consequently, reduction or destruction of enzymes or a loss in their power could account for the symptoms of aging. The little work that has been done along these lines so far does indicate that with age the efficiency of the enzymes decreases. However, most of the research on enzymes has been concerned with identifying them in various chemical reactions and with studying how they work. Discovery of how enzymes change with age therefore would seem to offer pioneering opportunities to scientists.

The foregoing ties in closely with what has been learned about nutrition in relation to later years. Nutrition experts are interested in learning how well food (including oxygen and water) is absorbed and used by the body as well as the kind and amount of food consumed. They believe that what we eat is closely related to the rate at which we age. By the time many persons reach the age of sixty, their bodies are poor in calcium, iron, protein, vitamin A and the B vitamins, and these deficiencies have been built up over years. Calcium alone is a sort of jack-of-all-trades in the body. It is essential to bones and teeth, in heart action and blood clotting. If the blood does not get all the calcium it needs from food, the blood takes what it needs from bones until they become as brittle as toothpicks. Anemia results, with a loss of muscular strength and lowered resistance to infections, if the body does not get enough protein. Without iron, the best source of which is red meat, the blood cannot carry life-sustaining oxygen to the tissues.

"The thin rats bury the fat rats"

The foods the older person does *not* need so much of are sugars and starches and fats. Cake and candy and gravy merely add pounds of fatty tissue that must be supplied by extra miles of blood vessels, adding undue strain to the circulatory system. As the famous experiments on rats conducted by Dr. C. M. McCay, at Cornell University, proved, the "thin rats bury the fat rats." Ordinarily, a rat reaches full growth at 4 months, it is elderly at 2 years and dies before 3. By feeding a group of rats a diet that was low in calories but had enough vitamins and minerals, the period of growth was extended from 4 months to as long as 1,000 days. In one experiment, the last senile survivor of a group receiving the usual kind of food died at 965 days, though at the same age the animals with a low-calorie diet were still bright young adolescents. Their whole life span was the equivalent of a human life span of from 100 to 150 years, and they were seldom diseased and were much more energetic in every way than their brothers on the usual diet.

Overweight in later years carries other penalties. In addition to the strain on the circulatory system, it is associated with gall-bladder disturbances, hernias and diabetes. In fact, 80 per cent of the cases of diabetes in adults are connected with overweight.

What, then, *is* a good diet for the later years? In general, it should include a high level of proteins, a low level of fats, only a moderate amount of sugars and starches, and higher than average amounts of minerals and vitamins. In terms of breakfast, lunch and dinner, this means: eat plenty of vegetables, fruits, lean meats, fish, cottage cheese and eggs; eat lightly of cereals, fats and sweetened fruits; shun concentrated sweets and alcoholic beverages. One quart of milk a day will supply calcium needs. (It is the thinner part of milk, the whey, that contains calcium.) However, ordinary milk contains considerable fat and other undesirable ingredients for older people. There is now a milk on the market, called "geriatric special milk," that is high in protein, low in fat and rich in calcium, iron, vitamins and other wanted materials.

As we said earlier, there is still tremendous variation among individuals in the rate at which they age and the course aging takes. No two persons who have lived a century or more ever give the same recipe for longevity. One centenarian will say that he has never smoked; and another, with impish relish, will state that he has smoked like a house afire all his life. So we cannot get away from the fact that some persons inherit stronger constitutions than others. Long life runs in certain families. At the same time, if many of the chronic illnesses that afflict the aged were detected early, they could at least be brought under control if not cured. Unlike the diseases that are more prevalent in youth, which are of short duration, the chronic diseases, as the name implies, are lingering and often result in complete invalidism. The best safeguards against them are thorough periodic examinations, even when the person feels well.

Aging affects the mind as well as the body. There seems to be a very slow decline in learning activities after the age of twenty; at least the speed of learning decreases. But if the factor of speed is removed from intelligence tests, older persons do as well as younger ones, with one important exception. This is that anything that is entirely strange or that upsets established habits is likely to be far more difficult for the elderly. Perception has become slower and the mind cannot unlearn old ways as easily as when it was younger. However, because of the greater grasp gained through experience, many persons past sixty, and even some over eighty, actually surpass in particular skills or capacities the average person of their own sex in the prime of life. The skills that are kept in practice do not decline like those left unexercised. Verdi composed FALSTAFF at the age of 80; Oliver Wendell Holmes was in service on the Supreme Court bench at 90; Titian painted Christ Crowned with Thorns at 95. Granted that these were exceptional men, it is still true that years alone need not dull the highest powers of the human organism.

It follows that for society to discard the skills of men and women just because they have reached a certain birthday is a tremendous waste of assets. It is also a tragedy for many of the workers themselves, not only from an economic standpoint but also because it is a symbol of the end of independence and purpose in life. Many of these people later become truly unemployable simply because they have not been employed.

As far as income is concerned, various union-management programs and social-security provisions alleviate the situation somewhat, but these are not final answers by any means. Under private pension plans that pay flat monthly sums on retirement, businesses that hire older workers face a heavier future-pension liability. This may result in a refusal to hire older workers. One plan has been put forth by which the question of retirement would be decided by the worker as he reaches the age of 62 to 68, at which his pensions would become payable. From 68 to 72 retirement would be decided by the union and management; and after that it would be at the option of management.

What needs to be done is to judge the worker on the basis of health, ability to produce, emotional attitudes and personality, and not on his calendar age. Aptitude tests for those over sixty have been suggested, as have factories especially built for older workers who may have a slower reaction time and less stamina but perhaps more patience. At that, many older workers hired in wartime proved that they were capable of splendid production records.

Greater value should be placed on maturity

From the days of the pioneers in North America, the highest value has been set on the period of youthful vigor. We must change this attitude and place greater value on maturity if we are to cope effectively with the economic problems of the aged.

Interwoven with the basic requirements of good health and economic security for satisfactory later years are many other considerations. Older people need opportunities for play as well as work; they need to give affection and have it returned just as much as younger human beings do. Nor does the creative urge lessen with age. In fact, there is no better insurance against boredom in old age than to have acquired an avocation or a hobby in youth or middle age. Many communities are meeting the need for recreational activities by developing special clubs for the aged. A typical center of this sort is self-governed and is open all day. The members work at various arts and crafts, chat together, edit their own magazine and plan monthly birthday parties and other entertainments. Many older persons have found new interests, and often new friends of their own age, in adult-education classes. Special housing projects for the aged have also been developed.

In this article we have been able to touch on only a few of the many aspects of geriatrics. Questions have been raised that perhaps only years of research can answer. Nevertheless, since so many of us today can look forward to a long life, it is reassuring to know that science has brought up its big guns and is using them in the battle for healthy, active later years.

See also Vol. 10, p. 277: "Geriatrics."

A BIT OF ELECTRONIC HISTORY

The Development of the Revolutionary Audion

BY LEE DE FOREST

[*Editor's note: The development of the three-electrode audion by Dr. Lee de Forest has been called the most important achievement in the history of electronic research. In the following chapter the inventor tells the story of that achievement.*]

IT WAS my rare good fortune to be young when Thomas A. Edison was nearing the peak of his marvelous career. In that period the individual inventor had free scope for his activity. American laboratories were not yet crowded with thousands upon thousands of highly trained engineers; United States patent numbers ran in the hundreds of thousands, not in the millions, as is the case today.

When I left Yale University with a Ph.D. degree earned by my studies on the subject of hertzian waves,* I found that the field of wireless telegraphy was wide open to newcomers. The only well-known experimenters in that field were Sir Oliver Lodge and the Italian Guglielmo Marconi.

Quite naturally, therefore, I plunged into wireless telegraphy research with youthful zeal. I was not encumbered with a mass of precedents, nor was I overburdened with financial resources!

I early resolved, "come hell or high water," to achieve an envied position in that well-nigh virgin field by inventing outstanding wireless transmitting and receiving devices. Even at that early date I foresaw that these devices would give rise to a vast world-embracing industry, one certain to have untold influence over our future civilization. Already, as my notebooks and diaries record, I foresaw that wireless telephony would ultimately supplement, if not supplant, the telegraph; that the human voice, and possibly music, would replace the time-honored dots and dashes of the Morse code.

Early in the course of my graduate studies at Yale I became interested in the possible application of hertzian waves to communication without connecting wires between distant points. For my doctoral thesis I made a study of the passing of the waves along parallel wires and their reflection from the ends of these wires. As a detector of these feeble waves I employed the Branly coherer.** I fully realized at that time the defects of the coherer: Its lack of sensitivity and the fact that it had to be tapped periodically in order to restore such sensitivity as it possessed. I obtained my Ph.D. degree in 1899. A year later, when I began to work for a wireless telegraph concern in Milwaukee, I decided to perfect a novel type of detector — an electrolytic anticoherer that I had previously developed in Chicago. With this device one could use a telephone receiver to listen to Morse dots and dashes from a distant spark transmitter. My anticoherer proved considerably more sensitive as a detector than the coherer. Since it required no tapping device to restore it to a receptive condition, messages could be transmitted much more rapidly than with the coherer.

While I was diligently toiling in my rented room at night to improve my detector, I observed an astonishing phenomenon. I was using a small spark coil and

* *Editor's note:* See Index, under Hertzian waves.

** *Editor's note:* See Index, under Coherers.

spark gap as my sending device for testing the detector. One evening I noticed that the light from the Welsbach mantle above my table would be dimmed to a marked degree while the sparking device was in operation; when the sparking stopped, the lamp light would brighten again. The phenomenon was repeated with complete regularity as I manipulated the spark coil.

This was an entirely unexpected reaction. The combustion of the hot gases in the meshes of the Welsbach mantle was in some mysterious fashion partially interrupted by the electric waves generated by the spark gap of the coil. I was certain that I had made an accidental discovery of immense importance. I had come upon an entirely new kind of "wireless detector" — a heated-gas wave responder. My amazement and delight were boundless.

Shortly thereafter I was permitted to carry on my detector research in the laboratory of Armour Institute,* in Chicago. I tried to demonstrate to Professor Clarence Freeman of the Institute how hertzian waves affected the combustion processes in a finely divided gas flame. The professor was skeptical. He happened to know more about the effect of sound waves on a gaseous flame than I did. He directed me to remove my little spark-gap coil to an adjoining closet and to close the wooden door. The spark no longer affected the gas burner, proving that the phenomenon I had noted was acoustic only.

My disappointment was profound; so was my chagrin at having proved so ignorant in the field of acoustics. But I clung doggedly to the belief that hot gases could be made to respond in some manner to hertzian waves. It was not until three years later that I was able to justify my belief. In the meantime I sought to perfect my electrolytic anticoherer; it was essential to my plans for creating an American wireless telegraph company that would bear my name.**

* *Editor's note:* It was combined with the Lewis Institute to form the Illinois Institute of Technology in 1940.

** *Editor's note:* This company, founded in 1902, was the American De Forest Wireless Telegraph Company.

In 1903, I resumed my heated-gas detector research, first making a careful study of J. J. Thomson's DISCHARGE OF ELECTRICITY THROUGH GASES. I inserted two platinum-wire electrodes in the flame of a small Bunsen burner. The cathode (negative electrode), coated with calcium salts, was connected to the negative pole of a small dry-cell battery; the anode (positive electrode), to the positive pole of the battery. By suitably locating the two wires in the gas flame (keeping the anode wire in the cooler part of the flame), I was able to detect wireless telegraph signals sent out by a Marconi spark transmitter from a ship in the harbor.

I realized at once that a fluctuating detector such as a gas flame was not a practical device and had no commercial possibilities. It was evident that the cathode would have to be heated electrically rather than by a gas flame.

The filament of a miniature electric lamp, connected to a small battery, would be a most suitable cathode. The glass envelope of the lamp would shield the hot cathode completely from the air-current fluctuations that had proved so annoying when I used the gas-flame detector. The other electrode — the anode — would also be housed within the little lamp.

My faithful, resourceful laboratory assistant, C. D. Babcock, chanced to know William McCandless, a manufacturer of miniature incandescent lamps. He induced McCandless to seal a small plate of platinum in one of his lamps. The bulb was exhausted, as was usual at that time, by means of a mechanical air-pumping device. At first, since I was under the impression that a gas medium was essential between cathode and anode, McCandless was asked not to exhaust the bulb to the degree ordinarily called for in the manufacture of his lamps.

With this new kind of diode (two-electrode tube), my experiments proceeded rapidly. The device required two batteries, one of six volts to heat the carbon filament, the other, of higher voltage, connected to the anode and to the inevitable headphone. I named these two the A and B batteries

Thos. Coke Knight Associates

The author with the grid electron tube that he invented. The development of this tube,
called an audion, represented an all-important milestone in the history of electronics.

— terms that are still used in electronics. Bab (as I called my assistant) and I soon found that the low vacuum that McCandless had produced in this special "lamp" permitted only a low potential to be impressed between the anode and the cathode — some ten or twelve volts. Any higher voltage would cause a blue arc in the gas contained in the bulb, and the response to a distant wireless signal would be greatly reduced.

In these early experiments the outside receiving antenna wire was connected directly to the anode lead of the bulb. The "earth wire" (the ground) was connected to the base of the bulb and, through the base, to the cathode. I had employed the same type of connections in the original gas-flame detector.

With the new detector, wireless reception was becoming pretty reliable and we two experimenters were quite excited. We told no one what we were doing.

McCandless rather reluctantly prepared additional tubes for us. This time he exhausted them more fully, reaching the degree of evacuation ordinarily employed in his commercial lamps. The new tubes, or "lamps," as I still called them, permitted me to use up to twenty or twenty-two volts without blue arcing. With this voltage the received signals were definitely stronger; we could obtain messages from more distant transmitters.

It was during this early stage of evolution of my "gaseous detector" that Babcock suggested the name "audion" for the tube, because we could *hear* signals in our headphones.* On account of the low vacuum then employed, the current traversing the space between filament and anode was ionic rather than electronic.

I realized that when I used my two-electrode detector some of the high-frequency energy intercepted by the antenna passed to earth through the B battery and telephone receiver circuit, instead of being concentrated on the ions between the plate and the filament. To avoid such loss of precious energy, I wrapped the cylindrical tube with a tin-foil covering; I then connected the tin foil to the antenna or to one terminal of the high-frequency tuner.

The tin foil, therefore, became a third electrode — a control electrode. I immediately noted a distinct improvement in the tube's response to wireless signals, much to my joy. It proved that I was on my way to a greatly improved wireless detector.

It seemed to me that if this third, or control, electrode were located *within* the tube its control action over the current through the rarefied gas from filament to anode plate (or in the reverse direction) would be even more effective. So Babcock cajoled McCandless into building me still another kind of "lamp," one with a thin nickel plate, about five eighths of an inch square, to match the anode plate; this additional plate was to be located on the opposite side of the carbon filament from the anode. Both plates were to be about three sixteenths of an inch distant from the filament. Of course this required an additional wire lead, sealed, as was the anode lead, through the upper end of the tube. The new arrangement marked a distinct

advance; the test signals were stronger than when the control electrode was wrapped outside the tube.

Next, I reasoned that this new control electrode would be even more effective in controlling the gaseous current from filament to anode if it were located between the filament and the anode plate. Obviously it should be abundantly perforated to afford the ionic current direct passage from anode to cathode. Bab therefore punched a new nickel plate full of small holes; and this was inserted in still another tube.

With the perforated-plate control electrode located midway between the "cold plate" (anode) and the filament cathode, our test signals came in with most gratifying strength and clarity. I now had the true three-electrode detector — the father of the grid tube, or triode.

It was rather a difficult job to prepare a perforated plate. Why not merely bend a piece of nickel wire back and forth so as to form a grid? We tried out a grid of this sort, and it proved to be most satisfactory.

After Babcock had shown my grid sketches to McCandless and persuaded him to construct two samples, I myself went down to the lamp factory and collected them. From there I walked back eagerly to my laboratory, carrying the world's supply of grid electron tubes in my pocket. This was in October 1906. In 1952 the Sylvania Company ** brought me from Los Angeles to New York and there presented me with the *billionth* grid tube it had manufactured! It was neatly housed in a block of lucite.

I tested the first grid tubes with a new assistant, young Jack Hogan. We connected the tubes one after the other to our receiver circuit, using a small antenna rigged to the flagpole on the roof of the building where we had our laboratory. The results elated me; thenceforth Mc-

* *Editor's note:* Audion comes from the Latin *audire*, "to hear."

** *Editor's note:* The full name of the company is Sylvania Electric Products Inc.

Candless was asked to construct only the grid-type tube.

I lost no time in applying for a broad patent covering the grid-type audion. I had consistently applied for patents on each type of detector as it was constructed and tested, beginning in 1903. These patents constitute an indisputable official record of the manner in which my inventive ideas and the experiments resulting from them developed step by step from the gas-flame detector to the grid-type audion. The development of the latter type of tube marked the dawn of the Electronic Age.

The experiments that led to the development of this tube were carried out in various places. My first New York laboratory was a small room, two flights up, at 27 Thames Street; it was there that I toiled with the gas-flame detector. Later, when I was experimenting with the two-battery gaseous diode, I occupied a large room in the penthouse on top of 42 Broadway, adjacent to the powerful transmitter

An early form of the revolutionary audion. Note the wire leads, sealed through the upper end of this tube.

Brown Bros.

station of the American De Forest Wireless Telegraph Company. There I had access to the large receiving antenna and could easily substitute my "mystery box" for the electrolytic detector ordinarily used there. I took delight in mystifying the operator, who would don the headphones connected to my little box. "My God, Doc!" he would exclaim, "What have you got there? These signals are ten times as loud as what I've been receiving!" Then I would disconnect my "mystery box" from his circuit and disappear into my laboratory.

After I had been "fired" from the organization I had built up, I was forced to hire laboratory space in the top floor of the Parker Building, at the corner of Fourth Avenue and Nineteenth Street. There I stored, in a pasteboard shoe box, the various types of tubes made in 1904 and 1905.

Not long after I developed the grid tube, a professional magician, Carl Anderson by name, visited me to see if I could design an effective amplifier for his act. The amplifier would boost the subdued voice of an assistant standing in the aisle among the audience and speaking into a small microphone concealed in his vest and connected to a metal plate fastened to the sole of his shoe. An insulated network of wires leading to the amplifier and telephone receiver back stage was to be laid under the carpeting of the theater. A man standing back stage would overhear what was spoken in the aisle and would relay the message in a whisper to the magician on the stage.

I was fascinated by this plot. At once I realized the possibilities of my audion as an amplifier of telephone currents. To be sure, I failed to meet Anderson's difficult requirements, but I lost no time in filing a patent covering broadly the essentials of an evacuated, three-element tube as a telephone amplifier, or repeater. This patent — No. 841,387, granted January 15, 1907 — proved to be one of the most important ever issued by the United States Patent Office.

Early in 1907, I began to broadcast from the Parker Building laboratory, using a little carbon-arc transmitter, and soon had a small audience of "crystal and cat's whisker" hams. Some of them visited me, curious to see my transmitter. Proudly I showed them my grid-type audion. Soon I was besieged by would-be buyers; they offered me as much as $8.00 a tube. Mc-Candless soon realized that my audions offered him a new and promising source of revenue.

About this time, with my approval, he altered the shape of the glass envelope, making it spherical instead of cylindrical. He found that the spherical type was easier to make, to provide with electrodes and to exhaust. It sold like hot cakes. We were quite unable to meet the demand.

I was in Chicago early in 1908 when fire destroyed the Parker Building. The shoe box containing the priceless collection of historic tubes was destroyed in that catastrophe. Only the various patents that I filed remain to attest to the accuracy of the recollections that I have set down here.

I strove continuously to increase the sensitivity of the tube and to add to its life. I had early inserted two filaments so that when one burned out the other could be readily connected to the base, thus doubling the life of the tube. Next I had Mc-Candless insert two anode plates, one on each side of the filament and connected together as a single element. This necessitated a double grid; the two grids, one in front of each anode, were connected in parallel.

When tantalum wire became available, it was used instead of the original carbon as the filament; it proved distinctly superior as an electron emitter. Later tungsten was used instead of tantalum as the filament material. To increase the emissivity of the tungsten filament, a mature but enthusiastic ham, one Dr. Hudson, wound the tungsten filaments with tiny coils of tantalum. The Hudson filament audion was soon in great demand.

In 1912 a maker of X-ray tubes in San Francisco re-exhausted some of Mc-Candless' audions to an X-ray vacuum; in this type of tube I was able to apply 250 volts of B battery to the anode without any blue arcing due to ionization. At the same time I developed the feed-back circuit, which made the tube an oscillator.

My tube made transcontinental telephony possible. In 1915, the Western Electric Company, using the tube as a telephone repeater, established radiotelephonic communication between Arlington, Virginia, and the Eiffel Tower in Paris, and also between Arlington and Honolulu, in the Hawaiian Islands.

Since that time the electron tube has gone on to a multitude of new triumphs. The electronic industry has revolutionized our modern civilization. It has brought about stupendous developments in communication (making possible radio and television), industry, warfare, medicine, navigation and many other fields of human endeavor.

I feel that my life has been rich, indeed, because of my contribution to these developments. Dr. I. I. Rabi, a Nobel Prize winner in physics, has called my audion "so outstanding in its consequences that it almost ranks with the greatest inventions of all time" (ATLANTIC MONTHLY, October 1945). A great technician, Charles F. Kettering, has observed that "perhaps the most important event in the history of electronics occurred when a young experimenter named Lee de Forest inserted a third electrode in the form of a grid between the cathode and anode of a vacuum tube. The spectacular growth of electronics to an enormous industry employing over a million workers and benefiting untold millions of people in all parts of the world may be said to have begun with that event" (RADIO-CRAFT, January 1947).

Of recent years the transistor, invented and perfected in the Bell Telephone Laboratories, has begun to take over many of the duties of the electron tube. It offers many advantages: it is small; its weight is negligible; it requires only the B-battery supply; its power requirement is modest indeed. The transistor — a three-electrode device wholly electronic in nature — is obviously an offspring, and a most ingenious one, of my grid electron tube.

See also Vol. 10, p. 281: "Electronics."

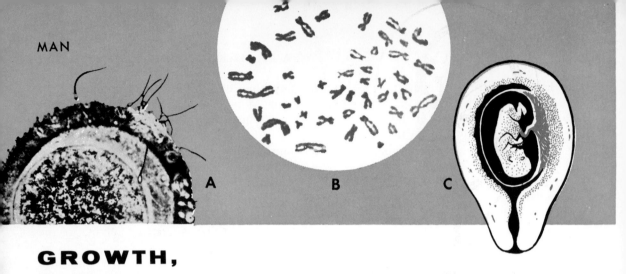

GROWTH, DEVELOPMENT AND DECLINE

A Survey of Human Life

BY BENTLEY GLASS

A. Human egg cell surrounded by spermatozoa (male sex cells). One of these has penetrated the egg cell and fertilization is taking place.
B. Human chromosomes in the nucleus of a cell.
C. Fetus at the end of three months.
D. Fetus at the end of five months.
E. A thriving baby, who has adjusted to the outer world.

THE Chinese, who reckon that a person is already one year old at birth, are nearer right than we are. Birth is not the beginning of life; it is only a change in the environment that surrounds the individual. It is a rather radical change, to be sure. Yet the growth and development that go on after birth are but continuations and alterations of processes that went on before that time in the embryo and fetus.* The saga of human life really begins when the egg is fertilized in the womb.

At this real commencement of life, the inherited pattern of a person's growth is fixed, once and for all. Each of the two parents has contributed to the child a set of twenty-three chromosomes — threadlike bodies within each cell. These microscopic threads carry the thousands of genes that determine one's hereditary nature. There are genes for long life, for the color of the eyes, for the shape of the nose, for the size of the feet; and innumerable others.**

Since the father and the mother each provide one chromosome for each of the twenty-three different chromosome pairs in the human cell, there must be two of every kind of gene, one inherited from the mother and one from the father. But the two genes of any single pair need not be exactly alike. A gene that promotes normal blood clotting, for example, may be paired with one, derived from the other parent, that is unable to provide normally clotting blood. When the two genes in a pair are different, one usually dominates. If a person has one gene for normal blood clotting and one that is ineffective, his blood will clot normally. It will not do so if both genes of this particular pair are ineffective.

Genes and chromosomes, then, are present in the fertilized egg from the very beginning, and they control development all through life. Some produce their effects early, others only in later years. Thus diabetes, as a general thing, shows up after the age of forty. The genes that make a long life possible may not reveal their presence until even later.

The genes represent the potentialities of all growth and development. Yet, since no human being can develop in a vacuum, normal growth and development can take place only in a normal environment. There must be an adequate supply of many sorts of essential things — water, mineral ele-

* *Editor's note*: The unborn young is called the embryo for the first two months or so of its growth. Thereafter it is known as the fetus.
** *Editor's note*: The genes are made up of nucleoproteins (nucleic acids plus proteins). The most important nucleic acid, DNA (an abbreviation of deoxyribonucleic acid), consists of four kinds of molecules. They can be arranged in various ways; and the varied arrangements account for the differences in the genes. See Index, under Deoxyribonucleic acid.

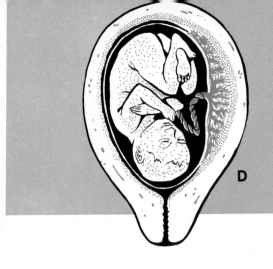

D

E

The life of a human individual really begins when a male sex cell of his father fertilizes a female sex cell (egg) of his mother (A) in her womb. Each parent contributes a set of chromosomes (B) to the individual; these carry the genes that determine his hereditary nature. The unborn young grows in the mother's womb for nine months or so (C, D). Then comes the crucial event of birth, when the newborn babe comes into the world. If all goes well, he adapts to his new environment (E).

A: from "Ovum Humanum," by L. B. Shettles, publ. 1960, Hafner
B: L. B. Shettles, Columbia University
E: Gerber Baby Foods

ments, fats, sugars and starches, proteins, vitamins, oxygen and so on. There must be adequate elimination of waste substances of all kinds. There must be careful control and regulation of physical and chemical conditions, such as temperature, acidity and the concentrations of substances. Changes in any of these factors may, and indeed will, alter considerably the processes of growth and development.

When we speak of an inherited characteristic, we generally have in mind one that is not readily altered by ordinary changes in the environment, or that we do not yet know how to change — such as eye color, for instance. However, this does not mean that the characteristic cannot be changed at all or that we ought to be fatalistic about such things. For example, diabetes, which in former times was a fatal disease, is inherited — that is, there is a gene responsible for it. Yet injections of insulin will keep it under control; in fact, some day we may even learn how to prevent it from developing at all. So an inherited condition may well be cured or prevented.

The environment of the still unborn baby is highly regulated and very uniform, compared to its surroundings after birth. Hence the features of growth before birth are much the same for everybody, except

for the differences that result from different genes. After birth, however, persons show more and more variation because of differences in surroundings, food, care, training and so on. So, although all a person's potentialities depend upon the genes he has at birth, what will come of these potentialities depends more and more on his special environment as he grows up.

The growing process in the human body

In its simplest sense, growing means adding living matter to the matter that is already present in the body. Every person is made up of a vast number of tiny cells. Because there seems to be a limit to the size that a single cell can have, a body's growth comes about through the multiplication of the number of cells. Each cell divides into two, and then these grow to their full size and divide in their turn. It takes many successive divisions of the cells to produce a baby from a fertilized egg. By that time, the growing process is almost finished, at least by comparison with what has gone before. The weight of the human body, from fertilized egg to newborn babe, has multiplied 2,500,000,000 times; it increases only twenty times or so from newborn babe to adult.

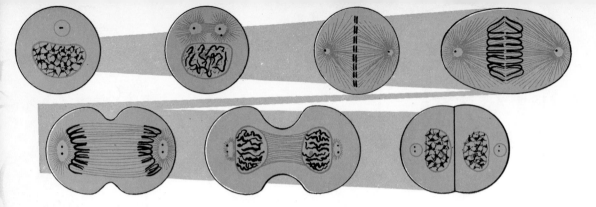

The growth of a human body comes about through an increase in the number of cells, due to mitosis, a form of cell division. Various stages of mitosis are shown above. At the end of the process, in place of one cell, there are two daughter cells.

Both brain and voluntary muscle have their full quota of cells before birth, and no new ones are produced after that time, even to replace those that happen to die. In the liver and various other organs, cells practically stop dividing by the time of birth, although they can again begin to do so if a part of the organ has been destroyed. In only a few structures are the cells constantly multiplying in number. Among them are the skin and the blood cells. Red blood corpuscles, for example, can serve as carriers of oxygen for only about eighteen weeks before they break down. Such enormous numbers of them are required that, in the bone marrow where they are produced, some two million of them must be turned out every second of the day and night.

The two main stages of growth and development

There are two main stages of growth and development. First, there is an automatic period. This is followed by a functional stage during which development depends upon contact, strain and stress, exercise and learning and all sorts of experiences. Birth does not represent a dividing line between these two stages, for many automatic steps in development occur after birth. Among them are the change from blue to brown eye color in many persons during childhood, or the regular changes of adolescence in everybody. On the other hand, many functional steps in development are taken before birth. When the unborn baby kicks and turns within its mother's body, it is beginning to make its bones and muscles strong through use.

Because some people are well fed and some are half-starved, because some stay well and others get sick, because of all sorts of other circumstances, there will inevitably be differences in people's growth and development. These will be heightened by differences in the genes they inherit, for no two people, except for identical twins, have identical genes. Some tend to grow faster, others slower; some, fatter, others leaner.

Growth in infancy and early childhood

Look well at the newborn babe — this tiny bundle of reflexes, just come into the

The Boys' Club of New York

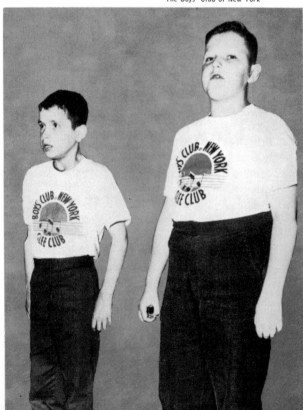

outer world to grow and develop into a man or a woman. It is from five to eleven and one-half pounds of helplessness; yet it is destined to become skillful in movement and to learn to think. Nothing is more marvelous than the story of its growth.

The very first adjustment after birth has to be that of breathing. The supply of oxygen from the mother has been cut off for some little while, and death will come quickly if the breathing is not started. For several weeks, the breathing muscles of the chest, abdomen and diaphragm have been practicing a bit at inflating the lungs. The shock of the cold outer air as the baby emerges from the mother's body usually provokes a cry that draws air into the lungs. Breathing then starts. For the first year, breathing depends mostly on the abdominal muscles, and it is shallow and rather rapid. After the baby commences to sit up, the thoracic, or chest, muscles begin to assist with the breathing, and it grows deeper and slower. Between three and seven years of age, the adult pattern of breathing is set.

The smaller a body, the more surface it has in proportion to its bulk, as is shown in the diagram on this page. Since greater

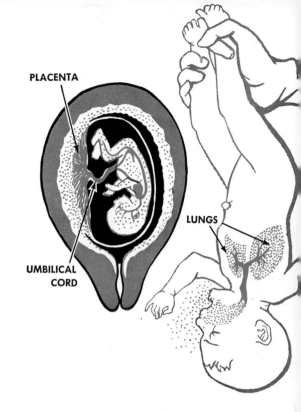

PLACENTA

UMBILICAL CORD

LUNGS

The fetus obtains its oxygen from the mother, by way of the placenta (and the umbilical cord), and the waste product carbon dioxide also passes through the placenta in the opposite direction. Hence the placenta acts like a lung. The real lungs are very small until the newborn babe starts to respire; they then expand considerably.

The photographs at the left shows strikingly the different rates of growth in different individuals. The eleven-year-old boy, at the right, is a year younger than the other youngster; yet he is taller and much huskier.

The smaller a body is, the more surface it has in proportion to its bulk. An examination of the two cubes, A and B, will show why. The bulk (volume) of cube A, whose side is 2 inches, is 2^3, or 8, cubic inches; its surface area is 4×6, or 24, square inches. (The area of one side is 2^2, or 4, square inches; there are six sides in the cube.) Cube B, with a side of 3 inches, has a volume of 3^3, or 27, cubic inches, and a surface area of 9×6, or 54, square inches. Obviously, the smaller cube, A, has more surface in proportion to its bulk than the larger cube, B. As the text points out, a baby would have far more surface area in proportion to its bulk than an adult, since it is so much smaller.

surface means more loss of heat, a baby, pound for pound, loses more heat than a grownup. This means that more food and lots of muscular activity — mostly kicking and squirming — are needed to produce the heat. Because of high heat loss and poorer control over heat production and body temperature, chills are a menace.

Body temperature is not well regulated until a child is about two years old, although in this respect each child is apt to follow his own rules. Temperature controls are connected with the development of the thyroid gland, which normally shows a steady rate of growth from birth to maturity. If a thyroid is underactive from birth, a child soon becomes a cretin. Growth is stunted to dwarf size; a puffy fullness of face, lips, nose, tongue and the skin in general is produced. The intelligence fails to develop normally and the child becomes practically an imbecile. Fortunately, if he can be treated with thyroid hormone early enough, he can overcome this condition.

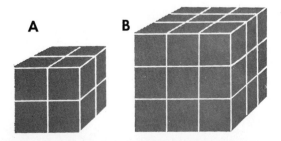

A B

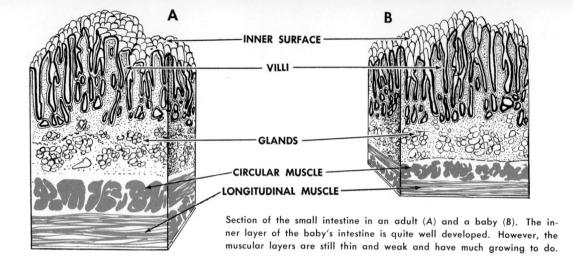

A — INNER SURFACE — B

VILLI

GLANDS

CIRCULAR MUSCLE

LONGITUDINAL MUSCLE

Section of the small intestine in an adult (A) and a baby (B). The inner layer of the baby's intestine is quite well developed. However, the muscular layers are still thin and weak and have much growing to do.

The development of the digestive system is another necessity in infancy. Everybody knows that a small baby cannot eat adult foods, although its digestive organs are further advanced than most of us realize. In the first two years, there is a fivefold increase in weight in the salivary glands. The stomach can hold from ten to twenty times more by the end of the second year. The pancreas and the glands of the stomach and intestines are all ready for action at birth, although their secretions are small in amount. At birth, the weight of the pancreas is only a little more than 4 per cent of its adult weight. At three years, it has already completed one-third of its growth.

The liver is relatively large at birth, and although its weight increases about ten times in all, the ratio of its size to that of the whole body diminishes by half. Its early growth indicates how important it is to the infant for storage and for the formation of bile. Both the large and the small intestines just about double in length between birth and full growth, with most of the increase coming in the first two years. At birth the inner layer of the intestines is well developed, so that digestion and absorption are well advanced. But the muscular layers of the intestines are still thin and weak and have a lot of growing to do.

The development
of the milk teeth

At birth, the milk teeth and most of the permanent teeth are already present in the gums, but eruption of the milk teeth does not begin until the sixth or seventh month, as a rule. Then the middle, or central, incisors start to come in, closely followed by the lateral incisors. The lower teeth break through the surface of the gums before the upper ones. The first molars erupt at 12 to 16 months and the canines afterward (16 to 20 months). The second molars are the last of the milk teeth (20 to 30 months). The buds of the permanent teeth already present at birth include all those that replace the milk teeth and the first permanent molars besides.

How the circulatory
system develops

The circulatory system has to make a considerable adjustment both at the time of birth and afterward. The lungs receive very little blood before birth, but afterward the whole flow must be passed through them. This is brought about by the closing of the passage between the right and the left auricles of the heart. Thereafter, all the blood entering on the right side of the heart is pumped to the lungs.

At first the baby's blood pressure is low and its heartbeat about twice as fast as that of an adult. The blood pressure slowly rises. The heartbeat becomes slower, losing about 25 beats per minute in the first two years, another 10 by the age of five, and gradually reaching the final rate of about 70 per minute. The heart, like the liver, is relatively large at birth. While its size doubles in the first two years and triples in four years, it does not keep pace with the growth of the rest of the body.

The kidneys are also relatively large at birth and are quite able to handle the task of excreting the body's wastes. Between birth and one year, they triple in weight. By the end of one year, they have already attained one-fourth of their ultimate size.

The transformation
of cartilage into bone

The skeleton and the voluntary muscles of the body have far more of their growth ahead of them than the heart, liver and kidneys. The bones of the newborn baby are hardly bones at all. Most of the skeleton is still formed of cartilage, which is softer than bone. Centers of transformation into bone, about eight hundred of them, arise in the cartilages before birth in some cases but in most cases after birth.

In the long bones of the arms, legs, hands, fingers, feet and toes the earliest center of bone formation in each bone is in the shaft.* Not long after birth, secondary centers arise in each enlarged end of the bigger arm and leg bones. These secondary centers of bone formation are known as epiphyses (pronounced *ee-pif'-i-seez*). The cartilage between the epiphyses and the main center in the shaft keeps growing rapidly. As a result, the epiphyses are pushed farther and farther apart. Only after the bone has reached its full length, toward the end of

* Editor's note: The shaft, in the bones mentioned here, is the long tubular structure that makes up the main part of the bone.

adolescence, does the cartilage stop growing, allowing the epiphyses to become firmly attached to the shaft.

The age of a child's skeleton can easily be told from X-ray pictures. For instance, at birth there are no epiphyses in the hand and no bones at all in the wrist, only cartilages. At one year, two centers of ossification (formation into bone) have appeared in the wrist and a number of epiphyses have developed in the hand. Between three and five years of age, all the epiphyses in hands and fingers arise, together with two more centers in each wrist. By five, only one center is still lacking. Girls develop the skeleton more rapidly than boys, though they lag behind in stature and weight in these early years.

The skull increases in
size rapidly during babyhood

On every little baby's head are soft places where the bones of the skull are still incomplete and soft cartilage covers the brain. During babyhood the skull increases in size very rapidly, the cranium, or brain case, doubling in capacity in the first year. The soft spots soon disappear as the cartilage is converted into bone. Even the biggest one has closed by the time the baby is a year or a year and a half old. The cranium reaches its full size by six years of age.

How the bones of the hand develop. At the left is shown the hand X ray (positive) of a girl of eight. At the right is another hand X ray of the same girl at the age of sixteen. She has now attained her full skeletal maturity.

Courtesy of Dr. Nancy Bayley; from the files of the Berkeley Growth Study, Institute of Human Development, University of California.

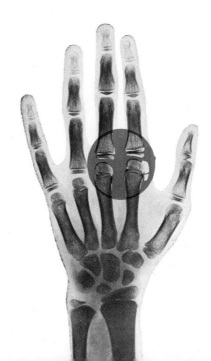
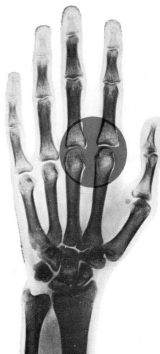

Meanwhile the bones that form the face grow quite slowly; their growth becomes rapid only as the teens are reached. That is one reason why young children's faces are so different from those of adults.

The rest of the skeleton passes through similar changes. At birth the spinal column is largely cartilaginous, but the ribs, needed at once for breathing, are already partly ossified. The typical curves of the spine develop after the baby begins to sit up and to stand. Just as the face and cranium grow at different rates in different periods, so other parts change in proportion, too, because of differences in their rates of growth. The arms and the legs both become longer in relation to the trunk.

The voluntary muscles of the body have a great deal of growing to do. Up to four years of age, they keep pace with the rest of the body. Then they begin to grow far more rapidly. Since no new fibers are added, this growth is largely due to an increase in the size of the single fibers.

At the start, the eye muscles and those needed for breathing are best developed, and the arm muscles are further along than those of the legs. The ligaments and tendons that attach the muscles to the bones are still poorly developed at birth; the connective tissues lack elastic fibers. It takes constant use to strengthen the muscles and their attachments and to bring about coordination between different muscles. But use is not enough. It must be supplemented by the development of the senses and the nervous system and by the acquisition of learning habits.

The baby's nervous system

The nervous system is far advanced physically at birth. The brain is already one-fourth full grown, although the whole body is only one-twentieth of its mature weight. In two years, the brain attains over half its full size, and at nine it has practically finished growing. The inner structure of the brain is also almost complete at birth, and no new cells are added to it thereafter. But the nervous system is very immature; perceptions, habits, learning processes and

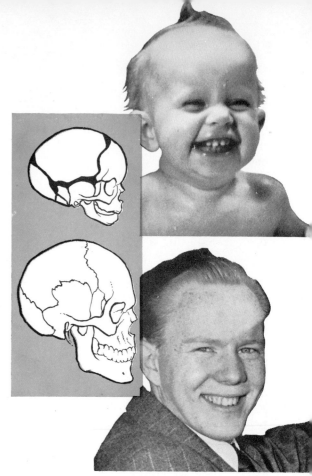

Harold M. Lambert, from Frederic Lewis

The diagrams show the striking differences between the skull of a baby and that of an adult. Note the far greater development of the nasal cavities and the jaws in the adult. Differences in skull structure are reflected in the baby and adult photos (of the same person).

memory are still to be established. Some reflexes, such as yawning, coughing, sneezing, sucking and swallowing, are present at birth or very soon thereafter. Other reflexes, however, have yet to appear. Hearing and the touch sense of the lips and tongue are well developed. It takes two or three months, however, before the baby can focus its eyes properly and see at all clearly. During this period the baby is wrapped up in itself; it is particularly interested in the state of its stomach.

The development of personality

When a baby is ushered into the brightness, noise and cold air of its new world, it is alarmed and confused. The sense of warmth and security that it develops as it snuggles against the mother and satisfies its hunger at her breast is the first great experience of life. It has been said that the shock of birth and the sense of security, or the lack of security, in earliest babyhood lay the foundations for the whole development of the personality.

The big job of the first year is to learn to co-ordinate and control bodily movements. First comes turning the head, and learning to focus the eyes and to co-ordinate them in following a moving object. After about two months comes the first smile.

In the second year, there is a widening understanding of self and of home relationships. Discipline comes to have significance as the baby learns to understand what "No" means. Control over the body functions of elimination generally improve rapidly at this age, although the bladder is still so small that urination must be frequent. The baby is eager to explore everything, but his attention is often and easily diverted. Between one and three years, he learns to focus his interest and attention on things, just as he

Changes in proportions of human body. A. Embryo, 2 months. B. Fetus, 7 months. C. Newborn baby. D. Child, 5 years. E. Teenager, 13 years. F. Young adult.

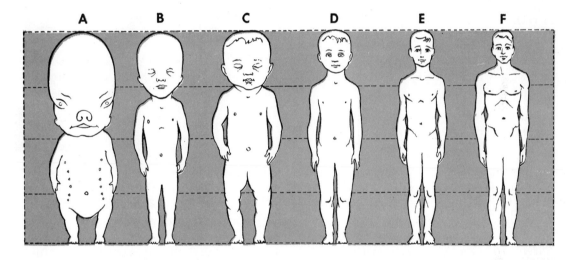

Next comes the co-ordination of hand and eye in grasping an object and transferring it from one hand to the other, in reaching out for it and in exploring it by turning it over and popping it into the mouth.

The cooing welcome that the baby gives strangers at three months of age is likely to be replaced by distrust and alarm at six months, for it can now distinguish between friends and strangers. Rolling over and sitting up comes in the second half year; creeping soon follows. Moving about while standing up and holding on and finally standing alone are the big achievements made toward the end of the first year or early in the second. Meanwhile a lot of experimenting in making sounds has been going on. Sounds that mean something — at least to the baby — may be used by the end of the first year.

had learned earlier to focus his vision. He becomes more sociable and dependent on the company of others, while at the same time he is learning in other ways to be more independent. Suspicion of strangers is likely to reach new heights, and worries and fears begin to enter his mind. He is learning fast, progressing from toddling to walking, from babbling gibberish to speaking sentences, from exploring his house to exploring the neighborhood. He learns to feed himself. He experiments with temper tantrums; new emotions, such as jealousy, appear.

From three to six, a new world opens up. Most of the contrariness of the baby just becoming aware of its own individuality now tends to disappear. Curiosity about all sorts of things is a dominant characteristic; this is the great age for questions. Keenness of observation increases steadily

Left-hand photo: Merrim, from Monkmeyer; right-hand photo: Hays, from Monkmeyer

Before and after puberty. The plump little girl shown at the left has become the slender maiden at the right.

at this period, as is shown by the skill with which the child can work out simple jigsaw puzzles. Imagination reaches a peak too. This is the age of make-believe, of imaginary playmates and fairy godfathers who may be as real to the child as his parents. Imitation becomes very prominent; a great deal of experience is gathered through imitating grownups. Playmates become an important part of life.

The school child —
from six to eleven

In the period from six to eleven, physical growth, rapid at first, slows down and nearly comes to a standstill. The child becomes less dependent on his parents and develops a sense of self-control.

School begins the formal education. Parents soon find that teachers have largely supplanted them as fountainheads of knowledge. Independence is expressed in bad manners and occasionally rebellion against parents' authority. Other children now set the pattern that Johnny and Mary strive to follow. Gangs and clubs spring up as the children strive to get their own community life organized. At this age, conscience develops into a profound emotional guide. There will be sudden onsets of good behavior in the usually heedless child, and attempts at neatness in the usually disorderly child. Compulsions are common. The child feels, for example, that he must not step on the cracks in the sidewalk or that he must touch every third picket in a fence.

Understanding and intelligence develop apace. Learning now becomes conscious; children begin to try to learn instead of learning without being aware of the fact.

In these years, they develop new reading skills. They cut down the number and duration of the pauses and backward movements that the eyes make as they move along the line of the page.

Puberty and
adolescence

Puberty is a time of rapid growth, culminating in sexual maturity; it is followed by adolescence, which lasts until adulthood is attained. Puberty and adolescence make up what is known as the period of maturation. The growth in this period, from about the age of eleven to nineteen in girls and from about thirteen to twenty-one in boys, depends in particular on increased hormone secretion by the pituitary and adrenal glands and sex organs. Sex hormones stimulate the rapid growth to mature size of the sexual organs, which have grown very little since infancy.

The very rapid increase in height at this age is due mainly to the lengthening of the legs, for the trunk grows chiefly in girth. Breathing capacity increases; the liver, heart and large blood vessels double in size; the blood pressure is stepped up. These changes make possible the effective use of the muscles, which also just about double in size, at least in boys, in this period. Physical strength increases rapidly, but lack of perfect co-ordination brings about temporary awkwardness. While the cranium changes very little, the bones of the face lengthen considerably. Childhood's visage fades away and the features of the adult emerge. The pores of the skin enlarge. Its oily secretion increases and there may be some trouble with pimples.

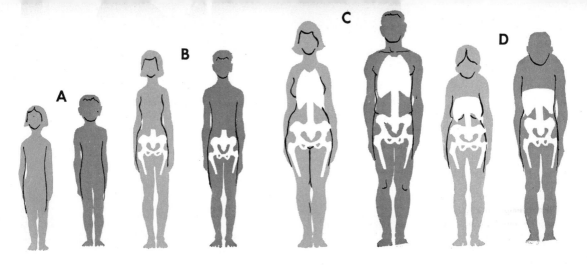

A pictorial study of growth, development and decline in a female (light green tint) and a male (gray). A. Childhood. B. Adolescence. C. Maturity. D. Old age.

The puberty of the girl usually begins at eleven years of age. She then shoots up three inches or so in a year, changing from a plump little girl to a slender young maiden quite abruptly. The gain of ten to twenty pounds per year in weight all seems to go into height. The first outward sign of sexual maturation is the enlargement of the breasts, accompanied by an enlargement of the pelvis. There is a change, too, at the shoulders.

Hair appears on the pubic regions and, usually somewhat later, in the armpits. At thirteen, on the average, the young girl has her first menstrual period, marking the first release from the ovaries of a mature egg cell. She may gain in height in the next few years: an inch and a half the first year, perhaps three-quarters of an inch the next and still less thereafter. Her growth, however, is rapidly slowing down. By the age of fifteen, the adolescent girl is about as tall as she ever will be.

Some girls enter puberty at eight or nine years of age and others only when they are thirteen or fourteen. These variations in development may lead to considerable self-consciousness and even alarm on the part of the child. She should be reassured, however, for there is no physical ill consequence from either precocity or delayed sexual maturation.

When the boy's puberty sets in, at about the age of thirteen, his shoulders broaden and his muscles fill out. He gains rapidly in height and strength. As his reproductive organs mature, spermatozoa, or sex cells, are formed and sexual glandular secretions begin. The pubic hair appears early, the hair in the armpits and the beard somewhat later. The larynx enlarges; the vocal cords lengthen considerably; the voice breaks and after a time it becomes about an octave deeper. At fifteen, two years after the onset of puberty, the boy's growth begins to slow down abruptly. He will gain another two or three inches, but more slowly. By seventeen most boys are full-grown in stature.

During puberty and adolescence, psychological development is not less pronounced than physical development. The sex hormones work a revolution in the attitudes of young girls and boys. Made self-conscious by the changes taking place in herself, the young girl tends to exaggerate her worries, perhaps to think that she is different and abnormal. She is likely to be sensitive to criticism. Sometimes she wishes to be considered as a grownup; sometimes she seeks the protection she enjoyed as a child. As boys of her own age outstrip her more and more in athletic activities, she tends to develop more interests of a passive nature. She is apt to become romantically attached to some older person, developing a "crush." The young lad who is maturing is also self-conscious, troubled and given to romantic daydreams. However, his psychological problems are usually less severe than those of a girl his own age.

The gulf between boys and girls of the same age

Because girls mature two years before boys, a sharp social gulf is created between girls and boys of the same age. In a given class at school, the average girl towers above the average boy and begins to be grown-up and romantic in her ideas and interests, while he "is still an uncivilized little boy who thinks it would be shameful to pay attention to her," to quote Dr. Benjamin Spock. It is best to allow for this situation in planning social affairs for youngsters by grouping young girls with boys a couple of years older.

Intelligence continues to develop rapidly in these years. The span and speed of visual perception, on which so much depends in learning, reach their peak at sixteen or seventeen years of age. So does memory; so do speed and accuracy in motor skills and habits. In fact, general intelligence, which shows a steady increase through childhood and early adolescence, improves little or not at all after the age of eighteen. The peak of mental development, at least in the simpler, more measurable qualities, is therefore reached even before the physical prime of the individual has been attained.

The period of
maturity and aging

The human body does not remain at the peak of its powers for very long. A slow but steady decline is the general rule for the years after maximum efficiency is attained. Aging is a normal, not an abnormal process; it cannot be said to begin at any one time. Because the peak of development comes at different ages for different parts of the body, there is no one age of highest efficiency in every respect.

Examples of different
peaks of development

For example, the ability of the lens of the eye to change its shape so as to focus on nearby objects declines steadily from babyhood up to the age of fifty, after which it remains about the same. The thymus gland reaches its maximum size at ten to twelve years of age. Then it rapidly diminishes in size, so that at the age of twenty it has only half the weight it possessed at its largest. It is actually degenerating while the repro-

ductive organs and the voluntary muscles are doing their most active growing.*

Muscular strength is greatest in the middle and late twenties, and an athlete is generally (not always) "old" at thirty. The muscles of the back relapse to the level of their power at twenty during the thirties. The biceps reaches this level in the forties, the handgrip in the fifties.

Changes in bone
structure in adulthood

The bones, too, undergo progressive changes. The bones of the skull, between the ages of twenty-five and thirty, make rapid advances toward complete fusion. Later, bones show an increase in mineral matter and a decrease in fibrous organic matter. They become more brittle; their power to regenerate and knit after a break is gradually impaired. But these changes are slow. Only after seventy does the atrophy ** of the jaws produce a marked change in the face, particularly after the teeth become loose in their sockets and fall out.

Other physical
changes in adulthood

The regulation of stable bodily conditions, such as body temperature and concentration of sugar in the blood, becomes more difficult with the years. The colloids of the protoplasm lose their capacity to take up water. They become less stable and less reactive as they shrink. The healing of wounds is slower; at the age of fifty they heal only half as fast as at twenty.

The digestive organs show the effects of age particularly in an increasing liability to various disorders. These include gastric ulcers (ulcers of the inner wall of the stomach), pernicious anemia, cancer and the formation of gallstones. The lungs do not appear to alter much with advancing age, except in elasticity.

There is a loss of elastic tissue in the skin, which becomes thinner as the fat stored

---

* *Editor's note*: It has quite recently been discovered that the thymus gland plays a very important part in the early months of life. It manufactures cells that help to build up resistance to disease.

** *Editor's note*: This type of atrophy is known as senile atrophy. It refers to the lessening in size, as a result of the aging process, of a bodily part that was formerly of normal size.

Ewing Galloway, N. Y.

Left: One of the characteristic changes due to old age is the atrophy (lessening in size) of the jaws. It is quite noticeable in the features of this senior citizen.

Below: Two old people performing work that requires great skill. One of them is repairing a watch; the other is working on the spruce-veneer fuselage of a light plane. An old person may surpass the average individual in the prime of life who works at the same task.

Column 2, upper photo: National Council on the Aging; lower photo, Black Star

A

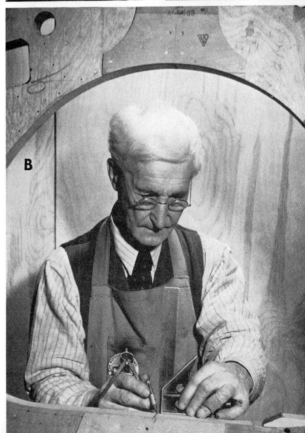

B

just beneath it is withdrawn. Folds and wrinkles appear. The hair becomes gray and sometimes falls out. The nails, especially the toenails, tend to thicken and become deformed as their rate of growth slows down.

The reproductive organs have a rather definite span of activity. In women, menopause, which comes at thirty-eight to fifty years of age, brings to a definite end the capacity to bear children. Like puberty, this change is brought about by the hormones, most probably those of the pituitary and adrenal glands. In men, over-all sexual capacity probably declines from early maturity on.

The heart and blood vessels are most important in the matter of aging because they are usually first to succumb. "A man is as old as his arteries." Breakdown of the circulatory system causes only 14 per cent of the deaths between the ages of twenty and thirty-nine, but it is responsible for almost two-thirds of all deaths after the age of sixty. Deposits of calcium in the heart and arteries, loss of elastic tissue, high blood pressure, heart strain, burst capillaries, blood clots, failure of circulation — these are the common steps in the process. Perhaps the brain and nervous system and the kidneys would hold out indefinitely if it were not for the failure of their elaborate circulations.

There is a steady decrease in the number of taste buds as one grows older, and the sense of smell also declines. Hearing above high C is impaired but below that it hardly changes at all, except in the really deaf. There is a loss of focusing power in the lens of the eye and a decline in the sharpness of marginal vision. In the brain and spinal cord, the white matter suffers more than the gray.

The outstanding thing about aging is the tremendous variation between individuals in its rapidity and its particular course. Long life runs in certain families. The genes undoubtedly play a big part in the later years, just as they do before birth and in infancy, childhood and adolescence. Even in the same individual, many of the processes involved in aging appear to be quite independent of one another. The nervous system, the respiratory system and the circulatory system may break down at different periods.

The decline in ability in learning activities

In almost all learning activities there is a slow decline in ability after the age of twenty, just as there is in physical skills and capacities. But if the factor of speed is eliminated from intelligence tests, older persons do just about as well in them as younger ones. In particular skills, many persons over sixty, and even some over eighty, may actually surpass the average persons of their own sex in the prime of life. We should point out, however, that older persons generally hold their own in a subject only if it fits the established patterns and habits of their minds.

Old age must be more moderate than youth, but it can also be rich and fruitful. That is why we should plan in our younger years those interests and activities that will keep us young at ninety. For when our interest in life is gone and only resignation is left, there will be little to keep us here.

Death — ringing down the curtain on the drama of life

What we call death represents the breakdown of the complex chemical and physical processes that exist in every living thing. It marks the end of the ceaseless interchange of substance between the living organism and its surroundings. It brings to a close, too, the constant outlay of energy that is characteristic of life.

Death is really gradual, not sudden. It is true that once a vital system has failed or an overwhelming shock has been experienced, in a few moments the damage done is irreparable. Yet certain tissues may remain alive for a considerable period thereafter. For example, for a number of hours after the death of a person, the still-living cornea of his eyes can be used to make windows in the opaque corneas of certain blind people.

Death may be premature. The lifespan may be shortened by such factors as the effects of war and peacetime accidents. It may be cut short, too, by disease.

It is true that medical science has made notable progress in the unceasing fight against the many ailments that have afflicted mankind. Consider the communicable diseases whose names once aroused dread: typhoid fever, cholera, typhus, smallpox, diphtheria, yellow fever and tuberculosis, among others. They have been conquered, or at least held at bay. Nutritional diseases have been unmasked and dealt with: scurvy, beriberi and pellagra need no longer be feared, if a proper diet can be provided. Progress has been comparatively slow, however, in combating other diseases, such as those involving the heart, blood vessels and kidneys. The cancers, too, have defied the best efforts of leading scientists. These unconquered killers are likely to strike at any age group, but they are the particular enemies of older people.

Perhaps, in another century or so, they too will be vanquished. Then, provided that he can escape the ravages of war, peacetime accidents and other untoward developments, the average baby will be able to live out the full term of the human hereditary life span. Nobody knows how long this span really is. The Russian scientist A. A. Bogomolets asserted that the average man should live to be 150, but he may have been unduly optimistic. The age of 80 may well become the average life span; we may possibly advance it somewhat beyond that mark.

Yet in the end we must all die, and others will carry on. Perhaps new minds, new hands, new eyes will develop more effectively than we the potentialities that lie in the ultramicroscopic genes.

See also Vol. 10, p. 276: "General Works."

THE STRANGE WORLD OF "SILENT" SOUND

Ultrasonics, a Powerful Tool for Man

BY HELEN MERRICK

AS YOU walk down a country lane at the height of summer, there is a constant drowsy murmur of humming bees, droning flies and chirping crickets. Even as you listen intently to these faint, thin noises, the busy little insects are making a great many other sounds that no human ears could possibly catch unaided. Scientists have succeeded, however, in inventing certain highly ingenious devices for detecting such "silent" sounds and also for producing them. The study of these sounds is called "ultrasonics" (from "ultra," meaning "beyond," and "sonic," meaning "referring to sound"). The name "supersonic" also used to be applied to sounds such as these. Nowadays, however, this particular word is used to describe speeds exceeding that of sound in the air — that is, about one mile every five seconds. Thus we refer to supersonic planes, which can fly faster than sound, and to supersonic wind tunnels, in which research on such amazing planes as these is carried out.

Since ultrasonic sounds differ in only one respect from sounds that we can hear, let us see first what any kind of sound is. For a sound to be created, three things must happen. First, an object must be made to vibrate, as when you press on an automobile horn, pluck a taut string or strike a tuning fork. The rapidity of the vibration — in more scientific terms, the frequency of the sound — depends on how elastic and dense the substance of the object is. No matter whether it is struck hard or lightly, the rate of the vibration for that substance will be the same and is its natural frequency. This gives sound its pitch — its high or low tone. The more rapid the vibration, the higher the pitch.

Second, as the object moves rapidly to and fro, it alternately pushes and pulls at the nearby molecules of whatever medium surrounds it, usually the air. (Sound cannot be created in a vacuum; there must always be some medium — a gas, a liquid or a solid.) When the molecules are

Sound waves of tremendously high frequency actually shake germs to pieces. The first shock of the waves makes air bubbles form (2) in a germ, which is then shattered (3) as the bubbles break.

Courtesy, Popular Mechanics Magazine

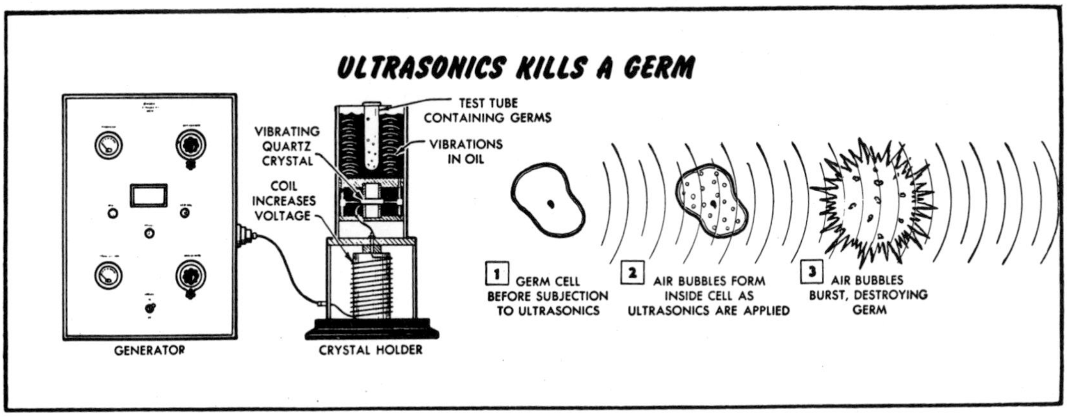

ULTRASONICS KILLS A GERM

VIBRATING QUARTZ CRYSTAL
TEST TUBE CONTAINING GERMS
VIBRATIONS IN OIL
COIL INCREASES VOLTAGE

1 GERM CELL BEFORE SUBJECTION TO ULTRASONICS

2 AIR BUBBLES FORM INSIDE CELL AS ULTRASONICS ARE APPLIED

3 AIR BUBBLES BURST, DESTROYING GERM

GENERATOR
CRYSTAL HOLDER

pushed together, they are compressed; when they are pulled, they are spread apart, or rarefied. Sound travels because this double motion is communicated to the molecules of the medium ever farther and farther away from the source of the sound. Thus a sound wave is a series of compressions and rarefactions in the medium through which it passes. Throughout the course of a sound wave's journey and regardless of its speed, the frequency of the sound remains the same as that of its source. The wave of motion might be compared to what happens when one pushes the last man of a long line of people waiting for a bus. The last man will have to push on the one in front of him and so on down the line. The initial push thus moves forward from the last man, "carried" by the people in front of him. The push travels along, but the line does not move.

Third, the sound waves must be received by some organ, such as the ear, or by some mechanical device. It is like the completion of a call when your friend answers the telephone. (Biologically speaking, sound is the translation in the brain of sensations received through the nerves of hearing.)

The difference between ultrasonic and ordinary sound waves is a matter of frequency. In fact, it is necessary to make this distinction only because human hearing has certain definite limits. The human ear is so constructed that even at its keenest it can register as sound only those frequencies that lie between 20 and 20,000 times, or cycles, a second. Any frequencies below 20 or above 20,000 are silent as far as man is concerned. Ultrasonics therefore is concerned chiefly with the frequencies above 20,000. Some study has been made of the lower range, but the upper one is by far the more important and interesting.

It must be remembered throughout this article that while there are certain similarities between the way electromagnetic waves (light, radio waves and so on) and sound waves behave, electromagnetic waves travel through space and sound waves can travel only through matter. Also, electromagnetic waves travel much faster, at about 186,000 miles per second.

To return to the insects, it is thought that some of them probably communicate with each other by means of "silent" sound — silent to us, that is. Certain grasshoppers produce sounds with a frequency of 40,000 cycles a second. Small mammals, such as cats, guinea pigs and rats, can hear frequencies up to 30,000 cycles and maybe even higher. Perhaps if you have a dog, you own a "silent" dog whistle. Its tone is pitched so high that you cannot hear it; but your dog, with a wider range of hearing than a human being, responds to the whistle.

Perhaps no members of the animal kingdom are so dependent on ultrasonic waves as bats. Experiments have proved that their inner ears are so delicately made that they can hear frequencies as high as 100,000 cycles. This was discovered in solving a problem that had long teased zoologists: What makes it possible for bats, nocturnal animals that fly about in the dark hours of the night, to avoid obstacles?

Westinghouse

Making ultrasonic waves "visible." To make the pictures (the room is then darkened), a photographic plate is put between the vibrating oil, in the jar, and a strong light. Exposures of up to a minute record the waves.

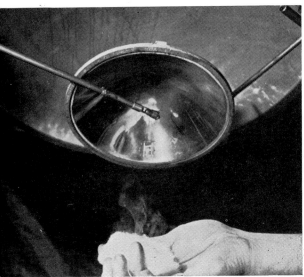

General Electric

Fire from a whistle! Ultrasonic waves from a whistle are focused to a point of high intensity. The waves stir the cotton so that a fire starts from friction.

After an apparatus was developed by which any high-frequency sounds the bats made could be detected, a number of the furry creatures were masked and made to fly through a room in which wires hung from the ceiling, only a foot apart. Provided the wires were not extremely thin, the masked bats avoided them with no trouble at all. The apparatus revealed that as a bat flies about, it produces ultrasonic cries, which are echoed back from any object in its path. As a bat approaches some hindrance, it may give as many as thirty to fifty of these cries per second; but if the path is clear, the cries are uttered at a lower rate.

As proof that these echoes guide the bats, they blundered helplessly into the wires and even into the walls of the room when, in addition to being masked, their ears were stopped or their mouths were gagged.

Later experiments showed that each cry consists of about 100 individual sound waves, which crowd together as each cry begins and then spread farther apart toward the end. The frequency of the waves ranges from 100,000, or even higher, at the beginning, down to 40,000 or even lower. It is likely that the longer inter-

vals between vibrations at the end of the cry are to help the bat hear the echoes from the first part. (This is really a kind of natural frequency modulation.)

However, you may well ask why the bat's cries need to be of such high frequency for the animal's purpose. The answer helps to explain one of the uses of ultrasonics to man, and has to do with the wave lengths of sounds.

We have said that a sound wave is a series of compressions and rarefactions. The wave *length* of a sound is the distance from one compression to the next. Consequently, the higher the frequency of the sound, the shorter its wave length must be. When the wave length is comparatively long, the sound flows around small obstacles, much as a large wave in the ocean surf flows around a piling, with hardly any disturbance, or reflection. (This ability of any kind of long wave — light and other electromagnetic waves as well as sound — to spread around an object is called diffraction.) But with short wave lengths, the sound is reflected back, or echoed, from small objects, just as little wavelets would be tossed back from the piling. In other words, short wave lengths are diffracted less. This is why the high-frequency cries of bats echo back from objects as small as wires, provided they are not hair-thin, and the animals are safely guided around unseen obstacles.

As long ago as World War I, a French scientist, Professor Paul Langevin, found a way to make use of this echoing quality of ultrasonic waves. At the naval base of Toulon, on the Mediterranean Sea, he and his assistants built an apparatus that could send strong bursts of high-frequency sound through the water. The waves traveled in straight paths without being diffracted. When they hit a submerged rock or a submarine, an echo returned. Here was a new way of detecting underwater perils.

After the war, two Canadian scientists applied Professor Langevin's discoveries further. They worked out a device for locating treacherous icebergs and hidden reefs.

Sonar, developed during World War II,

was a still greater refinement on the early ultrasonic underwater devices. In fact, sonar had a great deal to do with the conquest of the dread "wolf packs" of German submarines. Sonar works this way: When a pulse of ultrasonic waves is sent through the water from a ship equipped with sonar, an echo of the same frequency returns to the sonar apparatus from any solid object in the path of the waves. Since they are not diffracted, the direction from which the echo returns reveals the object's position. The distance, say, of a submarine from the ship can be calculated from the length of time that elapsed between the sending of the original pulse and the return of the echo. The device is able

sonic beam is aimed straight down toward the ocean bed, the depth of the water can be figured from the length of time it takes the beam to go down and be echoed back. Another peacetime use of sonar is on fishing boats because the device can locate schools of fish.

In 1951, Wayne M. Ross, a Seattle engineer, developed a simplified but even more sensitive sonar system. With this it is possible not only to locate schools of fish but also to identify the fish by determining the size of each school and the depth at which it is swimming. The device promises to be of even greater importance in the navigation of narrow, rocky channels because the echoes bounce back from hid-

A high-speed photo of the motions of water in a tank through which sound waves with a frequency of almost 3,000,000 cycles a second are being sent. The ultrasonic waves come from a crystal of barium titanate clamped between wires.

General Electric

to repeat the transmitting and receiving operations many times per second so that extremely brief periods of time may be measured. Submarine hunters were sometimes confused because the sonar system even picked up ultrasonic noises made by large schools of shrimps deep in the sea.

As you can see, sonar actually works on much the same principles as radar. The latter device is of no help under water, however, as it uses radio (electromagnetic) waves. These waves do not pass through water easily, whereas water is one of the best conductors of sound.

In peacetime, sonar is helpful to navigators in making soundings. If an ultra-

den shoals and from coastlines. The ultrasonic beam can be sent in any direction through the water, and the pattern of the echoes is displayed on a screen, much like radar. Ross's invention also records information in two other ways. The returning echoes are translated into audible sounds, over a loudspeaker, that a trained operator can easily recognize. A solid wall gives out a hard, clipped ping. Sounds from a smooth beach or a hidden sand bar are drawn out, as if someone were scratching granite with his fingernails. At the same time, automatic pen-and-ink records are made of all the echoes that appear on the screen and are heard over the loudspeaker. Thus there is a permanent record.

Scientists have discovered several ways of producing ultrasonic waves. The one most used for sending these waves through solids or liquids depends on a peculiar property of certain crystals, such as quartz, Rochelle salt (sodium potassium tartrate) and ammonium dihydrogen phosphate. This property is called the piezoelectric effect (from the Greek *piezein*, meaning "to press," plus "electric"). The effect was discovered by Pierre and Paul Curie in 1880.

The effect works like this: When pressure, which may be applied by weights, is brought to bear on such a crystal plate cut in a certain way, the plate becomes electrically charged. The amount of current generated is in proportion to the amount of pressure applied. Moreover, if the charged plate is stretched, the charge will be reversed. By alternate compression and stretching, therefore, an alternating current will be set up. The piezoelectric effect will work in the opposite way, once the natural frequency of the crystal is known. If an alternating current of the same frequency as a crystal's natural one is applied to the crystal, it will expand and contract in rhythm with the changes in the direction of the current.

The scientist produces ultrasonic waves by applying electricity to a crystal. Usually, small metal plates are attached to the crystal in an ultrasonic device and they move up and down in the same rhythm. This, in turn, sets up similar vibrations in the medium in which the apparatus is immersed — often a dish of light oil. With certain crystal cuts, their length or thickness is alternately increased and decreased and the waves travel in a longitudinal direction — that is, along the direction of the beam of sound. There are other crystal cuts from which the waves travel at right angles to the direction of the beam.

Another way of creating ultrasonic waves involves the use of an iron or nickel rod. A solenoid, which is a tubular coil for the production of a magnetic field and acts like a magnet, is placed around the bar. When a high-frequency alternating current is sent through the solenoid, the bar is magnetized in such a way that there are slight changes in the length of the bar in rhythm with the alternation of the current. The vibration thus set up in the bar is communicated to whatever medium surrounds it. Such a variation in length because of magnetization is called magnetostriction.

The strangest feature of these ultrasonic waves is that in the range of very high frequencies they have tremendous power. Ordinary sound waves have energy, but it is usually rather weak. Even a million persons all talking steadily for an hour and a half in a huge hall would produce only enough sound energy, converted into heat,

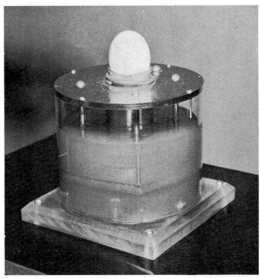

General Electric

A novel way of scrambling eggs. Waves with a frequency of 1,000,000 cycles pass to the water in which the egg rests, break down its contents and then cook it.

to produce a single cup of hot tea. Yet, in the experiments of Professor Langevin, of whom we spoke earlier, small fish swimming through the ultrasonic beams were killed instantly. When one of the professor's assistants held his hand in the path of the waves — only for an instant, you may be sure — he felt agonizing pain, as if his very bones were being heated.

What would we see in a laboratory where scientists are experimenting with ultrasonic waves? The focus of our attention

Sperry Products, Inc.

The box on top of the tread is an ultrasonic device called a reflectoscope. It is being used to test the take-up shaft on a giant power shovel. The reflectoscope sends short bursts of "silent" sound into the shaft, and any echoes that return to the instrument indicate cracks in the shaft.

would probably be a dish of oil. Immersed in it is a crystal, to which metal plates are attached, and from the plates there are wires connected with a maze of apparatus. As the electric current is turned on, there is an eerie hum. Watching the dish of oil, we see the surface suddenly begin to tremble. Slowly a mound forms and, so violent is the force, the oil bubbles and froths like boiling lava in an active volcano. At the climax, drops of oil fly up from the mound and make a little fountain that may be as high as twelve inches.

In one experiment, a small glass rod, pointed at one end, is dipped into the oil from above so that it touches the metal plates. If a piece of wood is held against the rod, smoke soon begins to spiral upward. After only a few moments, a hole is completely burned through the wood.

The evidence of great force is before our eyes, but where is it coming from? The metal plates move through distances of only ten thousandths of an inch. In fact, the plates appear to be still. Nevertheless, they change their direction at the enormous rate of hundreds of thousands of times per second. An object soaring out into space from the earth at such a rate of speed as this would be about a million miles away in ten seconds. It is the tremendous rapidity of the vibrations that gives such ultrasonic waves their power.

Both science and industry are beginning to harness this power. One device, called a reflectoscope, sends short bursts of "silent" sound into a metal object, such as a casting, searching out flaws. A quartz-crystal mechanism is placed against the object, with only a film of oil between them. The reflection of the high-frequency beam sent out by the crystal is picked up by the crystal, now acting as a microphone. Electronic switching is necessary (as in sonar) because the alternate use of the crystal as transmitter and receiver takes place in millionths of a second. Any flaws in the metal bounce back as echoes and flash on a cathode-ray screen, a kind of fluorescent screen, showing the location of the defects.

Flaws in automobile tires may be detected by a similar method. The tire is immersed in water between a transmitter of ultrasonic beams and a receiver. Rubber will transmit a sound wave from water with little reflection so that any echoes indicate a defect in the rubber.

Very high-frequency sound waves can make two liquids emulsify that ordinarily would not mix. For instance, alloys of iron and lead, aluminum and lead, aluminum and cadmium and so forth can be made to mix in the liquid state and kept mixed until they solidify. New bearing materials are being produced in this way. Another application of this effect is to make photographic emulsions stable and of the same consistency throughout.

Another use of high-power ultrasonics is to make solid and liquid particles in mist, dust and smoke clump together. Factories that make lampblack, for instance, an ingredient of varnishes, paints and some kinds of ink, are installing ultrasonic generators in the flues of their smokestacks. Ultrasonic vibrations push the lampblack particles together and they then drop down the chimney instead of escaping into the outside air. Around airports and harbors, ultrasonic waves can disperse fog and mist. It is even possible that large industrial areas overhung with palls of smoke and soot may be cleared by the same method.

Some kinds of chemical reactions are speeded up under the influence of ultrasonic waves. They may also break up long-chain polymers — molecules with a highly complicated structure.

In the United States in 1950, a patent was issued for an ultrasonic device that helps blind persons to avoid obstacles. The waves are sent out from a mechanism in a walking cane. Echoes from any object ahead are picked up by a receiver in the cane. Their energy is converted into electric pulses that travel by wire to the per-

Ultrasonic "bath" that cleans shaver heads for electric razors. The heads move through a trough filled with cleaning fluid. At the same time ultrasonic waves, generated by a crystal, are sent through the fluid. These clean small openings and corners where the fluid alone would be ineffective.

General Electric

son's ear where, by means of a kind of microphone, the pulses are translated into audible sound.

Quicker and better laundering is another very practical use of ultrasonics. The intense vibrations break down the attraction between dirt particles and fabrics and shake the grime loose.

Ultrasonic waves, as we have seen, can affect living organisms, although the reason for the effects is still rather obscure. Seeds, for example, have been treated with "silent" sound. In one case, it was reported that potato plants so treated blossomed a week ahead of time and their yield increased 50 per cent above the crop from untreated plants.

Living cells may be literally shaken to pieces by ultrasonic waves. Germs may be killed, and milk exposed to waves of high intensity is pasteurized in a few seconds. We pointed out that small fish had been killed by swimming through the ultrasonic beams produced by Professor Langevin's apparatus. Frogs and mice

Raytheon

In this machine, an oscillator, milk is homogenized by "silent" sound. It breaks up fat globules, making the fat mix with the milk itself.

have also been killed by silent sound.

In some kinds of surgery, ultrasonic beams may eventually take the place of knives and scalpels, performing an operation much more quickly and with less danger to the patient. Experimenting with dogs and rabbits, surgeons have shattered gallstones into fragments in fifteen seconds by the use of intense beams. The tissues around the gallstones were not damaged, and the fragments of the stones were easily eliminated through natural channels. In these operations the beam was transmitted through water, which is elastic. Living tissue is as elastic as water — it seems to "roll with the punch" — stretching without splitting under the impact of the beam. Only the solid gallstones were cracked into tiny pieces. Kidney stones might be treated in the same way.

Brain operations, usually long and delicate, are another ultrasonic possibility. At Columbia University, quartz crystals for ultrasonic generators were cut so that the waves could be focused on a particular spot inside the brain of a dog or a cat. The energy of such a concentrated beam can be as much as 150 times as forceful as that of an unfocused beam, and it can destroy a selected brain area in a few seconds.

Ultrasonic waves can have peculiar effects on human beings besides those we have mentioned. Just as a piece of wood is burned if it is held against the glass rod touching the vibrating metal plates, so would your fingers be burned if you touched the rod. Even stranger and rather frightening is the fact that if a person stands in the path of ultrasonic waves of high intensity, he feels confused and depressed and loses control of his movements. A mathematician reported that for several days after she had been subjected to such waves, she was unable to solve even the most simple problem in arithmetic.

Though the existence of ultrasonic waves has been known for many years, the development of the science of ultrasonics is fairly recent and received its greatest impetus after World War II. It is likely that we are only on the threshold of the mysterious house of "silent" sound.

Foam can be made to collapse by subjecting it to ultrasonic waves produced by a device called the Gulton Multiwhistle. Upper left: the device has been put on top of a plexiglass cylinder partly filled with a chemical foam. Upper right: the foam has been dissipated. The Multiwhistle has a number of applications.

Below: an experimental ultrasonic seam-welder is welding together two strips of aluminum. A continuous weld is made by passing the strips between two metal wheels that vibrate at the rate of 20,000 cycles per second. The crystals of the two strips are intermixed and a strong molecular bond is formed.

Upper photos, Gulton Industries, Inc. Lower photo, Westinghouse

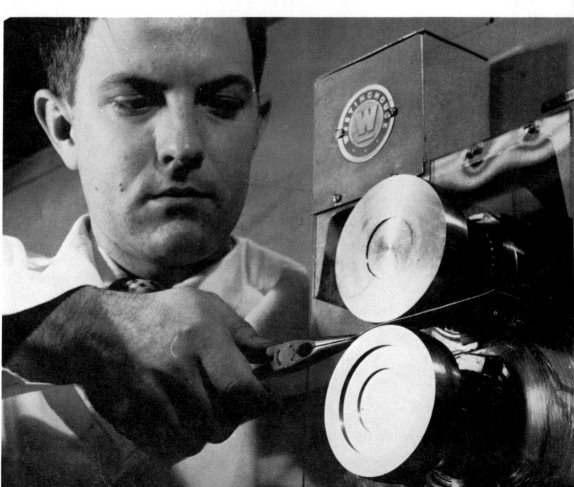

Two unusual aquariums. The elaborately furnished one at the left, with a Japanese motif, was arranged by R. Ferdinand of the Aquarium Stock Company, Inc. Below is shown an aquarium that was set in a room divider. It looks like an animated painting with an oval frame. A hinged panel at the back of the divider can be opened outward, thus giving easy access to the tank.

All photos on first three pages by W. A. Schwarz, with the co-operation and assistance of the Aquarium Stock Company, Inc., New York, N. Y.

HOME AQUARIUMS

Fish Pets and How to Care for Them

BY THOMAS GORDON LAWRENCE

A HOME aquarium to house fishes can be set up at moderate cost and will give you many hours of pleasure. You will be able to raise guppies, which bear living young, and egg-layers such as the familiar goldfish, the good-natured zebra danio and the gorgeous fighting betta. You will be able to study at your leisure such fantastic specimens as the African mouthbreeder, whose young make their home in their father's mouth.

How to select and maintain your aquarium

You can make an aquarium from any nonmetal container that will hold water. However, do not use a glass bowl. You can never see the fish properly through its curved sides. If you fill the bowl more than halfway with water, there will not be enough water surface exposed to the air. The best aquarium is a rectangular one with a metal frame.

An all-glass aquarium in one piece is very likely to break, and it is then a total loss. If you break one glass in a metal-frame aquarium you can have the glass replaced or replace it yourself. You can see the fish and plants better because each glass is the same thickness in every place. Varying thickness of the glass in an all-glass aquarium distorts your view of the fish.

Your aquarium should have a glass or plastic cover. This may save the lives of fish that jump out of the water from time to time. It protects the water from dust and checks evaporation. It also protects fish from sudden changes of temperature.

There are two important points to keep in mind. The first is that no metal should ever be in contact with aquarium water, because it is liable to produce poisonous substances. The second point is that the aquarium should be at least as wide as it is deep. Fish breathe oxygen that the water absorbs from the air. In a deep, narrow tank, the water might not absorb enough oxygen.

Two chief causes of death among aquarium fishes are overcrowding and over-

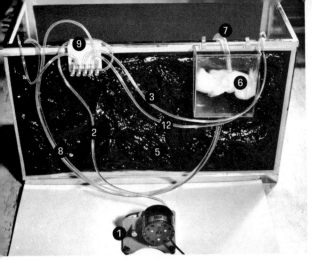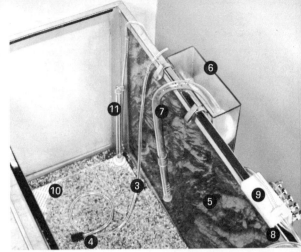

Two views of an aquarium. At the left, above, is a rear view, showing an installation for aerating and filtering water. At the right, above, is an overhead view of the same aquarium. The different parts of the installation are indicated by numbers, as follows: (1) Piston air pump. (2) Main air line from pump to five-gang valve. (3) Tubing through which air is pumped into tank from five-gang valve. (4) Air stone—a porous stone through which air is released into tank. (5) Three-dimensional rock-and-cave background, taped to outside of tank. (6) Hi-Speed filter box, containing charcoal and glass wool. It hangs outside tank. (7) Intake stem tubing; it draws water from tank into filter box. (8) Return stem tubing from bottom of filter into tank. (9) Five-gang valve. It consists of five valves mounted in series on lucite hanger. (10) Under-gravel filter plate (filter set under gravel in tank). (11) Under-gravel stem for item 10. (12) Air line from under-gravel stem to gang valve.

feeding. A safe rule is to allow one gallon of water for each inch of fish. "Inch of fish" here means inch of body length exclusive of the tail. If we apply this rule, a 5-gallon tank would be fit for two 2½-inch goldfish or five one-inch guppies. With aeration (bubbling air through the water), you can safely double the number of fish; if the aeration is kept up day and night you can put in three times as many. The colder the water, the more fish you can keep in a tank provided, of course, that they are the kind (such as goldfish) that can stand cold water. Cold water holds more oxygen than warm water does. When the water gets warmer, the fish need more oxygen and give off more carbon dioxide and other waste products. You could easily keep a hundred small goldfish in a 20-gallon aquarium at a temperature of 40° F. If the temperature in this crowded aquarium went above 80° F., the fish might die.

Water plants and fish both need light. If you do not use artificial light, place your tank where it will get light from the north most of the day. More than a small amount of direct sunlight may injure the fish. If leafy plants grow well in your aquarium, you will know that it is getting enough light. If there is too much, the water will become greenish because of the growth of the microscopic green plants called algae.

Green water may hide your fish; by cutting off light, it may harm the plants you put in the aquarium. You can check the growth of algae by cutting down on the light, since these plants make their own food in the presence of light.

You may have been told that the fish in aquariums get oxygen from the plants. The fact is that most of the oxygen the water contains has been absorbed from the air. The greater part of the carbon dioxide breathed out by the fish escapes into the air. Have you admired the stream of silvery air bubbles rising from an aerator in your own or a friend's aquarium? Although these bubbles do give off some oxygen to the water, their main value is to help carry off carbon dioxide. They also help to keep the temperature of the water more uniform so that you will not have a layer of warm water at the top and a layer of cooler water at the bottom.

Plants for the aquarium

When you buy fish, you will notice that often there are no plants in the tanks that contain fish for sale. It is easier to catch the fish and to keep the tanks free from parasites and small animals that devour young fry. However, to have a beautiful, natural-looking aquarium, you must set out

Live plants are a most useful and also decorative addition to an aquarium; plastic plants, as real as life, lend added color to it. A few examples of live and plastic plants are shown in these two photographs, as follows: (1) Water sprite (live). (2) Cabomba (live). (3) Azalea foliage (plastic). (4) Hornwort (live). (5) Red and green privet (plastic).

some plants. They give off substances that improve the water and they absorb waste matter given off by the fish. Many fish like to chew certain plants, which serve as fresh "greens" and are good for them.

To let the plants root themselves and to show off your fish to the best advantage, cover the bottom of the aquarium with one or two inches of clean sand or fine gravel. Bird gravel is excellent. Slope the sand instead of spreading it flat and even over the entire area. You might leave a small section at the front of the aquarium with no sand at all. The sand you buy at a fish store is cleaner than that which you gather by a lake or river. Never use sea sand in a fresh water tank! In any case, wash sand or gravel by putting it in a pail and pouring warm water over it while you stir it. Stir until the water runs off without carrying dirt or sediment.

You can add rocks for decoration and to help hold the roots of plants. Don't use limestone, soft sandstone or any soft rock that may slowly dissolve. Slate is excellent; so are rounded quartz pebbles. If you like, you can put in your aquarium the divers, castles and wrecked ships sold by pet stores. Do not use objects with sharp corners and edges that may injure fish when they brush against them.

When you fill a tank with water, put a piece of heavy paper over the sand and then place a dish over the paper. This prevents disturbing the sand. In adding water to a going aquarium, hold a saucer under the surface and slowly pour the fresh water into the saucer. Instead of this, you may spread a sheet of paper over the surface of the water.

Place your plants at the back and sides of the aquarium to form a background for the fish. If you have a large tank, set one or two plants near the center. Some beautiful and useful plants are *Elodea* (also called *Anacharis*), *Vallisneria* (eelgrass), *Cabomba, Cryptocoryne, Sagittaria* (resembling eelgrass but with thicker leaves) and the sedge *Eleocharis,* which looks like grass. Spatterdock, or cow lily, is an attractive plant, with its large leaves; it should be set by itself near the center of the tank. All of these plants will take root in the sand. Goldfish eat the tiny floating duckweed. Four handsome and rather strange floating plants are *Riccia, Salvinia* (a tiny mosslike fern), water lettuce and frogbit. Threadlike *Nitella* makes a jungle for young fish to hide in. Two plants that grow well, floating or rooted, are *Fontinalis* (water moss) and water fern, which sprouts new plants from its leaves.

After planting, let the tank settle for at least three days before you put any fish in it. Let the water age so that chemical adjustments may take place between the plants and the water.

A balanced aquarium

If you had a completely balanced aquarium, you would never need to clean it. Bacteria, algae and tiny animals would consume all dead leaves and waste prod-

The fish in your aquarium will need live food from time to time. The best is the tiny crustacean *Daphnia,* in the upper photograph, right. It is sometimes called the water flea. The fish will also relish a meal of mosquito larvae. One of these is shown in the lower photo.

Carolina Biological Supply Co.

ucts. The plants would use minerals from the animals; the fish would eat leafy plants, algae and small animals. I once kept such an aquarium for three years without removing anything. Your best policy will be to remove dead leaves and sediment whenever the fish begin to stir things up so that the tank looks unattractive. You can use either a dip tube or a siphon.

What about snails? Your aquarium can get along perfectly well without them, but the snails themselves, their eggs and their young are all interesting. They will eat food overlooked by your fishes, dead fish and the green or brown scum of algae that sometimes forms on the glass. They may benefit water plants by eating algae that grow over the leaves. You yourself can take care of algae that grow on the glass by scraping them off with a rubber scraper or a single-edged razor blade.

Food for fish
in an aquarium

Do not give your fish too much food. If you feed fishes every other day and never give them more than they will eat in fifteen minutes, you will not go far wrong. Remove all uneaten food with a dip tube. You can use prepared food sold by fish stores; most fish relish dried shrimp broken into small particles. Many fish will also eat cooked, dried and ground-up cereal and spinach, lettuce and other greens. Do not expect any aquarium fish to live entirely on vegetable matter.

Fish need live foods from time to time. The best is *Daphnia,* called the water flea because it "jumps" through the water. Newly hatched brine shrimp are ideal for many fry. Brine shrimp will not live long in fresh water, but fish swallow them before the water can kill them. If you feed *Tubifex* worms to your fish, wash the worms in a stream of cold water first. If a few *Tubifex* escape the fish, they will dig their heads into the sand and wave the other end of the body in the water. Mosquito larvae, *Cyclops* (a tiny crustacean related to water

fleas and shrimp) and fruit flies are all gobbled up by fishes. Leave small pieces of ripe banana peel in empty milk bottles and fruit flies will appear as if by magic.

Favorite types
of aquarium fish

The common goldfish is the easiest fish to keep. It can live through the winter in ponds under solid ice. Fancy goldfish are not so hardy as the common variety; yet some fancy types can also stand a considerable amount of abuse.

The best tropical fish to begin with is the guppy. It can bear a temperature as low as 60° F., while many tropicals soon die below 70° F. It thrives best at 72° F. Guppies are famous for giving birth to living young—and for eating their children. Many other fish eat their young, but not so many people have seen them doing it. The small, slim, lively males are very colorful; the females are a sober, grayish olive-green.

The molly is a beautiful live-bearer that can stand even lower temperatures than the guppy can. Mollies come in many colors, from light blue-green, with orange or red fins, to pure black. The popular platies and swordtails are also live-bearers (fish that bear living young). Male swordtails have part of the tail fin prolonged like a sword. Platies and swordtails may be red, blue, green, yellow, partly black or mixed in color. As all species interbreed, you can experiment by crossing different

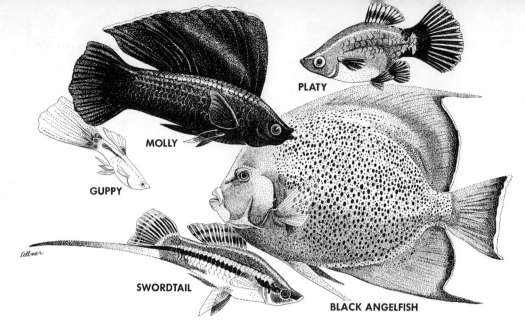

PLATY

MOLLY

GUPPY

SWORDTAIL

BLACK ANGELFISH

kinds. Both platies and mollies do better if you put a little salt in the water before you put in the fish.

Most kinds of fish lay eggs, but it is usually harder to raise egg-layers. Some egg-layers that you might want to start with are the zebra and pearl danios and the beautiful angelfish. Tetras, barbs, rasboras and *Panchax* are all interesting. Bettas (fighting fish), gouramis and paradise fish breathe air as well as water and they constantly force air bubbles through their gill covers when at the surface. The males use these bubbles to build nests for the eggs.

Diseases and natural
enemies of aquarium fishes

Fish may suffer from chill or over-heating, from poisons in the water or from diseases due to bacteria, fungi or animal parasites. Fungus growth looks like cotton wool; it often occurs when a fish's skin has been damaged. Very alkaline water may encourage fungus growth. To treat fungus, take the fish in your wet hand and touch the growth with mercurochrome or iodine solution diluted in the proportion of one part mercurochrome or iodine solution to ten parts of water.

Tail and fin rot are caused by bacteria that eat away the fins and may invade the body. Antibiotics offer the best treatment; 50 milligrams of aureomycin per gallon of water should remove all symptoms. Aureo-mycin or terramycin should also cure most cases of mouth rot.

Ich is a common disease among tropical fish. In ich, tiny parasitic animals cause the fish to be covered with white spots. You can place methylene blue or quinine-hydrochloride solution in the tank to kill the parasites. Methylene blue is harmless to fish, but you must change the water after using quinine.

Leeches, flukes, water bugs and some other animals will kill baby fish. Leeches and flukes are also dangerous enemies of larger fish. If you introduce wild water plants, you should inspect them to make sure none of these enemies are present.

Books about
aquarium fishes

Here are some interesting books about aquarium fishes:

H. R. AXELROD and L. P. SCHULTZ, *Handbook of Tropical Aquarium Fishes;* Mc-Graw-Hill Book Co., New York, 1955.

H. R. AXELROD and W. VORDERWINKLER, *Aquarium Fishes* (paperback); Sterling Publishing Co., New York, 1960.

H. R. AXELROD and W. VORDERWINKLER, *Saltwater Aquarium Fishes;* T. F. H. Publications, Jersey City, 2nd edition, 1963.

ANTHONY EVANS, *Care and Breeding of Goldfish* (paperback); Dover Publications, New York, 1957.

W. T. INNES, *Exotic Aquarium Fishes;* E. P. Dutton and Co., New York, 19th edition, 1959.

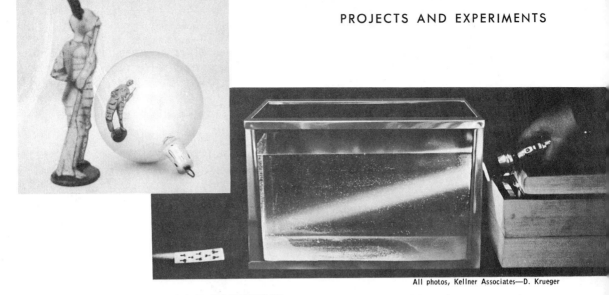

All photos, Kellner Associates—D. Krueger

EXPERIMENTS WITH LIGHT

How It Is Given Off and How It Travels

BY HARRY MILGROM

MOST of what we know about the world around us is brought to us in the form of light messages that enter the eyes. These messages carry information concerning the size, shape, distance, state of motion and color of objects in the immediate vicinity. Light messages also come to us from the depths of outer space — from planets, stars and far-flung galaxies.

Light is a form of energy. It does not move instantaneously, but has a measurable speed; it travels through a vacuum at the rate of about 186,000 miles a second. The light that reaches us at sunrise has sped for something like 8 minutes across 93,000,-000 miles of almost empty space to reach the earth. Unlike sound, light does not require any medium, such as air or water, for its transmission; as a matter of fact, it travels best through a vacuum.

How light is produced

Light is generally given off by very hot substances. Light a match, a candle, a pocket lighter and a piece of paper in turn. In each case, the heat generated by the chemical action of burning causes the carbon particles in the vapors that are produced to glow with a yellow light. Grasp a piece of steel wool with a pair of pliers, ignite it and hold it over a metal pan. Because the wool is in a shredded state, its temperature will become very high, and it will give off yellow-white sparkles as it burns.

The color of a heated, incandescent (glowing) body changes as its temperature goes up. At about 900° F., a dull red color appears. As the incandescent body grows hotter, it turns orange, then yellow, then white and finally bluish-white (at about 2,900° F.). A rapid method of finding out the temperature of a mass of molten iron in a furnace is to look at it through the instrument called an optical pyrometer ("fire-measurer"). To get a temperature reading of the molten mass, the viewer matches the color of its surface with the appropriate shade on a color scale in the instrument.

Turn on an electric bulb. The flow of electricity through the filament of the bulb heats the thin wire and causes it to glow with a white light; generally the bulb becomes too hot to handle. Plug in an electric broiler or toaster; the electricity will cause the wire of the broiler or toaster to get hot and glow red.

In the preceding experiments, light was produced by developing sufficient heat, either by the burning of a fuel or the flow of an electric current. Light is not always generated by heating effects. Obtain a neon bulb or a neon test lamp in your hardware store. Connect the bulb or lamp to the house circuit and note the characteristic orange-red glow of neon gas. Note, too, that the bulb or lamp does not become hot. When an electric current flows through the neon gas, the atoms of neon are stimulated to emit this orange-red light. Other gases that glow when they are electrically excited are mercury vapor (blue-green), argon (pale blue) and a mixture of the two (deep blue). These gases are sealed in variously shaped glass tubes to make colorful display signs.

Turn on a fluorescent lamp and touch the center of the glass; you will note that this type of light, too, is not accompanied by high temperature. Electricity flows through the mercury vapor in the lamp and generates invisible ultraviolet light. The ultraviolet, in turn, strikes the chemical coating on the inside of the glass to produce a relatively cold, visible light. Different coatings produce different hues. They are often used in neon signs to obtain more color variety.

Take a roll of friction tape, made with a cloth base,* into a dark closet. Pull a length of tape rapidly away from the roll. At the junction between the roll and the strip you are pulling out, you will see a line of light. This is a form of "cold" light that is generated by the forceful separation of the tape molecules.

Collect some fireflies in a glass jar. You will note that the fireflies produce nearly cold light from luminous organs on

* Masking tape and plastic electrical tape will not be satisfactory for the purpose of this little experiment.

the underside of the abdomen. These organs contain a fatty tissue called luciferin; it produces light flashes when it is acted upon by the enzyme luciferase. The insect controls the emission of this light, which is thought to be a signal between the sexes.

Some chemical effects of light

If you expose your face to bright sunlight (taking care not to look directly at the sun), you will feel an instant sensation of warmth. The infrared portion of the sunlight is mainly responsible for this heating effect. To test some of the effects of sunlight, cut out a cross of cardboard. Place the cross on a sheet of blue construction paper and expose the paper to bright sunlight. After ten to fifteen hours of exposure to winter sunlight (somewhat less to summer sunlight), the uncovered part of the paper will be faded by the light. When you remove the cardboard, a blue cross will be seen on the paper — the original blue color of that part of the paper not affected by sunlight. Evidently sunlight causes some substances to become lighter.

It darkens other substances, including your skin. If you are going to be in bright summer sunlight for several hours, attach two strips of adhesive tape, forming a cross, to the back of your hand. The skin of your hand will be tanned everywhere except where it has been covered by the tape. Place a key on a piece of photographic contact-printing paper and expose the paper to sunlight or to the light of a 100-watt bulb. The uncovered part of the paper will darken

1. A shows a small length of rope hung on a longer rope. The latter is attached to a fixed object. As the free end of the big rope is flipped with an up-and-down motion, a wave will travel along the rope and will knock off the small rope (B).

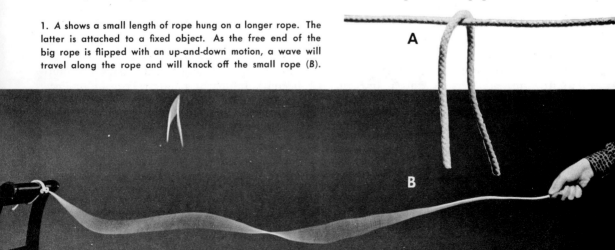

2. How light is transmitted from a flashlight through the air of a room.

3. How light travels from a flashlight through the water in a tank.

in a period of less than two minutes, clearly showing the outline of the key.

In all these cases, color changes have been brought about by chemical changes caused by light. One of the most important chemical changes in which light plays a part occurs in the process called photosynthesis (see Index). In this process, food is manufactured in the green leaf.

How light travels
from place to place

At one time physicists believed in either one or the other of two theories concerning the nature of light. Those who upheld the so-called wave, or undulatory, theory maintained that light travels in waves, just as sound does. Other scientists, followers of the emission theory, held that light was made up of particles that were emitted from luminous bodies and that were reflected or refracted (turned from a given course) by other objects. Today physicists use either the wave theory or the emission theory (also called the corpuscular theory), depending on the phenomena they are studying.

To explain certain phenomena, such as polarization and diffraction, scientists assume that light travels in the form of electromagnetic waves. (See Index under Polarized light and Diffraction.) To picture this wave motion of light, attach one end of a rope, about five feet long, to the back of a chair and hang a small length of rope near the fastened end (Figure 1). Use a cotton clothesline or a heavy cotton cord; if a plastic clothesline is used, it should be warm so as to be flexible. Stretch the rope and flip the free end with an up-and-down movement of your hand. A wave will travel along the rope and will transmit enough of

the force of the flipping motion to knock off the small rope. Of course, no part of the big rope actually travels from one end of itself to the other. Something like this takes place when light streams away from a source. The light source, corresponding to the moving hand, sends out impulses. These spread out in the form of waves, which strike different receivers to produce different effects.

The emission theory of light is used to provide an explanation of such phenomena as reflection and refraction. According to this theory, light consists of a stream of projectiles, called photons. A light source, like a machine gun, produces a rat-a-tat of photons which fly through space. This photon stream bounces off certain surfaces to produce the familiar effects of reflection. When the direction of the photon stream is changed as it goes from one region to another — for example, from air into glass — it is said to be refracted. A mirror reflects light; a magnifying glass refracts it.

Light can move in and through many different substances. Turn on a flashlight and direct it at the walls of a room (Figure 2). The light will travel in straight lines through the air in the room and land on the opposite wall. Next, turn a flashlight on one side of a tank filled with water (Figure 3). The light will travel through the glass and through the water and show up on the other side of the tank. Light travels fastest in a vacuum; a little slower in gases; slower still in liquids (for example, water or ethyl alcohol); and slowest of all in solids (such as glass).

Rays of light travel in straight lines (unless they are refracted, or bent). This can be shown as follows. Poke a pinhole through the center of the bottom of an empty cylinder-shaped cereal box. Put a screen of wax paper in place at the other end of the box, as shown in Figure 4A; hold the screen in place with a rubber band or tie it with a string. Let some dripping from a lighted candle fall in the center of an ash tray or the lid of a jar or can, and set the candle firmly upright on the dripping. Aim the pinhole end of the cereal box at the lighted candle. Because light travels in a straight line, the flame of the candle will appear inverted on the wax-paper screen, as in Figure 4B. Light rays from the top part of the candle have traveled through the pinhole, as shown, and have hit the bottom part of the screen; light rays from the bottom part of the candle have hit the top part of the screen. The apparatus you have made is called a pinhole camera.

Some effects
of reflected light

A number of things are visible because they are luminous: that is, give off their own light. The sun, a burning candle, a glowing electric light and a flashing firefly are all luminous. But most things are visible because they reflect all or some part of the light that comes to them from a luminous source. You see the words on this page by reflected light. Take the book into a closet where there is no light, and you no longer obtain a light message.

Stand near a window through which sunlight is entering. Reflect the light with a variety of different surfaces so that it strikes a piece of white paper set on the adjoining wall. Try aluminum foil, wax paper, glass, cloth, different colors of construction paper and cellophane in turn as the reflecting material. Use the same size square in each case. Which surface, the light-colored or the dull, dark surfaces reflect the light better? Can you apply your conclusions to the way that common mirrors are coated on the back?

Hold the palm of your right hand up against a large mirror. The image is the same size as your hand, but it is reversed; your right hand appears to be a left hand in the mirror. The mirror reverses all objects in this way. In his notebooks the renowned Italian engineer, sculptor and painter Leonardo da Vinci wrote from right to left; when these pages are brought up to a mirror, the writing is reversed and appears, as does ordinary writing, as if written from left to right. This is called mirror writing.

As you move your hand away from the surface of the mirror, the image seems to recede to the same extent in back of the mirror. Since the mirror is a flat surface and has no depth, the image produced in this way is not really where it seems to be; therefore it is called a virtual (apparent) image. Many striking optical illusions can be created by means of virtual images.

Stand a piece of transparent window glass in an upright position, as illustrated in

4. In A, we see the pinhole camera described on this page. Note the wax-paper screen and the lighted candle set up in front of the pinhole. B shows how the inverted candle flame will appear on the wax-paper screen when the room is dark.

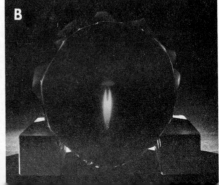

Figure 5. Place a lighted candle six inches in front of the pane and a tumbler filled with water six inches on the other side of the pane, as shown. If you look at the tumbler through the pane, you will apparently see the candle burning under water.

Cut the base of a cardboard box as shown in Figure 6. Fold and fasten the different sections together with Scotch tape (Figure 7); note that you will have to add a section. Mount the walls, or sides, at a right angle to the bottom (Figure 7). Put four mirrors in position by means of tape (Figure 8). The reflecting surfaces should all face toward the inside of the box. The mirrors must be placed exactly at an angle of 45° with reference to the sides of the box and with reference to the path of light (indicated in Figure 8). Next, put a cardboard top on the box, keeping it in position with Scotch tape (Figure 9). You will then be able to see an object even if you put a book or other solid thing between the object and your eyes (Figure 9).

Give a high polish to both sides of a silver or stainless-steel tablespoon. Look at your image in the inside part of the bowl; the image is small, upside down and egg-shaped. Move your eye closer and closer. The upside-down image will grow larger, fill the surface of the spoon and then reappear enlarged and right side up. Turn the spoon around. You will see an image that is smaller, right side up and egg-shaped. No matter how close you get to the spoon, the image will remain much the same.

If one uses silvered plastic material, it is possible to produce images in a variety of shapes, such as those in Figure 10. As the shape of the mirror changes, the kind of image it produces also changes. Irregu-

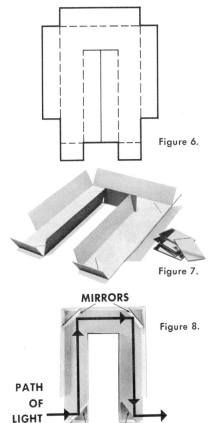

Figure 6.

Figure 7.

MIRRORS

Figure 8.

PATH OF LIGHT

MIRRORS   Figure 9.

6-9. Making a device that lets one "see" through a solid object. Figure 6 shows how to cut out the cardboard described in the text. In Figure 7, the cardboard has been cut out; mirrors have been obtained. The mirrors are in position in Figure 8; arrows show the path of light. The completed device is seen in Figure 9. A toy car has been set in front of one mirror (A) and it appears in the opposite one (B). Apparently the viewer sees through a book.

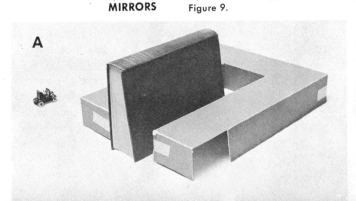

A

B

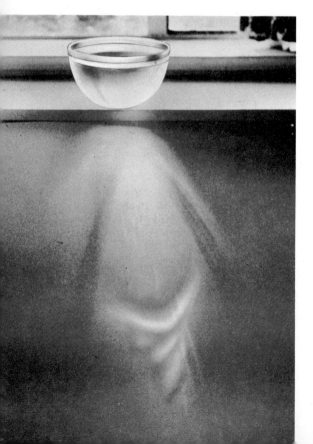

10. The image of the toy knight is distorted in the silvered plastic mirrors.

11. Sunlight passing through a glass bowl filled with water produces the colors of the spectrum on a sheet of white paper.

larly shaped mirrors are used in amusement parks to entertain people with their highly distorted images.

The hidden colors in white light

Fill a glass bowl with water and set it at the edge of a table, as shown in Figure 11. The sunlight will hit the water, will be refracted and will then emerge from the bowl to strike a piece of white paper. The white sunlight will be broken up into the colors of the visible spectrum: red, orange, yellow, green, blue and violet. Turn on the fine spray of a garden hose with sunlight in back of you and shift your view until you see a rainbow made up of the same sequence of colors. Use cut glass, beveled mirror edges, rhinestone jewelry, diamonds and other materials to see how they separate white light into its constituent colors.

One method used to learn more about the objects in outer space is to study the light that comes from those that are luminous. An instrument called a spectroscope separates the colors that make up the visible spectrum. In the spectroscope, a triangular glass prism (or a ruled grating consisting of parallel lines very closely placed on a glass) separates the colors in relation to their wave length. For example, starlight is gathered by a telescope and passed through a spectroscope. The light is then examined under magnification for color clues. Colors can give us information about a star's temperature, composition and speed of motion.

Cover the front of a flashlight with red, green, blue, orange and yellow cellophane in turn, and then turn on the light. (You can use the colored cellophane wrappings of lollipops for this purpose.) In each case the white light passes through the cellophane and comes out colored. Red cellophane, for example, absorbs all the colors of white light except red, which it transmits. Each of the other squares transmits its particular color in the same way. The color of a transparent object, then, depends on the way in which the object transmits light. The color effects of a material such as stained glass are due entirely to the transmission of light.

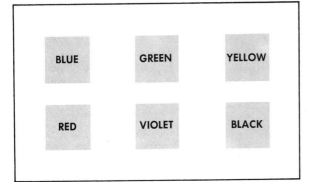

| BLUE | GREEN | YELLOW |
| RED | VIOLET | BLACK |

12. How to set up the squares of differently colored paper described in the text.

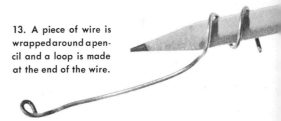

13. A piece of wire is wrapped around a pencil and a loop is made at the end of the wire.

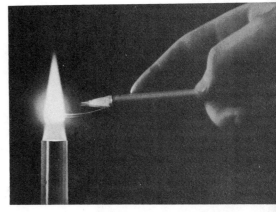

14. A crystal or two of table salt is set on the loop shown in Figure 13. The salt is heated in a Bunsen-burner flame and gives off yellow light. Nichrome wire was employed in this experiment.

The color of an opaque object, on the other hand, depends on the way in which the object *reflects* light. Paste six squares of differently colored construction paper on a white background, as in Figure 12. If you take the squares into a dark closet, you will find that the colors cannot be distinguished without a source of light. Now shine the white beam of a flashlight on each square in turn. The blue square will absorb all the colors of the white light, except those in the neighborhood of blue, which it will reflect. How about the other squares?

The effect is more striking if monochromatic light (that is, light of just a single color or wave length) is used. If the kitchen in your home is provided with a gas range, you can perform a striking experiment with monochromatic light. It will have to be in the evening, because before it will be over, the room will have to be made dark. First wrap the end of a piece of iron wire around a pencil, as shown in Figure 13. (Nichrome wire, obtained from your science teacher, would be even better than iron wire.) Make a very small loop in the free end of the wire, as in Figure 13. Heat the end of the loop over a gas flame in the range. When the flame above the loop has become nearly colorless, place a crystal or two of table salt on the loop and heat the salt in the flame. You will now have to make the room dark by pulling down the blinds and turning out the light. You will then note that the hot table salt gives off yellow light that is very nearly monochromatic. Bring the sheet of colored squares used in the previous experiment near the hot salt, and note the color that each square seems to have. Can you explain the appearance of each?

If you do not have a gas range in the kitchen, you can perform the experiment in the darkroom of your science laboratory in school if this room is provided with a Bunsen burner. You will have to adjust the burner so that its flame will be as nearly colorless as possible. Once this has been done, you can carry out the experiment as has just been described. (See Figure 14.)

Refraction—changing the direction of the pathway of light

Obtain a clear plastic or glass container, about four inches tall and four inches in diameter, and fill it two-thirds full of water. Add a few drops of milk to the water in order to make it cloudy. Prepare a circle of aluminum foil to use as a cover and cut a small slit (one inch by one-eighth inch) out of the foil near one side of the circle. (See Figure 15A.) A corner of the foil cover is to be lifted and smoke is to be introduced into the space above the water. The cover is then to be put back in place. The room is to be darkened and the beam of a flashlight is to be directed toward the slit, as is shown in Figure 15B. To keep the light beam from being too

15. Demonstrating the refraction of light. In A we see a glass partly filled with water to which a few drops of milk have been added. The glass has been covered with aluminum foil in which a slit has been cut. Smoke has been introduced into the space above the water. In B a flashlight beam, directed through the slit in the aluminum foil, changes direction as it hits the water.

16. Put a penny in the bottom of a cup; then lower your head until the penny appears as only a sliver (as in the photograph) and then disappears from view.

17. Without raising your head, fill the cup with water. The penny will seem to float into view though it has not moved—an effect of refraction.

diffuse, wrap a ring of heavy cardboard around the lens end of the flashlight so that it will form a hood, extending an inch or so beyond the lens. Keep the hood in place with Scotch tape.

The light beam will be seen to change direction as it enters the water, as shown in Figure 15*B*. Whenever light moves on a slant from air into a medium (such as water) in which it travels more slowly, it changes direction in this way. This is an example of the refraction of light.

Put a penny in the bottom of a teacup. As you lower your head, less and less of the penny will be seen (Figure 16); finally it will disappear from view. No light can now reach your eye from the penny and it will stay hidden. Without moving your head, fill the cup with water. The penny will appear to float into view (Figure 17). Of course the penny has not changed its position. But the rays of light reflected from the penny are refracted enough as they pass into the air from the water so that they reach your eye.

Hang a sheet of white paper on the wall, opposite a bright window. Bring the convex lens of a magnifying glass near the paper. Move the glass slowly away from or toward the wall until a clear image of the window scene appears on the paper. The distance from the glass to the clear image is the approximate focal length of the lens. The lens gathers light from the large area of the window and, by refraction, assembles the rays on the paper to form a small, upside-down image of the scene. The human eye and the camera use such a lens arrangement to compress large scenes into images that can fit on the small, light-sensitive retina at the back of the eye, or the small, light-sensitive film at the back of the camera.

Stand on one side of a darkened room. Hold a burning candle a little more than a focal length away from your convex lens and move the lens back and forth until a clear image forms on the opposite wall. You will find that the inverted image of the candle is much larger than the candle in your hand. The movie projector also blows up tiny pictures into huge ones.

See also Vol. 10, p. 289: "Physics, Experiments in."

GUIDANCE IN SPACE*

How the Flight of Missiles and Space Vehicles Is Controlled

BY E. P. FELCH

THE success of space programs and of long-range military missiles depends to a large measure upon ability to guide vehicles in space. We may think of space guidance as an extension of air navigation to the regions beyond the earth's atmosphere. Air navigation itself represents the addition of a third dimension — altitude — to the more familiar two-dimensional problem of finding one's way on the earth's highways, waterways and oceans.

There are various complicating factors in space. For one thing, speeds are so great that control or even intervention by humans is seldom practical. Precision acquires tremendous importance. An error of one-tenth of a degree in direction or a few miles per hour in speed can cause an Intercontinental Ballistic Missile (ICBM) to miss its target by several miles or can spell failure for a lunar mission. Even short detours must be avoided, for the precious supply of fuel seldom exceeds the minimum quantity required to complete a mission. Obviously, if adequate control is to be provided, it is necessary to determine position and speed in space with utmost accuracy. This is extremely difficult. Since there is no air in space, altitude cannot be measured by air pressure and speed cannot be determined by air flow. More complicated procedures, seldom used in conventional means of travel, must be employed.

* Article prepared with the assistance of Western Electric and Bell Telephone Laboratories, Inc.

Bell Telephone Laboratories, Inc.

Above: using a radio-command guidance system to track a *Titan* missile during a launch at Cape Kennedy, Florida.

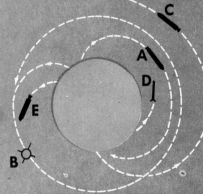

1. Objectives of space flight: A. Missile flights to definite target points on earth. B. Satellites set in orbit around earth. C. Space missions to moon or to other celestial bodies. D. Missiles used to intercept other missiles. E. Space vehicles rendezvousing with man-made objects that are already out in space.

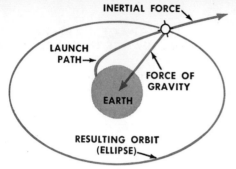

2. How a satellite falls into orbit. As it proceeds in its launch path, the satellite is acted on by the force of the earth's gravity and enters on an elliptical orbit.

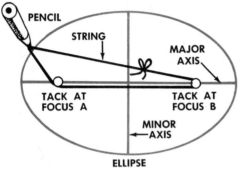

3. How to draw an ellipse, as explained on this page.

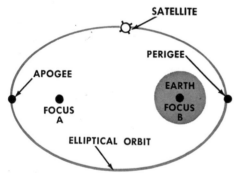

4. The elliptical orbit of a satellite around the earth. The apogee is the point that is farthest from the earth; the perigee, the point of nearest approach to the earth.

The objectives
of space flight

Flights in space have or will have various objectives, which may be summed up as follows: (1) Missile flights to definite target points on the earth (Figure 1, *A*). (2) Satellites set in orbit around the earth or the sun (Figure 1, *B*). (3) Space missions to the moon or planets, with vehicles proceeding to the vicinity of these celestial bodies or actually landing upon them (Figure 1, *C*). (4) Missiles employed to intercept other missiles, as in the *Nike-Zeus* Anti-Missile Defense System (Figure 1, *D*). (5) Space vehicles rendezvousing with objects already in orbit (Figure 1, *E*). This will be for inspection purposes or for the assembly of space stations or of particularly large vehicles capable of missions that could not be accomplished by craft launched directly from the earth.

An analysis of flights in
space — trajectories and orbits

Paths flown in space are usually called trajectories if they lead directly from one point to another. The path of a ballistic missile from launcher to target is a trajectory; so is the path of a space vehicle traveling from the earth to the moon. Flight paths which form closed circuits around the earth or around any other body in space are called orbits.

An object coasting in space is said to be in free flight; this means that neither propulsive force, nor friction nor pressure is acting upon it. Under these circumstances, the flight path of the vehicle is determined by gravitational forces of the earth, moon, sun or planets acting upon it, and by its own inertia. Inertia is the property which causes a body at rest to remain at rest and one in motion to remain in motion in a straight line at a constant velocity. This tendency toward motion in a straight line, when modified by gravitational forces, results in motion in a closed orbit, forming what is called an ellipse (Figure 2).

It is quite easy to draw an ellipse. Set two tacks in position about an inch apart on a board, and slip a loop made from about 4 inches of string over the tacks. Draw the loop taut with the point of a pencil. By moving the pencil clockwise while keeping the string taut, you can draw an ellipse, as shown in Figure 3. Each point where a tack has been placed is called a focus of the ellipse. The major axis is a straight line from one side of the ellipse to the other, going through each focus; the line perpendicular to the major axis and midway between the foci (plural of focus) is called the minor axis. A circle is simply an ellipse in which the two foci fall on the same point and in which the major and minor axes are equal in length. The eccentricity of an ellipse — the measure of how much it de-

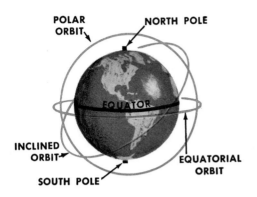

5. Types of earth orbits: equatorial, polar and inclined.

parts from the circular form — is related to the difference in length between the major and minor axes. The greater this difference, the greater the eccentricity. In the case of a circle, of course, the eccentricity is zero.

One focus of the orbital ellipse is always located at the effective center of gravitational force. For orbits within a few thousand miles from the earth, this may be considered as the center of the earth (Figure 4). The location of the other focus is determined by the position and motion of the vehicle at the instant when all propulsive thrust comes to an end.

The trajectory of a ballistic missile, after the termination of its thrust, is really a portion of an elliptical orbit whose minor axis is smaller than the earth's diameter. Hence a missile, or at least its warhead,

re-enters the earth's atmosphere and intercepts the earth at the target.

Various other terms are applied to orbits. The perigee is the point of nearest approach to the earth; the apogee, the point farthest from the earth; the orbit period, the time required for a single complete revolution. (See Figure 4.) The period is related to the size of the orbit; the larger the orbit, the longer the period. Here are some mean (average) altitudes and corresponding periods of earth orbits that are very nearly circular: 100 miles * — 90 minutes; 1,000 miles — 2 hours; 6,000 miles — 5 hours; 23,000 miles — 24 hours; 240,000 miles — 28 days. The last pair of numbers relates to the orbit of the moon.

An orbit girdling the earth in a plane parallel to that of the equator is called an equatorial orbit; the so-called polar orbit is at right angles to it, passing directly above the North and South poles. The orbits at intermediate angles are termed inclined orbits. (See Figure 5.)

Since orbits are fixed in space, rather than in relation to the earth, the rotation of the earth causes satellites in polar and inclined orbits to pass over different portions of the earth on successive orbits (Figure 6). Since the motion of the earth is precisely known and since that of satel-

* That is, statute miles. A statute mile is 5,280 feet. Confusion sometimes arises with respect to the dimensions of orbits, ranges of missiles and so on, because these figures are given in terms of nautical miles. The international nautical mile is 6,076 feet.

6. Successive passes of a satellite in an inclined orbit around the earth. Number 1 represents the first pass; 2, the second pass; 3, the third; 4, the fourth. The satellite was launched from a point in Florida; it went into orbit at the injection point indicated in the above diagram.

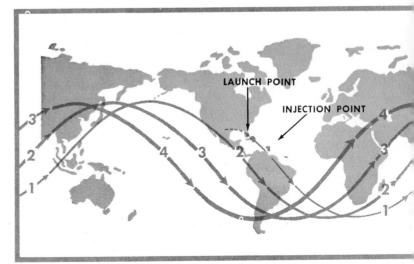

lites may be precisely measured after a few orbits, it is possible to predict the paths of orbits over our planet.

Most orbits more than 100 miles above the earth are exceedingly stable; however, there are exceptions. For example, satellites with unusually large size-to-weight ratios, such as the *Echo I* 100-foot balloon, are significantly affected by the pressure of the sun's radiation and by collision with the few molecules of air present even at altitudes of 1,000 miles. In the course of 6 months, the apogee of the *Echo I* balloon satellite changed from 1,050 miles to 1,350 miles, while its perigee changed from 950 miles to 600 miles. Satellites at altitudes of 100 miles or less encounter so much of the earth's atmosphere that the orbit altitude decreases rapidly until the satellite either returns to the earth or burns up from air friction in the lower layers of the atmosphere, just as many meteorites do when they pass through the atmosphere.

For many satellite missions, such as the *Echo* communications experiment and the *Tiros* weather-photography projects,* a constant altitude above the earth is desired. In these cases, a near-circular orbit is sought. However, even a perfectly circular orbit would not achieve uniform altitude above the earth since the earth is not a perfect sphere. Not only is it somewhat flattened at the poles but it is also slightly pear-shaped, as was first revealed by observations of the *Vanguard I* satellite, launched in 1958 during the International Geophysical Year.**

Guidance systems
for space vehicles

Thus far, we have been concerned mainly with what is called celestial mechanics — the physical laws that govern the behavior of any body in space. These laws of celestial mechanics represent both tools and limitations for the designer of guidance systems for space vehicles.

The three basic elements common to most guidance systems are (1) sensors;

(2) computers; (3) flight controls. Sensors are measuring instruments which determine the actual paths of space vehicles. Electronic computers* compare these paths with predetermined paths stored in their electronic memories, and they compute appropriate corrective orders for steering and controlling the thrust of engines. Flight controls accept these orders and put them into effect.

Guidance systems for missiles and space vehicles may be divided into two general categories: radio and inertial. We shall discuss them in turn in the pages that follow.

Radio-command
guidance systems

In a radio-command guidance system, such as that which has been developed by Bell Telephone Laboratories (Figure 7), the sensor is an amazingly precise radar located on the ground. It measures the position of the space vehicle in terms of angles of elevation and azimuth (Figure 7), and also in terms of range. The angular measurements are precise to a hundredth of a degree, while range is measured to within a few feet, even at hundreds of miles.

These precise measurements of position are passed by the radar to a UNIVAC ATHENA digital computer, also on the ground. This computer derives flight-path data which are compared with programmed data previously stored in its memory. Based on almost instantaneous comparison, corrective orders are generated by the computer and transmitted over the radar beam to the missile.

In the missile, a simple radar receiver receives these control orders, which are in the form of a code. It also triggers a radar transmitter (or beacon), which returns a strong signal to the ground to facilitate accurate tracking. A decoder connected to the receiver decodes the control orders and transmits steering and engine commands to the flight controls.

One of the principal advantages of the radio-command guidance system is the light weight of the guidance equipment that is carried aboard the missile. Remarkable reduction in weight is achieved through the use of the Bell Laboratories' invention called the transistor (see Index).

We pointed out that the decoder sends steering and engine commands to the flight controls. The steering function of the flight controls is accomplished by a device called an autopilot (Figure 8). This contains three spinning gyroscopes, or gyros. A gyro, like a child's spinning top, will maintain a fixed position in space even though its support is moved through space. The three gyros of the autopilot provide stable frames of reference for the three possible kinds of motion of the vehicle: (1) yaw, or right-and-left motion; (2) pitch, or up-and-down motion; (3) roll, or rotation about the long axis of the vehicle. All possible maneuvers of the latter are simply combinations of the three motions. In the absence of corrective orders, the autopilot aligns the flight path of the vehicle with the axes of the three gyros. Corrective control orders steer the missile by moving it with respect to the gyros in the autopilot. Steering is usually accomplished either by swiveling the main engine or the smaller auxiliary

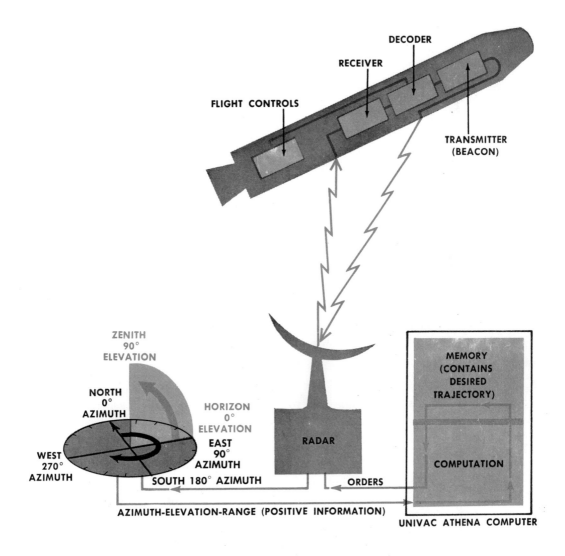

7. Above is a simplified diagram of the command guidance system of Bell Telephone Laboratories. It is described in these two facing pages.

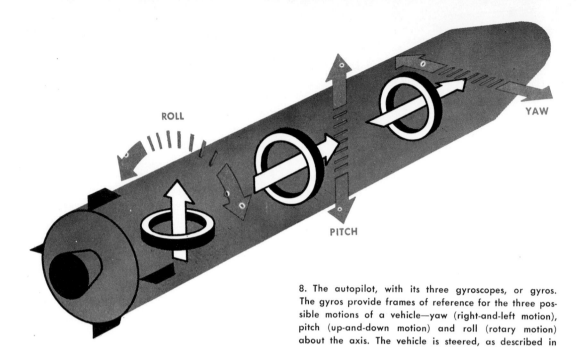

ROLL

YAW

PITCH

8. The autopilot, with its three gyroscopes, or gyros. The gyros provide frames of reference for the three possible motions of a vehicle—yaw (right-and-left motion), pitch (up-and-down motion) and roll (rotary motion) about the axis. The vehicle is steered, as described in the text, with respect to the gyros in the autopilot.

engines or else by controlling jets of hot or cold gases, aimed in appropriate directions.

The velocity or speed of a vehicle is controlled by shutting off the engines to terminate the thrust at the precise moment when the desired velocity is attained. It is extremely important to determine precise velocity for both ballistic missiles and space vehicles. Intercontinental ballistic missiles must achieve a speed of about 20,000 feet per second, or 13,500 miles an hour, to reach a target 5,000 miles away. A satellite must reach a speed of about 25,000 feet per second in order to orbit the earth. To travel beyond the grip of the earth's gravity, so-called "escape velocity" of greater than 36,000 feet per second must be attained.

Inertial-
guidance systems

Inertial-guidance systems employ inertial devices as sensors. In such systems, the sensors, together with the computers and flight controls, are carried aboard the space vehicle.

Gyros are employed as directional references in inertial systems in much the same way as they are in autopilots, as described above. To determine the path of a vehicle in space, velocities must be measured accurately along the directions determined by the gyros.

The inertial devices employed for measuring velocities are called accelerometers. These actually measure acceleration, or change of velocity, which can be converted into velocity. To understand the principle of operation of an accelerometer, let us consider the behavior of a weight suspended by a string from the ceiling of an airliner (Figure 9). While the airliner is at rest, the weight will hang straight down. When the airliner accelerates on taking off, the weight will swing to the rear; when it decelerates on landing, the weight will swing forward. The distance which it swings is a measure of the acceleration or deceleration, as the case may be. If the length of the string and the distance of the swing of the weight are known, it is possible to calculate the acceleration of the airliner in terms of feet per second per second. If the duration in seconds of a given acceleration is known, the speed in feet per second may be computed. Furthermore, the distance traveled may be calculated quite simply if the duration of a certain velocity is known.

The airborne computer of an inertial system has at its disposal all needed information on the direction and velocity of a vehicle's flight path. It can compute corrective control orders which it passes directly to the flight controls.

VEHICLE MOVING IN THIS DIRECTION

WEIGHT

VEHICLE
SPEEDING UP

VEHICLE
SLOWING DOWN

VEHICLE
AT REST OR
AT CONSTANT SPEED

9. The principle of the accelerometer, in inertial guidance systems. It is explained on the preceding page.

Hybrid
guidance systems

In addition to pure radio and pure inertial systems, there are hybrid systems which employ some of the techniques of both. Various types of sensors are sometimes used to supplement radar and inertial devices. Among these are "horizon sensors," which provide a reference with respect to the earth's horizon; "sun seekers," which align themselves with the intense radiation from the sun; and "star trackers," which can recognize and aim at star patterns.

Velocities are sometimes measured precisely by Doppler radar techniques. These depend upon the same phenomenon which, applied to sound waves, causes the whistle of an approaching locomotive to increase in pitch, or frequency. (See Index, under Doppler effect.) Radio waves behave similarly; the change in frequency is an exact measure of the relative velocity between the radio transmitter and the radio receiver. The transmitter is in the space vehicle, the receiver on the ground; or vice versa.

Guidance in powered
phases of flight

Most space vehicles are powered only during a portion of their flight. The rest of the time, they are in free flight, or coast-ing. While a vehicle may be tracked during any part of its flight, it is obvious that active guidance can be applied only as long as power is available to modify the course of the vehicle. Future missions to the moon or planets may employ different types of guidance for the ascent, mid-course and terminal part of the flight.

Command guidance systems have already achieved many striking successes. Thus the Bell Telephone Laboratories system, aboard the Douglas *Thor Delta* vehicle, has served to place in orbit *Echo I*, the *Tiros II, III, IV* and *V* weather satellites, the *Explorer X* and *XII* space probes and the *OSO, Ariel* and *Telstar* satellites.

Guidance systems are most versatile. For example, in order to place the hundred-foot *Echo I* balloon satellite in a nearly circular orbit about 1,000 miles above the earth, it was necessary to resort to what is known as the coasting orbit technique (Figure 10). The third-stage rocket and the payload (that is, the *Echo* balloon) were placed in an elliptical orbit with an apogee of about 1,000 miles. As apogee was reached, the third stage was fired, and the payload was placed in the desired orbit.

10. How the *Echo I* satellite was put in a nearly circular orbit. The three-stage rocket and its payload (the *Echo* satellite) attained an elliptical orbit with apogee at 1,000 miles. When apogee was reached, the third stage was fired and the payload entered the desired orbit.

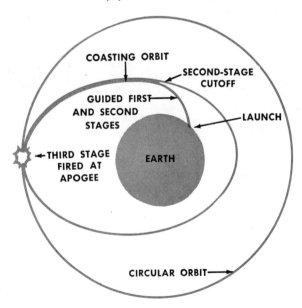

COASTING ORBIT

SECOND-STAGE
CUTOFF

GUIDED FIRST
AND SECOND
STAGES

LAUNCH

THIRD STAGE
FIRED AT
APOGEE

EARTH

CIRCULAR ORBIT

Australian News and Information Service

The dingo, above, is a notorious sheep-killer; it roams over the plains of Australia in large numbers.

THE DOG FAMILY

Wolves, Coyotes, Jackals, Foxes and Others

WHEN we speak of dogs, we generally have in mind the domesticated dogs, numbering over a hundred different breeds and ranging from the big Newfoundland, with its magnificent, shaggy coat, to the diminutive hairless Chihuahua. In this article, however, we shall use the phrase "dog family" in the sense of "animals belonging to the family Canidae." In this group we find not only our domesticated dogs but also wild dogs and other wild animals such as wolves, foxes and jackals.

The dog family has had a long and interesting history. One of the earliest direct ancestors of the group was the small animal known as *Cynodictis,* which lived about 35,000,000 years ago. It had a long and flexible body and a long tail. Its legs were rather short and the claws on its short, spreading feet were probably retractile, or capable of being drawn back. *Cynodictis* was in all likelihood a tree dweller and had a larger brain, in proportion to its size, than other primitive carnivores. From *Cynodictis* and other animals similar to it have descended the modern members of the dog family.

The early dogs gradually departed from the tree-dwelling habits of their forebears and took up their abode in open plains country, where they hunted in packs. They had to be fleet in order to capture other plains animals. Hence in time their legs were lengthened; they developed endurance for running great distances. The "thumb" of each forefoot and the "big toe" of each hind foot were greatly diminished in size; the claws became blunt and could no longer be drawn back. Like the cats, the dogs possessed shearing teeth, or carnassials, for slicing through tough fibers in the flesh of their prey. But, unlike the cats, the members of the dog family retained certain molar teeth for chewing hard foods. Because of these grinding teeth, the members of the modern dog family have a much more varied diet than the cats. The European red fox, for example, adds snails, insects and berries to its usual diet of small rodents.

There were several different lines of dog evolution. One line led to the now-extinct hyenalike dogs that once roamed over the plains of North America. Another branch terminated in large bearlike dogs, which also have become extinct. The wolves, jackals, foxes and domesticated dogs formed still another group — our modern dog family — which is now distributed throughout the world. The Cape hunting dog of Africa, the hunting dog, or dhole, of India, the Malay wild dog, the Siberian wild dog and the Brazilian bush dog are other representatives of the family.

Of the various kinds of carnivores, the dog family has undoubtedly been of the greatest direct service to mankind. Other land carnivores, such as the weasels, bears and big cats, keep in check the rodents and large hoofed mammals that, if left unmolested, would destroy a great deal of natural and cultivated vegetation. But the dog family has given man his most important animal ally and companion — the domesticated dog. This friend of man has been taught to draw his master's carts and sleds and to guard his home and his other domestic animals.

According to the most commonly accepted theory concerning the development of domestic dogs, these supremely useful animals were probably descended principally from wolves, with various other wild canines contributing to the strain.

Ewing Galloway

The gray, or timber, wolf, which ranges from northern Mexico to the Arctic Circle. This wolf is a large animal; the male has an over-all length of five feet and may weigh as much as a hundred pounds.

U. S. Fish and Wildlife Service

The red wolf is found in Texas and several other places.

prietorial sense. When they have taken up their quarters somewhere, they resent intrusion. A cat will attack another cat, and sometimes even a dog, for trespassing upon his home. Had some earlier cat shown the same spirit, "watchcats" might have been second to none by now as guardians of our possessions and of our loved ones.

We must view the wolf, therefore, with a certain amount of respect for the boon that he has indirectly conferred upon the human family. Since it takes a thief to catch a thief, the first wolves that became domesticated were the best animal friends that man could have had for the preservation of his flocks and his children from the rest of the wolves.

If domesticated wolves turned traitor to their fellows, it was perhaps because most animals acquire what may be called the pro-

Amer. Mus. of Nat. Hist.

The coyote, above, looks a good deal like a small wolf.

Amer. Mus. of Nat. Hist.

Jackals feed on garbage and on the leavings of hunters such as lions and tigers. They also hunt effectively in packs.

This sense of proprietorship is as strongly developed in the dog as in any other animal. The humblest mongrel may be a hero in defense of its master's home — an instinct that finds its counterpart in the wolf's resentment at intrusion on its den.

The wolf that early man had trained in his habitation from the time it was a puppy would fly at other wolves that it might meet on the prowl near the home of its master. To be sure, sooner or later it might desert and might rejoin its friends of the forest. But its cubs, born in the master's house, would be more likely to remain faithful to him and to become the trusted guardians of his abode.

At one time wolves abounded in many lands where they are now scarce or unknown. They were not exterminated in England until the sixteenth century; they continued to dwell in Ireland and Scotland until the eighteenth. In the United States wolves are practically extinct east of the Mississippi but still flourish in considerable numbers in the wilder regions of the western United States south to Mexico. They are also found in the northern and western Canadian forests and in Newfoundland and the Hudson Bay region. The true wolf (*Canis lupus*) occurs over almost all of Europe except for the thickly populated areas. It is found in large numbers in Russia and Scandinavia; in Asia it is distributed widely to the borders of India and the plains of China. Wolves are not found in South America or Africa.

The wolves of North America include the gray, or timber, wolf, the black wolf of Florida, the red wolf of Texas, the dusky wolf of the central plains and the coyote. In the Arctic regions we find the arctic wolf, which is nearly pure white, with a tail that is tipped with black.

The gray wolf, or timber wolf, of North America is much the same animal as that of Europe and Northern Asia. Zoologists believe it to be a single, wide-ranging, adaptable and variable species. A full-grown specimen measures five feet from the muzzle to the end of the tail, which is itself about sixteen inches long. The height at the shoulders is about twenty-seven inches; the weight is something like a hundred pounds. These are the measurements of a large male; the female is considerably smaller. The normal color

The northern Rocky Mountain wolf ranges from Idaho to Southern Alberta, preying on smaller animals.

A. W. Ambler, from Nat. Audubon Soc.

seems to be a yellowish white. The under-fur is gray; many of the hairs are tipped with black. The animal has brown markings on the ears and about the face. The coat is quite variable, however, in color; it may be blackish in southern specimens.

The gray wolf once ranged throughout the whole continent. Civilization has driven it to the northern and western regions; it abounds on the plains, in the Rocky Mountains and western coast ranges and throughout the forests of northern Canada. It was the especial enemy of the buffalo in the old days; it is now the scourge of the deer, moose and caribou. In the western United States it raids cattle and horse ranches; constant war, aided by government bounties, is waged upon it.

The coyote is the
jackal of the New World

The coyote is a smaller sort of wolf; it takes the place of the jackals and other wild dogs of semitropical parts of the Old World. Though it is often referred to as the "red wolf," the coyote shows the same gray tone as the gray wolf and has the same darker varieties. The real distinction is that of size; the coyote, or prairie wolf, is much smaller than the gray wolf. The length of a large male is rarely more than four feet, of which some fourteen inches represents the tail. The weight averages about twenty-eight pounds for the male and twenty-four for the female. Coyotes vary greatly; some systematists have established a dozen or more species and subspecies in North America.

These small wolves are restricted to the open regions of the West. Though they formerly ranged eastward into Indiana, they are now found almost exclusively in the thinly settled plains beyond the Mississippi. Here they live in burrows and feed on mice, gophers, rabbits and the like.

The never-ending
campaign against the coyote

They are a pest to sheep herders and poultry raisers. It is impracticable to reduce their numbers effectively by guns or traps. A widespread poisoning campaign has been going on against the animals for a good many years. It is doubtful whether this campaign is justified, since the coyote is an important check on vegetation-destroying rabbits and other rodents. Perhaps the best protection against the animal is wire fencing; the coyote hesitates to jump over or dig under this fencing.

Wolves are found in
open country and forests

The wolves of both Europe and North America frequent open country and forests. They make their lair in rocky caverns or in the decayed trunk of a fallen tree; occasionally they make burrows in the ground or enlarge those begun by other animals. The wolf has no friend among men, for he lives, when near settled areas, at the expense of our flocks. Wolves sometimes attack humans but, generally, only when they are in large packs or are ravenously hungry. They seem to be more of a menace to man in the Old World than in the United States and Canada. There have been authentic cases of travelers being attacked and devoured by wolves, but a good many of the stories that have been told about the man-eating propensities of these animals have been highly exaggerated.

How the wolf has
survived in a hostile world

Every year the wolf's chances for survival in a hostile world must become more precarious, for civilization and wolves are incompatible. They have successfully resisted extermination thus far partly because they can exist on a highly varied diet and partly because they are exceedingly prolific. The female wolf brings forth from six to ten cubs at a time; she rears them with rare fidelity.

The legend of children reared by wolves is a favorite one. The ancient Roman historians solemnly averred that Romulus, the supposed founder of Rome, and his brother Remus were rescued and suckled by a she-wolf after they had been taken from their mother and thrown into the Tiber River by the tyrant Amulius. Few people take legends of this kind seriously today.

The common wolf is found along the borders of India; but the chief representatives of the wolf tribe in that country are two smaller animals — the woolly wolf of the Himalayas in Kashmir and the Indian wolf of peninsular India. The brown-colored Indian wolf, which dwells on the plains, seldom hunts in large packs.

Africa is the home of the jackal

In Africa, the wolves are replaced by jackals. The Abyssinian wolf, red jackal or cuberow, is a wolflike animal but slightly smaller and rather long-legged. It is bright reddish-brown with white underparts, and the bushy, foot-long tail is tipped with black. The Egyptian, or wolflike, jackal is another large canine similar to the wolf. Its home is North Africa, where it hunts small antelopes, mice, hares and ground birds; it also makes forays against livestock. The animal is buffy-brown or pale reddish in color.

Jackals are generally smallish, foxlike creatures and are richly colored; they have bushy tails. They are lithe animals fortified with great endurance that allows them to travel quite rapidly over long distances; their highest running speed is something like thirty-five miles per hour. They are, for the most part, confined to Africa, but the Oriental, or common, jackal inhabits not only northeastern Africa but also southeastern Europe, ranging eastward to India, Ceylon and Burma.

Jackals serve as scavengers

Jackals eat all types of foods — in fact, almost anything at hand. Carrion and refuse and even buried bodies are not overlooked. They have a keen sense of smell for nosing out all sorts of refuse, and they serve admirably as scavengers in and about villages and towns. Jackals hunt in packs, mainly on the plains, bringing down antelopes and sheep and goats. Solitary hunters will forage for mice and other small mammals. Vegetable food also makes up part of their diet. It is a known fact that jackals follow lions, leopards and tigers and eat what remains after these mighty hunters have had their fill. Lions, at least, apparently do not resent the company of jackals, as long as these wild dogs refrain from seeking a share of the meal until the lions have finished eating.

The common jackal is small, standing about fifteen inches at the shoulder. It is a pale reddish or yellowish color with dirty-white underparts. Like other jackals, it hides by day and begins to roam about at dusk and on through the night. Packs of these animals enter villages where they do scavenger service but also make general nuisances of themselves by thieving food from unguarded shops and homes. A female common jackal has a gestation period (period of carrying her young in the womb) of around nine weeks. There are from five to eight young in each litter.

Side-striped and black-backed jackals

The side-striped jackal is similar to the common jackal in color but wears a slanting, light-colored stripe on each side between the ribs and the hip. Its muzzle is more pointed, and its tail is usually tipped with white. This jackal occurs from Zululand to Ethiopia, Morocco and Senegal. The handsome black-backed jackal of west Africa is a reddish animal with a yellowish-white belly. The back is black, spotted with white or gray. The legs of this jackal are somewhat short compared to those of the common and side-striped species. All of these jackals possess a gland at the base of the tail, which discharges an offensive-smelling secretion.

Australia's native dog is the dingo, or warrigal, a medium-sized, reddish-yellow animal; it is the only large wild mammal not a marsupial, or pouch-bearer, on this continent. In more populated areas where dingoes readily cross with domestic dogs, a pure-bred dingo seldom is seen. Small packs of wild dingoes, pairs or even lone individuals hunt in the forests and open plains. Dingoes find lairs in crevices among rocky outcrops, in ground burrows or in hollow logs. From five to eight pups are borne by the female at one time.

Both photos, N. Y. Zool. Soc.

Two untamed members of the dog family are shown on this page. Above is the Andean wild dog (*Pseudolopex culpaeus reissii*), found in mountain areas in South America. Left: the Cape hunting dog, belonging to the genus *Lycaon*. It is a strong beast with powerful jaws; it hunts in big packs numbering from twenty to sixty.

We are not entirely certain about the manner in which this animal came to be an inhabitant of Australia, where the other mammals are so different. It was first maintained that the dingo was a recent importation. This theory was apparently upset when fossil remains of the animal were found in Australia, together with the fossils of marsupials, in Pleistocene deposits going back about a million years. Did this discovery indicate that the dingo evolved independently, together with marsupials and the other bizarre animals of the island continent? Modern biologists are inclined to believe that the dingo was an importation after all; that it was introduced to Australia by prehistoric man in the early days of his history.

When the first white settlers came to Australia, the dingo did not represent much of a problem. However, the introduction of the rabbit changed the situation. Rabbits increased and multiplied, and so did dingoes, which preyed upon the rabbits and therefore were assured a staple food supply. Unfortunately, the dingo also turned its attention to sheep and poultry and became a menace to the stock raiser and farmer. It still has a most unsavory reputation as a sheep killer.

The native Australians hunt the dingo for its flesh, of which they are very fond. Often, too, they seek out the young dingoes in their lairs, and they bring up these wild dogs as domestic animals.

The dhole is a remarkable hunter

Among the most remarkable hunters of the dog family is the dhole, the wild dog of India. This animal has a distinctive red coat. It weighs about forty pounds and has an over-all body length (including the tail) of almost four feet. In India, dholes are found in wooded areas; where there are no forests, as in Tibet, the animals dwell in the open country.

Dholes hunt in packs. Possessed of a keen sense of smell, they can track a quarry by scent until they are close enough to keep it in view. One of the dholes acts as a sort of "master of the hunt"; it keeps in the lead and prevents the pack from scattering by emitting a series of sharp yelps. When the quarry, say a deer, is at bay, the dholes circle it and soon overpower it. Deer are the favorite prey of these animals; they also attack wild pigs, goats and even such animals as bears, leopards and fully grown buffaloes. Popular belief in India has it that dholes sometimes successfully hunt tigers. There is no conclusive evidence that this is so. It is believed, however, that tigers and dhole packs will sometimes fight over the possession of a slain deer or goat.

In the breeding season, the packs are dissolved for a time. The female dhole has from four to six pups, after a gestation period of about nine weeks. When the young are weaned, they are fed with partly digested food that the mother has regurgitated (cast up from its stomach).

The African, or Cape, hunting dog

The African, or Cape, hunting dog (genus *Lycaon*) also hunts in packs. It is a strong animal, with a big head, powerful jaws and long legs; its over-all length is approximately four feet. The large, oval ears are rather like those of the hyena; for this reason, the animal is sometimes known as the "hyena dog." The basic color of the coat is generally tortoise-shell, mottled with yellow, black and white patches. Occasionally one comes upon black specimens of this striking-looking animal.

Cape hunting dogs hunt in packs numbering from twenty to sixty. They prey on antelope, hartebeest and other animals dwelling in the brush country south of the Sahara Desert. The dogs are always on the move and travel great distances; they communicate with one another by musical calls. When surprised, the Cape hunting dog barks angrily. At night, it sometimes makes chattering sounds.

Although Cape hunting dogs can swim very well, they will not cross deep water, possibly because of the danger of lurking crocodiles. It is said that antelopes sometimes make good their escape from a pursuing pack by jumping into the water of broad rivers or ponds.

In the breeding season, the dogs generally take over aardvark dens, which they clean and enlarge. The gestation period is about two months; there are from two to six pups in a litter. As in the case of the dholes, young Cape hunting dogs are fed regurgitated food when they are weaned. They are allowed to take part in the hunt at an unusually early age.

The cunning and bold fox

The fox, with which our chapter closes, has long had a reputation for cunning and boldness; it is skillful in eluding its enemies and in tracking down its prey. It has an unusually varied diet. Among its victims are hares, rabbits, pheasants, partridges, rats, mice, moles and lambs. It sometimes varies the diet with frogs, beetles, worms, shellfish and crabs.

This animal has had a long and prosperous career. It is almost world-wide in distribution; in fact, it is easier to enumerate the places where it is not found than those in which it lives. There are no foxes in South America, Australia and various islands. They are found in great numbers in the northern half of the New World, as well as in Africa, Asia and Europe. The fox can adapt itself to a wide variety of climates. It is an important member of the fauna in the desert lands of Asia, in the frozen wastes of the arctic regions and in the steaming tropics.

The eastern gray fox (genus *Urocyon*) is noted as a bold and cunning hunter.

N. Y. Zool. Soc.

The fox is bred for the sport of fox hunting in Great Britain and some parts of the United States; it is also raised for its fur in the United States and Canada. In the wild, the animal is ruthlessly hunted down; it is trapped, shot or poisoned. Yet it manages to survive because of its ability to adapt itself to changing conditions. As a result, it has acquired a merited reputation for cunning.

The best-known of all the foxes is the common red fox, which is to be found throughout the Northern Hemisphere. The red fox is about as large as a small dog. Its fur is rusty red; its tail, which is tipped with white, is long and bushy; it has a long nose. The red fox lives in wooded areas or in rolling country.

The gray fox is grayish in color; it has a longer body than the red fox and also a longer tail. It is to be found from the Great Lakes east to the Atlantic, in the southern United States and in Mexico. Mainly nocturnal, it preys on rabbits and mice and, occasionally, on birds and reptiles; sometimes it eats fruits and berries. It often climbs up a tree when pursued.

The arctic fox is particularly remarkable for the seasonal change in its coat, which is dark brown to slate in the summer and white in the winter. (One variety, called the blue fox, is smoky gray or bluish drab throughout the year.) This fox is found in northern regions in open, treeless lands. In the summer it hunts along the coasts or in open plains; in winter it often seeks its prey out on the ice. The arctic fox sometimes follows the polar bear, feasting on the seal leftovers of the bear.

Other foxes dwell in widely separated areas. Among others, there is the Tibetan sand fox with unusually short ears and tail; the small Indian fox; Rüppell's fox, found in Egypt; the African sand fox, ranging from Senegal to the Nile region; and the South African silver fox.

The smallest of the foxes is the fennec, which is found in the Sahara. The fennec is a graceful little animal about fifteen inches long, with a tail measuring seven inches; its large, erect ears give it an alert appearance. It burrows into the sand and can dig so rapidly that it can escape most of its pursuers. It is nocturnal in its habits. It spends the day curled up in its burrow; at night it comes out to search for water and for its prey.

See also Vol. 10, p. 275: "Mammals."

All photos, Kellner Associates—Dick Krueger

EXPERIMENTS WITH MAGNETISM

A Mysterious Force That Serves Man in Many Ways

BY NELSON F. BEELER

PROBABLY all of you who read this chapter have picked up various objects of steel and iron with a small magnet. The strange effects of magnetism have been known for centuries and have puzzled many generations of learned men. Modern science has not entirely solved the mystery of magnetic attraction and repulsion, though it offers a fairly satisfactory theory to account for it.* But even though we do not completely understand magnetism, we have put it to work for us in many different ways. We use it in the electric generator, the telephone, the telegraph and a host of other devices.

In the pages that follow we are going to examine certain effects of magnetism, and we shall show how to make working models of some of the devices in which it plays an important part.

* The theory is discussed in the article Magnets Large and Small, in Volume 7.

For our first experiment, we shall need two bar magnets. Support one of them in a stirrup, or holder, of thread, as shown in Figure 1; then suspend it in a place as far as possible from objects of iron and steel. The magnet will swing back and forth, eventually coming to rest in what is approximately a north-south direction.* The end, or pole, that points to the north is called the north pole; the other is the south pole. (Actually, "north-seeking" and "south-seeking" would be more appropriate terms.) In some magnets, the north pole is marked "N" and the south pole "S." If your two bar magnets do not have these markings, you can add them after having tested the magnets.

Now suspend one of the magnets as you did in the first experiment and bring the second one toward the first (Figure 2). The suspended magnet will start moving while the other is still at a distance from it. Which poles will be attracted to one another? Which ones will repel one another? Note that this action-at-a-distance is brought about because of the presence of a magnetic field of force outside of each bar magnet.

It is interesting to "map" a magnetic field of this sort. Sprinkle some iron filings on a cardboard or, better, a sheet of window glass, and put a bar magnet right below it. If you tap the cardboard or glass, the lines of force will become evident. Draw these lines, and compare them with the "map" of the lines of force in Volume 7, page 164.

Some of the magnetism of a bar magnet can be transferred to unmagnetized steel objects such as a sewing needle. Stroke the magnet with a needle several times; each stroke is to be from the center of the magnet to one of the ends (Figure 3). You will now be able to attract small objects of steel or iron with the needle. The steel will retain the magnetism that you have put into it. You can also magnetize a soft iron nail, but it will soon lose its magnetism.

A large magnet acts like a lot of little magnets all lined up in the same direction; each of the molecules is a magnet in itself. To show that this is indeed so, magnetize a steel needle by stroking a bar magnet

* We shall explain later why the magnet does not point exactly to true north.

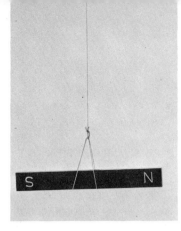

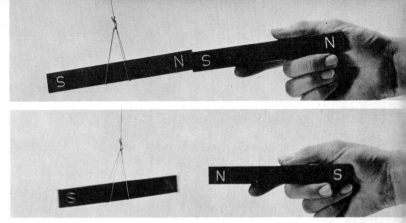

1. Suspending a bar magnet so that it will move freely about a pivot.

2. What happens when you bring (1) unlike poles together; (2) like poles together?

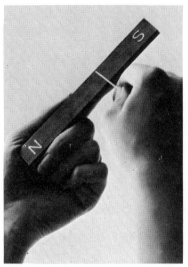

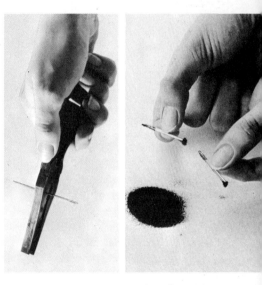

3. If you rub a steel needle several times from the center to the end of a magnet, it will become magnetized.

4. If a magnetized needle is cut in two with a pair of cutting pliers, each half will act like a magnet.

with it. Now cut the needle in two with a pair of strong cutting pliers or a hacksaw (Figure 4). Test the ends of the two pieces by trying to attract small iron or steel objects with them. You will find that you have really made two magnets out of a single needle magnet. If each of the two pieces is cut in two, you will have four magnets.

One theory holds that an unmagnetized needle is made up of great numbers of little magnets whose poles point every which way. No magnetic field is produced around the needle. When you magnetize the needle, you line up its molecules so that their north poles all point in the same direction, and a magnetic field is set up. Not all metallic substances can be magnetized. Iron, co-

balt and nickel and various alloys of these metals and of aluminum show particularly strong magnetic properties.

For your next experiment, place the cork lining from a pop-bottle cap on water in an aluminum or glass dish or bowl. Set a magnetized needle on the cork, as in Figure 5. You will note that the needle will swing around until it reaches what is approximately a north-south position. You have made a simple compass.

The compass needle moves because the earth has a huge magnetic field around it — a field that is strongest at the so-called magnetic poles. In the Northern Hemisphere, the North Magnetic Pole is about a thousand miles from the true North Pole. The compass needle points to the North

5. How to make a compass. If you put a cork in a container of water and set a magnetized needle on the cork, the needle will point approximately north.

7. Making a dipping needle that shows the angle of dip.

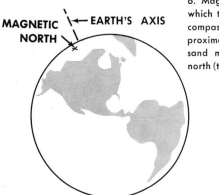

MAGNETIC NORTH ← EARTH'S AXIS

6. Magnetic north, to which the needle of a compass points, is approximately a thousand miles from true north (the North Pole).

Magnetic Pole and not to true north (Figure 6).

Apply a magnetic compass to the top and bottom of iron and steel objects that have been standing in place for a long time. Try a radiator, a steel door frame, a steel window frame, a metal stove or an iron fence or gate. You will find that these objects have become magnetized and will affect your compass. This is because the earth's magnetic field has been acting on them. Large metal ships have to be brought into port and demagnetized from time to time. Otherwise, the iron parts of the ship become so strongly magnetized that they begin to have a definite effect on the ship's compass.

Provide yourself with two knitting needles. Push the first one through a cork. Then push the second one through the same cork, at right angles to the first and very close to it in the cork so that the apparatus will not be too heavy on its underside. (See Figure 7.) Set the first needle on two glass tumblers, as in the figure. Adjust the second needle by moving it through the cork in one direction or the other so that when it comes to rest, it will be parallel to the tabletop. Magnetize this needle. Note that when it is magnetized, it is no longer parallel to the tabletop but that it points downward at a

certain angle. This is because it is attracted in such a way by the lines of force coming from the earth that it forms what is called the angle of dip. This angle varies as one moves from north to south about the earth.

Your bar magnet is called a permanent magnet because it retains its magnetism. Another important kind of magnet displays magnetic properties only as long as electric current is passing through it. This device is called an electromagnet. You can make one with very simple equipment (Figure 8). Wrap a single layer of No. 18 insulated bell wire around a large nail. You should have at least 40 turns of wire. Leave about a foot of wire loose at each end. Connect the end of one of the loose wires to a pole of a dry cell. Place several paper clips or thumbtacks on the table. Hold one end of the nail near the clips or tacks and bring the other end of the wire in contact with the other pole of the dry cell. (Do not *connect* the wire to the cell.) When the current flows in the coil, the nail, which is the core of the coil, becomes a magnet and will pick up the clips or tacks you have put on the table. If you remove the loose end of the wire from the pole of the dry cell, the nail will no longer be a magnet and the clips or thumbtacks will fall from it.

Wrap another layer of wire around the nail; you should now have about 80 turns of wire. Test it to see how many paper clips it can pick up now. You will find that the more turns of wire there are, the stronger the magnet will be. Now try two cells instead of one (Figure 9). Your magnet will be still more powerful.

You can play various games with your electromagnet; here is an example. Mark a line across the middle of a large square piece of cardboard and make the outline of a goal at either end, as in Figure 10. Place some thumbtacks, with the sharp end pointing upward, along the middle line. Two people, each provided with his own electromagnet, can play this game. Each player turns on his switch at the word "Go" and attempts to put as many tacks in his opponent's goal as possible. As soon as he has deposited a tack, he must turn off the current, so that the tack will remain where it is. He turns on the current again as soon as his magnet is within range of the tacks that have not yet been taken. The person who draws most tacks into his opponent's goal is the winner.

We are now going to make working models of a few devices in which the electromagnet plays an important part. The first of these devices is a telegraph key (Figure 11). Drive an eightpenny nail (nail

8. Electromagnet consisting of a nail, a wire and a dry cell.

9. Two cells will make a stronger electromagnet.

10. How to play the electromagnetic game described on this page.

No. 1 in the figure) into a board. Wrap about 80 turns of No. 18 insulated bell wire around the nail and connect one end of the wire to a switch * at some distance from the coil. After you have done so, run a wire from the other end of the switch to a dry cell and connect it to one of the poles of the cell. Connect the unattached end of the coil on the upper board to the other pole of the dry cell, as shown in Figure 11. Cut a strip from a tin can; make it about eight inches long and an inch wide. Bend the strip as shown in the diagram and mount it so that the end barely misses the nail. The strip can be held onto the board with tacks if you first make holes in it with a nail. Set a second nail (nail No. 2 in the figure) in place on the board. This should be a twelvepenny nail. Bend the head of this nail with a pair of pliers so that it rests about 1/16 inch above the strip.

When you close the switch, the nail with the coil around it will become an electromagnet. It will attract the tin-can strip which will strike the nailhead above it and make a sharp sound. When you release the switch, the natural spring of the metal moves it away from the nailhead and it will strike the other nail. A quick closing and opening of the switch will result in two sounds close together: a dot in the Morse code (see Index). If you take more time in opening and closing the switch, the sounds will be farther apart and will represent a dash in the code.

In your simple telegraph set, the electromagnet nail will become permanently magnetized to a slight degree after a while; so will the metal strip. When this happens,

* The article on Experiments with Electricity in Volume 6 shows how to prepare such a switch.

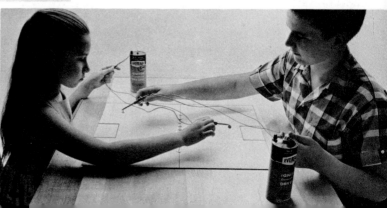

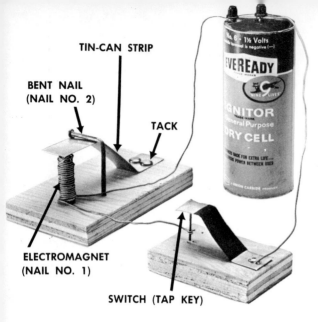

TIN-CAN STRIP

BENT NAIL
(NAIL NO. 2)

TACK

ELECTROMAGNET
(NAIL NO. 1)

SWITCH (TAP KEY)

11. A simple telegraph sender.

the strip will hold fast to the nail even when you open the switch. If you put a piece of cellulose tape over the end of the strip, the leftover magnetism will not be strong enough to keep your telegraph set from operating.

The telegraph works because the operator can turn the magnetism on and off by means of a switch or key. In a buzzer, the magnet is turned on and off automatically by a so-called "make-and-break" device. You are now going to make a buzzer, complete with make-and-break unit (Figure

12). The electromagnet is made of three to five tenpenny nails wrapped with 80 turns of No. 18 bell wire. Attach one end of the wire to a pole of a dry cell, which is then to be connected in series with another dry cell, as shown in Figure 12.

The sounder is a piece of metal cut from a tin can; it should be about three inches long and a little less than an inch wide. Bend the end of the sounder around a large nail so as to make a cylinder of this end. Slip a thin finishing nail through the cylinder and drive the nail into a board. The sounder should be free to move back and forth, with the finishing nail as a pivot. Attach a wire to the top of the finishing nail. Connect the other end of the wire to a switch. Another wire is to lead from the switch to the dry cells, as shown.

Cut another piece of metal from a tin can and bend it as shown in Figure 12. Drill a hole in the bent strip and insert a stove bolt, held in place by nuts, as is shown in Figure 13. Adjust the bolt so that it makes contact with the sounder strip. Connect the unattached end of the electromagnet wire to the bolt. A rubber band, attached to a finishing nail, as in Figure 12, will keep the sounder in place against the tip of the bolt.

When you close the switch, current will flow through the coil and the bolt. Since the tip of the bolt touches the sounder, current will also pass through the sounder and back to the dry cells through the switch. As soon as the current flows in this way, the coil-wrapped nails become a magnet. The latter attracts the metal sounder, which moves away from the tip of the bolt, thus

12. A buzzer that really works.

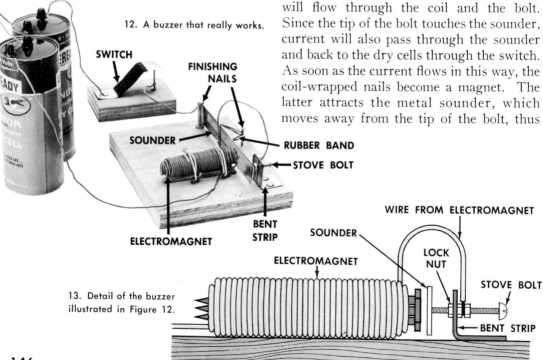

SWITCH

FINISHING NAILS

SOUNDER

RUBBER BAND

STOVE BOLT

ELECTROMAGNET

BENT STRIP

WIRE FROM ELECTROMAGNET

SOUNDER

LOCK NUT

ELECTROMAGNET

STOVE BOLT

BENT STRIP

13. Detail of the buzzer illustrated in Figure 12.

breaking the circuit. But once the circuit is broken, the electromagnet will no longer attract the sounder; the rubber band will then pull the sounder back against the tip of the bolt. This will again close the circuit and the electromagnet will again attract the sounder. In this way the circuit is made and broken automatically again and again. The metal sounder will move back and forth so rapidly that a buzzing noise will result. If the end of the sounder were equipped with a tiny hammer that would strike a gong, it would make a ringing sound. You would then have an electric bell.

The amount of electricity flowing in a wire can be measured by a device called a galvanometer. (It was named after Luigi Galvani, an Italian pioneer in the study of electricity.) You can illustrate the principle of the galvanometer by a simple piece of apparatus.

Mount a 6-inch-square board with 2-inch nails in each corner on a base as shown in Figure 14A. Make a shelf for a compass in the center of the upright board by gluing it to a block of wood, as in Figure 14A. Do not use nails to fasten the shelf, since they would affect the compass needle. Place a small compass on the shelf and rotate the apparatus until the compass needle is parallel to the plane of the upright board. Wrap one loop of wire around the four nails and connect the ends of the wire momentarily to a dry cell as shown in Figure 14B. Note the deflection angle of the compass.

Now wrap a second loop of wire around the four nails and connect the ends to a dry cell as before. Note the difference in the deflection of the compass needle. Try three, four and five turns, noting the angle of deflection of the compass needle each time. What do you conclude is necessary to have a very sensitive galvanometer that would measure small currents? Try reversing the connections on the dry cell. How does this change the deflection of the compass needle? (To save the life of the battery, a flashlight battery may be put in series with the wire loop.)

In a working meter of this type, the coil is sometimes mounted in such a way that it will turn as current flows through

it. It carries a pointer which moves across a dial to indicate the quantity of current that is flowing. In other meters, the coil works against a spring. The amount of tugging that the coil does against the spring becomes a measure of the quantity of current passing through the meter. In either case, although the electric current cannot be seen, it makes itself apparent by the magnetic field it produces.

The electric motor is one of the most useful devices based on the electromagnet. Perhaps the best way to find out how it works is to make a motor yourself. You will need only a few odds and ends: four nails, a cork, a piece of glass tubing and three pieces of bell wire each about four feet long. Figure 20 shows how the finished motor will look.

The cork is used for the rotor, or armature, which is the moving part of the electric motor. The rotor must be able to turn with very little friction. To provide a smooth bearing, insert a piece of rounded glass tubing into the cork. A glass medicine dropper can be utilized for this pur-

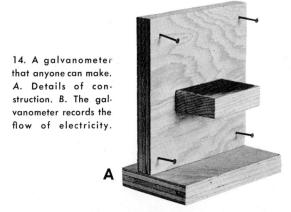

14. A galvanometer that anyone can make. A. Details of construction. B. The galvanometer records the flow of electricity.

A

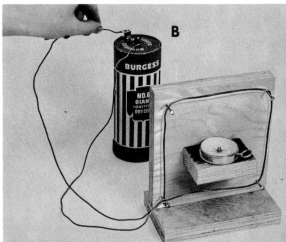

B

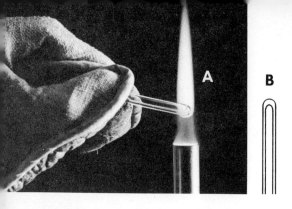
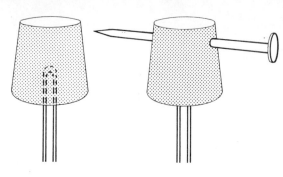

15. Preparing a glass tube (A) that will fit in the electric-motor rotor. Keep the tube turning in the flame. Its end will become dome-shaped, as in B.

16. The tube (cross section) shown in place in the rotor.

17. A tenpenny nail is to be pushed through the rotor.

pose. Holding it in an asbestos mitt, melt one end in a gas flame (Figure 15A) until it softens to form a dome shape like the one shown in Figure 15B. Some ball-point pens have a cap to keep the pen from leaking ink into your pocket. One of these may be substituted for the glass tube, if its surface is smooth enough.

Now insert the glass tube in the cork, as shown in Figure 16. A corkscrew is most convenient for making the hole in the bottom of the cork in which the glass tube will fit; but a hole can also be gouged out with a thin file or a pocketknife. The hole should be just large enough to allow the glass tube to fit snugly in the cork.

Push a tenpenny nail through the cork (Figure 17). Wind one of the four-foot pieces of bell wire around the rotor (Figure 18). Start along the side of the glass tubing, leaving 2 inches of wire below the cork. Bring the wire up the outside of the cork and then wind it clockwise around one of the protruding parts of the nail; you should have about 40 turns of wire on this part. When you have finished winding the wire around this portion of the nail, pass it across the top of the cork. Wind it around the other protruding part of the nail in the same direction as before — that is, clockwise. Again, you should have about 40 turns of wire; leave 2 inches of wire below the cork. Look at the drawing carefully; be sure to apply the wiring as shown. Unless it is wound correctly, the motor will not run.

Cut the insulation from the 2-inch lengths of wire that run along the glass tube. Wind and bend each wire as shown in Figure 19A. Hold the bared and bent

wires firmly in place against the tubing by means of adhesive tape or friction tape (Figure 19B). The two bent wires should be opposite one another on the glass tube. They make up the part that is called the commutator.

Now make the field coils. Drive two twelvepenny nails into a piece of pine board, as shown in Figure 20. Place the nails just far enough apart so that the armature can turn freely between them. Now drive a third nail — a finishing nail — up through the bottom of the base board, exactly midway between the two twelvepenny nails. The rotor is to be put in place upon the finishing nail and is to turn upon it. Figure 20 shows how the three nails of this section are to be arranged.

Wind one of the four-foot pieces of wire around one of the twelvepenny nails. You should have between 125 and 150 turns of wire. Leave a fairly long length of wire to connect the motor to the dry cells that will supply the electric power. The other end of the wire from this coil should have the insulation removed. Staple the wire to the board so that the bare end just touches one of the bare wires of the armature when it is in place. The bared wire is called a brush. When the motor operates, current will flow through the brush from the coil and into the rotor.

The second coil is now wound around the other twelvepenny nail in the opposite direction to the first — that is, if the direction of the first coil was clockwise, the direction of the second is to be counterclockwise. When they are wound in this way, current moves in one direction in the first coil and in the other direction in the second

148

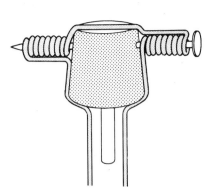

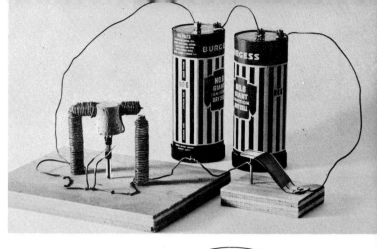

18. How to prepare the winding of the armature.

19. Details of the commutator.

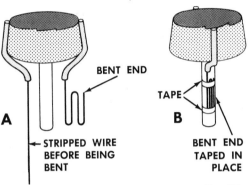

A — STRIPPED WIRE BEFORE BEING BENT

BENT END

TAPE

B — BENT END TAPED IN PLACE

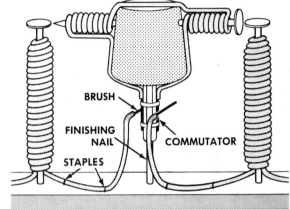

BRUSH

FINISHING NAIL

STAPLES

COMMUTATOR

20. The diagram shows the completed motor. The loose end of the wire at either end is connected to a switch and to two dry cells set in series, as shown in the photo above.

coil. This is so that the top of one nail will be a north pole, while the top of the other will be a south pole. When you have the motor assembled and ready to run (Figure 20), it will be well to check with a compass to see that you have made unlike poles of the tops of these nails. If they are like poles, your motor will not work when current is sent flowing through it.

Place the rotor on the finishing nail. If it does not spin easily, file the supporting nail to make it fit better in the glass tube. Attach the two long wires from the field coils to cells, and then the motor will be ready to run. You will have to move the brushes slightly perhaps before the motor will start. You may also have to tinker with the two wires running up the sides of the glass tube on the rotor. As we pointed out, they must lie exactly opposite each other across the glass tube.

This is how the motor works. When electric current from the dry cell passes through the first field coil, the coil becomes a magnet. The current then passes through the brushes and the armature, and the armature also becomes a magnet. Its north pole will be opposite the north pole of field coil

No. 1. Since like poles repel one another, the armature will make a half turn. Each brush will then make contact with another segment of the commutator and the poles of the armature will be reversed. What was formerly the north pole of the armature will now become the south pole; since it is opposite the south pole of field coil No. 2, it will make another half turn as the two poles repel each other. The process will be repeated again and again, and the armature will keep spinning.

The induction coil is another device that works because there is a connection between magnetism and electricity. You know that when a current moves in a coil of wire, a magnetic field appears. The reverse is true, too. When a magnetic field moves across a coil, a current begins to move in the coil.

You can show that this is true by making an induction coil and hooking it up to your galvanometer. To prepare the induc-

149

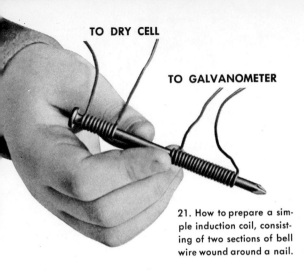

21. How to prepare a simple induction coil, consisting of two sections of bell wire wound around a nail.

tion coil, wind ten turns of bell wire around a nail near the head end of the nail (Figure 21). This will be your primary coil. Leave the ends of the wire long enough so that they can be connected later to a dry cell. Wind a coil of twenty or more turns on the other portion of the nail. This will be the secondary coil.

Connect the ends of the secondary coil to your homemade galvanometer. Attach one end of the primary coil to one pole of a dry cell. Touch the other pole for just an instant with the wire from the other end of the primary coil. The contact between wire and pole should be as brief as you can make it. As the free end of the coil touches the pole of the dry cell, current will flow for an instant in the primary coil. Notice that the compass needle in the meter moves when this happens. This indicates that current must be flowing in the secondary coil, too, during that instant. The coils are not connected by wire but they are wound on the same core of magnetic material — the nail.

When the primary coil becomes a magnet, the magnetism floods down the nail and across the turns of the secondary coil. It is this moving magnetism which causes the current to flow in the secondary. If you left the primary coil attached to the dry cell, the current in the secondary coil would die down immediately. It flows only when the primary current is changing. Tap the free end of the primary coil against the unconnected pole of the dry cell several times in quick succession. The compass needle will dance in time with it.

You can step up, or increase, the voltage by an induction coil. The number of turns in the secondary coil as compared to the number in the primary one will determine how much the voltage will be changed. If there are ten turns on the primary coil and twenty on the secondary, the voltage on the secondary coil will be twice that on the primary coil.

There is a very important induction coil in every automobile. It has a great many turns on the secondary and only a few on the primary. The storage battery of a car supplies at most 12 volts; the generator only a few volts more. But a voltage in the thousands is produced in the secondary coil. This makes a spark jump at the spark plug and ignites the gas in the cylinder.*

Like the induction coil, a transformer is made up of two coils and a core. The two devices are different in one important respect. The primary of the induction coil in a car operates on direct current. The current passing through the coil is alternately turned on and off by a make-and-break mechanism such as the one in your buzzer. The transformer, however, does not have a make-and-break device. Instead, its primary is attached to alternating current. The electricity in the primary flows first one way and then the other. This changing current produces magnetism, which floods the secondary coil of the transformer.

The toy transformer used with electric trains has fewer turns on the secondary than on the primary; it steps down, or reduces, the voltage. The current in the ordinary home line flows at 110 volts. It is reduced to 6 or 12 volts to make it possible to operate a toy train safely.

We have considered only a very few of the fascinating things that can be done with magnetic fields. The basic principles involved in the simple devices you have built apply also to complex machinery and gadgets of all kinds. It is a far cry, of course, from your toy electric motor to the big motors of subway trains, or electric locomotives or the elevators of buildings; but they all run because a magnetic field has been put to work.

See also Vol. 10, p. 289: "Physics, Experiments in."

* For an account of an automobile's ignition system, see the article called The Automobile, in Volume 2.

THE MILKY WAY

A Modern Interpretation of This "Band of Light"

THE Milky Way is one of the most striking sights in the night skies. It is too faint to be seen in bright moonlight or amid the myriad lights of our large cities, but on moonless nights in the country we can easily make out the outlines of its cloudy track of light across the heavens. If we peer at it through a powerful telescope, we realize that the Milky Way represents the combined light of vast numbers of stars, which cannot possibly be made out individually by the unaided vision.

We know today that the Milky Way forms part of a vast system of stars, to which our sun belongs. In bygone ages, however, it was a celestial puzzle that filled men with a sense of mystery or destiny. It was explained in various ways in Greek and Roman mythology. Some writers called it the highway of the gods, leading to their abode on Mount Olympus; others held that it sprang from the ears of corn dropped by the goddess Isis as she fled from a pursuer. Still others believed that the Milky Way marked the original course, later abandoned, of the sun god as he sped across the skies in his chariot.

In medieval times the Milky Way became associated by pilgrims with their journeys to various sanctuaries. In Germany, for example, it became known as Jakobsstrasse, or James' Road, leading to the shrine of St. James at what is now Santiago de Compostela, in Spain. In England it was called the Walsingham Way; it was associated with the pilgrimages to the famous shrine of Walsingham Abbey. The pilgrims of those days did not seriously believe that the Milky Way had anything to do with their travels. They saw it lying overhead, a misty path in the heavens, and their belief in the universal kinship of all things caused them to find comfort in its presence. This simple affection for the heavenly bodies, which we find in Chaucer and many other medieval poets, is not nearly so widespread now that we know how far away the stars are.

The best time to see the Milky Way is on an autumn or winter evening (in the country, as we pointed out); it is then highest in the heavens and therefore its light is least affected by our atmosphere. It is seen to stretch like a vast, ragged semicircle over the skies of the Northern or the Southern Hemisphere. Actually, it traces a rough circle, for it is continued in the other hemisphere.

The path traced by the Milky Way is full of irregularities. Its average width is about twenty degrees, but it varies considerably, both in width and in brightness. Its track lies through the constellations Cassiopeia, Perseus and Auriga, in the Northern Hemisphere. It passes between the feet of Gemini and the horns of Taurus, through Orion just above the Giant's club and through the neck and shoulder of Monoceros, here entering the Southern Hemisphere. It passes beyond Sirius into Argo and through Argo and the Southern Cross into Centaurus. Here it divides into branches separated by a dark rift, in a way that suggests the divided course of a river around an island.

The divided course of the Milky Way extends over one-third of its entire length — that is to say, one hundred and twenty degrees in the heavens. Finally the divergent branches are reunited in the Northern Hemisphere in the constellation Cygnus. The brighter stream passes through Norma, Ara, Scorpio and Sagittarius, along the bow of Sagittarius into Antinous, then

Amer. Mus. of Nat. Hist.

The southern part of the Milky Way, showing the nebula called the Coal Sack somewhat below the center of the photograph.   The bright objects at the right and top of the Coal Sack are two of the stars that form the Southern Cross.   The other two stars are too faint to be identified in this photograph.

enters the Northern Hemisphere again. Passing through Aquila and Vulpecula, it arrives at Cygnus and reunites with the branch that left it in Centaurus. From Cygnus, the starry stream of the Milky Way, now single, passes through Lacerta and the head of Cepheus to Cassiopeia, from which we traced its course.

Irregularities in
the Milky Way

It is evident that the Milky Way is by no means a simple stream of stars. Even with the naked eye, one can make out something of its irregular detail when the atmosphere is unusually clear and there is no moon. When viewed under such conditions through a good telescope, the Milky Way is a truly exciting spectacle.

Its general effect has been likened to that of an old, gnarled tree trunk, marked here and there with prominent knots. The Way is riddled with dark passageways, which are linked together by shimmering arches. As details become clearer in a telescopic view, we see that at one point the Milky Way may consist of separate stars scattered irregularly upon a dark background. Elsewhere, there are gorgeous star clusters, sometimes following one another in a long, processional line. In certain places, the stars seem to collect in soft clouds, presenting the appearance of drifting foam as the telescope sweeps over them. Or else they seem to be connected in some mysterious fashion, forming straight lines. These may run parallel to one another or else radiate from a common point. In certain instances, curved lines are formed.

Milky Way Stars that
are embedded in nebulosity

In many places, the track of the Milky Way is engulfed in nebulosity, in which a great many stars may be embedded. In Norma, a complicated series of nebulous streaks and patches appears; these cover the Scorpion's tail and spread faintly upon and beyond Ophiuchus, as if to meet a corresponding series of nebulosities sent off from the region of Cygnus in the Northern Hemisphere. The latter series is very bright; it runs south through Cygnus and Aquila and disappears from view in a dim and sparsely starred region. From Cassiopeia, a "feeler" extends to the chief star of the constellation Perseus.

A powerful telescope reveals the presence of a great many dark rifts in the Milky Way. Sometimes these are parallel; sometimes they radiate like branches from a common point; sometimes they are lined with bright stars. In certain places, they are quite black, as if utterly void of content. In others, they are slightly luminous, as if powdered with small stars.

The dark area
called the Coal Sack

Large, dark areas occur here and there. The most famous of these is the so-called Coal Sack, which is near the constellation called the Southern Cross. Just before the Milky Way divides into two branches in the southern constellation Centaurus, it broadens. It now becomes studded with a collection of brilliant stars, so that this is one of the most resplendent areas in its whole course. Right in the center of this host of bright stars, near the four stars that form the Southern Cross, is the inky black "cavern" of the Coal Sack. It seems to be an opening into a great void, where neither stars nor any other material things could exist.

Other dark areas
in the Milky Way

The Coal Sack is by no means unique. There are many similar black areas in the Milky Way, though they are generally less clearly defined and less striking in appearance. The American astronomer Edward E. Barnard described one of these, in the constellation Sagittarius, as "a most remarkable, small, inky-black hole in a crowded part of the Milky Way, about two minutes in diameter, slightly triangular, with a bright orange star on its north preceding [northwesterly] border, and a beautiful little star cluster following."

As we pointed out, the starry band that extends across the heavens is really part of a galaxy, or system of stars, which in-

Yerkes Observatory

Diagram of the Milky Way system, or galactic system, edge on, showing the stars and globular clusters. Note the position of the sun, about two-thirds of the way from the center to the outer edge.

cludes many stars and star groups in other parts of the sky. Every star that can be seen with the naked eye belongs to this vast system, which is often called "our galaxy," because our own star, the sun, belongs to it. Sometimes the old name "Milky Way" is applied to our galaxy.

Various methods have been used to solve the mystery of its structure. The German-English astronomer Sir William Herschel (1738-1822) decided to attack the problem by making a survey of the stars. He called his method "star gauging." It consisted of counting all the stars visible in a reflecting telescope with an 18-inch mirror and a field 15 inches in diameter.

The Dutch astronomer J. C. Kapteyn developed an even more elaborate method for determining the distribution of stars in the heavens. He selected 206 areas distributed uniformly over the whole sky, and he urged astronomers to determine the apparent magnitudes (see Index) and other

data for all the stars in these regions. A number of the great observatories of the world have taken part in the co-operative venture suggested by Kapteyn. Two outstanding contributions to the program are the Mount Wilson Catalogue and the Bergedorfer Spectral-Durchmusterung (Bergedorf Spectral Catalogue). The latter gives not only the apparent magnitudes but the spectral classes of stars.

Another way of determining the dimensions and structure of our galaxy is to observe the positions and distances of some of its members, such as the Cepheid variables. We can find the approximate distances of these star groups from the earth if we know their apparent magnitudes.

The radio telescope has provided information about our galactic system. We discuss this telescope in some detail in Radio Astronomy, in Volume 9.

We can also obtain a clearer idea of our own galaxy by comparing it with others.

Our chief difficulty in determining the extent and structure of our galactic system is that we are embedded in it so deeply that we cannot obtain a comprehensive idea of it through mere observation. It is almost as if we were required to make a map of New York City, for example, from a vantage point somewhere in a crowded section of the Bronx. However, we do have a clear over-all view of a number of other galaxies, such as the magnificent spiral galaxy in Andromeda (Messier 31) and the Whirlpool in Canes Venatici (Messier 51). They enable us to draw a number of plausible conclusions about our own galaxy.

The "wheel" of the Milky Way

An analysis of our galactic system by the methods described above indicates that most of the stars it contains are crowded into a sort of wheel with a pronounced hub. When viewed edge on, the wheel and hub would look something like the drawing on the preceding page. Of course, since the sun, our own star, is located within the wheel, we cannot see the latter edge on; we infer its shape from other galaxies whose form we can make out. Our sun does not lie anywhere near the center, or hub, of the wheel. It is at a distance of about two-thirds of the way from the center to the outer rim. The wheel as a whole is inclined at an angle of 62° to the plane of the celestial equator (see Index).

Ours is a spiral galaxy

The stars that make up the wheel show a distinct spiral pattern, for ours is a spiral galaxy. The spiral arms spring from a nucleus or core at the center of the galactic system. In the spirals, we find individual bright stars, star clusters, bright nebulae and a great deal of obscuring matter, made up of dust particles and various gases. The general haze it causes has been detected through the dimming and reddening effects that it produces.

The name "dark nebulae" has been given to the masses of obscuring matter which have no stars nearby to illuminate them. Some of them can be seen quite clearly with the naked eye and have been known to astronomers for a long time. The earlier astronomers thought of them as gaps in the starry firmament. The dark nebulae, or masses of obscuring matter, cut off our view of vast numbers of stars that lie beyond them. To obtain an approximate idea of the form of the galactic system, certain observers have counted the stars in various parts of the sky and have then made allowances for the obscuring matter.

The Coal Sack is not an optical illusion

It is this matter that is responsible for the large dark areas such as the Coal Sack. At one time, the attempt was made to explain the Sack as an optical illusion. This hypothesis never seemed too convincing, however. One could not explain away the sharp distinctness of its outline, its huge size, its utter darkness and the even brightness of the starry edge surrounding it. The existence of obscuring matter, now fully proved, offers a simple explanation of what was once a mystery.

The clouds and wisps of bright nebulosity that we mentioned before are the bright nebulae, whose atoms have been excited by the hot stars in the vicinity or else reflect the light of nearby stars. We have described these effects in the article The Astronomer Explores Space, in Volume 9.

It is interesting to note that the most luminous stars are all to be found in the wheel of our galaxy. The so-called galactic clusters are also confined to this area. They are sometimes called "open clusters." They consist of groups of hot stars, each group consisting of several hundred stars. They excite the atoms of dust or gases in their vicinity; hence they are embedded in bright nebulae. Among the best-known of the galactic clusters are the Pleiades, the Hyades and Coma Berenices.

The spectacular globular clusters

The globular clusters differ in many respects from the galactic clusters. For one thing, each one contains hundreds

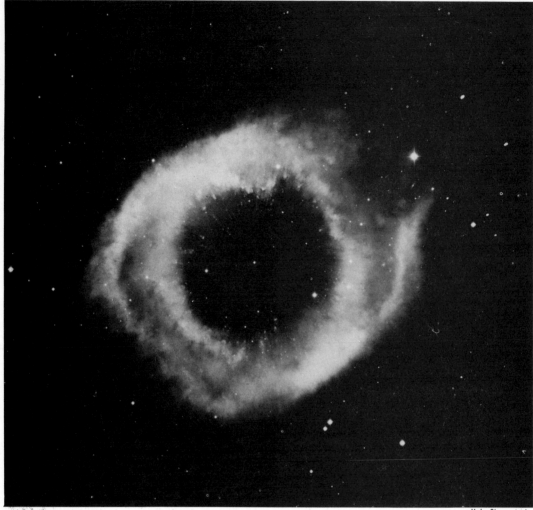

Hale Observatories

Planetary nebula in Aquarius (48 NGC 7293), photographed in red light.

of thousands of stars, or perhaps even millions, compactly and symmetrically grouped; they are comparatively free of gas or dust. The globular clusters are distributed through a roughly spherical region bisected by the plane of the galaxy. The American astronomer Harlow Shapley held that the main aggregation of stars in the galactic system is arranged in the form of a disc (or wheel) and that this disc is enclosed in a roughly globular haze of stars — globular clusters and others.

The core or nucleus of the galactic system is so heavily obscured by clouds of dust that astronomers have a very imperfect idea of its structure. It has never been successfully observed photographically.

With the aid of the radio telescope, however, we have been able to determine the existence of a heavy concentration of stars in this area. It has been estimated that these stars account for something like a half of the total mass of our galactic system.

The rotation
of our galaxy

The entire galactic system is rotating around an axis which is at right angles to the wheel of stars. Outside the dense nucleus of the galaxy, the speed of rotation increases and the period decreases with greater distance from the center. Corresponding differences in period and speed

A CELESTIAL COLOR DISPLAY

In these pages, certain familiar nebulae and galaxies are shown in their glowing natural colors. Their faint light was focused by the powerful 200-inch Hale telescope and the 48-inch Schmidt telescope of the Palomar Observatory on extremely sensitive color film for exposures as long as four hours. William C. Miller, research photographer for the Mt. Wilson and Palomar Observatories, took all the pictures shown here and he also supervised their processing.

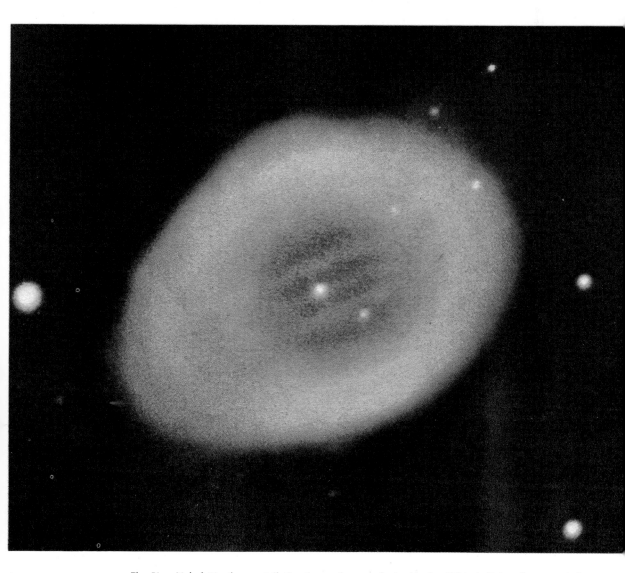

The Ring Nebula in the constellation Lyra, photographed with the 200-inch Hale telescope. Only high-speed color film, exposed in a powerful telescope such as the 200-inch, could record the delicate hues of the nebula in a single exposure. The blue star at the center of the ring is the source of powerful ultraviolet light, which causes the gases in the ring to radiate in their characteristic colors by fluorescence. Hydrogen, oxygen and nitrogen are the principal gases in this well-known nebula.

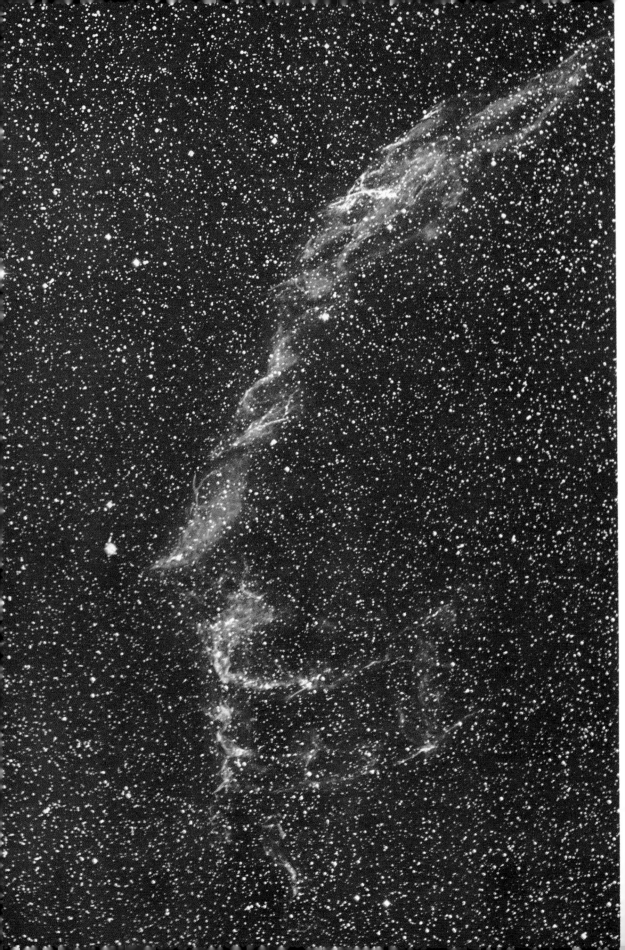

Left: the Veil Nebula in Cygnus, photographed with the 48-inch Schmidt telescope. The colors radiated by these clouds of gas result from the rush through space. The gas was ejected from an exploding star more than 50,000 years ago with an initial velocity of nearly 5,000 miles per second. The velocity has been slowed down to about 75 miles per second because of collisions with gaseous atoms in interstellar space. The force of these collisions ionizes the gas of which the nebula consists, and makes it glow with its characteristic colors. Because of the decline in speed caused by these collisions, the nebula will no longer glow in another 25,000 years or thereabouts.

Below: the Great Galaxy in Andromeda, photographed with the 48-inch Schmidt telescope. This great island universe of stars is located at a distance of nearly 2 million light-years from the earth. (A light-year is about 6 trillion miles.) The outer portions of the galaxy appear blue because the dominant stars in that region are relatively young and extremely hot stars. The central region is of a reddish color because there the predominant stars are older and cooler.

All photos by William C. Miller—Palomar Observatory; copyrighted by the California Institute of Technology

The Crab Nebula in Taurus, photographed with the 200-inch Hale telescope. Here we see the result of a stellar explosion which occurred nearly 5,000 years ago. Because of its distance from the earth, this explosion was not observed until nearly 4,000 years later. On the morning of July 4, 1054, a star was seen by Oriental astronomers to flare to such extraordinary brightness that it was visible in full daylight; it showed with a brightness a hundred million times greater than before. After several months the star slowly faded from sight. In the place where that star flared, modern telescopes now reveal this large cloud of gas, still expanding at a rate which causes the diameter to increase by nearly 70,000,000 miles each day. High-energy electrons, still dashing about as a result of the explosion, make the center of the nebula glow with nearly white light. At the same time, they cause the wispy filaments of gas to shine with characteristic colors.

156-f

The Great Nebula in Orion, photographed with the 200-inch Hale telescope. This great cloud of glowing gas is visible to the naked eye as the middle star in the sword of Orion, one of the best-known constellations in the winter sky. The delicate curtains of gas are caused to shine by several very hot blue stars in the center of the nebula. At its densest part, the nebula is more rarefied than the best vacuum that could be obtained on earth. However, because of its vast size, its mass is ten times that of our sun.

The North American Nebula, photographed with the 48-inch Schmidt telescope. This vast body of gas derives its name from its resemblance to the continent of North America. Its atoms absorb ultraviolet light from stars, and the energy is reradiated in visible colors. The relatively dark areas are caused by an intervening cloud of opaque dust, blotting out not only the light of the nebula but also the many stars that form the background.

can also be noted in the planets that revolve about the center of the solar system — that is, the sun. The nearer a planet is to the sun, the faster it revolves around it and the shorter the period of revolution. For example, in the case of Mercury, the planet nearest the sun, the period is only 88 days, compared with the period of 248.43 years of the planet Pluto, whose orbit is farthest from the sun.

Obviously, then, the stars between the sun and the galactic center go around the

The Horsehead Nebula, a dark nebula in the Milky Way, photographed in red light.

Yerkes Observatory

center of the galaxy more rapidly than the sun does and eventually overtake and pass it. On the other hand, the stars between the sun and the outer rim of the wheel revolve around the center of the galactic system more slowly than the sun does. As a result, they lag farther and farther behind our star in their voyaging.

The dimensions
of our galaxy

In the foregoing pages, we have given you some idea of the form of our galaxy. Astronomers have also sought to discover its dimensions in space. Recent researches, based on the study of special types of variable stars (the Cepheid variables) and on the visual, photographic and spectrographic analysis of the globular clusters, have revealed how extraordinarily great these dimensions are. This is natural enough when we consider that our own solar system, vast and complex though it is, is an infinitesimally small part of the galactic system. The estimates of the dimensions of our galaxy vary considerably, but even the smallest are almost overpowering when we seek to grasp their significance.

Modern estimates of the
size of the galactic system

According to one of the "small" estimates, made by the American astronomer Heber D. Curtis, a pulse of light, starting from one edge of the galactic system and traveling at the speed of over 186,000 miles a second, would take from 20,000 to 30,000 years to reach the other edge. According to another American astronomer, Harlow Shapley, whom we mentioned previously in this article, Curtis's estimate, staggering though it may seem to us, is far too modest. If we are to accept Shapley's figures, we must tax our imagination still further. He maintains that it would take about 100,000 years for a pulse of light to travel from one confine of the galaxy to the other. In other words, the diameter of the galactic disc is approximately 100,000 light-years.

We must remember that our galaxy, of which the solar system forms a part, is only one of millions of other isolated galactic systems. Each of these island universes contains millions upon millions of stars. The nearest — two irregular ones, called the Magellanic Clouds — are about 100,000 light-years distant from us.

Beyond that distance, the heavens are studded with galaxies. When we reach the limit of vision — two billion light-years — with our greatest telescope, the two-hundred-inch giant of Palomar Mountain, the island universes show no signs of thinning out. It is clear that this telescope, gigantic though it is, has not yet reached the extreme "edge" of the universe. With improvements in instrumental and photographic technique, it may be possible to extend our horizons still farther.

Why we shall probably never
see the "edge" of the universe

Whether we shall ever approach the "limit" of the universe is extremely doubtful in view of a certain sobering consideration. If the universe is five billion years old, as some scientists believe, our horizon will be limited to a distance of five billion light-years from the earth, even if the universe actually extends far beyond that distance. The reason is that the light from galaxies beyond that mark will not have been able to reach us since the beginning of time. The same is true of the radio waves emanating from such galaxies, since light waves and radio waves travel at the same speed — roughly, 186,280 miles per second. Such galaxies, therefore, will remain a deep mystery to astronomers, unless some hitherto unsuspected method of detecting them is discovered.

Astronomical problems as a
standing challenge to scientists

Our astronomers are not discouraged by the apparent limit set upon their efforts. For, imperfect as our knowledge of the heavens will always be, the order and beauty they reveal will continue to justify the most devoted attention of scientists. The study of the heavens will always afford an ever widening field for the exercise of man's ingenuity and persistence.

See also Vol. 10, p. 268: "Milky Way."

A section of the Milky Way, near the border of the constellations Sagittarius and Scorpius. This portion of the Milky Way has a quite irregular underlying network of dark nebulosity.

Emission nebulosity in the Milky Way near the Eta Carinae Nebula, in the southern skies. The photo was taken at the Boyden Station of Harvard Observatory, Bloemfontein, South Africa.

Both photos, Harvard College Observatory

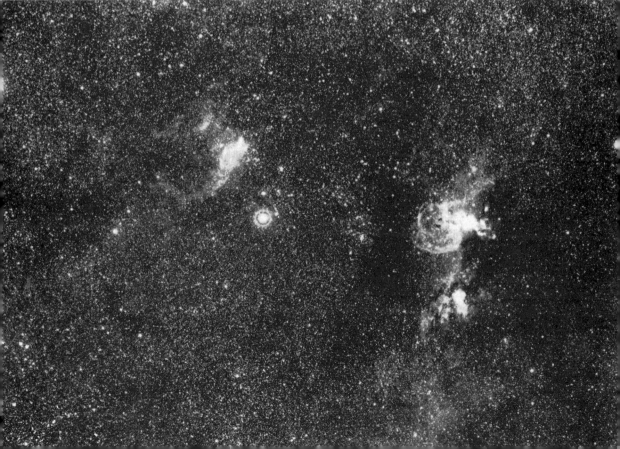

N. Y. Zool. Soc.

Eastern chipmunk.

E. P. Haddon, U. S. Fish and Wildlife Service

Gray squirrel.

Ott, Nat. Audubon Soc.

Ground squirrel.

Golden mouse.

Maslowski and Goodpaster,
Nat. Audubon Soc.

THE GNAWING ANIMALS

Rodents, Pikas, Rabbits and Hares

THE rodents, or gnawing mammals, are the most abundant of all mammals in number of species and number of individuals. The land areas of the globe teem with these comparatively small-sized creatures. For the most part they are ground dwellers; many live in colonies, constructing a vast system of underground chambers and galleries. There are, however, tree-living rats and mice, squirrels and porcupines; beavers and water rats are partially aquatic. All the rodents are herbivorous, or vegetation-eating animals, though some rats sometimes eat flesh also.

At one time rabbits, hares and pikas (or picas) were classified as rodents; but many of their physical features and their evolutionary history show definitely that they are not typical rodents. Their habits are quite rodentlike, however, and it is convenient to consider them together with the true gnawing mammals.

One of the most curious things about the rodents and the hares and rabbits is the regular fluctuation in the number of individuals of various species. For a period of several years there will be a noticeable scarcity of rats, mice, voles and rabbits. Then, during a single season, the populations of these animals will increase tremendously. This cycle of decrease and increase is due to various factors — the abundance or scarcity of carnivorous animals; the reproduction rate of the plants upon which the rodents feed; the competition with other vegetation-eating species; disease; the amount of solar radiation reaching the earth. Obviously, when the number of rodents increases, our natural and cultivated vegetation alike is in danger of quick destruction.

Typical rodents have a pair of large, chisellike front teeth, or incisors, in the upper and lower jaws; the bases of these teeth curve far back in the jaw bones. The molars, or cheek teeth, are often high-crowned. Since the canine teeth and some premolars are absent, there is a large gap between the incisors and the molars. Consequently the incisors can operate freely and food can be easily transferred from the front of the mouth to the grinding teeth. The strong muscles that control the lower jaw are adapted for moving it backward, forward and crosswise, thus making a thorough grinding action possible.

The rodent's skull is long and low; the brain is comparatively small. Often the hind legs are much longer than the forelegs; in some species the animal hops about on its two hind legs. The front limbs are flexible, allowing the paws to bring food to the mouth.

The rodents have had a long evolutionary history. Judging by fossil evidence, they were already flourishing something like 50,000,000 years ago. Apparently they descended from an insectivorous, or insect-eating, mammal resembling the shrew of today. The rodents did not include any giant species among their numbers. The largest rodent known was a beaver that lived in North America about 1,000,000 years ago; it was nearly as large as a black bear. Today the largest member of the group is the South American capybara, which is the size of a small pig.

The rabbits, hares and pikas evolved quite independently. No fossil has ever been discovered that could be considered as a definite connecting link between these animals and the true rodents.

The incisor teeth of the gnawing animals are a distinguishing feature. These teeth are the animals' most precious possession for without them no rodent could survive. The teeth are self-sharpening. This is accomplished by a very ingenious arrangement. The outer surface is made up of a very hard enamel; the inner surface is composed of a softer bone material which wears away at a much faster rate. The difference in rate of wear causes the tooth to be always sharp. The same principle is used in the ax and chisel.

The rodent incisor is in constant use and we might expect that it would be worn away to a tiny stump in a relatively short time. The fact is, however, that the tooth is constantly growing from a permanently active pulp. In some species, this is also true of the molars.

Another interesting adaptation in the rodent's mouth is the method of filtering out unwanted food such as foreign substances and chunks of food that are too large and would choke the rodent. As we have already mentioned, the typical rodent has a gap in its teeth between its incisors and its molars. The cheek fur extends inside the mouth and fills up this gap in the teeth. The hair serves as a screen or filter for the food.

The squirrel family comprises one of the largest groups of rodents. In addition to the various species of tree squirrels, it also includes the ground squirrels, such as the woodchuck, marmot, chipmunk and prairie dog. The tree squirrels include a large number of tree-dwelling species; all have a long, bushy tail and a habit of hoarding quantities of seeds and nuts.

The European pine squirrel is the main species found in Europe and northern Asia. It is usually red in color and has long tufts on the tips of its erect ears. The male is about eight inches long. The European pine squirrel changes its coat with the seasons; it has a grayish-brown coat in the cold months instead of the red one it displays in the other months.

A close relative of the European variety is the North American red squirrel. It is found particularly in the eastern half of the United States and Canada. The eastern gray squirrel is rather more familiar because it is partly domesticated. It is like the red variety except for the color of its coat and its slightly larger size. The western gray squirrel is larger than its eastern cousin, has a handsomer tail and has longer and finer fur. In the southeastern United States we find the fox squirrel, the largest in that area. It owes its name to its fox-colored coat and the sharp, foxlike face.

Central America and Mexico have a large number of especially gay-colored varieties of squirrels. Two well-known species are the variegated squirrels and the Humboldt's squirrels. In South America, one finds the flame squirrels, the common pygmy squirrels and the midget squirrels.

The Orient has some of the most interesting of squirrel species. The common Oriental squirrel is most widely represented; 320 different forms have already been named. One also finds in Asia the largest squirrel of them all, the Oriental giant squirrel, which attains a length of over three feet.

Africa, too, has its share of squirrels. The oil palm squirrel is the largest on the so-called "dark continent," growing to over a foot in length. At the other extreme, there is the West African pygmy squirrel, which has a head and body length of perhaps less than three inches. There are more than fifty varieties of squirrels in Africa.

Marmots are ground squirrels. They are too heavy-bodied and clumsy to climb trees; instead, they make their homes in burrows in the ground. The American species of marmots are called woodchucks or groundhogs. These are hibernating mammals, reaching a maximum weight of fourteen pounds. The hoary marmot, so named for the color of its coat, is a native of the high mountains. Related to it is the European or Alpine marmot, found in Europe and eastern Asia. There are also Chinese and Kashmirian varieties of marmots.

One of the most interesting ground squirrels is the prairie dog, once one of the most abundant species inhabiting the great plains of North America. Before so many of them were poisoned by American cattlemen, they could be found in giant underground prairie-dog cities, which often extended for many miles.

The eastern chipmunk of North America is easily recognized by the characteristic five stripes running down its back. This tiny rodent is often surprisingly tame. There is also a somewhat different western variety. It is more slender and has a longer tail and more stripes.

The pocket gopher is a well-known burrowing rodent; it constructs miles of tunnels during its lifetime. The name "gopher" is derived from the French word *gaufre,* meaning "honeycomb"; the reference is to the maze of the animal's tunnels. The "pocket" part of the gopher's name refers to its cheek pouches, which extend from its cheek back to the shoulders and are used to carry quantities of food.

Thus far we have dealt only with the tree-dwelling and burrowing rodents. Some rodents take to the water. The best-known of these is the beaver, the largest rodent in the Northern Hemisphere. It attains a length of four feet, including the tail, and it averages something like thirty-five pounds in weight. With its webbed hind feet and its broad and flat tail, it is as much at home in the water as on land. Its thick, soft fur is much esteemed by furriers. Beavers live in colonies in shallow streams. Each family consists of two parent beavers and their young up to the age of about two. Two-year olds leave their families in the spring in order to mate and to form their own families.

The beaver is an accomplished engineer and is particularly renowned for its dambuilding. A beaver pair engaged in new dam construction will begin by felling trees near the proposed dam site. The two will work together at the task, gnawing through each tree close to the roots. Once the tree is down, the limbs are removed and the trunk is cut into lengths that the beaver can handle. The branches and logs are then hauled to the damsite. They are put into position in the water and held in place with rocks and mud. When the dam has

been raised to the desired height, the whole is plastered with mud.

Next, the beaver constructs its lodge on the dam itself, or on a bank of a stream or on an island. The structure is of wood and twigs, heavily plastered with mud. When these materials freeze in the winter, they cause the lodge to become an almost impenetrable fortress. The chamber of the lodge has a stout arched roof; it is lined with twigs, vegetable fibers and mud. There is a straight tunnel running down into the water; through it the beaver brings up supplies that have been stored on the bed of the stream. Various other tunnels lead from the chamber into the water.

The nucleus of a beaver colony is the original pair of permanently mated adults. When the young grow up, they usually settle at the damsite, building another lodge and co-operating in the task of keeping the dam in repair.

In the colder months of the year, beavers feed on the supplies amassed on the stream bed and consisting of logs that have been cut to manageable lengths. The logs are taken into the chamber one at a time; the animals devour the bark and then throw out the rest. In the warmer months, the beavers feed on the roots and stems of plants growing in the water. They also go ashore and feast on the bark of trees.

Certain rodents are gliding experts. The flying squirrels represent a familiar species of this type. They are able to glide because of a special adaptation — a loose membrane or fold of skin extending between the hind and front limbs. When the animal stretches out its limbs, the membrane is pulled tight, thus producing a wide gliding surface. The scaly-tailed squirrels of Africa are also able to glide.

The kangaroo rat is a tiny rodent with highly developed hind legs. It moves by leaping forward much as a kangaroo does. It is not a true rat.

One of the most curious rodents is the porcupine. This normally unaggressive creature rarely meets a violent death, because of its almost impenetrable defenses. The porcupine's quills are both a defense mechanism and a deadly offensive weapon.

Philip Gendreau

Amer. Mus. of Nat. Hist.

A dangerous pest — the Norway rat, also known as the gray rat and brown rat. It is found almost everywhere.

A flying squirrel. Actually, this little rodent glides, rather than flies. Loose folds of skin between the front and hind legs are stretched tight when the animal extends its limbs and provide a wide gliding surface.

N. Y. Zool. Soc.

Below is a large beaver dam on the Powder River, in Wyoming. The dam is several feet high and it is many feet long. At the right is a fine specimen of a beaver.

Philip Gendreau

When threatened, the porcupine immediately raises them and turns its back to the foe. Some species will rattle their quills in warning just like a rattlesnake. If the warning is not heeded, the animal charges backward forcing its quills into the threatening animal. Contrary to widespread belief, the porcupine cannot shoot its quills. Quills are not simply annoying; they can kill. There is a case on record of the finding of a dead panther whose head had been run through by seventeen quills; two of them had penetrated the brain. Once a quill enters an animal's body, its end opens up into several small barbs which prevent the quill from being removed. It will actually work its way deeper into the victim. There are a number of species of porcupines on all the continents except Australia. The largest is the African crested porcupine, which measures over three feet in length.

The Muridae, or mouselike family, constitute the largest group in the rodent order. It is in this family that we find the multitudes of rats and mice. The true rat alone has over 550 subspecies. In addition, there are the rice rats, cotton rats, roof rats, water rats, tree rats and so on ad infinitum. There are also innumerable kinds of mice. There are more than 130 subspecies of the common mouse. There are also field mice, pygmy mice, harvest mice and white-footed mice, to name only a few. Besides the exceedingly numerous mice and rats, there are a great many other kinds of rodents. There are the hamsters, voles, lemmings, meadow mice, muskrats, gerbils, bandicoot rats, jerboas, dormice and jumping mice. This sketchy list will give some idea of the vast number of rodents. In this article, we shall describe some of the more interesting ones.

Hamsters are natives of Europe and Asia. The animals that are now being bred in North America as pets were imported from there. The name comes from the High German word for "weevil." Like weevils, hamsters are burrowers; they venture from their underground homes only at night. They subsist mainly on grain and tubers. The European hamster is somewhat larger than the golden hamster that is so well-known in this hemisphere.

The curious rats called gerbils live in the hot, arid desert regions of Asia, Africa and southern Russia, often many miles from water. Communities of these animals make their homes in underground tunnels and burrows. Ratlike in many ways, they differ from most rats in having rather large ears and in leaping instead of scurrying along. The gerbils resemble the jerboas closely; it has, in fact, been suggested that "gerbil" is just another form of the name "jerboa."

Jerboas are nocturnal in habit. They are buff-colored little rodents with rabbit-like ears and kangaroolike locomotion. They have been credited with attaining speeds of up to forty miles per hour.

The dormouse is an Old World species. The name, coming from the French *dormir* ("to sleep") is very appropriate. The dormouse is one of the hibernating mammals, sometimes sleeping for six months at a time. During the months that it is active, it sleeps throughout the day and ventures forth at night. Thus, the animal may spend as much as three-quarters of its six-year life in slumber. The dormouse closely resembles the squirrel. It has a long, bushy tail, soft, silky fur and large, round eyes. It also resembles the squirrel in its habit of spending a good deal of time in the trees. Like the squirrel, it feeds on nuts, acorns and seeds.

The muskrat is trapped in North America for its fur. It is the most valuable fur-bearing animal in North America if one considers the total number of skins that have been sold. However, the animal is so prolific that there is no danger of the species being wiped out. It gets its name from a secretion of its hind quarters; this secretion gives off the odor of musk and is used in the manufacture of perfume. The muskrat is usually found in the marshy regions of North America.

The rats are the most widespread members of the order of rodents. They are found almost everywhere that man himself dwells. From our point of view they are thoroughly obnoxious creatures. The fight against them has been a continuous one and, so far, a losing one. In North America, despite all of man's efforts to destroy them, there are more rats than people. The rat found in most temperate regions is the Norway rat. It first came to America, probably about 1775, on the ships of man. The Norway rat ranges from five inches to almost a foot in length and its scaly tail is even longer. Its color varies from gray to grayish-brown and it breeds as often as twelve times a year. Although it lives mainly in the filth and squalor of city slums, where it spreads disease, this rat may be found almost anywhere. In South America and southern Europe, a black, slightly smaller variety is more common. Both species are believed to have originated in China and spread west.

The rat has managed to thrive, despite man's utmost efforts to destroy it, because it is fertile, hardy, voracious, adaptable, cunning and courageous.

One of the most curious patterns of behavior to be seen in the rodent family is the famous death march of the Scandinavian lemmings. The lemming is a small, heavy-bodied rodent, usually about six inches in length and covered with soft, golden fur. The Scandinavian variety has patches of glossy, black fur behind the shoulders.

A colony of lemmings is formed on a mountain slope where it soon grows to considerable proportions depending upon the severity of the weather and length of the warm season. If spring comes early and there is a long warm season, the colony grows rapidly. Because of the longer season, there is more vegetation to feed the increased number in the colony, but a long summer also tends to scorch the vegetation because of the excess heat. This reduces the normal supply of food — a serious condition because of the greater number of lemmings. Suddenly, as if a signal had just been given, the rodents start migrating toward the sea. The instinct seems to affect all the lemmings in the colony. It is unknown whether or not some of them remain behind to start up the colony again, but if they do, they have not been observed, although the lemmings appear again after a few years.

Like a huge army, the host of lemmings begins its march. Many will perish before they even catch sight of their destination. As they move along, all the food encountered by the long column is

A chinchilla that is being raised for breeding purposes. Chinchilla fur is very valuable; a coat made from 80 to 120 pelts may be worth $30,000. The animal is a native of South America.

Ewing Galloway

H. L. Dozier, U. S. Fish and Wildlife Service

The coypu, or nutria, is a large rodent found along the streams of southern South America. The animal sometimes attains a length of two feet, not including the tail, and may weigh nine pounds or more.

eaten by those in the front ranks, while the rest go hungry. Ordinarily, they do not swim, but now they cross rivers and fjords without hesitation. Huge numbers of lemmings are eaten by birds of prey, foxes and even ordinarily herbivorous animals, such as the caribou. When the lemmings reach the sea, they plunge in and drown en masse. Little is now known as to why lemmings behave this way. Some authorities interpret the phenomenon as a kind of population control, because it occurs when the weather is mild and the lemmings are reproducing very rapidly. There are distinct changes in the blood, for example, of Alaskan lemmings, such as increased concentrations of hormones and other biochemicals, during these reproductive periods. It is possible that one of these blood chemicals attacks the nervous system, causing violent behavior.

Another puzzle, of a similar sort, arose in connection with the coypu (*Myocastor coypus*), or nutria, as it is called. A large, muskratlike aquatic rodent, characterized by partially webbed hind feet, it inhabited the river banks of South America. In the late 1840's, hunting coypus for their pelts was so popular that peons were not willing to work on the sheep and cattle ranches during the summer months. In order to protect the ranchers from ruin, coypu hunting was made illegal. As a consequence, the coypu swarmed everywhere. It even changed its dietary habits and its habitat, establishing itself on dry land many miles from the nearest river. Then, quite suddenly, for no known reason, the coypu became practically extinct. In recent years, it has been introduced to the United States, where it now appears to be thriving.

Among the few valued representatives of the fur-bearing rodents are the South American chinchillas, small burrowing animals with bushy tails. Chinchilla fur, which is much prized for its beauty, is silky in texture and in color a delicate pearl grey, darkly mottled on the upper surface and dusky white beneath. In mountainous districts of Peru and Bolivia, the chinchilla is well-known, and here the little creature attains its greatest size, ten inches or more with an imposing ten-inch tail. The chinchilla is also found along the coastal area of Chile, while one variety, the viscacha, is native to the Argentine pampas. Chinchillas live in large colonies, remaining in their capacious burrows by day and venturing out at twilight to forage and work on their tunnel-building enterprises. As in the Peru of the Incas, the chinchilla is still hunted for the adornment of man.

Other species of South American rodents are the agouti and its close relative the paca. Both resemble over-size guinea pigs, the agouti measuring from eighteen to twenty inches in length, while the paca is more than two feet long. These rodents are long-legged, with broad flat nails like rudimentary hoofs on their toes. Their ears are small; their tails insignificant. They are not gregarious like the colony-loving chinchillas but dwell in pairs or little groups. One variety of paca lives in the lowlands of Mexico and the eastern part of South America. Another inhabits the lofty forests of the Andean foothills in the countries of Ecuador and Venezuela.

A third group of southern rodents is the cavies, small plump creatures with short legs. Perhaps the best-known cavy is the domesticated guinea pig. It is probable that this object of childish interest and boon to the medical profession derives from the Peruvian cavy of the Incas. The cavy lives chiefly in marshy districts where it can burrow in the soft earth. However, the rock cavy of Bolivia dwells in colonies amid the high and rock-strewn slopes of the great mountain range of the Andes, in South America. The rock cavy is highly prized by the Indians as a delicacy.

Another member of the cavy family is the Patagonian cavy. At first glimpse, this elongated cousin of the rat might be mistaken for a large hare. The Patagonian cavy is eighteen inches in length and has a two-inch tail. It stands over a foot in height at the shoulder. The creature is colored a light cinnamon on the flanks and breast, while the belly is white. This cavy, with a smaller relative, the dwarf Patagonian cavy, inhabits the salt wastes of the Argentine.

The largest of the rodents is the capybara, found in the South American forest regions from Guiana to the River Plate. The size of a year-old pig, the capybara attains a length of more than forty inches and stands twenty inches at the shoulder. The capybara is a strong swimmer and forages in herds of from fifty to one hundred. Herbivorous, constantly on the move, denuding whole districts of fruit and vegetation, this rodent can be a considerable nuisance to man.

As we pointed out, pikas, hares and rabbits were once classified as rodents. Actually, for all their rodentlike appearance, they belong to another order — that of the

N. Y. Zool. Soc.

A South American spotted paca.

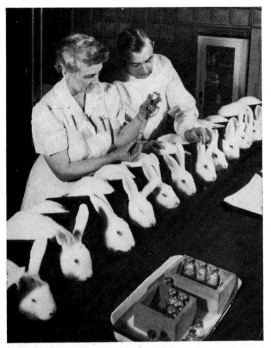

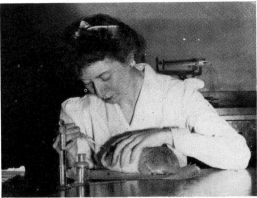

U. S. Civil Service Commission

This technician is testing the effect of a medicinal preparation on a guinea pig. Guinea pigs are frequently used for experimental purposes.

Commercial Solvents Corp.

Trying out the effectiveness of a penicillin preparation on a group of rabbits. Experiments like these have brought about many great advances.

Borden Co.

The cages shown here house white rats, on which different food combinations are being tried out.

Lagomorpha (lagomorphs). The hares and rabbits make up the family of the Leporidae. The pikas are assigned to a separate family, called the Ochotonidae. The pika looks very much like the rabbit or hare, except that its ears are shorter and it is tailless. It is very much smaller, too, averaging about six inches in length when fully grown. Unlike other lagomorphs, it lives high up in the mountains, where it makes its den in rocky crevices.

Although its shrill, bleating call is frequently heard in elevated regions, the animal itself is hard to find. Its coloration makes it blend into the background. When it moves, the motion is so quick that it simply disappears from one spot and then appears motionless at another (or so it is claimed by naturalists who have observed it). To survive during the long, barren mountain winter, it harvests bits of vegetation and dries them in the sun. If the drying is threatened by rain, the entire pika colony will turn out and work until late into the night, seeing to it that the winter's fodder is safely stored away. Since spring in the mountains comes late, the first litters are not born until May. Other litters continue to arrive until the month of September. In North America, the pika is found from the Rocky Mountains westward. In Europe and Asia, it is widely distributed east of the Ural mountains.

The hares and rabbits look very much like one another but differ in some important respects. The young of the rabbit are born, blind and hairless, in underground burrows. The hare is born above ground in an open nest and has hair and vision from birth. The true hare averages something like twenty-five inches in length. The ears are about seven inches long. The tail appears quite short, but this is partly because much of it is hidden in the fluffy fur of the belly. The true rabbit is somewhat smaller than the true hare. It averages eighteen inches in length and from three to six pounds in weight.

Both rabbits and hares have conspicuously long, powerful hind legs. Their hopping gait leaves a very distinctive track. In front are the two large prints, side by side, made by the hind paws. Slightly to the rear, one behind the other, are the smaller prints of the front legs, which are used more for balance than for propulsion. For short distances, the powerful hind legs of the rabbit enable it to travel at speeds up to 45 miles per hour.

The animals commonly called rabbits in North America are really hares. The true rabbit is native to Europe. It is now found wild in Australia, although it was completely unknown there until brought over by Europeans. Having no long-standing natural enemies, the rabbit prospered in that huge island-continent. It did such damage to crops that foxes were introduced to control it. Although they did reduce the number of rabbits, the foxes themselves soon became a major pest.

Jack rabbits abound in the western part of North America. They are generally about two feet long and have prominent ears. They are a serious pest in many areas. In Harney county, Oregon, bounties were paid on a million jack-rabbit tails in a single year. The black-tailed, or California jack rabbit is responsible for more crop damage than any other hare in North America.

Among the other large hares of the Western Hemisphere are the snowshoe rabbit and the arctic hare. The snowshoe rabbit, or varying hare, as it is also called, wears a brown coat during the summer; this changes to white during the winter months. The name "snowshoe" is derived from the fact that a heavy coat of hair develops on the foot during the winter months and forms a sort of snowshoe. The arctic hare, largest of all hares, lives farther north and keeps its white fur all year round.

The cottontail occurs in great numbers in North America. It has both harelike and rabbitlike characteristics. It looks very much like a true rabbit; its young, like those of the rabbit, are hairless and blind at birth. However, they are not born underground but in nests, as are the hares. Cottontails thrive in brush and along the fringes of various cultivated areas. They are very widely hunted and millions of them are shot every year.

See also Vol. 10, p. 275: "Mammals."

THE WILY WEASEL FAMILY

Small Flesh Eaters That Roam the Wilds
in the North Temperate Zone

THE members of the weasel family include not only weasels but also martens, wolverines, minks, skunks and other animals. For the most part they are slim and lithe carnivores, noted for their courage and their strength. Most of these animals dwell in the North Temperate Zone; representatives of the group are sometimes found in other regions.

Like the other flesh eaters, such as the cats, dogs and civets, the weasels evolved from small, tree-dwelling carnivores called miacids, which lived something like 40,000,000 years ago. Among the descendants of the miacids was a weasel about the size of a black bear, which dwelt in America about 25,000,000 years ago. Today the members of the weasel family range in size from the tiny weasels proper through the medium-sized skunks and martens to the wolverines, which may weigh as much as thirty-five pounds.

Modern members of the weasel family have evolved but little in some respects from their earliest known ancestors, the miacids. They usually have long, lithe bodies and stocky legs; they have specialized teeth for shearing tough meat fibers. Some still dwell in trees; others do not. The badger, for example, has become a burrowing animal; the otter is well adapted for life in or near the water. Many weasels have scent glands; those of the skunk are particularly well developed. Certain members of the group have coats of thick, dense fur, greatly esteemed by furriers.

Among the most striking members of the weasel family is the American marten, often called the American sable or pine marten. This handsome animal, which is an expert tree climber, is about twenty-five inches long. It is colored a rich light brown above; its underparts are darker brown, as are its legs and bushy tail. Its wide throat patch is whitish or buff. Martens shun man; they are found in the denser forests of Canada from the Atlantic to the Pacific, in some parts of New England and some regions of the West.

Martens hunt squirrels in the treetops; on the ground they track down rats, rabbits and grouse. They are active during the winter, traveling easily over the snow as they hunt for mice and shrews. They often supplement their meat diet with insects, honey, berries, fruits and nuts. The female marten builds a grass-lined nest in a tree hollow or burrow, in which from one to five young are born in April. The chief enemies of the marten, besides man, are the horned owl, the lynx and the fisher, also a member of the marten group.

The fisher, or pekan, is a large, powerful animal that looks like a bulky, long-nosed cat. It measures about a yard in length; its long and soft fur is much darker than that of the American marten. The fisher dwells in the northern part of the United States and in Canada. It is sometimes found in the neighborhood of swamps and other bodies of water, where it preys upon fish, frogs, muskrats and beaver. It also hunts at night for rabbits, squirrels and other small mammals. The fisher is one of the few animals that venture to attack the porcupine; it turns this rodent over and pounces upon its underside, which is not protected by the formidable quills. The trapper regards the fisher as one of his chief enemies in the wilds, since it steals the bait in his traps and devours any animals that are caught in them.

A still worse nuisance to the hard-worked trapper, however, is the animal called a glutton in northern Europe, and wolverine, or carcajou, in Canada, where it lives from ocean to ocean. The wolverine is the largest of the weasel family, equaling, in the adult stage, a small bear, and differing from other weasels in that it stands fairly high upon the leg. The wolverine not only robs the trapper of both his bait and catch, but such are its skill and craftiness that it is almost impossible to trap. It shows an uncanny guile in eluding

Both photos, N. Y. Zool. Soc.

This streamlined grison hunts small mammals and birds. When it is attacked or cornered, it exudes an unpleasant odor like that of a skunk.

the artifices of the despairing trapper. It preys upon all small mammals, including the beaver, and will pull down a wounded or sick deer.

Like certain birds and other animals, the wolverine combines an enormous appetite with a strange kleptomania. It will creep into the unoccupied tent of the trapper and carry away and bury every portable article that is available, from a gun to a blanket. All things considered, the wolverine is a considerable trial to the trapper, and there is no animal that has made a more audacious and cunning fight against man and his arts in the wilds than this lord of the wily family of weasels.

Two smaller but very troublesome and savage members of this family are the tayra and the grison, both belonging exclusively to Central and South America. The first matches the otter in size; the grison has more the proportions of the marten. Both frequent hollow trees, clefts in rocks, and the deserted burrows of other animals; both prey upon various small mammals and birds. Eggs play an important part in the diet of the tayra and the grison.

The true weasels are distinguished from the martens by their small size, slimness of body, preference for the ground — although all are good climbers — their close fur, comparatively short tails and change of coat in winter. Several species inhabit the Old World, of which the most familiar is the stoat, or true ermine weasel; and several others belong to North America. One species ranges south along the Andes. The specialists tell us of more than twenty species in North America, but the best-known are the common, or short-tailed, weasel, the northern long-tailed weasel, the little

The wolverine is unequaled by any animal in courage and craftiness and is reputed to be the most powerful mammal of its size in existence.

"mouse-hunter" of the northern plains, and the bridled weasels of the Pacific Coast, marked with a black band across the face. The Arctic coast has another species very like the stoat. But wherever the weasel is encountered, he is found to be "a keen, agile, indomitable hunter, within his powers a being of the highest type of effectiveness."

Photos, N. Y. Zool. Soc.

Striped skunk, most common North American skunk.

A handsome species, the small, slim spotted skunk.

New York weasel, showing its partial winter pelage.

The short-legged, heavy-bodied American badger.

What these powers are may be seen in the following paragraph from THE FUR-BEARING ANIMALS by Dr. Elliott Coues:

"Swift and sure-footed, he makes open chase and runs down his prey; keen of scent he tracks them, and makes the fatal spring upon them unawares; lithe and of extraordinary slenderness of body, he follows the smaller through the intricacies of their hidden abodes and kills them in their homes. And if he does not kill for the simple love of taking life, he at any rate kills instinctively more than he can possibly require for his support. Yet which one of the larger animals will defend itself or its young at such enormous odds? A glance at the physiognomy of the weasels would suffice to betray their character.

N. Y. Zool. Soc.

The savage, weasel-like tayra of South America.

The teeth are almost of the highest known raptorial character; the jaws are worked by enormous masses of muscles covering all the side of the skull. The forehead is low and the nose is sharp; the eyes are small, penetrating, cunning and glitter with an angry green light. There is something peculiar, moreover, in the way this fierce face surmounts a body extraordinarily wiry, lithe and muscular."

The fur of all the true weasels — in summer, light or dark brown above and white, sulfur yellow or orange, according to species, underneath — becomes white in winter wherever the climate is such that snow re-

mains continuously from late November to April. South of the line of permanent winter snow the weasels keep their brown color all the year round, but along this climatic borderland they may become partially white. This white winter fur of the weasel is the valuable ermine — in the Middle Ages a symbol of royalty and later so distinguished a mark of high office on the judicial bench that we still refer to a judge as a "wearer of ermine." The weasels are of great service in destroying vermin in the fields and in hunting rats and mice about barns and storehouses, and these activities should be encouraged.

The mink, on the other hand, is of little value to man except for its fur. Minks exist in northern Europe and are encountered over all the North American continent. The mink is larger, stouter and more robust than the weasel, and its fur is everywhere thicker and darker brown in color, except for the white patch of variable extent on the chin. It lives on the ground, making its den in holes under or inside old stumps, in crevices among loose rocks or in burrows usually opening in the bank of a stream or lake. It hunts its food mainly along stream and lake shores where it preys upon frogs, fish (especially eels) and other aquatic animals. Though trapped by amateurs and professionals alike, it holds its own in even the most thickly settled regions. Mink farms are now a successful enterprise throughout the northern United States and Canada. The mink are as carefully raised and tended as though they were precious children, which in a sense they are. Most mink skins for fur coats now come from these farms.

The polecats, apart from certain structural differences, are distinguished from the martens by the fact that while the martens are inoffensive in the matter of odor, they are equipped with a notoriously evil fluid for confusing their enemies. The name "pole," or "pol," as it was originally spelled, may be a corruption of the French poule, a hen, in reference to the animal's great weakness for poultry. Or it may come from the old English word foumart, or "foul marten," referring to its fetid odor.

Polecats of various species are common in many parts of Europe and Asia. Though they vary in size and coloration, their habits are similar and their odor is equally abominable. It represents a truly remarkable feat for man to have bred the domesticated type of polecat that we call the ferret so that the offensive odor has largely disappeared. The ferret differs from the polecat in the color of its fur, which is generally yellowish white; its eyes are pinky red. The animal is about fourteen inches long, not including the tail. It is exceedingly prolific; each year the female brings forth two broods, each numbering from six to nine young. The ferret is often used to kill mice and rats. It also serves to drive rabbits from their burrows; the rabbits are shot by hunters as they emerge. According to the ancient Roman naturalist Pliny, his countrymen also used ferrets for rabbit hunting.

Like all members of the weasel tribe, ferrets are extremely bloodthirsty. They must be muzzled before being turned into a rabbit's burrow; otherwise they would kill the quarry and drink its blood. They are not muzzled, however, when they are used to kill rats, since these rodents are fierce fighters when cornered.

Courage and power are attributes of every member of the weasel family. In none do we find these qualities more pronounced than in the ratel, which is found in India and Africa. Ratels are badger-like in build, with rudimentary ears. They have powerful claws, with which they can dig themselves into the ground with great rapidity in time of danger. They are generally black in color, with a whitish crown and a somewhat grizzled back. The ratel is fond of honey, and for that reason it is sometimes called the honey badger. It also eats rats, frogs, birds and insects, particularly termites. It is very fond of poultry and occasionally raids chicken coops.

The ratel has been successfully tamed and it generally makes an amusing pet, always astir and turning somersaults in order to attract attention. However, it is likely to bite upon the slightest provocation, apparently not from malice but fear.

The mink at home. Famous for its fur, it is trapped by professionals and amateurs.

Amer. Mus. of Nat. Hist.

Sea otter on Amchitka Island, in the Aleutians. The animal is now scarce.

Fish and Wildlife Service

Specialization reaches a high point in the American skunk, whose defense mechanism is about as effective as can be found in the animal kingdom. The skunk is equipped with scent glands from which it can, at will, discharge a most nauseating fluid. This fluid, which is clear yellow in color, is dispensed in a fine spray that can carry as far as ten feet at the time of discharge; air currents can distribute it even farther. The spray is strongly acidic; if it gets into the eyes it may bring about temporary blindness. The sickening odor of the fluid is so penetrating that it generally puts to rout any attacker.

The skunk uses its defensive spray only when it is sufficiently provoked or actually attacked. It gives due warning of its intent by turning its back to its tormentor and raising aloft its long, bushy tail. The animal relies so thoroughly upon its defensive weapon that when it is attacked, it almost never tries to escape or to defend itself with its sharp teeth.

Skunks are black in color, with white longitudinal stripes or white spots. They are found in many areas of the United States and Canada. They commonly avoid deep forests and waterless deserts, preferring slightly wooded or open country. They eat insects, mice, frogs and snakes; sometimes they raid poultry yards. They hibernate during the winter; in the spring the female has a litter of from six to ten young.

The badger is another interesting member of the weasel family. The American badger, distributed over the western and central parts of North America, is a clumsy, heavy-bodied animal, with a much flattened body and a short and bushy tail. It is silvery gray and black above and yellowish white below. The animal is a powerful digger; it is equipped with strong front legs and inch-long claws. It digs burrows in the ground for its own use. It also digs out the burrows of other animals, such as gophers, prairie dogs and the like; it travels from one burrow to another in its nightly hunting and devours the hapless occupants. Because it keeps down the population of small mammals that attack the crops, the badger performs a real service to mankind. The fur of this animal is of considerable value; its bristly hairs are widely used for artists' brushes and shaving brushes.

The European badger is quite similar in appearance to the American variety, but differs somewhat in certain details; for one thing, the head is much more slender. Its general habits and food are much like those of the American badger. Badgers are also found in Asia and Africa.

The cruel "sport" of badger baiting

The European badger was formerly used in the cruel "sport" called badger baiting. A captive badger was placed under a partly overturned barrel or similar receptacle, and dogs were allowed to attack the animal. After they had dragged it out, it was put back and a second set of dogs went after it. The "sport" was prohibited in Great Britain about the middle of the nineteenth century. Early settlers in the western part of the United States also practiced badger baiting to a certain extent, though it was not so popular as in England. The "sport" gave rise to the familiar term "badgering," in the sense of "teasing" or "harassing."

Next we come to the otter, which is perfectly at home in the water. It has the webbed feet, short and thick fur and small ears that are characteristic of aquatic mammals. Its legs are short; its rudder-like tail is fairly long. The otter is a handsome animal, with dark brown fur. An excellent swimmer and diver, it feeds almost entirely on fish; however, when fish are scarce it will hunt crayfish, frogs and muskrats. It is found in Europe, Asia, South Africa and the Americas.

The otter rests for the greater part of the day in its den, near the stream in which it fishes; it comes out at night to feed. This den is usually an excavation in the bank or in a log lying close to the stream. If the den is in a bank, the entrance is often to be found below the water line.

The otter is noted for its playfulness

The otter is a playful beast; it loves to slide down river banks or snowy hummocks. It can be tamed quite easily, and it makes a most playful and affectionate

pet. It can be taught to follow its master like a dog and to enter and leave the water at a word of command. It can even be trained to catch and bring in fish. It is first taught to retrieve things on land and afterward to fetch and carry a dummy fish thrown into a river. Then comes the catching of a real fish cast into the stream.

The true sea otter of the North Pacific is quite a different kind of animal; it is more seallike than the otter. It is particularly well adapted for a marine habitat. It is about four feet in length; it has a flat and broad head, very short ears, a flexible body, short limbs and a short tail. Its front feet are small; the hind feet are webbed to form flippers, suggesting somewhat those of the seal. The molar teeth are provided with grinding surfaces for crushing the shells of various kinds of shellfish. In addition to shellfish, the sea otter feeds upon salt-water fish of various kinds, cuttlefish and sea urchins.

N. Y. Zool. Soc.

American otter (*Lutra canadensis*).

The sea otter is rapidly vanishing from the earth. Today the animal has become so scarce that a sea otter's skin fetches a very high price. In North America the animal is now restricted to a few scattered areas from Vancouver Island north to the coast of Alaska and the nearby islands. Attempts to domesticate the animal have met with scant success. The young that are captured generally refuse food.

See also Vol. 10, p. 275: "Mammals."

THE
LAND
THEORY OF
COLOR
VISION

A Revolutionary
Mid-Century Development

UP TO the year 1955, most physicists would have found no difficulty in answering the questions "What is color?" and "How are impressions of color registered in the brain?" In that year, however, Dr. Edwin H. Land * began a series of epoch-making experiments that cast doubt upon the old established ideas concerning color vision. He proposed a new theory that ranks with the most striking scientific developments in recent years.

* Dr. Land is the inventor of the Polaroid Land Camera (see Index) and the founder and head of the Polaroid Corporation.

To have some understanding of the Land theory of color vision, we must first consider the older notions. Of course, the existence of color was known to men of every age. It was not, however, until Sir Isaac Newton carried out a series of classical experiments in 1666 that scientists began to learn something about the physical basis of color.

Newton passed a narrow beam of light into a darkened room, and then placed a triangular glass prism in the path of the beam. He found that the glass broke up the white light of the sun into all the colors of the rainbow. Instead of a single color — white — there was a band of colors: red, orange, yellow, green, blue, indigo and violet. This color band was what we call today the visual spectrum of light.

The fact that light is refracted, or bent, when passing through various materials had been known before Newton's day. Newton showed that different colors are refracted through different angles. Red is bent least; then come, in order, orange, yellow, green, blue, indigo and violet. To prove that all these colors were derived from white light, Newton used a second prism to bend the colored rays back again to their original paths. They then produced white light.

Next, he tried to recombine only parts of the spectrum. He used a slotted board to cut off all but the desired bands. When he combined two such bands, letting the

Sir Isaac Newton showed in a famous experiment in 1666 that white light consists of a combination of colors. He passed a narrow beam of sunlight through a triangular glass prism in a darkened room. The prism broke up the white light of the sun into bands of various colors: red, orange, yellow, green, blue, indigo and violet.

VIOLET

INDIGO

BLUE

GREEN

YELLOW

ORANGE

RED

Below is a diagram of the camera used in Dr. Land's experiments. In a typical experiment, he took two pictures of the same scene, one through a red filter and the other through a green one. He obtained two black-and-white transparencies. The one taken through the red filter was called the long record; the one taken through the green filter was called the short record.

The two records were inserted in a double-lens projector. The long record was projected through a filter transmitting red light; the short record through a neutral filter, transmitting white light. When the two images were superimposed on a screen, the objects in the transparencies appeared in a full range of color.

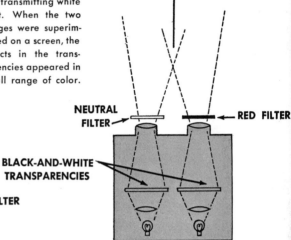

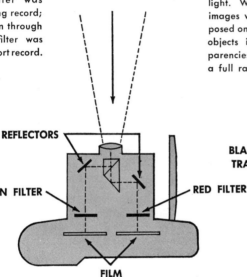

NEUTRAL FILTER →   ← **RED FILTER**

BLACK-AND-WHITE TRANSPARENCIES

REFLECTORS

GREEN FILTER

RED FILTER

FILM

rays mix on a screen, a third color appeared. Generally, it matched a color lying between the two bands in the spectrum.

In the years that followed, many additions were made to the theory of color. Thomas Young (1773–1829) and other investigators established that light is a wave disturbance * and that the different regions of the visual spectrum correspond to different wave lengths. The wave lengths decrease as we pass from the red end of the spectrum to the violet end.

The Englishman James Clerk Maxwell (1831–79) and the German Hermann von Helmholtz (1821–94) worked independently on the problem of color. They found that they could match most colors by mixing wave lengths chosen from the red, green and blue bands of the spectrum. Hence red, green and blue came to be called the primary colors. This formed the basis of the so-called three-color theory of color vision.

According to this theory, one group of cones in the retina of the eye (see Index, under Rods and cones of eye) is particularly sensitive to the red region of the spectrum (that is, to the wave lengths corresponding to the red region). A second group of cones is particularly sensitive to the blue region; a third to the green region.* The sensation of color results when wave lengths in varying degrees of strength strike the cones of the retina and are registered in the brain. The colors (or color) that an observer sees depend on what wave lengths are proceeding from a particular place — a flower garden, say.** They also depend on the relative strengths of the wave lengths.

* There is nothing in the appearance of the cones to indicate this; they look alike under the microscope.
** It should be borne in mind that the wave lengths actually originate from the source of light — the sun or some form of artificial illumination. As they strike a given object, part of them may be absorbed by it; those that are not absorbed are reflected from the object and strike the retina of the observer. See the article, The World of Color, in Volume 7.

* Actually, light has a dual nature. In many ways, it behaves like a series of waves. However, in certain respects it behaves more like a stream of small particles of energy. See the discussion on the first page of the article An Introduction to Optics, in Volume 7.

Dr. Land's experiments have cast doubt on this classical theory of color vision. In a typical experiment, he took two pictures of the same scene, one through a red filter and the other through a green one. The red filter passed the long wave lengths corresponding to the red area of the spectrum; the green filter passed the shorter wave lengths corresponding to green. In both cases he obtained black-and-white photographic transparencies. There was no color in either of them; there were only darker and lighter areas. The transparency obtained with the longer wave lengths (red) was called the long record; the one obtained with the shorter wave lengths (green) was called the short record.

Next, Dr. Land passed two beams of yellow light through the transparencies. One of these beams was from one end of the yellow band of the spectrum; the other, representing a different wave length, from the other end of the yellow band. As the beams fell on a screen, forming a superimposed pair of images, one would have expected to see the screen illuminated in yellow. Yet the objects pictured in the transparencies appeared on the screen in different colors — red, gray, yellow, orange, green, blue, black, brown and white. These colors were pallid but could be clearly recognized. (For the experimental setups, see the figures on the preceding page.)

In other experiments, instead of working with two wave lengths from the yellow area of the spectrum, Dr. Land chose wave lengths from entirely different areas to light up the screen; again he obtained full color. In fact, if he illuminated the long-wave and short-wave transparencies — the long and short records — with beams representing almost any pair of wave lengths and superimposed the images, he would obtain a colored image. If he sent the longer wavelength beam through the long record and the shorter wave-length beam through the short record, he would obtain most or all of the colors in the original scene and in their proper values. If he reversed the process, the colors would be reversed, so that the reds would show up as blue-greens and so on.

Light sources for the records did not have to be restricted to pairs of wave lengths, each from a definite part of the spectrum. For much of his work, Dr. Land used white light including all lengths of visible light as the source for his short record. These combined wave lengths are short, compared, say, to the wave lengths corresponding only to the red area of the spectrum. Hence the white light could transmit the short record, while the long record could be transmitted by the red. On the other hand, since the combined wave lengths of white light would be longer than the wave lengths of cyan (blue green), white could be used for the long record if cyan were used for the short record.

Dr. Land found that the full-color image was not achieved when the colors in the photographed scene represented a systematic pattern, such as a checkerboard arrangement. There had to be a random dispersal of the different colors in the scene.

From these experiments and many others, Dr. Land has derived a new theory of color vision. He holds that the colors we see in images do not arise from the effects of certain definite wave lengths upon the retina of the eye. Rather, the sensation of color results from the interplay of longer and shorter wave lengths which can be selected from many different parts of the visual spectrum. Any pair of wave lengths that are far enough apart * can be used to produce the long and short records. Many different combinations of beams can produce the full range of color.

Does this mean that the classical theory of color vision is all wrong? Dr. Land maintains that the work of most previous investigators had little to do with color as we normally see it. They were interested chiefly in *spots* of light and particularly with pairs of spots which they tried to match. They did not take into account the "surrounds" — the areas that surrounded these spots. In other words, men such as Maxwell and Helmholtz dealt only with a restricted area of color sensation and not with all of it.

* The minimum distance required is very small in some parts of the spectrum.

THE NATURE OF MOLECULES

An Experimental Approach

BY EARL J. MONTAGUE

THE stuff of which all things, living and nonliving alike, consists is called matter. In many of its forms, according to scientists, matter is made up of units called molecules. A molecule is the smallest particle of a given substance that has a stable independent existence under present conditions. Molecules are thought to be building blocks of liquids and gases and many solids.

The molecule is not the smallest subdivision of matter, for it can be broken up into smaller units, called atoms.* Each molecule of water, for example, is believed to be made up of an atom of oxygen and two atoms of hydrogen, as is indicated by the chemical formula H_2O. If a molecule of water were to be broken up into the atoms of which it consists, these atoms would not be water. They would be atoms of two gases, oxygen and hydrogen. Furthermore, these gases are entirely different from water.

In another chapter — How Molecules Behave, in Volume 1 — we set down some basic facts about molecules. In this article, we are going to approach the subject in a different way. We shall bear in mind that we have only indirect evidence that molecules exist and that they possess certain properties. We are going to examine some of this evidence at first hand through a series of simple experiments, and we shall be able to make up our own minds about how convincing the evidence is. We shall also learn something about the properties of molecules through this experimentation of ours.

* *Editor's note*: Atoms can be subdivided into even smaller units, called subatomic particles. These include electrons, protons and neutrons. See Index.

Evidence supporting the theory that molecules exist

The electron microscope has enabled us to see and photograph certain giant molecules, including those called viruses (see Index). Generally speaking, however, we know of the existence of molecules only through indirect evidence. Thus there are certain situations that can most readily be explained if we assume that molecules exist. We are going to examine some of these situations in the present article. *Before proceeding, we urge that the reader go over the chapter on The Tools of the Chemist, in Volume 4. It provides information on chemical apparatus, on the measurement and handling of solid and liquid chemicals, on safety precautions and on other important matters.*

Weigh exactly 50 ml (milliliters) of ethyl alcohol and 50 ml of water. Mix the alcohol and the water and measure both the weight and volume of the mixture. You may be surprised to find that though the weight is equal to the combined weights of the ethyl alcohol and water, the volume is less than 100 ml (that is, 50 ml + 50 ml). How can we account for this curious fact?

Let us compare this situation with an apparently similar one in which we would mix a bushel of marbles and a bushel of sand. The weight of the mixture would be equal to the combined weights of the bushel of marbles and the bushel of sand; but the volume would be less than two bushels. It is clear why this is so. The individual marbles and the individual grains of sand are not of the same size; hence the sand grains can fill the vacant spaces between the mar-

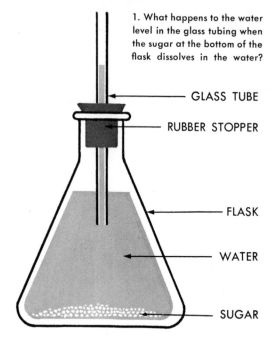

1. What happens to the water level in the glass tubing when the sugar at the bottom of the flask dissolves in the water?

GLASS TUBE

RUBBER STOPPER

FLASK

WATER

SUGAR

bles. That is why the combined volume is less than two bushels. If we bear this example in mind, and if we assume that alcohol and water are made up of particles (molecules), we have a possible explanation of the way they behave when they are mixed. It should be pointed out here that we have not *proved* that alcohol and water are made up of particles. We have assumed that such is the case (and a useful assumption it is!) in order to explain our observation of the alcohol-water mixture.

Consider now the apparatus shown in Figure 1. It consists of a flask, a one-hole rubber stopper and a piece of glass tubing. (A test tube may be substituted for the flask.) The bottom of the flask should be covered with sugar to a depth of about 1/8 inch. Fill the flask with water. Insert a glass tube through the hole in the stopper; then put the latter in the flask. The stopper should be placed in the flask so that the water level is about half way up the tube. Mark the level of the water in the tube. Then at intervals compare this level with the original level. Does the volume of the combined sugar and water change as the sugar dissolves in the water? Is this situation similar to the one in which alcohol and water dissolved in each other? It is important not to touch the flask or the

stopper once they are in place. It might bring about changes in temperature and pressure that might affect the level of the water.

The experiment can be repeated using table salt (sodium chloride), hypo (sodium thiosulfate), ammonium chloride, sodium acetate or any other substance that is soluble in water. Can we explain what happens in all these cases by assuming that we are dealing with small particles of different sizes?

There is reason to believe that in some solids molecules are linked together in a definite and repeated pattern to form the structures called crystals. A fascinating way to study crystalline patterns is to grow your own crystals. For this purpose, you will have to use glass jars that have been very carefully cleaned. You can begin with the compound called alum (potassium aluminum sulfate).

Prepare a supersaturated solution of alum and water,* using hot distilled water as the solvent. Allow the solution to stand until it is cool. Then pour the clear liquid into a clean jar and cover the latter with a cotton handkerchief or other kind of cotton cloth. As the solution stands in the jar, crystals will gradually form on the bottom. Wait until some of these crystals grow to about 1/4 inch in diameter.

Editor's note: In a supersaturated solution, there is an excess of the dissolved substance (the solute) in the dissolving substance (the solvent).

2. Growing a crystal. It is suspended from a matchstick set across the top of a glass jar. See the text.

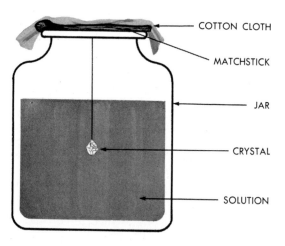

COTTON CLOTH

MATCHSTICK

JAR

CRYSTAL

SOLUTION

Provide yourself with a second clean jar, and set a matchstick across the top of the jar, as in Figure 2. Select a crystal from those on the bottom of the first jar and attach the crystal to the matchstick by means of thread. Next, pour the solution from the first jar into the second one so that the suspended crystal is well below the surface. Cover the jar with a piece of cotton cloth so as to keep out dust and other impurities. Allow the suspended crystal to remain undisturbed until it grows to the desired size. If you wish to preserve the crystal after it has been removed from the solution, place it in mineral oil or in carbon tetrachloride.

Many different kinds of crystals can be produced in this way, though it may be difficult to grow some to a large size. You might try to grow crystals of copper sulfate, potassium ferricyanide, sodium nitrate, potassium sodium tartrate (Rochelle salt), potassium chromium sulfate (chrome alum) and nickel sulfate. Some of these compounds are poisonous. To be on the safe side, do not handle the crystals with your hands. Always wash your hands immediately if you come in contact with the solution or the crystal.

You are all familiar with the crystals of sodium chloride (table salt). Try to grow a sodium-chloride crystal. Some interesting phenomena will occur in this case. Unfortunately, it is very difficult to grow large crystals of sodium chloride.

After you have grown several crystals of one kind, examine them closely. Are the angles between the flat sides (planes) the same? Do not let the size of the surfaces deceive you. Using a single-edge razor blade and a small hammer, attempt to split the crystals. (You may use the handle of a screwdriver instead of the hammer.) Are there certain angles at which smooth, flat planes are produced? How do the angles between these planes compare with the angles between the planes of the original crystal?

You will find useful information about crystals in the article called Symmetry Unlimited in Volume 5 of THE BOOK OF POPULAR SCIENCE. The theories of crys-

talline structure are discussed in CRYSTALS AND CRYSTAL GROWING, by Alan Holden and Phylis Singer, published by Doubleday and Company, Inc., New York City, 1960. This book contains many ideas for experiments with crystals.

Molecules in motion

We have examined evidence supporting the notion that matter is made up of small particles, which in some cases are linked together to form crystals. We shall now do some more experiments that will lead us to further understanding of these particles. We shall begin with an examination of the assumption that they are constantly in motion until the temperature falls to a point, called absolute zero, at which all motion stops.* This is in accordance with what is called the kinetic-molecular theory of matter (see Index).

In the next experiment, you are to use an evacuated glass tube containing mercury and small particles of colored glass floating on the mercury (Figure 3). Your science teacher may have the device in question. Heat the mercury gently by holding the device with a test-tube holder over a candle flame. An interesting thing will happen. Can you explain it by assuming that mercury is made up of small particles which

* Editor's note: Absolute zero is −273.16° C. or −459.69° F.

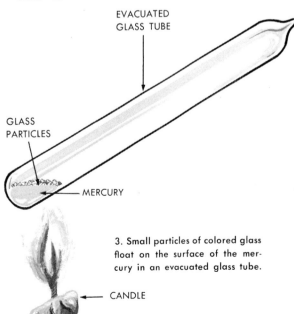

EVACUATED GLASS TUBE

GLASS PARTICLES

MERCURY

3. Small particles of colored glass float on the surface of the mercury in an evacuated glass tube.

CANDLE

evaporate as the mercury is heated and that these particles move about rapidly?

The perpetual dance of particles is called the Brownian movement, after Robert Brown, a Scottish scientist, who called attention to it in 1828. The Brownian movement can be observed by means of an apparatus that is found in most school science laboratories. Figure 4 shows how it operates. The basic element of the apparatus is a box, called a cell, with a glass window at the top. There is a lens on one side of the cell; a rubber bulb fitted into an opening in the cell on another side; and a tubular opening on still another side. The inside of the cell is illuminated by a strong light passing through the lens at the side of the cell. (A slide projector can be used to supply the light.) The interior of the cell can be viewed through a microscope whose objective lens is brought close to the glass window at the top of the cell.

First, the rubber bulb is squeezed and is held in the squeezed position. A lighted match is then put near the tubular opening of the cell. The match is blown out and at the same time the pressure on the rubber bulb is removed. As a result, smoke from the match is drawn into the cell. The microscope is then adjusted so as to bring the smoke particles into view. They will be seen to be in violent motion. Remember that you are not seeing individual molecules colliding with one another. Such molecules would be far smaller than the smoke particles and would be invisible under your microscope. We assume that what you are

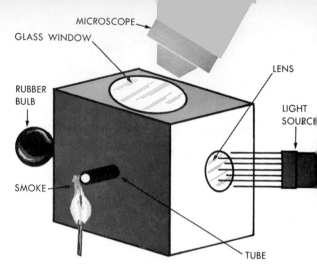

4. With the apparatus illustrated in the drawing you can observe Brownian motion. It is explained in the text.

observing is a series of collisions between the visible smoke particles and the invisible molecules of the air. The motion of the smoke particles may be attributed to an uneven distribution of collisions with air molecules.

The motions of gas molecules can be demonstrated strikingly by means of another piece of apparatus. Provide yourself with a glass tube about 2 feet long and ½ inch to 1 inch in diameter. Attach a small wad of cotton to each of two rubber stoppers that are to fit securely in each end of the tube, as shown in Figure 5. Before putting the stoppers in place, moisten one cotton wad with concentrated ammonium hydroxide (NH_4OH). This will provide a source of ammonia (NH_3) gas, since ammonium hydroxide is a solution of ammonia in water. Moisten the other wad with concentrated hydrochloric acid (HCl). This will provide a source of hydrogen chloride

5. Cotton wads have been attached to the stoppers at either end of the glass tube. One wad has been moistened with ammonium hydroxide, serving as a source of ammonia gas; the other wad has been moistened with hydrochloric acid, serving as a source of hydrogen-chloride gas. What happens when ammonia molecules and hydrogen-chloride molecules meet in the tube?

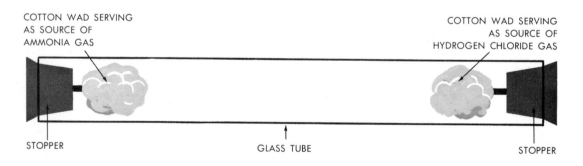

COTTON WAD SERVING AS SOURCE OF AMMONIA GAS

COTTON WAD SERVING AS SOURCE OF HYDROGEN CHLORIDE GAS

STOPPER

GLASS TUBE

STOPPER

(HCl) gas, since hydrochloric acid is a solution of hydrogen chloride in water.

Insert the two stoppers in the ends of the glass tube. When ammonia and hydrogen chloride gases come in contact, they form a white solid — ammonium chloride (NH_4Cl). The formation of this solid will be indicated when a white ring appears in the tube. Observe carefully the first appearance of the ring and its position. In order to explain the formation of the ring, we can assume that molecules of ammonia and hydrogen chloride moved from the cotton wads toward the center of the tube. Here again is evidence supporting the notion that molecules move.

Did the white ring first appear in the middle of the tube? If not, then ammonia and hydrogen chloride molecules do not move with the same speed. Which ones have the higher average speed?

The molecules of gases can exert pressure in some unusual situations. Let us examine one of these; it is illustrated in Figure 6. Obtain an unglazed or porous cup from a science laboratory, and fit it with a one-hole stopper. Insert a short glass tube in the stopper and connect a rubber tubing to this glass tube. A U-shaped glass tube (a manometer) is then attached to the other end of the rubber tubing, as shown in Figure 6, and is partly filled with colored water. The manometer will indicate any change of pressure in the porous cup. Suppose that the gas pressure in the cup increases. In that case, the water level in the farther arm of the manometer will rise and

the level in the nearer arm will fall. If the pressure in the porous cup decreases, the level in the nearer arm will rise and that in the farther arm will fall.

Support a beaker with the open end down and with the porous cup inside it, as indicated. Prepare some hydrogen in a gas generator of the type described and illustrated in Volume 4, pages 349–50. Hydrochloric acid and zinc are made to react in this generator; hydrogen gas is formed. Introduce the hydrogen gas in the beaker. Note the behavior of the water levels in the manometer.

Next, place the beaker upright and invert the porous cup in the beaker. Substitute limestone (calcium carbonate) for zinc in the generator. The reaction of hydrochloric acid with calcium carbonate will produce carbon dioxide gas (CO_2). What happens to the water level in the manometer when carbon dioxide is introduced into the beaker in this way?

You may want to use other gases in the same apparatus and to observe their behavior. Can we explain this behavior by assuming that molecules of different gases move at different speeds? Do hydrogen gas molecules move (diffuse) into the cup faster than air molecules move out of the cup?

If there is evidence that gas molecules move, it seems reasonable to assume that molecules composing a liquid are also in motion. Let us see if we can back up this assumption. Place some water in a large bottle or beaker. Add several drops of colored vegetable dye (or any colored sub-

6. An interesting experiment demonstrating the movement of gas molecules.

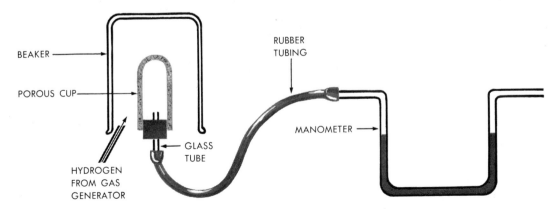

BEAKER

RUBBER
TUBING

POROUS CUP

GLASS
TUBE

MANOMETER

HYDROGEN
FROM GAS
GENERATOR

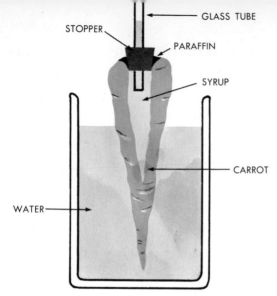

GLASS TUBE

STOPPER

PARAFFIN

SYRUP

CARROT

WATER

7. This apparatus was designed to demonstrate that the molecules that make up a liquid move. See the text.

stance in solution) to the undisturbed water. Observe the water for some time. Does the dye spread throughout the water?

Assemble the apparatus shown in Figure 7. Carve out a hole in one end of a carrot, being careful not to split the sides of the carrot. The hole should be just large enough at the top so that it can be effectively closed with a one-hole rubber stopper. Before the stopper is put in place, the hole in the carrot is filled with syrup or concentrated sugar solution. A long glass tube is inserted into the rubber stopper (see Figure 7) and the stopper is then inserted in the carrot hole. Place melted paraffin around the top of the carrot in order to keep the stopper firmly in place. Now put the carrot in a beaker of water, as shown in Figure 7, and allow the water to stand for several hours. Is there any similarity between the situation that develops and the one in which hydrogen diffused into a porous cup?

The size of
molecules

It is certainly impossible at the present time to measure directly the size of molecules. Yet, by noting the behavior of certain substances, we can determine the approximate size of molecules with a fair degree of accuracy. In the following experiment, illustrated in Figure 8, we shall use oleic acid ($C_{18}H_{34}O_2$). This acid spreads out in a very thin layer when placed on

water; if a very small quantity is added to the surface of water in a container, the acid will form a film one molecule thick.

Make a 0.2 per cent solution of oleic acid in alcohol. Prepare a fairly large flat tray by cleaning it thoroughly. Pour distilled water into the tray to a depth of ¼ inch, and allow it to stand quietly for several minutes. It is important that the water be motionless. Hold two blackboard erasers above the tray of water and clap the two together, thus depositing a thin layer of chalk dust on the surface of the water. (Talcum powder may be substituted for chalk dust.)

Using a burette or medicine dropper, permit one drop of the oleic-acid solution to fall upon the surface of the water in the tray. Support the burette near the water in order to avoid a splash. The alcohol in the solution soon evaporates or dissolves in the water, leaving a thin circular film of oleic acid on the surface, as shown by the chalk dust pushed aside. Measure the diameter of the circle and calculate the radius. (The radius is half the diameter.) It will probably be necessary to make an estimate of the diameter because it is highly unlikely that a perfect circle will result when the drop of oleic acid is added to the water.

Now count the number of drops of oleic-acid solution in one cubic centimeter released from the burette. This will enable

8. How to measure a single molecule, as explained in the text. The molecule in question is that of oleic acid.

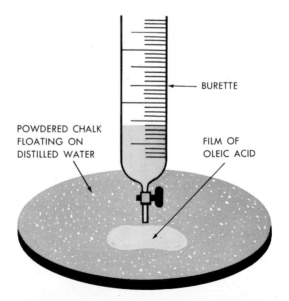

BURETTE

POWDERED CHALK
FLOATING ON
DISTILLED WATER

FILM OF
OLEIC ACID

you to determine the volume of one drop of the solution; simply divide one cubic centimeter by the number of drops. Calculate the volume of oleic acid contained in one drop, remembering that in the solution there were 0.2 parts of oleic acid to 99.8 parts of alcohol. The volume of oleic acid in one drop will represent the volume occupied by the film of oleic acid on the surface of the water. The thickness of the film can be calculated by employing the following relationship:

$$\text{Thickness} = \frac{\text{Volume of film}}{\pi\,(\text{Radius of film})^2}$$

If the layer of oleic acid is one molecule thick, then the thickness just calculated should be the length of one molecule. How many molecules lined up end to end would make a total length of an inch? How many oleic-acid molecules would there be in one drop? If our experimental procedures and the necessary assumptions are correct, we can get a clear idea of the very small size of a molecule.

Attraction
between molecules

Fill a drinking glass with water until the surface appears level with the top of the glass. Set paper clips, one after the other, very gently on the surface of the water and see what happens. What enables the water to bulge above the level of the glass without overflowing? A reasonable explanation would be that the surface of the water remains unbroken because the molecules forming the surface are held together by their mutual attraction.

It also seems reasonable to assume that while molecules may attract one another, the degree of attraction will vary according to the molecules. Some idea of the relative attraction between molecules may be obtained by a relatively simple experiment. Obtain or prepare two glass tubes, approximately 40 inches long and ¼ to ½ inch in diameter. Each of the tubes is to be closed at one end. Fill each tube about ¾ full with water. Cover up the open end with your finger and place the tube, open end down,

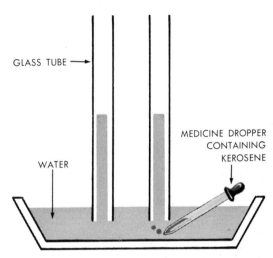

9. Showing that the degree of attraction between molecules will vary according to the different molecules.

in a tank of water, as shown in Figure 9. Use burette clamps to hold the tubes firmly in place.

Next, measure the level of the liquid in each tube. Partly fill a medicine dropper with kerosene. Squeeze the bulb of the dropper gently, forcing one drop (and only one drop!) under the first tube. Did the level of the water in the tube change? If so, measure the change. Continue adding kerosene, drop by drop, until no further change in the water level is noted. Compare this level with the level of the water before kerosene was introduced. Repeat the general procedure in the other tube; but this time add drops of gasoline, ether or benzene. In which tube did the water level change the most? Can you explain the results of this experiment? Is it reasonable to assume that the less the attraction the liquid molecules at the base of the tubes have for each other, the more the water level in the tubes should drop? If this is so, the relative attraction between molecules of any two liquids could be determined. You will find that the results will be far clearer if you use mercury instead of water in this particular experiment.

Record the time it takes for a single drop of each of the liquids used in the experiment to evaporate. Is there any relationship between the rate of evaporation and the amount the water level in the tubes falls?

A factor that is believed to affect strongly the attraction between molecules is the unequal distribution of electrons in certain molecules. The unsymmetrical distribution causes one side of such a molecule to have a partial positive charge and the other side to have a partial negative charge. A molecule of this type is called a polar molecule. Polar molecules have unusually high boiling points and other properties characteristic of substances in which strong attraction exists between the molecules of the substances.

The molecules of water are polar, as can be demonstrated by the following experiment. Allow water to flow in a very small stream from a faucet. The flow from a burette would be even better, since it can be easily regulated. Rub a comb or other hard rubber object against wool, causing the comb to acquire a charge of static electricity. Bring the comb close to the stream of water (Figure 10). The water will then be attracted to the comb. If you have used a burette containing water for the experiment, empty out the water. Then fill it in turn with carbon tetrachloride, alcohol, ether and other liquids in place of water. Will these liquids show the same behavior when the charged comb is brought close to them? You will find that the results may surprise you.

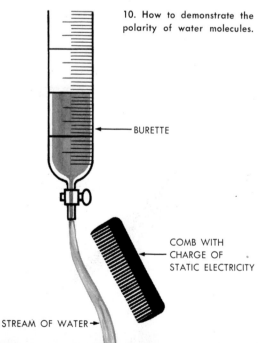

10. How to demonstrate the polarity of water molecules.

BURETTE

COMB WITH CHARGE OF STATIC ELECTRICITY

STREAM OF WATER →

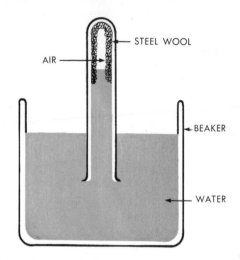

STEEL WOOL

AIR

BEAKER

WATER

11. As the oxygen in the air above the water in the test tube unites chemically with the iron in the steel wool, what happens to the water level in the test tube? Why?

Separation of molecules

The attraction between molecules varies with different substances. As a result, these substances have different boiling temperatures, vapor pressures, surface tensions, rates of evaporation and chemical properties. It is possible to use our knowledge of such differences to separate mixtures of molecules.

We can, for example, separate oxygen from the other gases contained in air * by making use of one of its chemical properties — its ability to combine chemically with iron. Place some moistened steel wool in a test tube and invert the tube in a beaker of water. Suspend the tube with a burette clamp in such a way that the opening of the tube is about one inch below the surface of the water (Figure 11). Mark the level of the water in the test tube. The gases of which air consists will be imprisoned above the water in the tube. Allow the tube to stand for 24 hours; then note the level of the water. The oxygen in the air above the water has been combining chemically with the steel wool; how has this affected the level of the water?

Alcohol and water are different substances; it is reasonable to assume, therefore, that they can be separated from each other. Add 10 ml of ethyl alcohol to 10 ml of water in a 125 ml flask. The flask should

* *Editor's note:* Air is a mixture of nitrogen, oxygen, argon, carbon dioxide, water vapor and other gases.

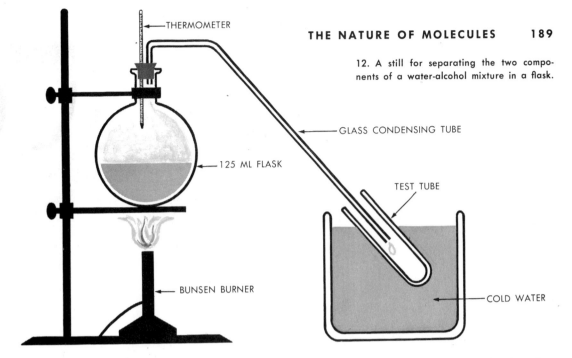

THERMOMETER

12. A still for separating the two components of a water-alcohol mixture in a flask.

GLASS CONDENSING TUBE

125 ML FLASK

TEST TUBE

BUNSEN BURNER

COLD WATER

be attached to a ring stand by means of a burette clamp, as shown in Figure 12. Stopper the flask with a two-hole stopper containing a thermometer and a glass condensation tube (Figure 12). The condensation tube is to lead to a test tube, which is to be immersed, as shown, in a beaker of cold water. You have prepared a still (see Index). Heat the flask with a Bunsen burner. Collect 5 ml of liquid in the test tube. What is this liquid; why did it boil off first? Continue collecting 5-ml samples until almost all the mixture has boiled off. Observe the temperature changes; note the difference in the samples collected. We are able to

separate water and alcohol in this situation because they have different boiling points. Neither the alcohol nor the water has been changed chemically in the process. (In the preceding experiment, the oxygen *was* changed chemically, since it combined with the iron in the steel wool, forming iron oxide.)

Now place some wood splinters in a test tube and connect the latter to a condensing tube, as shown in Figure 13. Heat the wood strongly. Liquid will collect in the condensation tube. See if you can separate different substances in the liquid mixture, using the procedure followed in separating alcohol and water.

13. Wood splinters are put in a test tube, and this is connected to a condensing tube. The wood is heated and liquid will collect in the condensing tube. Will this liquid be a mixture? How can you demonstrate this?

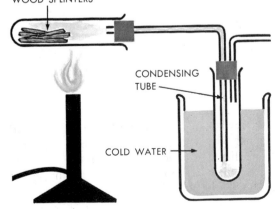

WOOD SPLINTERS

CONDENSING TUBE

COLD WATER

Summary. The situations that we have examined in this article can all be explained by assuming (1) that the substances with which we experimented were made up of molecules; (2) that molecules are in motion; (3) that molecules attract one another. We have not actually proved that molecules exist, but we have shown by means of our simple experiments that there is much evidence to support this belief. This is the general method used by scientists to substantiate their theories. Theories arise as a result of experimentation or observation; then further experimentation or observation is required to show whether or not the theories are tenable.

IRRIGATION AT WORK

How Man Brings Water to His Farm Lands

BY ELIZABETH RUBIN

AGES ago, man learned that if he is to have an adequate food supply, he must find a way to make crops flourish not only where rain is plentiful but also where it is insufficient for plant growth. By bringing water to thirsty fields in canals or ditches or by flooding, he has greatly increased the productivity of his lands and has multiplied the population that they can support. Such artificial watering of farm lands is called irrigation.

Hundreds of millions of acres throughout the world need some form of artificial watering. As a rule, if a region receives less than ten inches of rain a year, it will not produce crops of any kind without irrigation. Where there is between ten and twenty inches per year, some plants can be grown without irrigation, but the yield is generally low and uncertain. To get a dependable harvest, the farmer in these areas must supplement rainfall with some artificial watering method or other.

In certain areas, the total rainfall per year is apparently adequate; yet not enough falls during the growing season. The farmer cannot always manage his crop planting to take advantage of the seasonal rainfall distribution. For instance, in southern California, Italy and Israel, most of the rain comes during the winter months, while the summer is dry. In India, too, nearly the whole year's supply of rain pours down during the rainy season, called the monsoon. Unless irrigation is practiced in these areas, crops will wither and die in the fields during the long, dry growing season. Starvation may be the

only reward the anxious farmer gets for his labors.

Even in humid areas, periods of drought may occur. If water is stored and supplied as desired, the farmer can provide his fields with just the right amount at the most favorable time. Besides, a much greater variety of crops can be grown in nearly any area with irrigation than without it.

Different soils have different water-holding capacities. Rainfall that would suffice in one area would be quite insufficient in another. Sandy soils, for example, often do not retain water well; it may percolate too deeply to be reached by the root tips. On the other hand, close-packed clay soil may not absorb a sufficient amount of moisture before the water runs off.

In the case of certain crops such as rice, a great quantity of water is needed. The land must be flooded to a depth of several inches during most of the growing season. For such crops, artificial watering is often required.

Irrigation has been practiced for thousands of years, but only within the last century or so has man applied scientific knowledge to the watering of arid land. Increasingly, population pressures have made it necessary to cultivate dry areas, and so irrigation has become more and more imperative. We must remember that land surfaces are limited in extent, since three-fourths of the world's area is covered by water.

In many parts of the world, drainage goes hand in hand with irrigation. Some

Bureau of Reclamation—U. S. Dept. of the Interior

Looking along the All-American Canal toward the Imperial Dam. The canal irrigates a vast expanse of irrigable land in Southern California. Water for the canal is provided by the Hoover Dam, which stores up the flow of the Colorado River. The Coachella branch of the canal turns out at the left. U.S. Highway 80 crosses the canal in the foreground.

method of draining excess irrigation water must always be provided. Otherwise the soil becomes waterlogged — that is, both large and small air spaces in the ground will be clogged with water.

It is fascinating to study the progress of irrigation through the ages. The first civilizations developed in regions where the people could grow a dependable food supply with the aid of irrigation. In these areas — the Nile River valley, Mesopotamia and certain parts of Persia, India and China — great rivers annually flooded the lands through which they flowed. These inundations provided a natural system of irrigation which early farmers partially controlled. Through irrigation, a greater amount of food could be produced by fewer men's efforts.

Ancient Egypt was a pioneer in the artificial watering of the land. Her paintings and sculpture show that by 3000 B.C. the Egyptians were engaged in large-scale, highly skilled irrigation work. The nearly rainless country is cleft by the Nile River. In ancient times, the desert came down close to the marshes that edged the river. The people filled the marshes and built mud walls to keep out flood water. Then they dug canals that cut across miles of land to lead the river's waters to the fields. When flood waters were dangerously high, they were diverted and stored, thus accomplishing both irrigation and flood control.

Mesopotamia, the country between the Tigris and Euphrates Rivers, is almost rainless, like the Nile country. Yet this area, too, was famous in ancient times for its rich crops and advanced civilization. The Sumerians, early inhabitants in that region, built embankments to control the flood waters of the Euphrates. They drained the marshes bordering it and dug irrigation canals and ditches.

The Sumerians were conquered by the Babylonians, who further developed the irrigation system. Since the Euphrates did not flow uniformly, the early Babylonian engineers built dams to store flood waters in reservoirs, thus obtaining a dependable water supply. This was accomplished by 3500 B.C. These huge reservoirs covered 650 square miles and were 25 feet deep when full. They had the disadvantage of retaining all the silt contained in the water stored in them. As a result, they eventually became filled up. This problem was not solved until comparatively modern times.

Other ancient lands applied irrigation methods. The Chinese, whose civilization goes back several thousand years before the birth of Christ, constructed extensive works to provide artificial watering. Ancient stone causeways used for this purpose in the region of the Hwang Ho River still exist. The vast and heavily populated lands of India, dependent on a very fickle rainfall, early developed irrigation methods to serve agriculture. Wells, reservoirs and canals were in use before 300 B.C. The Romans also were skilled in applying irrigation methods.

Prehistoric cliff dwellers in Mexico and the southwestern United States built check dams * to provide water for their fields and ensure good crops. These early planters developed a higher civilization than did the hunting Indians. Remains of the Inca civilization in the high mountains of Peru show terraced fields, some with more than fifty "steps" climbing the mountainsides. Stone causeways several miles long wind among the terraces and show that irrigation was highly developed.

In ancient times, several mechanical devices served for irrigation purposes — devices still used in some parts of the world. The shadoof was probably first employed in Egypt. It consisted of a long tapering

* Check dams are barriers placed across a stream to hold back the flowing water and make the level higher. This makes it easier to draw off the water for irrigation purposes.

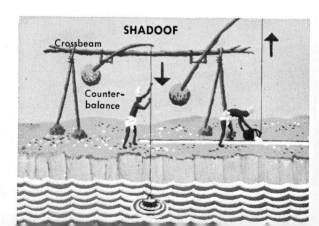

SHADOOF
Crossbeam
Counter-balance

pole supported on a crossbeam. From the thin end a bucket on a rope was hung; the thick end carried a counterbalance. The worker pulled down on the rope to fill the bucket. The counterbalance then dropped and raised the water so that it could be emptied into a trough leading to the field.

The sakia, or Persian wheel, used animal power. An ox or donkey was hitched to the end of a long lever, which was attached at its other end to the hub of a horizontal wheel. The lever formed the radius of a circle in which the animal was forced to walk. As the animal moved around, of course, the horizontal wheel turned also. It forced another wheel, vertically placed and to which it was geared, to move around. The vertical wheel was connected by means of an axle to a second vertical wheel, which was partly submerged in a well. This wheel had buckets lashed to it; as it turned, the buckets would first scoop up water from the well and then empty it into a channel leading to the fields.

In the third century B.C., Archimedes of Syracuse invented the device known as Archimedes' screw. It consisted of a bent tube spiraling around one end of an axis. This apparatus was immersed in water in a slantwise direction, one end rising to the riverbank. When the device was rotated, water rose up through the internal screw and flowed onto the soil.

When the Roman Empire succumbed to surges of barbarians from Northern Europe, the highly developed Roman irrigation works were allowed to fall into disrepair. Yet certain features of the Roman methods were retained and were incorporated in later systems. In England, water-

meadow irrigation, for example, goes back to the Roman occupation. In this method, the fields are left flooded during the winter and thus the soil is prevented from freezing. In the spring the fields are drained and early planting is possible.

The feudal system of medieval times did not favor the spread of irrigation methods. The division of the fields into serfs', lord's and fallow strips interfered with any but the smallest scale irrigation plans. In all Europe during this period, only monks connected with certain large abbey lands practiced effective irrigation. They used the records they had kept of Roman systems.

When the Mohammedans swept over the lands along the Mediterranean, they introduced more skillful irrigation in all the countries they conquered. The Crusades brought Europeans in contact with these methods. With the breakdown of feudalism, artificial watering was applied increasingly throughout Western Europe. So few peasants remained after the plagues of the fourteenth and fifteenth centuries that larger-scale irrigation systems, which made more efficient use of manpower, had to be put into practice.

The age of steam marked a new era in irrigation progress. Pumping engines driven by steam made much more water available for the fields. In the eighteenth and nineteenth centuries, several huge irrigation projects were launched. Among these was the Aswan Dam, on the Nile River. Work on this structure, among the largest in the world even today, was begun in the middle of the nineteenth century, though it was not completed until the early twentieth century.

ARCHIMEDES' SCREW

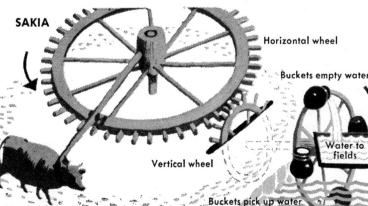

SAKIA

Horizontal wheel

Buckets empty water

Water to fields

Vertical wheel

Buckets pick up water

Snow

Rain

Dam

Tunnel and pipeline

Canal

THE SOURCES OF IRRIGATION WATER

The sources of irrigation water

The source of most irrigation water is the rain and snow that fall in the uplands. Some of this precipitation makes its way to surface streams; some of it seeps through the ground, forming what is called ground water (see Index). In either case, it can be made available for irrigation purposes. Of course, ground water will have to be pumped up to the surface before it can be put to use.

The heaviest flows from melting snow usually occur in spring or early summer. Therefore, storage works must be developed to control flooding and to retain the excess water for use during the time of low stream flow. To get a dependable water supply, engineers or farmers build dams to impound flood water in storage basins. There are sluices, or gateways, in the dam; these can be closed sufficiently to raise the water level until it flows into the storage basin as desired. The stored water can then be led through canals to the fields.

In large-scale irrigation by means of dams and canals, water is led from rivers and streams to farms that may be hundreds of miles away. Gigantic and complex systems of dams are built to hold back water in huge man-made reservoirs. These dams furnish billions of cubic feet of water every year for irrigation. Since they store water during times of peak water level in rivers, they help prevent floods. Many of them serve other purposes, too, such as storing up water for human consumption and industry, and generating electricity. Some of them maintain a minimum depth of water for ships that ply the rivers. (See the article on dams, in Volume 8.)

As the dams raise the level of the water in the river, the water may be diverted to reservoirs, lakes or huge tanks where it is stored. From here, canals convey it to the point where it is needed. For irrigation, it is distributed by smaller canals and ditches to the fields.

Sometimes aqueducts are used instead of canals to carry the water from the dam to the point where it is needed. It may be necessary, in other instances, to tunnel through a mountain in order to get the water to the desired place. The longest

irrigation tunnel in the world is the United States' thirteen-mile-long Alva B. Adams Tunnel, which was driven through the Rockies under the Continental Divide.

Much irrigation water is pumped from wells that tap the natural underground reservoirs that we call ground water. Various types of pumping systems are used to bring the water to the land. These vary from one or two horsepower engines to huge plants whose power is measured in thousands of horsepower. The pumps may raise the water from depths of several hundred feet. Ground water is not usually carried for long distances, but is put to use for the most part on the farms or districts where it is pumped up. One-third of all the irrigated land in the United States receives water from underground sources.

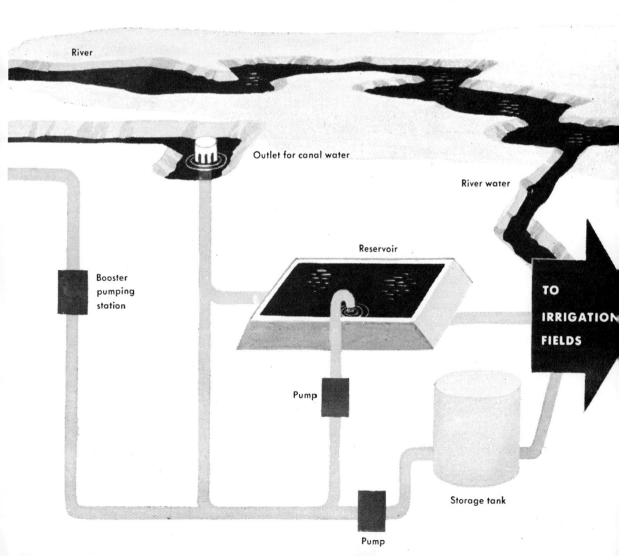

BASIN

Methods used to irrigate the land

The three most common methods in the United States and Canada for applying water to individual farms are surface irrigation, sprinkler or overhead irrigation and subsurface irrigation.

Surface irrigation. In one form of surface irrigation — flood irrigation — the water completely covers the surface of the land. The slope of the fields must be carefully considered when this method is used, in order to prevent the soil from being washed away or eroded. Flooding methods are best used where water is plentiful.

There are several types of flood irrigation. In "basin irrigation," water is quickly ponded on a flat area surrounded by little dikes. A large quantity of water is used, and it continues to percolate into the soil for some time after the supply is shut off. This method is especially suited for orchards, where several trees may be included in one basin.

Flooding from "contour ditches" is used on rather steeply sloping or hilly lands. The water is allowed to flow as a sheet downslope between closely spaced field ditches which are dug across the slope, following the contour of the land. The ditches distribute the water through frequent openings in their embankments; each captures the runoff from the ditch above and redistributes it.

Border-strip irrigation is usually applied to land having a regular slope in one direction. First the fields are leveled to a uniform grade. Then low earth ridges are formed between parallel strips of land. The strips usually run down the slope. A large quantity of water is introduced at the head of each border strip and a sheet of water, confined by ridges, advances down the slope, entering the soil as it goes. This method is recommended for close-growing crops that are not damaged by temporary flooding, such as alfalfa and pasture grasses.

If the incline is great, flooding cannot be used unless the land is planted to dense cover crops. On the steepest slopes, sprinkler irrigation must be applied to start the crop; then careful flooding can be used after the young plants have attained a certain size. Checks may be provided on sloping land. These are small low ridges cut across the slope to hold the water in the plots in between the ridges. A very small stream of water is introduced to prevent erosion. The size of the plot in all the flood methods is determined by the size of the water stream available and the rate at which the particular soil soaks up water.

Crops are sometimes grown on terraces. These are long, level, raised platforms of earth which cut across the slope of land with a steep incline. The face of each terrace is supported by a wall or by turfing. The terraces are placed one above the other.

Another important type of surface irrigation is furrow irrigation. It is especially suited to row crops such as cotton, corn and sugar beets. It is chosen when the stream of water to be used is small, and when land surfaces are uneven. It also serves where the land is steeply sloping. The water is derived from a main ditch or canal leading from the source of supply. Smaller ditches, called laterals, leave the main ditch at regu-

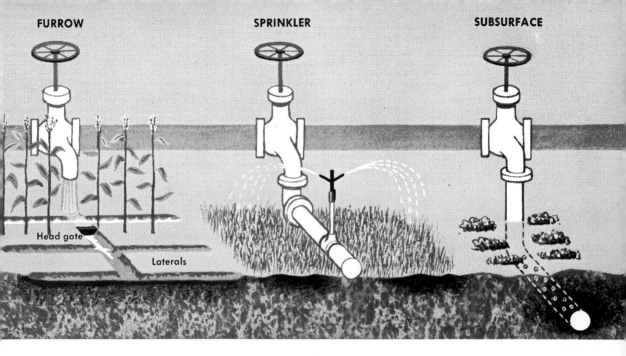

FURROW   SPRINKLER   SUBSURFACE

Head gate

Laterals

lar places and lead the water to the fields, where water is run into the furrows. Special gates — head gates — regulate the flow into the laterals. In the farm ditch, canvas checks, or dams, raise the level of the water to divert and spread it through the furrows. The water soaks into the soil and spreads out into the crop root areas between furrows.

The different types of surface irrigation may be 50 to 90 per cent efficient, depending on the soil type and the crop grown, as well as on the nature of the land.

Sprinkler, or overhead, irrigation. In this method, water is sprayed into the air, falling on the land and crop in minute drops. It is adapted to most soils and all major crops (except those which must be flooded, such as rice), and it provides excellent control of the water. Sprinkler irrigation can be used with less preliminary land leveling than any other artificial watering method. Light applications for new seedlings and small plants can be made easily. Farming operations are not hindered by field irrigation ditches and dikes, as in other systems. Sprinkler irrigation is especially suitable to permeable soils that erode easily under other methods of irrigation. Small constantly flowing streams are the best source of water for this method.

Sprinkling usually requires the greatest initial investment and labor requirements may be high. Light, portable aluminum pipe with special coupling is employed. This permits quick dismantling for use in another field. A circular sprinkling apparatus, similar in design to the common lawn sprinkler, has been developed in some areas.

Subsurface irrigation. In this method, the farmer must create an artificial water table beneath the surface of the land, within reach of the plant roots. He accomplishes this by either of two methods. In some areas, water is turned into widely spaced supply ditches, and soil conditions favor a rapid lateral flow of water through the soil. In other places, it is necessary to apply water beneath the surface of the land through various kinds of pipes and conduits. The roots of the plants get moisture by the upward capillary movement of the water. (See Index, under Capillarity.) When using these subsurface methods, the depth at which the water is maintained must be controlled carefully; otherwise root growth may be retarded or stopped. The soil must be highly permeable for satisfactory subsurface irrigation, and only water containing low percentages of mineral salts should be used.

The subsurface system can often be put into effect on soils that are difficult or expensive to irrigate by other methods. It can be applied to uneven land and does not interfere with tillage practices. Installation costs on subsurface projects are usually low; so are labor requirements.

Crops such as strawberries, which are planted on raised beds, are irrigated most effectively by this method. Water is run in ditches between and below the beds. It soaks into their sides and reaches the roots from underneath. The top of the bed remains dry; this reduces the spoiling of the fruit by mildew.

Problems to be considered in attempting irrigation

Many problems confront the farmer after he has made his decision to irrigate. First, of course, he must decide on the most suitable method of irrigation for his land. He may have to use more than one irrigation method. This is because different crops require different irrigation systems, as we have seen, and the land is seldom planted to a single crop year after year. For example, rotation from crops that grow in rows to crops that "cover" the land is common on today's farms.

No matter what system is used, the farmer must know something about how much water is needed to grow the desired crop, how much his soil can store and how much of the stored water can be used by the crop before it is necessary to irrigate again. Next, he must prepare the land by leveling, digging ditches, installing connections for sprinklers and so on. The nature of the land influenced the choice of the irrigation method much more in the past than now. New earth-moving equipment makes it possible to modify the land surface to almost any condition desired.

Of course, the water supply is important in choosing a system of irrigation. It should be kept in mind that more water must be available than will actually be used by the crops, since no method is perfectly efficient. Where irrigation water is short in supply, the farmer must plan his crop to utilize whatever rainfall is available after his irrigation source is dry. On the other hand, uncontrolled application of water may cause erosion. The very water that is so essential to crop growth can also ruin both the crops and the land. The farmer must know his soil. Some farm lands resist erosion, others do not. Land that has not

been properly leveled will erode most rapidly under irrigation.

Good drainage is required for top yields. Most crops will not live very long in standing water, especially during hot weather. Lands once waterlogged often will not become fully productive again. If the water table is high, excess salts may accumulate in the surface soil. Artificial drains will help provide the leaching needed to prevent the accumulation of salts; under extreme conditions, the water may have to be treated before it can be used.

The rate at which water infiltrates the soil is influenced by the size of the particles in the soil and by its depth. Water percolates rapidly in some soils, slowly in others. The proper length of the furrow or plot and the size of the stream for each furrow is determined by the percolation rate. To further complicate matters, the amount of water required by crops varies under different physical and climatic conditions. A farmer who is going to apply irrigation methods must consider all these problems.

Experience, of course, is the best guide in such matters. There are other ways, too, of eliminating guesswork in water management. The study of weather data is important; it makes possible short-range prediction of drought and rainy weather. Various government agencies supply useful information. In the United States, for example, the irrigation guides provided by the Department of Agriculture are exceedingly helpful.

Irrigation systems of modern countries

The United States, India, Pakistan, China and Russia lead in number of acres under irrigation. Together these countries have more than two-thirds of the world's artificially watered areas.

In the United States, there are at present 29,500,000 acres under irrigation. Nine-tenths of this is in the seventeen western states and the three southern states of Arkansas, Louisiana and Florida; but every state practices irrigation in some way.

The eastern part of the country has adequate rainfall, on the whole. The early

Bureau of Reclamation, U. S. Dept. of Interior

All-American Canal, looking upstream from Station 1180, in the Pilot Knob rock section in Arizona.

settlers who established colonies in this region generally did not have to apply irrigation methods. After the Americans won their independence, enterprising farmers and others made their way westward. In the course of time, they penetrated the arid and semi-arid lands of the West, where irrigation became a "must." The Mormons started the nation's first modern irrigation project in Utah in 1847, and this was followed by a number of others.

In 1902, the United States Congress passed the National Reclamation Act. It provided that the Federal government was to take over multipurpose projects, involving irrigation, water supply and the like, if these were too large and difficult for private enterprise. The Bureau of Reclamation of the Department of the Interior now has charge of all multipurpose projects except those located on Indian reservations. Those are handled by the Bureau of Indian Affairs.

The five biggest dams constructed since the passing of the Reclamation Act are the Grand Coulee in Washington, the Hoover in Arizona and Nevada, the Shasta and Friant in California and the Marshall Ford in Texas. They are all multipurpose and rank among the largest concrete structures in the world.

Hoover Dam, which stores up the waters of the Colorado River, is an excellent example of a multipurpose dam. It furnishes water for Los Angeles, power for southern California and irrigation for hundreds of thousands of acres of desert land. Some of the water it dams up flows through the 80-mile-long All-American Canal, which serves 200,000 acres of irrigable land in the Imperial Valley of southern California. Before the dam was built, the Colorado River flooded the farms of this valley and also the Yuma valley in Arizona when the mountain snows melted; it became a sluggish stream in the long dry summer. Now that the stream has been tamed, it has transformed vast areas of the Southwest into superb farmland and has provided a steady supply of water for other purposes too.

In Canada, irrigation is practiced chiefly in the prairie provinces where extensive farming operations are in progress.

Alberta, Canada: building the St. Mary's River Dam, creating a reservoir for irrigation purposes.

Editorial Associates

Deane Dickason, from Ewing Galloway

Egypt's Aswan Dam, on the Nile River, provides water for hundreds of thousands of acres of land.

About one million acres are irrigated. Many Canadian irrigation systems are small, and have been built by private individuals or companies. The government grants licenses to those who wish to develop such systems and in some instances has provided loans for this purpose. Two of the largest Canadian undertakings are in Alberta, where the waters of the Bow river are utilized to irrigate 100,000 acres of farm land. This is accomplished by means of the Horseshoe Bend Dam and a reservoir near Bassano.

India's agriculture is quite dependent on irrigation, for her rainfall is concentrated in a few weeks of the year. This country has practiced irrigation for thousands of years and has one of the world's largest irrigated acreages; yet much remains to be done. There have been scant harvests, crop failures, floods and famines in some part of India almost every year.

When India became a sovereign republic in 1950, irrigation became one of the top-priority projects of the new government. Two five-year plans have already put many millions of additional acres under irrigation. At present some 61,000,000 acres are artificially watered, and many new projects are being planned. Much of the work is done by private individuals; the government provides technological advice and, to a certain extent, financial assistance. Here, as in other Asian countries, centuries-old-traditions and primitive labor practices have begun to yield to modern methods.

Irrigation has been the basis of Egyptian agriculture for thousands of years, for rain is very scarce. At present 5,400,000 acres of cultivated land are under irrigation. The Nile River provides the sole water supply; its flow is carefully regulated. The government is now building the Sadd-El-Aali project on the Nile, south of the Aswan Dam. This project will store water to be used in the development of two million additional acres of agricultural land in Egypt. It will provide hydroelectric power and raw materials for nitrogenous fertilizer production; it will also give complete flood protection and improve navigation and drainage conditions.

Israel has a rainy winter and a dry summer. Irrigation is necessary, therefore; it is a major concern of the government. Irrigation increases the crop yield four to six times in northern and central Israel; in the arid region called the Negev, it can make the difference between no crop at all and two bumper crops a year. The Yarkon-Negev project is the largest irrigation undertaking carried out so far in Israel; it aims to make the Negev more productive and populous. Israel now has 275,000 acres under irrigation and is producing 65 per cent of the food her people consume. Her ultimate aim is to be entirely self-sufficient in this respect.

Today, as in the past, food is the most pressing problem in China. The weather is very unpredictable, and millions used to starve when droughts and floods ruined the crop. It seemed that there was always either too much or too little water. For many centuries the Chinese have sought to control floods and to irrigate the land by means of dikes and canals. China is now embarked on an intensive program of agricultural improvement and new irrigation projects, some of considerable size, are now under construction.

The total figure for lands under irrigation in China is now 142,550,000 acres. The new irrigation projects have been hindered by the countless grave mounds that dot the landscape and house the remains of generations of ancestors. In some places, these mounds take up 9 per cent of the cultivable area. It has been necessary for the government to move many of them to other localities to accommodate new projects. It is now illegal to build grave mounds; the dead must be cremated.

For thousands of years, Pakistan has been ravaged by floods, which have led to waterlogged, saline and barren land. Efforts to obtain a balance between too much and too little water are centuries old. In recent years efforts have been intensified and aided by modern engineering. The arid Sind province contains one of the largest projects in the world — the Lloyd Barrage system on the Indus. It has seven huge canals carrying water to 5,000,000 acres of farmland. 30,500,000 acres are now being irrigated in Pakistan and many new projects are under construction.

Nature has portioned out water with an erratic hand in the Soviet Union. For example, on the eastern Black Sea coast, yearly rainfall is almost 100 inches; in the lower reaches of the Amu-Darya River near the Aral Sea, it falls to a low 3 inches a year. Compare this with the range in the United States. Alabama, the rainiest state, receives 67 inches a year; Arizona, the driest, 7 inches.

In pre-Revolutionary Russia, irrigation efforts were on an insignificant scale for such a huge country. The method by which farmland was divided interfered with any large-scale plans. Peasant holdings were cut up into tiny individual tracts, which were further divided as each generation inherited its elders' property. Individual plots were separated by ridges which were not cultivated. Backward methods, resistance to change and stifling ancient customs hampered agricultural progress.

After the Revolution, centuries-old irrigation systems in these primitive regions were modernized and expanded. Private property in land was abolished. Many peasants were put to work on huge collective co-operative farms run on modern lines. The collective farms, having larger land areas, could plan suitable irrigation projects.

By 1940, the total area of irrigated lands was 17,500,000 acres, as against 10,000,000 acres in 1913. In 1953–54, a program was adopted to develop virgin and long-fallow lands, mainly in desert areas. Huge dams are being built to provide power and irrigation, especially in Kazakhstan and Western Siberia. Hundreds of thousands of acres of once-barren desert are becoming productive. One of the largest completed projects is in Central Asia — the 200-mile Grand Ferghana Canal. It carries water to 874,000 acres of land which had lain fallow for centuries.

The imposing Tilaiya Dam, in India's Damodar Valley.
Ewing Galloway

Constructing a pipeline to the Negev region of Israel.
Israel Office of Information

THE CATS

Lions, Leopards, Jaguars, Pumas and Lynxes

THE cats, such as the lions, leopards, jaguars, pumas, lynxes and cheetahs, are outstanding among the carnivores, or flesh-eating animals, for their unsurpassed combination of muscular power and grace of movement. These magnificent creatures generally catch their prey by clever stalking or ambush. They must be on the alert and tensed to make the kill; at the same time they must remain quiet enough not to frighten the intended victim. Sometimes they pursue game that they can neither smell nor see; in such cases, they rely upon such clues as footmarks.

The long and lithe bodies of the felines are wonderfully specialized for their hunting activities. A good deal of the weight is carried on the forelimbs. Strong back and hindleg muscles enable the animals to make sudden leaping movements as they pounce upon the prey. The cats are not as sturdy runners as the dogs; instead of steadily pursuing the prey for minutes at a time, they make a rapid dash followed by a sudden leap. They walk on their toes; there are five toes on each front foot and four on each hind foot. The toes are armed with long, curved and sharp claws, which can be drawn back, or retracted, when not in use. The claws are primarily utilized for seizing prey.

A cat's head is large, while the face is relatively short and rounded. There is a bony crest at the top of the skull for the attachment of the powerful temporal muscles that manipulate the lower jaw. The animal has highly specialized teeth. The front teeth, or incisors, serve to bite off flesh; the fangs, or canine teeth, are efficient daggers for piercing and slashing. Finally, there are modified teeth, the carnassials, which shear flesh and cut tough

sinews. The carnassials, which are shaped like cutting blades, are a combination of the last premolar of the upper jaw and the first molar of the lower jaw. As the jaws close, the upper blade moves past the lower one; the action suggests that of a pair of scissors. On the tongue there are backward curved prickles that provide a rough surface for rasping particles of meat from bone.

The cats' early ancestors were primitive carnivores called creodonts. From the creodonts evolved a group of forest-dwelling hunters known as the miacid carnivores. The miacids were small; they had long bodies and tails and short, flexible limbs. The cats, civets, weasels, dogs and probably the seals and walruses as well are direct descendants of the miacids.

The first cats appeared about 35,000,-000 years ago. These animals developed formidable upper fangs, up to 8 inches in length — which accounts for their popular name of saber-toothed tigers. Saber-tooths of various kinds roamed the forests and plains of all continents except Australia; they died out when the thick-skinned mammoths and other herbivores on which they preyed became scarce.

The modern felines probably evolved from early saber-toothed cats, whose canine teeth were smaller than those of most saber-tooths. While the saber-toothed cats were specialized for killing large, heavy and slow-moving herbivores, the true felines became adapted for hunting and catching agile animals, both large and small. Today, felines are among the most widespread and successful of mammalian groups. They inhabit all major land areas, with the exception of Australia, Madagascar and the oceanic islands. They are, however, most abundant in tropical regions. The smaller cats are adept at climbing trees; the large felines live mostly on the ground.

Cats are pretty much nocturnal hunters, and their large eyes are perfected for gathering the maximum amount of reflected light available to them at night. In open country, especially, cats use their eyes to detect quarry. The sense of smell is fairly well developed, but not relied upon to the extent that it is by other carnivores such as dogs and weasels. The sense of hearing, on the other hand, is exceptionally keen, and is probably the greatest single asset to a cat's hunting. The facial vibrissae, or whiskers, are feelers associated with touch. No doubt, they warn a cat when it is getting into too thick underbrush, especially at night. A cat's brain is large and well developed, making the animal extremely alert and controlling its highly co-ordinated actions.

Cats prey on any animal that they are quick enough to catch and strong enough to subdue. Although felines are almost exclusively meat eaters, the lion and leopard are known to feed on vegetable matter, principally fallen fruit. The lion will also eat rotten wood and ashes, when the animals it preys on become scarce.

As a rule, cats are solitary animals, traveling and hunting alone. Only at mating time do they accompany one another; of course, the young follow their mother until capable of hunting for themselves. The lion is an exception; though it may live alone, twenty or thirty individuals frequently associate in a troop.

Cats usually restrict their movements to a home range inhabited by game upon which they can prey and having suitable hideaways where they can retreat when not hunting. The size of the range depends on the species of cat. The bigger cats of necessity must have ranges that support the larger hoofed mammals on which they chiefly feed. For example, the home range of a lion troop may extend from 12 to 20 miles. A cat will expand its home range or move to another area if game becomes scarce.

To know feline nature, you need but watch a house cat. It yawns most superbly and stretches elegantly; it grooms itself in a self-indulgent way. On soundless footfall, its supple body moves across a room. It rolls on the floor and often sleeps on its back, paws relaxedly drooped in the air. A sound makes it instantly alert, ears rotating toward the source. Anger it, and it transforms from a gently purring pet into a spitting, courageous terror — nose wrinkled, ears back, claws extended, fur erect and tail lashing. In truth, the house cat is a wild cat that has found it convenient to live amidst the society of man. It is only partly domesticated.

Fifty distinct species of cats are known

There are at least fifty species of cats, belonging to the family Felidae (mammalian order of carnivores). They are built on much the same plan, but for convenience are divided into the big and the not so big cats. The cheetah, however, is a somewhat different kind of feline.

Those designated as the big cats (genus *Panthera*) include the lion, tiger, leopard, jaguar, clouded leopard and snow leopard. They are good-sized carnivores. The lion and tiger may weigh up to 500 pounds, stand 3 feet at the shoulder and measure 7 feet from nose to rump; the leopard measures around 5 feet and weighs a hundred pounds; the 5-foot jaguar weighs about 250 pounds. These cats hunt under a variety of conditions; they vary their behavior to fit the terrain and the quarry that they are pursuing.

The lion needs no description; it is the big carnivore of Africa, inhabiting grassy plains, scrub land and semideserts south of the Sahara. A few lions still live in India, in the Gir Forest. Once lions were common in eastern Europe and the Middle East but were hunted to extinction. The wanton slaughter of the lion (or any animal for that matter) is unforgivable. It is true, however, that sometimes a lion finds man's domestic animals easier to catch than its natural prey; rarely it becomes a man-eater.

The lion, and tiger, too, develop a taste for human flesh because of various reasons. There is a possibility that certain individuals lose their innate fear of man by killing a person in a chance attack. Once successful in man-killing, these cats become less cautious toward man and find him easier prey than the hoofed animals on which they usually feed. A period of famine, too, may embolden a big cat and make it a man-eater. This is especially true of a female that is nursing young; it may account for the fact that most man-eaters are females. In the Ruindi-Rutshuru plain, in central Africa, man-eating reached epidemic proportions among lions around 1860 and again between 1904–09. This behavior was probably caused by decreased numbers of topis and kobs, antelopes common to the area. Man-eating among tigers and leopards was pronounced in Gujarat, India, during the famine of 1901–03.

A lion will not ordinarily attack a human being unless suddenly startled, bothered, wounded or diseased. Aged individuals are known to stalk and kill a person, but usually old ones exist on scorpions, insects and small rodents.

The reproductive and social behavior of lions

Sexually mature lions have no particular breeding season. When a female is sexually receptive, the sexes pair off and mate. After a gestation period of 105-113 days, two to six spotted cubs are born. These young grow quickly, and within a year are capable of making their own kills. By the time the animals are two years old, they have become sexually mature.

Several female lions, with their young, commonly join in a loosely organized troop. This is usually dominated by a big male lion in excellent physical condition. Normally, fighting among lions is a rare occurrence, though frequently a lion who is a stranger to the vicinity is attacked; male lions may battle fiercely, with sometimes fatal results, for possession of a sexually receptive lioness. When the troop leader shows signs of a decline of power and speed, he is quickly driven out from the

Vernon O. Wahrenbrock

group by the younger, more vigorous males. The former leader then lives alone, depending more and more on smaller animals for food, until he dies. His remains are as often as not devoured by the members of his former troop. The fate of the old lioness is not so stern, for she is allowed to follow the group and feed on the kills of the others. Lions have a potential longevity of thirty years, but they seldom reach this age in the wilds.

Hunting is a co-operative affair. Normally the male circles a herd of zebra or

Family group of African lions. In nature, ten or more lions usually associate in a pride, or troop.

wildebeest or other antelopes while the females wait in ambush. The hunters keep in communication by emitting grunts every so often. The male gets upwind of the quarry so that his scent is carried to them. He singles out several individuals and tries to cut them off from the rest of the herd, meanwhile driving the prey toward the hidden lionesses. Once a victim is within range, there is little hope of its escape. The lioness makes a short, high-speed charge, up to fifty miles an hour, and usually leaps at the shoulder area of the prey. The powerful forepaw with claws extended violently wrenches back the victim's head as the teeth sink into the back of the neck. The kill is swift and clean. Of course, the male is perfectly capable of making the kill, but his function is usually that of co-ordinating the movements of his troop.

Having made the kill, the rest of the lions of the group join in the feast. The carcass is disemboweled, the heart and liver being especially prized. The big cats gorge themselves on the meat, but since these animals do not crush the bones nor pick them clean, there usually remains enough to feed the jackals, hyenas and vultures that always follow in the train of the great hunters.

Since lions gorge themselves to satiety, they may not kill again for several days. That they hunt only to satisfy hunger and not to kill wantonly is attested by the fact that lions are often seen resting near a herd of antelope or zebra or even walking through the herd.

African lions are found where herds of plains- and semidesert-dwelling hoofed animals congregate to feed and drink. In periods of drought, when the herbivores move to new feeding areas, the lions accompany them. During the hot season, they remain in shady places until well into the night before they commence to prowl. In wet or cold weather, the lions hunt in the late morning or afternoon.

In all truth lions cannot be described as bloodthirsty or cruel. They beautifully exemplify the hunter's way of life and deserve the protection that is given the hoofed animals that serve them as prey. Thanks to laws that now guarantee its existence, the African lion continues to stand in the animal world as a symbol of power, agility and nobility.

Tigers are the great feline carnivores of Asia

The tiger, the largest cat of Asia, differs considerably in habit from the lion. It takes an expert, though, to distinguish between the two when their skinned carcasses are laid side by side. Externally the tiger has a ground color of reddish orange striped with black. Some specimens may be dark, others quite light in coloration; rarely, one may be white with black stripes. Just as the normally maned lion is sometimes practically devoid of a mane, so the characteristically maneless tiger often possesses a fairly extensive growth of long hair about the cheeks and neck. Siberian tigers have a thick underfur and an overcoat of long pale hair.

A forest dweller, the tiger ranges widely from the Caucasus, Persia and Afghanistan, through India, China and southern Siberia, to Manchuria, Korea, Burma, Indochina, Malaya, Sumatra, Java and Bali. Though the animal is at home in the dense vegetation of the tropical rain forest, it suffers considerably from the heat. Consequently, it bathes and swims to cool off and rests by day in a shaded spot near water. When on the prowl, the tiger keeps to game trails, man-made paths and the banks of rivers, preferring not to make its way directly through heavy vegetation.

Tigers, unlike the social-living lions, are ordinarily solitary and independent creatures. However, the animals sometimes live in pairs for long periods. Mating takes place when the tigress is receptive, which may be at any time of the year. The gestation period lasts from 105 to 109 days, and one to five cubs are born; two is the usual number. The young remain with the mother until they are half grown. Tigers may live about nineteen years.

Tigers, especially older individuals, can subsist on locusts, fish, scorpions, crabs and frogs. Their usual prey, however, is larger game, including wild hogs, the four-horned and nilgai antelopes, water buffalo and various kinds of deer, such as the sambar, chital and muntjac.

The tiger's sense of smell is poor and its eyesight is not much better, but its hearing is superb; it is on this latter sense that the big carnivore mainly depends for hunting. Let a deer or antelope stand perfectly still in the forest gloom, and the tiger will pass it by or may even look in its direction without distinguishing it as possible prey. But if the animal flicks an ear, moves a leg or above all makes a sound, the prowling tiger becomes instantly transformed into the epitome of controlled and tensed destruction.

Creeping or stalking stealthily, and remaining unseen, unheard and unsmelt by its intended victim, the tiger gets to within rushing distance. Then, with several bounds it is upon the prey, hugging it with forepaws and biting it at the throat or the back of the neck. In the case of a water buffalo or nilgai, the tiger may first disable its victim by hamstringing it,

Amer. Mus. of Nat. Hist.

In spite of its formidable appearance, the tiger is usually a timid animal, making every effort to avoid man. Adult tigers are not good climbers but may ascend trees if threatened with danger.

The jaguar — the big spotted cat of the New World — is often called "tiger" or "el tigre." A retiring feline, the jaguar can climb trees much better than any of the other great cats.

N. Y. Zool. Soc.

The leopard of Africa and Asia, one of the most beautifully marked of the cats.

N. Y. Zool. Soc.

that is, by biting the back part of each hind leg. The big cat drags its kill to a sheltered spot and gorges itself on the meat. Afterwards, the tiger rests up nearby, returning from time to time to feed on the kill, even if the meat is quite rotten, until little remains.

Tigers themselves are not indestructible. Packs of wild dogs can kill them, the elephant can mash them to a pulp and the water buffalo can rip them open with a sweep of its heavy horns. Tigers are shy of man and try to remain out of his sight. Since they sometimes feed on cattle, horses, sheep and goats and infrequently become man-eaters, tigers have been extensively hunted over many areas.

The leopard, another big cat but somewhat smaller than the lion or tiger, is one of the most quick-reacting of mammals, becoming quite fearless and fighting savagely when cornered or otherwise bothered. Primarily a night hunter and shy of man, it will, nevertheless, maraud villages and farms, often in daylight, for domestic animals. Its wide range includes all of Africa, except the Sahara and Kalahari deserts, and extends east from the Black Sea through India to Burma and Malaysia and north through Manchuria. Those leopards that inhabit semidesert regions or rocky and treeless areas tend to be of large size and pale colored. In forested lowland and mountainous districts, leopards are usually smaller and have a darker coat profusely patterned with large black blotches and rings. In certain areas, some of the leopards may be glossy black.

A leopard frequently conceals itself by day in a cave or other den or in a densely leaved tree. It usually hunts by itself, but occasionally two of the beasts prowl about together. One to four cubs, but usually two, are born after a gestation period of ninety-two to ninety-five days. They remain in company with the female for the first six months of their lives. Leopards are known to have a life span of at least twenty-one years.

Since the leopard is an expert at tree climbing, it spends a great part of its time in trees where it hunts small arboreal creatures, especially birds and monkeys. It cleverly stalks its quarry through the forest canopy or waits in ambush until an unwary creature comes to within springing distance. Baboons are sometimes preyed upon, but these are most often sick or solitary individuals or stragglers cut off from the main troop; for baboons in concert can usually ward off the attacks of the leopard. Deer, antelope, pigs and young water buffalo are also natural game for this carnivore. The leopard often hides itself in a tree or rocky outcrop above a water hole or game trail, dropping upon its prey from the site of ambush. If the leopard's victim is so large that it cannot be entirely eaten at one time, the big cat drags the remainder of the carcass up into a tree, apparently as a protection against hyenas. One observer saw the remains of a young buffalo—over one hundred pounds of flesh and bone—hung by a leopard in the branches of an acacia tree; the branches were at an estimated height of twelve feet above the ground.

Photos, N. Y. Zool. Soc.

The puma, also known as the cougar or mountain lion, feeds on mammals and birds.

The snow leopard. This big cat dwells in the mountainous districts of central Asia.

A leopard has been seen to roll on the ground and engage in other "playful" antics, apparently arousing the curiosity of deer in the vicinity and causing them to approach within leaping distance of the cat. The author observed similar behavior by a house cat near a group of pigeons. The cat rolled and twisted on the sidewalk, in apparent nonchalance, advancing itself toward the wary but unstartled birds.

The clouded and snow leopards are not true leopards, but are leopard-size and decorated with dark blotches and broken rings. These attractive felines are secretive creatures, rarely encountered by man. The clouded leopard, which has an exceptionally long tail, is an arboreal carnivore of southeastern Asia's dense forests. The snow leopard lives in the high altitudes of central Asia — from the Himalaya to the steppes and Altai Mountains; its long, soft, plumelike hair covers a dense woolly underfur.

The jaguar is the
New World's big cat

The jaguar is patterned somewhat like the leopard, but is more stocky and has a larger head. It is the big feline of the New World, ranging from southern Arizona and New Mexico to lower Patagonia. Though the jaguar lives in arid country and mountainous regions, it prefers the tropical lowlands, especially near the banks of water courses. An expert at catching fish, the jaguar lies on a river bank or overhanging branch and with extended paw scoops up its prey as they come within reach. The cat swims well and will pursue turtles in the water. It attacks capybaras (rodents) and tapirs that come to drink. The jaguar also climbs trees after monkeys. In fact, the cat spends much time in the trees when the lowlands are flooded during the rainy season. The male jaguar seems to assist in providing food for the young and protects them from possible danger. The kittens, usually two or three, are born after a gestation period of approximately one hundred days.

The lesser cats (genus *Felis*) vary in size from the four-foot puma to the small house cat. They can scream or yowl but have not the vocal apparatus to roar as the big cats do.

Pumas (also called mountain lions, cougars, catamounts or painters) have one of the greatest distributions of any mammal — from Canada's northern forests to the southern tip of South America. Unfortunately these magnificent animals, which once occurred over much of North America, have been systematically slaughtered until they now inhabit only the more remote areas of the western third of the continent. Some may still be found in Florida.

The tawny, trim puma hunts deer, tapirs, small mammals and birds in all kinds of country, from swamps and rain forests to brushy and grassy plains and arid, mountainous terrain. Traces of the cat have been found at 18,500 feet in the Andean highlands. The puma's home range covers many square miles; the animal is known to travel as much as twenty miles in a single night. Like all cats, the puma is a stalker and springer, cautiously taking advantage of all possible cover to get within a bound or two of its prey before making a leaping charge. After eating its fill, the cat covers over any remains of the kill and returns to it a day or so later. Pumas are known to track humans, apparently out of mere curiosity. Fictional tales of pumas attacking man are complete nonsense; authentic cases of unprovoked assaults are indeed exceedingly rare if indeed they occur at all.

Active both by day and night, the puma dens up in a cave, a fissure in the rocks or in a clump of dense vegetation. One to four dark-spotted kittens are born some ninety days after the female mates, which may be at any time of the year. The young remain with their mother until over a year old. Sixteen years is a good life span for these cats.

The ocelot, a smaller feline, is handsomely marked with dark blotches, bars and small rings. It is a nocturnal animal of open country or closed forest and makes its den in a rocky cave or hollow tree. Ocelots are known to hunt in pairs, for small vertebrates of all kinds, especially snakes. They mark their hunting grounds by piling their excrement at certain fixed places. The animals are found from extreme southern United States to Paraguay.

Another American cat, the jaguarundi, is a uniformly gray or brownish red feline, having a slender, elongated body, short legs and a long tail. It is a thick-brush cat, somewhat larger than a house cat, of South, Central and southern North America. American cats also include the margay, of South and Central America, which is small but similar to an ocelot in shape and pattern, and several other spotted and striped cats, such as the Andean highland cat and the pampas cat.

In Africa and Asia are a number of small to medium-sized felines, similar in appearance but varying their behavior to fit the habitat in which they live. They inhabit forests, open woodlands, brushy plains and deserts, hunting all kinds of small game; some fish with their paws.

The golden cats are beautiful reddish-brown felines living in the mountainous forests of southern Asia and Africa. They may grow to the size of a small leopard. Among the smaller Oriental spotted cats are the leopard cat, rusty-spotted cat, marbled cat and fishing cat. Small felines that are desert-adapted, having pale, sandy colored coats, include the pallas cat of central Asia, the desert cat of southern Asia and the sand cat of the Middle East and Africa. Felines that are patterned like our domestic tabby cats include: the true wild cat, still found in remote areas of Europe; the black-footed cat, which

is an African feline only fourteen inches long excluding its six-inch tail; and the African wild cat, from which our house cats have probably been derived. In southeast Asia are two plain-colored, short-legged felines with stubby tails — the flat-headed cat, which is dark brown above and white below, and the Bornean Bay cat, which is a reddish chestnut all over.

A somewhat different kind of cat is the lynx. Its bobbed tail juts from high hindquarters that gently slope toward the lower forequarters. A ruff of hair is at the throat, and hair tufts project from the high, pointed ears. The lynx is heavy-

N. Y. Zool. Soc.

The northern lynx (above) is a heavy-bodied feline with large feet serving as snowshoes in carrying the animal over snow. Thick and long hair protects it from the rigors of northern winters. The cheetah (below) is a leopardlike cat, adapted for swift running in open country.

Amer. Mus. of Nat. Hist.

bodied and may measure up to forty inches in length. Its grizzled gray-brown pelage is rather long.

The relationship between the lynx and the snowshoe hare

Primarily a forest animal, the lynx lives in the great coniferous forests of circumpolar lands. Sometimes, if food becomes scarce, the cat will prowl far out on the tundra. Its large fluffy feet serve effectively as snowshoes that pace over the snow without sinking in. The animal climbs trees and swims well, but does its hunting on the ground. Though powerful enough to pull down a deer, the lynx preys chiefly on the snowshoe hare. In fact, lynxes are so dependent on these hares that when the latter decline drastically in numbers every seven to ten years, hundreds of lynxes die of starvation.

Lynxes are usually solitary, but several of the animals are sometimes seen traveling or hunting together. The lynx makes its lair in a hollow log, or in the shelter of a fallen tree or under an overhanging ledge. Here the two to four kittens are born to the female who mated some sixty days before in the late winter.

The bobcat, or bay lynx, is a smaller short-haired lynx that inhabits forests, swamps, deserts and mountainous country of the United States and Mexico. It hunts rabbits and other small animals and may travel twenty to twenty-five miles in a night over an irregular hunting route. A similar species, the Isabelline lynx, lives in southern Europe.

The caracal is a slender lynxlike cat

In the hot and dry areas of Africa and southern Asia is found the caracal, a slender lynxlike feline with a short reddish brown coat. Long plumes of hair tip its ears. A very agile and quick animal, the caracal can catch a gazelle and captures flying birds by leaping into the air and knocking them down. Another lynxlike feline, the jungle cat, prowls the forests, dry scrub and tall grass areas of eastern Asia and northern Africa.

The serval is a somewhat uncatlike feline of equatorial Africa. It is similar to a lynx in being long-legged, long-eared and short-tailed, but is a more slender animal with golden buff fur strikingly blotched with black. The serval lives in the long grasses of the open plains and is capable of running down small game.

The cheetah (genus *Acinonyx*) is even more uncatlike and has the distinction of being the world's swiftest ground-running animal. The cheetah's lithe, streamlined body is supported on long, slender legs. The head is small and the feet narrow; the stout claws cannot be drawn back into sheaths. Quick to break into a run, the cheetah can reach forty-five miles an hour in only two seconds after starting. At full speed the animal attains a little over seventy miles an hour, swifter than the fastest greyhound. Such a pace, however, cannot be long maintained.

How the cheetah catches its prey

The cheetah, like other cats, is a stalker. With belly to the ground it creeps to within a hundred yards of its quarry; then it puts on its enormous speed to run down the prey over a short distance. If the intended game — a rodent or a medium-sized or small antelope — is not caught within five hundred yards, the cheetah gives up the chase; it does not have the necessary endurance to be a long-distance runner. The running cheetah can agilely change direction and usually makes the capture regardless of how much the prey turns in its flight to escape; the cheetah's long tail serves as a rudder in these maneuvers. Cheetahs hunt by sight during the day or on moonlight nights.

The cheetah, which is as big as a leopard, is a plains and desert carnivore of Africa and India. Its sandy brown coat is profusely spotted with black. This beautiful and easily tamed animal has been semidomesticated in India for centuries; wild cheetahs that are young but experienced at making their own kills are captured and trained for hunting.

See also Vol. 10, p. 275: "Mammals."

THE RACES

OF

MANKIND

A Modern Approach to

an Age-Old Classification Problem

BY STANLEY MARION GARN

THE word "race" has been defined in various ways. It has been applied to citizens of a particular nation, to followers of a single religion and even to persons who speak a given language. It is true that national boundaries, a single faith, a common tongue tend to hold people together. But nations do not endure forever, religious affiliations change and new languages arise. Only in rare instances, where complete geographical isolation exists, does race actually coincide with nationality, religion or language.

The German scientist Johann Friedrich Blumenbach (1752-1840) maintained that mankind is made up of five races: Caucasian (white), Mongolian (yellow), Ethiopian (black), American Indian (red) and Malayan (brown). This system was generally accepted for a great many years. Then another classification was proposed. Mankind was divided into three races: Caucasian (white), Mongoloid (yellow) and Negroid (black). This three-way system is still followed in many elementary textbooks; but it is far from adequate. Over half a billion people cannot be properly described as either Caucasian, or Negroid or Mongoloid or as a combination of these three "races." Any classification that leaves out so many people obviously will not do. To be satisfactory, a race system should include the great majority of the more than three billion human inhabitants of the earth.

In the last century certain students of race affirmed that there once existed in Europe various "pure" stocks, or races — Nordic, Alpine, Mediterranean and so on — which were later merged by intermarriage. The supporters of the pure-stock theory selected individuals as "types" to serve as examples of these hypothetical pure races. By selecting "types" on the basis of outward appearance alone, they could assign brothers to different races. One brother could be Nordic, another Mediterranean, a third Alpine and so on. However, the mere fact that different types exist in a population does not necessarily show that a mixture of pure stocks took place in the past.

The so-called "types" that can be identified in populations do not necessarily indicate what the ancestral stocks were like. All races of men are normally very variable; all possible recombinations of genes *

---

* Genes are hereditary units; they form part of larger units, called chromosomes, which are found in each body cell. The genes are responsible for the development of individual traits. See the article Heredity in Man, in Volume 8.

can and do occur. There was variation in the past, just as there is at present. That is why it is difficult to believe that at a given time in the more or less remote past there was a pure Nordic race, say, all of whose members looked, thought and acted alike.

When modern students of race try to distinguish one race from another, they take it for granted that variations within a race exist. They take the average of measurable characteristics and the frequencies of various traits. They do not think of variations as deviations from rigidly fixed standards.

According to the modern viewpoint, a race is a distinct breeding population. It is largely isolated from other races; ordinarily it occupies a given territory. It differs from other races in the frequencies of a great many hereditary traits. The important differences between races are those that have a hereditary basis, not those imposed by the environment.

The different
criteria of race

In comparing races, the geneticist and the physical anthropologist make use of the various differences existing among human beings. The color, form and amount of hair, the color of the skin, the shape of the nose, mouth and eyelids are obvious criteria. Less obvious are the shape and size of the teeth, the proportions and size of the skull and face and the absolute and relative lengths of the long bones and the skeletal trunk. Such minor traits as the ability to taste PTC (phenyl-thio-carbamide; see Index) can be added to the more obvious physical differences.

The various blood groups or factors — OAB, MNS, Rh and so on — are extremely valuable clues in the study of race in man. They are based on the fact that the red blood cells of certain persons are apt to form clumps when the blood of these persons is mixed with the blood of certain other people. This phenomenon made blood transfusion highly dangerous until its nature was revealed. We know now that the agglutination, or clumping to-

gether, of the red blood cells is due to the presence of certain substances — agglutinogens, which may occur in the red blood cells, and agglutinins, which may develop in the serum. A certain kind of agglutinin will cause red blood cells to form clumps if these cells contain a certain variety of agglutinogen.

The blood of human beings has been classified in various ways on the basis of the agglutinogens and agglutinins that are present or absent in it. The first and still most important system — the OAB system — was worked out by an Austrian pathologist, Karl Landsteiner, in 1900. In this system the blood of human beings is classified into four groups — O, A, B and AB. There are also various subgroups within the main groups. We discuss the OAB blood-group system in the article The Life Stream, in Volume 3.

The blood-group system based on the so-called Rh factor is next in importance to the OAB system. The Rh factor is an agglutinogen found in the red blood cells of most people. It is also found in the red blood cells of the rhesus monkey; hence the term "Rh." Blood containing the Rh factor is called Rh-positive; blood without it is called Rh-negative. (See the article The Life Stream, mentioned above.) Other well-known blood-group systems are the MNS system, the Kell system and the Duffy system.

There are several reasons why such systems are particularly important for the student of race. For one thing, differences in blood groups are far less complex genetically than the more obvious differences in man, such as skin color, nose shape and form of hair. Besides, the blood groups are not affected, as are bodily form and skin pigmentation, by age or environment. Finally, many millions of individuals have been blood-typed. We therefore possess a wealth of data on the blood groups of many peoples who have never been studied from any other physical standpoint. Blood-group frequencies can be used, to some extent, to trace the amount of admixture between races and to indicate the extent of ancient relationships.

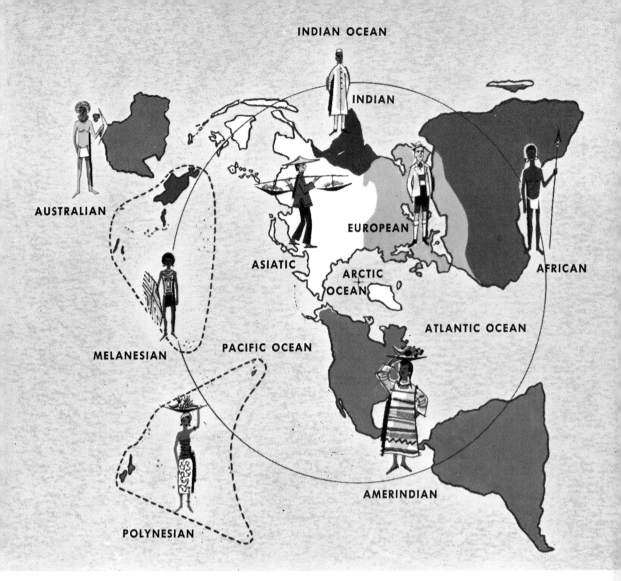

INDIAN OCEAN

INDIAN

AUSTRALIAN

ASIATIC

EUROPEAN

ARCTIC
OCEAN

AFRICAN

ATLANTIC OCEAN

MELANESIAN

PACIFIC OCEAN

AMERINDIAN

POLYNESIAN

Map showing the distribution of the eight geographical races. It should be pointed out that certain races have migrated extensively and are to be found in great numbers in various areas where they were formerly unknown. Sometimes they outnumber the original inhabitants. Thus the European race is now predominant in North America, which was formerly inhabited only by Amerindians. The African geographical race is also well represented there.

The eight geographical races

On the basis of all the criteria that we have described, students of race recognize a limited number of geographical races. Each consists of a number of local races.

The major continents and island chains serve as effective areas of isolation. It is easier to migrate within continents or along island chains than to cross the major oceans or to climb high mountains. Therefore the geographical races given below correspond reasonably well to the continents, subcontinents and island chains.

These are natural limits, as has been shown by studies on animals, and not artificial distinctions made by geographers.

In all, eight geographical races are generally distinguished, though the number is rather arbitrary in certain respects. The eight races may be described as follows:

(1) *European geographical race,* originally inhabiting Europe, western Asia and Africa north of the Sahara, and comprising at least six local races. It has migrated extensively in post-Columbian times (that is, following the discovery of the New World by Columbus). As a result,

EUROPEAN

AFRICAN

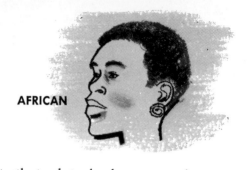

the Americas, New Zealand and Australia now make up secondary population centers of the European geographical race. In general, members of this race show a high percentage of blonds (up to 30 per cent in some localities). The skin reflects from 15 to 40 per cent of sunlight in the short-wave (violet) end of the spectrum. The males tend to have considerable body and facial hair. There are many local variations in the form and color of hair and in eye pigmentation. Average height ranges from moderate (about five feet five inches) to tall (five feet ten inches). The percentage of type A blood is high (up to 50 per cent). The Rh-negative blood type is more common (15-30 per cent) than in any other geographical race.

(2) *African geographical race,* inhabiting most of Africa south of the Sahara, and (until recently) north of the Kalahari Desert, in South Africa. Like the European geographical race, it has established itself in various other parts of the world, particularly the Americas. The members of this race generally have dark skin, which reflects less than 5 per cent of the sunlight that falls upon it. The hair of the head ranges from helical to extremely peppercorn.* There is an increased number of sweat glands and there are various other adaptations to extreme heat. The teeth and jaws commonly protrude; the

teeth tend to be larger, on the average, than those of other geographical races. The Rh-negative blood type is quite common. Members of the race are subject to the hereditary disease known as sickle-cell anemia. There are various dark-skinned populations in the Pacific, but they do not appear to be intimately related to the African geographical race.

(3) *Asiatic geographical race,* found in much of Asia, southward to Burma and Java, eastward to the Philippines and northeastward to Japan and the Kurile Islands. There are small local populations extending in a circumpolar band to the Aleutians, Alaska, northern Canada and Greenland. The following physical features are found in a great many members of the Asiatic geographical race: a fold of skin overlapping the inner angle of the eye (Mongoloid fold); straight, coarse, black hair; wide and projecting cheekbones; shovel-shaped incisors; extra sutures * in the skull. The trunk is generally large in relation to the legs. Blood type B is quite common (30 per cent or more). There is considerable evidence of special adaptation to the cold. Many of the populations in this race are quite small in stature.

(4) *Indian geographical race,* inhabiting the subcontinent of India, with many local races. The members of this race are generally dark-skinned, and have straight

* The hair of the head is said to be helical if the individual hairs are twisted in the form of a helix, or spiral. The term "peppercorn" refers to hair that tends to form little knots because it is both helical and short.

* Sutures are the lines of union, or seams, between the bones of the skull.

AUSTRALIAN

MELANESIAN

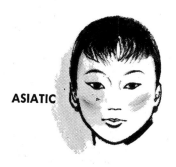

ASIATIC

INDIAN

to wavy black hair; however, they do not have the thrust-out lips or protruding jaws found in Africans. Indians often show resemblances to the Mediterranean division of the European geographical race. Blood group B is fairly common (30 per cent or more).

(5) *Australian geographical race,* limited to the island-continent of Australia. The members of this race are sometimes thought of as "primitive whites." They often have considerable hair on the face and body; occasionally blonds are found among them. Prominent brow ridges and other features suggest that the Australian aborigines may be an early Neanderthal-like population. The teeth are generally large. Blood group B is almost entirely lacking; blood group N reaches a world high (more than 90 per cent).

(6) *Melanesian geographical race,* made up of natives of New Guinea and the Melanesian Islands. Certain Negro-like characteristics, including dark skin and deeply waved, often moplike hair, are found among these peoples. However they differ considerably in blood-group types from the members of the African race. Among other things, Rh-negative individuals, common among Negroes, are not found in the Melanesian race. The different populations vary considerably; some are of pygmy proportions.

(7) *Polynesian geographical race,* covering a broad area from New Zealand to Easter Island. The Rh-negative blood type is absent; B is also rare. Shovel-shaped incisors and Mongoloid folds are found quite often. It is thought, therefore, that the Polynesians may be related to the Asiatic geographical race.

(8) *Amerindian geographical race,* inhabiting the Americas from southern Alaska to the tip of South America. It is obviously related to the Asiatic geographical race, as is shown by the prevalence of straight, black, coarse hair, Mongoloid folds, broad and projecting cheekbones and shovel-shaped incisors (up to 98 per cent). However, marked differences in blood-group frequencies suggest a very early separation from the Asiatic Mongoloids. Blood group B is virtually absent; so is the Rh-negative blood type. On the other hand the frequency of blood type M is extremely high (nearly 90 per cent). This sets off the Amerindian geographical race from all other geographical races.

This listing should make it clear that no simple system of three races or divisions of mankind would suffice to describe the geographical races. As a matter of fact, there may be even more than eight of them. The cases of the Bushmen-Hottentots of southern Africa and Ainus of Japan are rather doubtful. We list them below, among the local races. But the Bushmen-Hottentots may represent a distinct geographical population. As for the Ainus, they may be the remnants of a once-large, pre-Mongoloid population of eastern Asia.

POLYNESIAN

AMERINDIAN

The local races of mankind

Each geographical race, as we have pointed out, is made up of a number of local races. Some of these, such as the Lapp, are clearly recognizable. It is more difficult to distinguish between others; but it can be done by studying the averages and frequencies of various physical traits. In some cases local types (as opposed to races) have developed because of isolation and inbreeding.

Many local races that once flourished have become extinct. For example, various populations well known to Roman authors are now remembered only as place names. On the other hand certain important local races are of recent and hybrid origin. Among these are the Neo-Hawaiian (No. 24 in the list given below), the Ladino (No. 27) and the North American Colored (No. 28).

The following listing is really only a selection because there is not enough space in this short article to include all recognized local races. As a matter of fact, it would be impossible, on the basis of our present knowledge, to draw up a comprehensive list of local races, even if space permitted. Much of Asia, a great part of India and parts of South America and Africa are little known from the standpoint of racial genetics. Casual exploration, desultory missionary activity and journalists' trips provide little useful information about the lesser-known people of the world.

Here, then, is an abridged list of local population groups:

A. EUROPE

(1) *Northwest European:* the basic population of Britain, Scandinavia, the Low Countries and Germany. So-called Nordics are simply blonder and taller *individuals,* drawn largely from this population.

(2) *Lapp:* small people with very small teeth; aboriginal herders, hunters and fishermen. The northernmost extreme of the European geographical race.

(3) *Northeast European:* a regional population inhabiting the broad area from the Baltic to the Black Sea.

(4) *Alpine:* the peoples of central Europe and western Asia, tending on the average to roundheadedness.

(5) *Mediterranean:* the basic population dwelling on the Mediterranean coast and extending eastward to Afghanistan, Pakistan and India. Brunet, longheaded and slender.

B. AFRICA

(6) *Hamitic:* properly a linguistic term applied to intermediate Negro-Mediterranean populations of northern Africa and the Sudan.

(7) *Sudanese:* a dark-skinned population; lankier and with less protruding teeth and jaws than the Hamitic population.

(8) *Forest Negro:* the populations of west Africa and much of the Congo; heavyset and with protruding teeth and jaws.

(9) *Bantu:* a large population of east and south Africa, of recent introduction.

(10) *Bushman and Hottentot:* two related populations of south Africa. They are smaller in build than the Bantus; their skin is lighter; they have more extreme peppercorn hair. In the Hottentots there is excessive development of fat on the buttocks.

(11) *South-African Colored:* an increasingly complicated mixture, principally Northwest European (1) and Hottentot (10).

(12) *Negrito:* a series of tropical forest populations, widely spread from Africa to the Pacific islands and Australia.

C. ASIA

(13) *Hindu:* the slender brunet population of much of India. Hindus resemble darker-skinned Mediterraneans.

(14) *Dravidian:* the more rugged, "primitive"-appearing tribes and castes of southern India.

(15) *Turkic:* Central Asiatics, suggesting a European-Mongoloid blend. They are the Mongols who played such an important part in the history of Asia.

(16) *Tibeto-Indonesian:* a population inhabiting Szechwan, Tibet, Burma and other areas south and east through the Sunda Isles, in the Malay Archipelago.

(17) *Extreme-Mongoloid:* a stocky population ranging from the Amur to Japan, Siberia and the far northern part of North America. They are extremely well adapted to the cold.

(18) *North Chinese:* found in northern China; taller than the Extreme-Mongoloid. They show less evidence of adaptation to the cold.

(19) *Southeast Asiatic:* dwelling in southern China, Burma and Thailand; found also in Indonesia and the Philippines.

(20) *Ainu:* a small aboriginal population from northern Japan, antedating the Extreme-Mongoloid Japanese.

D. AUSTRALIA

(21) *Murrayian and Carpenterian:* two somewhat different populations of the Australian geographical race.

E. ISLANDS OF THE PACIFIC

(22) *Papuan and Melanesian:* a series of populations extending from New Guinea to the Fiji Islands. It properly includes separate Papuan, Melanesian and Micronesian populations.

(23) *Polynesian:* aboriginal inhabitants of Polynesia; they comprise a number of distinct populations. (See the data on the Polynesian geographical race above.)

(24) *Neo-Hawaiian:* a new Hawaiian population of Mongoloid, European and Polynesian origin.

F. THE AMERICAS

(25) *American Indian-Marginal* *: a large number of marginal, nonagricultural populations from the Great Plains and West Coast areas of the United States to Tierra del Fuego, in the extreme southern part of South America.

(26) *American Indian-Central:* a large number of more sedentary, agricultural populations from the southwestern United States to Bolivia and Peru.

* Marginal populations are those dwelling on marginal, or comparatively unproductive, lands.

(27) *Ladino:* a recently formed hybrid population of Mediterranean (No. 5 above) and American Indian-Central (No. 26) origin; found in Central and South America and the West Indies. The Ladino local population contains certain Negro elements in Brazil.

(28) *North American Colored:* the so-called "Negro" population of the United States and Canada. Contains little or no Indian ancestry.

The formation of the world's races

All available evidence indicates that human races were brought about by natural selection, working on isolated populations widely separated from each other and in markedly different environments. Even today peoples most distant from one another tend to show the most striking differences. But the differences between populations are not so great and so numerous as to warrant setting up different species for man, or for assuming different lines of descent.

In a general way human variation is similar to that found in other living things. Skin pigmentation tends to be greatest toward the tropics and wherever solar radiation is most intense (Gloger's rule). Body size tends to be largest among inhabitants of colder or temperate climates; it is smaller among peoples who live where it is hot and dry (Bergmann's rule). The limbs tend to be somewhat reduced in size, relative to body mass, in extremely cold areas and larger in hot deserts (Allen's rule). But these are not hard and fast distinctions. People have migrated a great deal; many have developed artificial protection (hats, light clothing, furs and so on) against heat or cold.

The fact that most race populations show considerable variation in skin color, hair form or blood type used to be considered as evidence against natural selection. We know now that a variable population is likely to adapt successfully to changing conditions of the environment. Polymorphism (variety of forms) fits into the general picture of natural selection.

Except for a very few isolated groups, no human population is pure in the sense that a laboratory-reared line of rabbits is pure. All of written history reveals a constant mixing of human populations; the formation of new races is still in progress.

Race and disease

Many diseases are hereditary; some of them are largely restricted to particular geographically related race populations. Sickle-cell anemia, mentioned above, is a prime example. It occurs quite frequently in south and central Africa, and to a lesser extent in Madagascar and the Arabian peninsula. Mediterranean anemia, a somewhat similar disease, is (as the name suggests) most common among peoples from the coastal areas of the Mediterranean.

There is also evidence that some of the metabolic diseases *, including diabetes and coronary heart disease, are more common among particular races. In this case it is not certain to what extent they are hereditary. They may be due, in part at least, to diet and the general way of life. Many diseases formerly noted only among whites in South Africa are now found increasingly among Bantus and Zulus, as they give up their tribal life and traditional foods and move into the large industrial South African cities.

Certain racial differences are shown in the reaction to injuries. Many Negroid peoples tend to form enlarged, raised scars,

* The term "metabolism" is applied to the chemical reactions involved in such vital processes as growth, reproduction and the replacement of worn-out tissue. The name also applies to the breakdown of organic substances and the release of the energy stored in these substances. Metabolic diseases interfere with these processes.

The skills needed by a primitive people for survival may be comparable to those required by dwellers in cities.

called keloids. There is at least some evidence that there are racial differences in susceptibility to dental caries (tooth decay). However, nutrition is also linked to the problem of dental decay in man.

Race, behavior and intelligence

At one time it was widely believed that members of different races were inherently different in behavior. We are familiar with the popular stereotypes of the "stolid" American Indian, the "inscrutable" Oriental and the "carefree" Negro. However, these stereotypes break down in a common cultural environment, such as is provided in the United States. Either there are no racial differences in behavior, or, if there are, they can be virtually wiped out by experiences in a different culture.

Racial differences in intelligence pose a knotty problem. What is measured by a test of intelligence is vocabulary, mathematical reasoning, reading ability and the solving of certain types of problems. All of these factors are related to the particular background and experience of the individual who is being tested. It would be silly to give the same kind of test to an Australian native living in the bush and to an American living in a big city. If an intelligence test is to be fair, all the persons taking it should have a comparable background. As a matter of fact, the more similar the background of members of different races, the closer their averages for intelligence tests prove to be.

Marked racial differences in intellectual ability need not be expected. The ability to think has been important in every type of culture. The skills necessary to survive in the Kalahari Desert or in tiny Eskimo settlements are comparable to those required by city dwellers today.

It is likely that any hereditary racial differences in behavior and intelligence that do exist are to be found in the response of the autonomic nervous system (see Index). Racial differences are also to be found in such minor abilities as tonal memory, the perception of distance, color discrimination and the like.

THE GROWING WORLD OF PLASTICS*

Moldable Materials That Serve Man in Countless Ways

BY MAURICE W. SCHWARZ

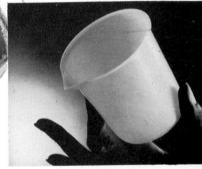

Photos: upper left and lower right, Eastman Chemical Products Inc.; center, Hooker Chemical Corp.

CHILDREN often play with plastic materials — substances that can be molded**; they make mud pies and sand castles and snowballs. In the remote childhood of mankind, as far back as the New Stone Age, men transformed plastic clay into pottery containers for their food and drink. Later, in the Age of Metals, they learned how to mold metals, such as copper, bronze and iron, and they fashioned weapons and tools. Waxes, pitch and gums were also worked for various purposes. Eventually, there were developed other moldable substances — glass, cement, concrete, plaster and rubber.

More recently, within the last hundred years or so, chemists, physicists and engineers, using all the resources of modern

* The assistance of the plastics industry in the preparation of this article is gratefully acknowledged. Information about melamine-formaldehyde and diallyl phthalate was supplied by American Cyanamid Company and Durez Plastics Division, Hooker Chemical Corporation, respectively.
** The word "plastics" comes from the Greek *plastikos*: "that which can be molded."

221

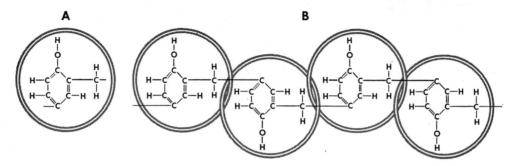

A   B

The synthetic resin phenol-formaldehyde is a long-chain polymer based on the unit, or "link," illustrated in A above. In B, we show a few links in the polymer chain. The letters C, O and H stand for the carbon atom, oxygen atom and hydrogen atom, respectively. The straight lines between atoms represent bonds. It is to be noted that each carbon atom has four bonds, each oxygen atom two bonds and each hydrogen atom one bond.

science, have created the wide array of moldable organic substances * we call plastics. These versatile materials serve in myriads of products, from golf tees to lifeboats; from moving-picture film to wall panels; from hosiery thread to towing rope; from overshoes to raincoats; from drinking straws to water pipe; from table tops to wrappings for meat and vegetables.

All plastics are moldable, but not all moldable substances are called plastics. Inorganic substances that can be molded, such as metals, clay, glass and concrete, are not considered to be within the scope of the plastics industry. Neither is rubber (natural or synthetic), which is one of the organic materials of industry.

Each one of the plastics consists of, or contains as a basic ingredient, an organic substance with high molecular weight (see Index, under Molecular weights) — a substance that can be softened and molded under applied heat or pressure, or both. It is made up of giant molecules called polymers, consisting of smaller molecules joined end to end like a chain. Sometimes the chains are crosslinked, in chain-armor fashion. In some cases, each link in a given chain is the same basic unit, or monomer. Some polymers, known as copolymers, are made up of a mixture of two or more monomers.

Most plastics are produced by the synthesis, or building up, of various chemicals, yielding a product in which carbon is combined with one or more other elements,

including hydrogen, oxygen, nitrogen, chlorine and fluorine. The term "synthetic resin" is often applied to such plastics because in some respects they resemble the natural substances called resins. These include shellac, rosin and amber and are noteworthy for the large size of their molecules. In preparing the synthetic resins, it is necessary to build up smaller molecules into polymers under carefully controlled conditions. Examples of synthetic resins are the alkyds, phenolic resins and vinyls.

Other plastics are derived from natural organic substances, which are modified chemically to produce the desired plastic material. The cellulosics, for example, are derived from cellulose, a carbohydrate found in plants and readily obtainable from cotton and wood pulp. Casein plastics are produced from the protein casein, which is formed when milk is curdled. The basic molecules of such natural organic substances are polymers. It may be necessary to reduce the size of these giant molecules in order to produce a material that is easy to mold on an industrial scale.

All plastics can be divided into two major groups, which are called **thermoplastic** and **thermosetting** and which differ in the way in which they are affected by heat. (*Thermē* means "heat" in Greek.) Thermoplastic materials become soft when they are heated and they harden when they are cooled. If heat is applied again, they become soft once more; they can be molded in the same shape or in a different shape. Only a physical change is involved and not a chemical change. Various substances that

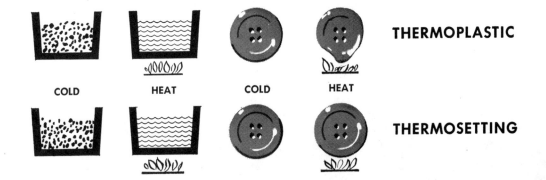

COLD   HEAT   COLD   HEAT

THERMOPLASTIC

THERMOSETTING

are not plastic also act in this way; paraffin wax is a well-known example. Among the thermoplastics are cellulose acetate and polystyrene.

Thermosetting plastics are affected quite differently when heat is applied to them. They first become soft and then hard, and after that they cannot be softened again by the application of heat. The permanent hardening, called curing, is a chemical change; it may be compared to the hardening of an egg in boiling water. The thermosetting plastics, or thermosets, as they are sometimes called, include the phenolics and the epoxies.

Pioneers in plastics manufacture

The first commercial plastic was Celluloid, a cellulose nitrate, which was created by John Wesley Hyatt, a young American printer, in 1868. He had entered the competition for a $10,000 prize, offered for the best substitute for ivory in the manufacture of billiard balls. Hyatt decided to try pyroxylin, a cellulose nitrate, as an ivory substitute. The English chemist Alexander Parkes had reported that if pyroxylin, which is brittle and cannot be molded, was mixed with camphor, the resulting substance was moldable. It did not stand up well, however, under hard use. Hyatt improved upon the mixing process, using pressure and heat, and he produced a plastic material that he called Celluloid. It could be formed, machined and rolled. It was not quite right for billiard balls, unfortunately, and Hyatt did not win the competition.

However, he found that Celluloid was a fine substitute for ivory in many products. In the early days, it was used for denture plates, collars, cuffs and shirt fronts. Later, it served for photographic film, automobile window curtains, hairbrush handles and many other products. Its position was unchallenged for years, but it lost ground rapidly after better plastic materials were developed in the present century.

In the last years of the nineteenth century, two German chemists, Wilhelm Krische and Adolph Spitteler, began looking for a composition material to replace slate for blackboards. They discovered that the action of formaldehyde on casein produced a hornlike substance — a plastic that lent itself to many uses. About 1900, the commercial production of casein plastic was begun in Germany and France. The product received the trade name of Galalith ("milk stone"; from the Greek *gala*: "milk" and *lithos*: "stone").

Another important name in the history of the plastics industry is that of Dr. Leo Baekeland, a Belgian-born American chemist. The eminent German chemist Adolf von Baeyer * had noted in 1872 that resinous masses were formed when various phenols and aldehydes reacted together. These resinous materials were not used in industry until 1909. In that year, Baekeland discovered how to control the reaction between phenol and formaldehyde so as to produce a useful phenolic plastic product, which was given the trade name of Bakelite.

* Von Baeyer received the Nobel Prize in chemistry in 1905 for his research on organic dyes and aromatic hydrocarbons.

It could be formed by casting, or by molding under heat and pressure. In solution, it could be used as adhesive to bond together layers of wood, cloth, paper and other materials. Bakelite was the first commercial synthetic resin.*

Since 1909, a great many other kinds of plastics have been developed and pro-

The first commercial plastic, Celluloid, was created in 1868 by John W. Hyatt, who had sought to develop a substitute for ivory in the manufacture of billiard balls. Celluloid found many uses. In the early days it served, among other things, for collars, cuffs and shirt fronts.

duced commercially. The plastics industry has been growing steadily and has invaded one field after the other. In the United States alone, several billion pounds of plastic materials are produced for many purposes every year.

* The trade name Bakelite is now applied to various groups of plastics.

How plastic materials are prepared

Generally speaking, the chemicals from which plastics are made do not occur as natural products; they must be manufactured from available resources, such as petroleum, natural gas, coal, limestone, salt, fluorspar, sulfur, water and air, to name some of the more important. The fact that we have to go back to such basic materials in the manufacture of plastics has led to statements such as "Nylon is made from coal, air and water" and "Polyvinyl chloride is made from coal, limestone, salt and water." This is certainly an extreme form of simplification; it suggests a magician's sleight of hand in breaking some eggs into a top hat and turning out an omelet. Actually, it is a complicated task to transform basic natural materials into chemicals and to change the chemicals into plastics. Elaborate equipment must be provided; skilled scientists and engineers and technicians must be employed.

To give some idea of just what is involved, let us outline the processes used in the manufacture of the synthetic resin melamine-formaldehyde, a combination of melamine and formaldehyde.

The starting materials for the manufacture of melamine are coal, limestone and air. We do not, indeed, use coal as an ingredient; we use coke, which is derived from coal. It is obtained, together with coal gas and coal tar, when bituminous coal is heated in a by-product oven (see Index). Similarly, it is not limestone that we use as an ingredient in the manufacture of melamine; it is lime, which is obtained, together with carbon dioxide, when limestone is heated in a kiln. Finally, we do not use air in our melamine-manufacturing process; we use nitrogen, which is derived from air. It is obtained by compressing, expanding and cooling the air, which becomes a liquid in the process. The liquid air is then distilled through a column to separate its nitrogen from the oxygen and other elements and compounds it contains.*

* *Editor's note:* Air consists largely of nitrogen and oxygen: approximately 78 per cent nitrogen by volume and 21 per cent oxygen. It also contains small amounts of argon, neon, carbon dioxide, water vapor and other gases.

The chemist sets to work with the three ingredients coke, lime and nitrogen. A mixture containing lime and coke is heated to a high temperature in an electric furnace and forms calcium carbide and carbon monoxide. The calcium carbide is heated with nitrogen in an electric oven, and the product is calcium cyanamide. This chemical reacts with water and acid to form cyanamide, from which dicyandiamide is obtained by treatment with alkalies. Finally, dicyandiamide is heated with ammonia and methanol (wood alcohol) to produce melamine. So now we have the melamine component of the melamine-formaldehyde resin.

The starting materials for the manufacture of formaldehyde are coal, water and air. First of all, the coal must be converted to coke, by the process we just described. The coke is then made to react with steam (derived from water) to form hydrogen and carbon monoxide. When these two gases are heated under high pressure in the presence of chromic oxide and zinc oxide or some other catalyst,* methanol is formed. Methanol vapor is then made to combine with the oxygen contained in air, with the aid of a catalyst; the resulting product is formaldehyde.

We now have melamine and formaldehyde, which are to be combined to form melamine-formaldehyde resin. For this operation, the chemist uses a stainless-steel reaction vessel, called a kettle; it is provided with an agitator, a steam jacket for heating and an electric instrument to record temperatures. Formaldehyde is run into the kettle first; it is followed by melamine and the necessary catalysts. After the mixture has been heated and stirred by the agitator for a sufficient time, the liquid resin is drawn off and pumped through a filter press to remove solid impurities.

The filtered liquid resin is then mixed in a stainless-steel double-arm mixer with sulfite pulp, which has been derived from wood by chemical and mechanical treatment.** The mixed resin and sulfite pulp are dried; they are then broken down by a

* Catalysts are substances that modify the rate of a chemical reaction without themselves undergoing any permanent change.
** *Editor's note*: We discuss the preparation of sulfite pulp elsewhere; see Index, under Sulfite process.

cutter and mixed with various modifiers, or added substances, by grinding in a ball mill (see Index). These modifiers include colorants (coloring agents), lubricants and catalysts. Some of the ground mixture is screened to remove fine particles; it is then packed in drums and sold as plastic molding powder. The rest of the mixture is compacted (pressed together) and is passed through a cutter to produce grains of the required size. After screening, it is packed in drums and is sold as granular molding compound.

This, then, is a very brief outline of the many chemical and physical processes required to transform coal, limestone, air and water into a typical plastic material — melamine-formaldehyde. Many other processes are employed in the manufacture of plastics. The addition of modifiers is of great importance in practically all cases, since synthetic resins by themselves are usually not suitable for molding purposes.

Colorants are almost always added. They include dyes and pigments, such as red cadmium, yellow cadmium and white titanium-dioxide pigments. Lubricants are also generally included; they make it easier to fill a mold when the molding powder is made into a finished or semifinished product.

Some plastics develop a charge of static electricity, which causes dust to adhere to their surfaces. Destaticizers are added to the plastic mixture to overcome this trouble. One type of destaticizer increases the electrical conductivity on the surface and thus allows the static charge to leak away. Exposure to sunlight causes some plastics to deteriorate. This is offset by the addition of substances called stabilizers, which include organic-acid compounds of barium, cadmium, calcium and zinc. Other modifiers serve as flame and fire retardants. Among these are organic phosphorus compounds, which are chemically combined with bromine or chlorine.

When suitable fillers, such as carbon black, clay, cotton flock, mica and wood flour, are mixed with plastics, they add bulk, make the material less porous and reduce the over-all cost. Plasticizers are other important modifiers. They are in liquid or solid form and they are added to plastic powder to improve flexibility and provide other desired properties. Something like four hundred organic compounds have been used as plasticizers.

After modifiers have been added as required, the plastic material may be made into a powder or may be prepared in granular form, as in the case of the melamine-formaldehyde resin described above. It may also be supplied in the form of pellets or flakes. As powder, grains, pellets or flakes, it is now ready to be made into finished or semifinished products.

Important groups
of plastic materials

There are thousands of plastic materials today and they are marketed under a profusion of trade names. For example, Herculoid, Nitron and Rowland CN are all trade names for cellulose nitrate; Beetle, Plaskon and Catalin are trade names for urea-formaldehyde; Styron, Lustrex and Bakelite Polystyrene are trade names for polystyrene. It would obviously be impossible in this short article to name each of the available plastic materials, to say nothing of the even more numerous trade names by which they are known. In the paragraphs that follow, we shall give a list of the more important plastic groups, with a brief account of the characteristics and principal uses of each group.

Acrylic resins. These thermoplastic materials are noted for their toughness, weather-resisting qualities and their clarity.

The transparent dome skylights shown in this photo were formed from single sheets of Plexiglas acrylic plastic.

Rohm & Haas Co.

Polymethylmethacrylate is the most important resin in this group; it is best known under the trade names Plexiglas and Lucite. It remains clear after weathering, resists impact, is available in a wide range of colors, burns slowly and does not become brittle when cold. It serves for display signs, automobile taillights, hairbrush backs, combs and costume jewelry. It is well suited for glasslike materials used in airplanes because of its clarity and its resistance to weather.

Alkyd resins are thermosetting; they are fine electric insulators and resist heat. They are widely used in paints, enamels, lacquers, adhesives and printing inks. Alkyd molding materials serve in light switches, fuses, automobile starters and electron-tube bases.

Allylic resins, which are thermosetting plastics, are excellent electric insulators and have low water absorption and low shrink-

Diallyl phthalate was used for this component of the telephone cable system from Newfoundland to Scotland.

Durez Plastics Division, Hooker Chemical Corporation

age during and after the molding process. They are molded to form automotive ignition parts, coil forms, distributor caps, commutator assemblies, electron-tube bases and other electric and electronic parts. The allylic resin called diallyl phthalate provides reliable electric insulation under the most adverse conditions of temperature and humidity. For this reason, it was used in the terminal equipment of the world's first transoceanic telephone cable system, extending 2,250 miles from Clarenville, Newfoundland, to Oban, Scotland. (It was put in operation in September 1956.)

Amino resins are a group of thermosetting plastics which include melamine-formaldehyde resin and urea-formaldehyde resin. Molded amino products are translucent or opaque; they are glossy, have good colorability and are free of odor or taste. Melamine-formaldehyde resin has low moisture absorption and gives a smooth hard surface; hence it is used in great quantities for tableware, buttons, cases for hearing aids and distributor heads. Urea-formaldehyde resin serves for radio cabinets and electrical devices, among other things.

Casein plastics are produced by the action of formaldehyde upon the protein casein, resulting in a tough, hornlike thermosetting material. Casein plastics have a brilliant surface polish and come in a wide range of colors. They are used for buttons, beads, game counters, pushbuttons, knitting needles and toys.

Cellulosics, derived from cellulose, form a group that includes cellulose acetate and cellulose acetate butyrate. These materials are thermoplastic; they are tough and are available in a wide range of colors. Cellulose acetate is the most widely used of the cellulosics. It serves for fibers (acetate rayon), photographic film, buttons, combs, eyeglass frames, lamp shades, vacuum-

Handsome chessmen molded of Tenite cellulose acetate. The pieces have been weighted in order to prevent tipping.

Eastman Chemical Products, Inc.

cleaner parts and floor polishers. Chessmen molded of this material are pleasant to the touch and sturdy; finger marks upon them may be easily wiped off. Cellulose acetate butyrate is superior to cellulose acetate in toughness and in low moisture absorption; it withstands weathering and is easily cleaned. It is an ideal material for automobile steering wheels, portable radio cases, tool handles and telephone handsets.

Epoxy resins are chemically resistant thermosetting plastics; they are durable, flexible and tough. They make fine protective coatings; they serve in primers, topcoats and varnishes and as lining materials for cans, drums, pipe, tanks and tank cars. Epoxy resins are outstanding as adhesives; they have also been employed increasingly for molding, casting and laminating. Laminates, or combined layers, of epoxy resins and glass-fiber products have found wide application in printed electric circuits, aircraft, ducts, housings, tanks and tooling.

Fluorocarbons excel in durability; they resist chemicals, flame, heat and weathering. These thermoplastics are flexible at low temperatures and are good electrical insulators. They are employed for valve seats, pump diaphragms, gaskets, nonlubricated bearings, electronic parts, wire coating and tubing.

Phenolic resins are thermosetting plastics that possess strength, rigidity and heat resistance. They have a natural brown color and are usually not provided in the bright array of hues that the public has come to associate with plastics generally. Phenolics are used for the items or parts where special mechanical or electrical properties are required, rather than colorability.

Distributor cap made of Durez phenolic resin. This plastic resists heat and it has fine electrical properties.

Durez Plastics Division, Hooker Chemical Corporation

Among other things, they serve for camera cases, distributor caps, electric-iron handles, switches, telephone handsets, electron-tube bases and washing-machine agitators. In liquid form, they are used as protective coatings and adhesives and in the manufacture of laminates.

Polyamide resins. Nylon is the best-known of these resins; it is not a single plastic but a group of plastics, each with its own physical characteristics and uses. The word "nylon" suggests fabrics, stockings or bristles to the general public; but these are but a few of the many uses of the nylons. Because nylon products are tough

Du Pont Co.

Gears molded of Zytel nylon resin. They are tough and have excellent resistance to chemicals and abrasion.

and resist abrasion, heat and chemicals, they are molded to make gears, bearings, cams for speedometers, business-machine parts and household appliances. They also serve for rifle stocks, rotary switches, silverware handles and slide fasteners.

Polycarbonate resin is resistant to chemicals, heat, impact and weather. This thermoplastic is used to make various parts for aircraft, automobiles and business machines and it has many applications in electrical and electronic devices. Like aluminum, brass and copper, but unlike most other plastics, it can be cold-forged (formed without the application of heat) by rolling, punching and drawing to make disks, caps and tubing.

Polyester resin compounds are thermosetting; they are excellent electrical insulators, have low moisture absorption and resist heat. They are widely used for coatings. The fiber-forming polyester known under the trade name Dacron has become important in the textile industry. When reinforced with fibrous glass, asbestos, synthetic fibers and other materials, polyester resins can be molded into various products which have outstanding electrical properties, are lightweight and strong and can withstand impact. These reinforced mate-

rials are used for building panels, electrical-equipment parts, boats, chairs, laundry tubs, luggage, awnings and many other products.

Polyethylene is among the most widely used plastics. This thermoplastic material is tough and can be made either rigid or flexible; it resists heat and cold, water and weathering and is an excellent electrical insulator. Its many applications include pipe and tubing, jacketing for electric cables, flexible ice-cube trays, rigid and squeezable bottles, tumblers, dishes, mixing bowls, brush handles and toys. In film and sheet form, it serves for bags for candy and food

The Hydraulic Press Manufacturing Co.

These thirty-gallon containers, molded of polyethylene, resist weathering and food acids and are easy to clean.

products, rainwear, balloons for meteorological observations, freezer bags and moisture barriers under concrete and in walls. High-density polyethylene is rigid, resistant to water, weathering, food acids and abrasion and is easy to clean; hence it is an outstanding material for garbage cans.

Polypropylene, a thermoplastic, combines many valuable properties. It is lightweight, tough, strong, resistant to chemicals, boiling water and cracking and is a good electrical insulator. It is employed in dishes, pipe and tubing, valves, bottles, battery boxes, refrigerator parts, household articles, textile-machinery parts and insulation for wire and cable. Polypropylene laboratory ware is impact-resistant, withstands temperatures up to 140° C. (284° F.) and resists chemicals and cracking.

Polystyrenes are thermoplastics which are noted for their clarity, hardness, gloss and electrical properties and are available in a wide range of colors. Their applications include kitchenware, food containers for refrigerators, instrument panels, wall tile, portable-radio housings, storage boxes, handles and toys. Polystyrene rigid foams are supplied in plank, block or sheet form and also as beads which can be processed into foams. They are used for packaging and as sandwich panels in the construction of buildings.

Silicones resist chemicals, heat, water and weather and are fine electrical insula-

Silicones Division, Union Carbide Corporation

Applying a silicone silver finish with an aerosol spray dispenser. The silicone provides a long-lasting finish.

tors. These thermosetting plastics are employed in aircraft and missiles, coil forms, connector plugs, generator insulation, switch parts and terminal boards. Silicones in liquid form are used to control foam in chemical products; they also serve as lubricants. They make it easier to release molded products; in the baking industry, for example, bread pans are coated with silicone so that the loaf may be readily removed from the pan after baking.

Urethan resins are thermoplastic materials. They form flexible, semirigid and rigid foams, which are tough and lightweight and resist moisture. Flexible urethan foams have proved useful in automobile and airplane seats, bedding, crash pads, furniture cushioning, interlining for clothing, protective packaging and thermal (heat) insulation for tanks. Rigid foam is serviceable as a core material for curtain-wall panels (see Index, under Curtain walls), low-temperature insulation and packaging. Urethan resins are also used in adhesives, bristles, coatings and solid plastic products.

Vinyl resins, or vinyls, are strong and tough thermoplastics which are good electrical insulators, resist chemicals and are

available in many colors. They are used for floor tile, handbags, phonograph records, raincoats, shower curtains, upholstery, garden hose, electric plugs and wire and cable insulation.

Monsanto Chemical Co.

The Tessera vinyl floor covering shown above is ultrastrong; it has virtually no seams to collect dust or dirt.

How plastics are processed

If a manufacturer is thinking of using plastics for the products he makes, he wants to be sure that one or more of these materials will serve his purpose better than wood, metal or ceramics. He will have to consider what his products are intended to do, how long they should stand up under wear, what properties they must have and how many he will be able to sell. He must also estimate at what price he can sell his merchandise in order to make what he considers to be a fair profit.

If he finally decides upon the use of plastics, he will have to make up his mind which plastic or plastics to select. Certain desirable properties are found in many different kinds of plastics: their resistance to rusting and tarnishing, their retention of their original color and their smoothness. Other properties are characteristic only of a few plastic materials.

If a manufacturer produces brush and mirror backs, costume jewelry, kitchen items, tableware and toys, he has a wide choice of plastic materials at his disposal. The situation is quite different, however, if he is going to manufacture electrical commutators, electronic parts, gears, pipe fittings or washing-machine agitators. He may be limited to one or two plastic materials, which will have to meet the requirements in Federal or industrial specifications.

When he has finally selected the appropriate plastic materials, the processor works them into finished or semifinished products. A number of different methods are available to him; here are some of the principal ones.

Compression molding. Thermosetting plastics are often formed by compression molding. Plastic molding powder or preforms (roughly molded forms) are set in the cavity of a heated mold. The mold is then closed by a plunger worked by hydraulic pressure. The molding compound within the mold first softens and flows and then hardens as the chemical reaction that crosslinks the polymers takes place. The mold opens and the finished piece is ejected from it.

Transfer molding. This is a variation of the compression-molding method. It is used for thermosetting plastics which are to have deep holes for inserts or when the mold cavity has a rather complicated shape. The plastic compound is first heated in a separate pot, called a transfer chamber. Then a plunger forces the softened plastic from the transfer chamber into the cavity of the mold.

Injection molding is used to process various thermoplastic materials and is particularly effective in the rapid production of more or less intricate pieces. Often the mold consists of a number of small cavities, connected by channels, called runners. Each cavity serves as the mold for a single small item — a button, say. Molding pellets flow from a hopper into the rear end of a heating cylinder where heating starts. A plung-

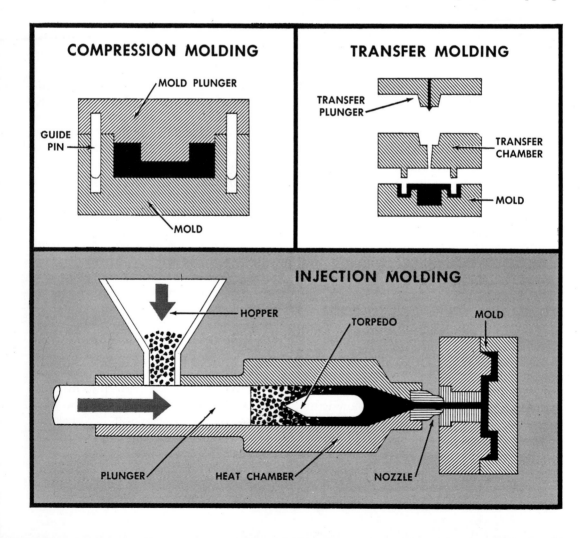

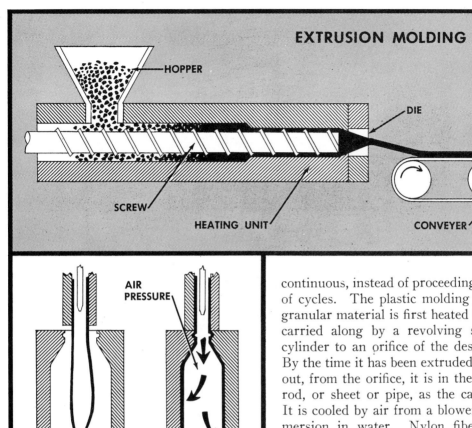

EXTRUSION MOLDING

HOPPER

DIE

SCREW

HEATING UNIT

CONVEYER

AIR PRESSURE

BLOW MOLDING

er pushes the plastic forward and it is forced into the space between the walls of the cylinder and a heated "torpedo" in the middle of the cylinder (see diagram). The material is melted by the time it reaches the cylinder nozzle and is injected into the mold. The fluid plastic completely fills the mold cavity (which has the shape of the desired product) or the cavities that are connected by runners. When the plastic has cooled and has become solid, the mold is opened and the finished piece is ejected. Plastic that has solidified in the runners between cavities is later removed. Most injection-molding machines are horizontal; some are vertical.

Extrusion molding serves to produce sheets, rods, tubing, pipe and filaments (such as nylon fibers used in textiles). No molds are required; the flow of product is

continuous, instead of proceeding in a series of cycles. The plastic molding powder or granular material is first heated and is then carried along by a revolving screw in a cylinder to an orifice of the desired shape. By the time it has been extruded, or pushed out, from the orifice, it is in the shape of a rod, or sheet or pipe, as the case may be. It is cooled by air from a blower or by immersion in water. Nylon fibers are extruded through thousands of tiny holes in a die, called a spinneret.

Blow molding is used for the production of hollow thermoplastic containers, houseware and toys. In one type of blow-molding machine, plastic material in the form of a tube is received from an extruder and is fed into the cavity of the mold. One end of the tube is then sealed off and after the mold has been closed, compressed air is blown through the other end of the tube. This causes the plastic to expand against the walls of the mold; it is cooled and hardened and then removed from the mold. Flexible polyethylene squeeze bottles are produced by blow molding.

Slush molding. Overshoes and other flexible objects are made by this method. A thermoplastic material in the form of a slush or slurry (a thin, watery mixture) is put in a warm mold. The slurry is then agitated and drained off, leaving a thin layer of material on the inner surface of the mold. This layer is fully fused by additional heat, cooled and then stripped off.

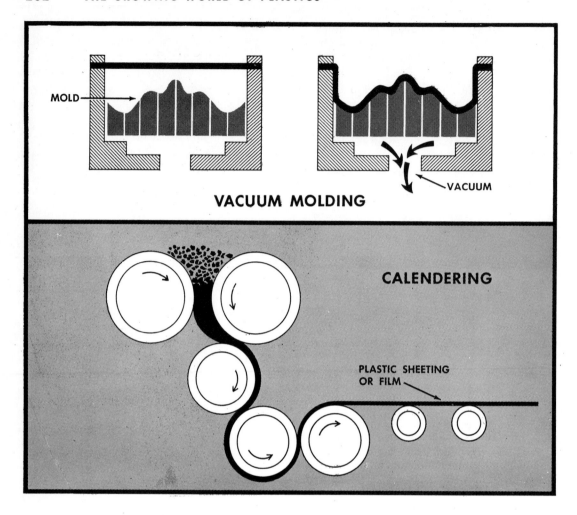

MOLD

VACUUM

VACUUM MOLDING

CALENDERING

PLASTIC SHEETING
OR FILM

Vacuum molding. Atmospheric pressure plays a part in this process. In one method, a sheet of thermoplastic material is clamped over a hollow mold, forming a tight diaphragm, and is then heated and softened. Air is now withdrawn from the space between the thermoplastic sheet and the inside of the mold. Atmospheric pressure then forces the sheet against the inner surface of the mold. After cooling takes place, the clamp is released and the piece that has just been formed is removed. In another type of vacuum forming, a sheet of thermoplastic material is placed over a raised mold. Heated and softened, the sheet drapes itself over the projecting parts of the mold. The edges of the sheet are then clamped tightly around the mold, and the sheet becomes a sealed bag. Air is withdrawn from the inside of the bag, and at-mospheric pressure causes the sheet to fit tightly against the mold.

Casting. Acrylic and vinyl thermoplastics and phenolic, polyester and silicone thermosets can be cast into shapes, film, sheeting, rods and tubes. Pressure is not necessary in this method.

Calendering. In this process, thermoplastic material is fed between several heated calenders, or rollers, and emerges as a sheet or film. Calenders are also used for coating cloth or paper with plastic material.

Laminating. Layers of wood, fabric or paper, impregnated with a solution of thermosetting plastic material, can be combined to form strong, rigid products. The material is treated with the thermoset solution and is placed between the platens, or flat plates, of a laminating press. Heat and pressure are then applied.

Foaming. To produce an expanded plastic in the form of a foam, several methods are used. Gas bubbles may be introduced into the melted plastic by chemical reaction, by heating a volatile liquid or by forcing gas in under pressure. Among the plastics that can be foamed are cellulose acetate, polyethylene, styrene, urethan, the vinyls, the phenolics and the silicones.

Reinforced plastics. It is possible to reinforce certain thermosetting plastics effectively by embedding various substances in them. The reinforcing materials include glass fiber, asbestos, cord, cloth and paper; they may be in the form of chopped fibers or of sheets. Pressure is applied to the reinforced plastic to mold it into shape.

Additional operations. In addition to all these operations, certain additional ones are often required. For example, the moisture content of molding powders must be reduced to a very small amount before the molding operation; in some cases preheating takes place. Some plastic products are preformed before compression molding. Sometimes they are annealed after molding — that is, they are heated and then slowly cooled. Annealing relieves stress that develops during molding and cooling. After the molding process, some excess plastic material is frequently attached to the molded part and must be removed in order to obtain a presentable finished product. This is often done by mechanical tumbling.

Using the processes we have just described, manufacturers turn out either finished products or semifinished materials, such as sheets and rods. Manufacturers called fabricators, employing various kinds of machine tools * and other equipment, will transform the semifinished materials into costume jewelry, boxes, display signs, television lenses, raincoats and a host of other products.

See also Vol. 10, p. 284: "Plastics."

* *Editor's note*: See the article Power-Driven Tools, in Volume 10.

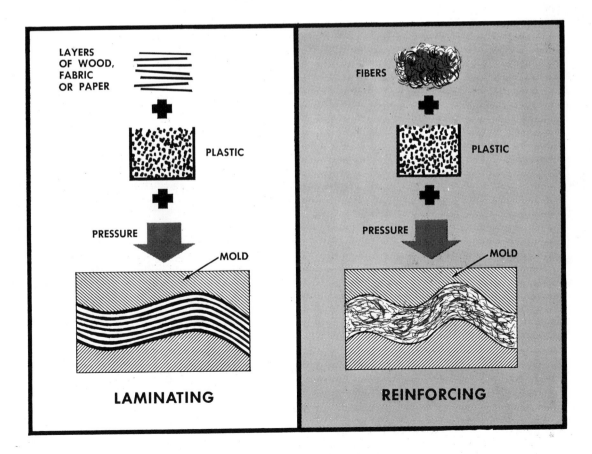

LAYERS OF WOOD, FABRIC OR PAPER

PLASTIC

PRESSURE

MOLD

LAMINATING

FIBERS

PLASTIC

PRESSURE

MOLD

REINFORCING

N. Y. Zool. Soc

A monarch of the widespread deer family, the stately male red deer, or stag, of Europe.

THE DEER FAMILY

The Story of the Widespread Antlered Herd

THE deer and their kin belong to the large group of even-toed ungulates, or hoofed mammals. Besides the deer family, this group includes pigs, hippopotamuses, camels, giraffes and cattle.

The earliest known deer were small creatures living in the forests of Eurasia some 25,000,000 years ago. Some of the early deer had no antlers but developed tusklike canine teeth. The heads of other primitive members of the family had bony projections that were covered with skin and that were never shed. As evolution proceeded, some species of deer increased greatly in size. They included a giant moose that lived in North America something like a million years ago. The Irish elk of Europe was another huge animal; it carried a ponderous set of antlers.

The members of the modern deer family are swift-footed and have keen vision and an acute sense of smell — valuable assets to animals that are constantly harassed by carnivores. All the animals of the group are herbivores, or plant eaters; they feed upon such vegetation as grasses, twigs, leaves and acorns, which they grind with their broad molar teeth, provided with ridged surfaces. They share with camels, giraffes and cattle the habit of ruminating, or chewing the cud. In this process, partially digested vegetable matter (the cud) is brought back from one compartment of the stomach for prolonged chewing before it is returned for further digestion to another compartment.

The males of various species of deer grow striking antlers of bone; so do the females of the reindeer, or caribou. The species that have antlers shed them every year. In this respect they differ from all the other members of the animal kingdom save one — the prongbuck, or pronghorned antelope. This is a rather curious animal. It is neither a true deer nor a true antelope (a member of the cattle family), but it shows striking similarity to both deer and antelopes. Let us pause for a moment to consider this unusual animal, which is to be found chiefly within the boundaries of the United States.

The prongbuck is a swift and graceful creature, colored a reddish buff or tan above and whitish below; it has a rump patch composed of white hairs. By means of muscles of the skin, the prongbuck is capable of erecting these hairs until they stand out conspicuously over the rump. Some prongbucks are solitary creatures; others gather in small bands during spring and summer, as they graze on the grasses, weeds, cactus and sagebrush of grassy or desert plains on the North American continent. In May or early June, two or three young are born to each doe.

The "antlers" of the prongbuck are really horns. A pair of bony projections grow from the top of the skull; the horns are sheaths that develop over these projections. Each of the horns has an ascending branch, which may be twenty inches in length; it also has one small hooked branch in front. The prongbuck's chief point of similarity, perhaps, with the members of the deer family is that it sheds its horns in the fall as deer do their antlers.

In other respects the horns of the prongbuck and the antlers of the deer are quite different. The deer's antlers originate upon a pair of bony knobs on the upper part of the forehead behind the eyes. Upon these knobs there develops each spring a velvety skin, which is most liberally supplied with blood vessels.

N. Y. Zool. Soc.

Mule deer with antlers still in velvet.

The velvety growths expand, elongate and finally branch out. The developing antlers are spongy and filled with the blood that brings nourishment needed for growth. In time the blood supply is cut off, growth ceases and the antlers harden. The velvety skin dries and comes off in shreds; the deer, by rubbing them against tree trunks and branches, helps rid its antlers of this dead skin. Antlers serve as weapons when the males contend against one another for possession of females during the autumn mating season. In late winter or early spring, the antlers are shed; new antler growths appear very soon afterward. Red deer and caribou are known sometimes to eat their shed antlers. This seems to occur when the animals lack calcium and other mineral salts in their diet.

About one hundred known species of deer occur widely distributed over North America, Europe and temperate Asia; several species inhabit South America and the Mediterranean coast of Africa. They are found in forested and mountainous areas, in brush country and swamps, on grassy plains and the arctic tundra.

Most species live in groups, the males normally joining the females only at breeding time. After mating, the males abandon their mates or harems to associate in herds of their own sex; some of the animals take up a solitary existence. The females, accompanied by young, form their own herds and roam together over a fairly well defined range, the size of which usually depends upon the abundance of food. These herds are led by an older, but still reproductive, female leader. Social behavior such as this is typical with the red deer. Eight months after mating, the females give birth to one or, rarely, two white-spotted fawns. The young closely follow the parent for some time.

Red deer are found in Europe, Asia and northern Africa. They prefer moorlands with running brooks or forests of deciduous trees interspersed with meadows. They are handsome animals, wearing a sleek reddish brown coat in summer, which becomes grayer during the winter; buff-colored hairs cover the rump. Standing from four to four and a half feet, the red deer is well proportioned and carries itself most gracefully. The male normally has six points, or tines, on each of its beautifully shaped antlers.

Red deer usually remain hidden during the day and roam in search of food at dusk. The amount of humidity seems to control herd movements. In very humid weather, the deer do not travel far; when the air is dry, they keep on the move. It is suggested that humid air favors the deer by carrying scents to a greater degree, thus allowing the deer to rely on the sense of smell for safety rather than on perpetual movement. A lone deer will rest with its back to the wind; it scans the area to the front of it and depends on its olfactory sense to tell of danger at the rear. Whenever a herd rests, one or several deer watch in all directions. With the wind blowing toward him, a person may walk to within a few hundred yards of a herd. The deer stare and uneasily stand their ground.

USDA Forest Service

N. Y. Zool. Soc.

The magnificent bull moose, the largest of American deer.

Young male moose. In time, its antlers may spread across more than sixty inches.

During most of the year, a number of stags associate in a loosely-organized band, which is not dominated by any particular group leader. The animals wander over a fairly definite home range, keeping to lower valleys and meadows during winter and ascending to higher elevations in summer. This range provides ample feeding areas and deer wallows, where the animals roll about in the mud.

Closely knit bands of females, or hinds, have their own separate home ranges. The lead hind controls the movements of her followers, usually consisting of her own female offspring and their calves. She is ever on the alert, watching, listening and

An antlered prince of America's West—the regal bull wapiti, or elk.

Canadian Govt. Travel Bureau

sniffing the air. When she senses danger, she emits a sharp, staccato bark. The group immediately stops all activity and becomes watchfully motionless until the lead hind resumes normal activity; if she moves off, the group follows.

During the brief rutting, or mating, season, in September or October, stags invade the hinds' home range and establish temporary harems. Each male gathers a number of females and fights for his position against contending stags. The animals vigorously push against each other with their antlers, measuring their relative strengths.

Each stag dominates his harem in the sexual sphere only; if danger threatens, the hinds gather around their accustomed female leader. After mating, the males go off to their winter range. They shed their antlers in early spring.

Fawns are born in the spring. The females never wander far away from their young and return frequently to nurse them. If an enemy comes near, the fawn "freezes" on the spot. This "freezing" plus the pattern of the little animal's spotted coat is usually ample protection against predators. When fawns get older they engage in a variety of amusing "games." Sometimes they race with one another; often such races take place around a hillock, which is the goal to be gained by the victor, as in "king of the mountain." Sometimes one fawn chases another until the pursued is "tagged"; then the pursuer is chased. Fawns also stage energetic sham battles.

The red deer of North America is the American elk, or wapiti. It is a magnificent creature with a shoulder height of five feet. Generally reddish brown, it has a pale yellowish rump patch and a short white tail. Wapiti usually live in the semi-open forests and highlands of the Rockies and rain forests of the Northwest. They move in summer to higher elevations, which are free from mosquitoes and flies, and come down into the valleys when winter snows are heavy. Browsed vegetation, low broken tree branches, bark scraped from aspen trees and mud wallows are signs of the wapiti's home range.

In September and October, bulls seek out cows, collecting as large a harem as they can maintain. The males announce their presence, as a challenge and threat to rival bulls, by bugling. This vibrant call begins in a low key, ascends to a high pitch and then abruptly drops in a scream. Fights between disputing bulls are savage, though they seldom end fatally. The rivals, with necks swollen and nostrils flaring, dash headlong at one another. Though the huge spreading antlers are sharply pointed, they inflict no deep wounds because the spread of the antlers serves as a basketwork to catch the onslaughts. Usually after two or three battering assaults, the weaker bull gives up and hurriedly quits the field. Infrequently, bulls so enmesh their antlers that extrication is impossible; the animals exhaust themselves and die of starvation.

A powerful bull may gather a harem of fifty or sixty cows, but he rarely keeps it. A contending male provokes the harem master into battle. While the fight rages, other bulls split up the harem into smaller groups and each moves off with one of these groups as his own.

At calving time, in the following May or June, when one to three calves are born, the cows are alone or in twos or threes. Shortly after birth, the white-spotted calves

dwelling in grassy plains and marshes in southeastern Asia; and the sikas, which are relatively small, spotted deer of eastern Asia, Japan and Formosa.

Probably the handsomest deer is the axis deer, or chital. This graceful, medium-sized animal sports a rich reddish coat that is vividly spotted through all seasons. Large herds roam the plains and foothills of India. Allied with the chital is the hog deer of southeastern Asia. This is a small, meager-antlered, thick-bodied creature. Chiefly nocturnal, it associates in twos or threes in grass jungles.

The small fallow deer is strikingly colored in summer with orange-red brown

Indian sambar deer—the large red deer of southeastern Asia.

N. Y. Zool. Soc.

stand and even take a few wobbly steps. As summer advances, young bulls and cows with their calves congregate in herds numbering several hundred.

Another red deer is the sambar of southeastern Asia and the Philippines. These large, forest-dwelling deer live singly or in small bands. The rusa is a small, light-brown colored sambar of Malaysia. Other deer closely related to the red deer include: the spotted red deer of the Philippines; Thorold's deer of western China and eastern Tibet, which has white cheeks, muzzle, lips and chest; the Indian barasingha, or swamp deer, whose bright reddish summer coat is sprinkled with white; Eld's deer,

profusely spotted with white. The male's antlers assume a palmate, or hand-shaped, structure at the top. Large herds are kept in a number of European parks. In their native Mediterranean region, fallow deer form small bands, which hide in thickets during the day. A larger relative, the Persian fallow deer, has smaller and nonpalmate antlers. It inhabits Asia Minor to Persia.

The deer called Père David's deer is unique in that it is unknown in the wild state. During the last century a herd of these medium-sized, heavily built deer were kept by the Chinese Emperor outside Peking. Several animals were brought to

Muntjacs, or barking deer, give a sharp dog-like bark as a mating call or cry of alarm.

N. Y. Zool. Soc.

Europe, and their ancestors are now bred in captivity in several parks and zoos.

Leaving the red deer and their allies, we consider next the musk and water deer. Musk deer, which live on wooded mountainous slopes of central and eastern Asia, are the most primitive of living deer. The male does not carry antlers but grows, in the upper jaw, enlarged canine teeth that form curved tusks. Musk deer stand twenty inches at the shoulder, are compact of body and have thick, sturdy legs, large hoofs, long and thick hair and a very short tail. A gland on the belly secretes an aromatic substance called musk, which probably serves to attract the sexes during the mating period. Except at this time, in January, musk deer are solitary, hiding by day and feeding and moving at night.

Water deer are similar to musk deer, the males being antlerless but having tusklike upper canines. Semiaquatic in habit, these deer frequent lowland marshes and brush country of China and Korea. Unlike other deer, female water deer bear litters of five or six young at a time.

Muntjacs, or barking deer, are a group of small, shy deer that inhabit forests of southeastern Asia. Males carry tusks and also short antlers, which grow from high pedicels (bases) on the forehead.

The largest and perhaps most impressive of deer is the moose, or European elk. This magnificent beast, which stands on extremely long legs, may measure as much as seven feet at the withers. The male has enormous spreading antlers. The snout is broad and rather pendulous, and a dewlap, or "bell," hangs from the throat. Moose are fond of willow thickets and forested uplands in the vicinity of lakes and rivers. They browse on leaves, buds and bark and also, in summer, consume water plants, which they get by wading or sometimes completely submerging under water. They are vigorous swimmers. Moose occur in Canada, Alaska, extreme northern United States and northern Eurasia.

During the autumn rut, bulls look for mates. They give a hoarse bellow, and a cow calls back with a softer, somewhat longer low. If another bull is near, the two rivals fight fiercely for possession of the female. The victorious male remains with the female for several days and then searches for another cow.

Usually two unspotted calves are born the following May. They nurse for some time and may follow the cow until just before next year's calving when the cow drives away her yearlings. Females and their calves go to high ground in the summer and descend to lower country as winter approaches. In winter, several moose may associate. They tramp down the snow in a sizable area, called a yard. Here they remain until the food supply is exhausted; then they move to another locality.

The powerful bull moose, using antlers and forefeet, can kill a full-grown bear. Man is practically its only enemy. Moose were formerly hunted with dogs; at night hunters in boats would use spotlights. Both practices are now outlawed.

The water deer and the moose are members of a large subfamily of deer, the hollow-toothed deer. This group includes the white-tailed deer, the marsh deer, the pampas deer, the guemals, the brockets, the pudus, the caribou and the roe deer. These animals are slightly different from the red deer that we have mentioned. Except for the moose, they are generally smaller in size, lighter in build, and their antlers are differently constructed. In the Virginia deer, which is a widely known white-tailed deer of the United States and Canada, the antlers sweep backward and then forward with all the tines pointing forward. These deer have large ears, and a fluffy white patch on the underside of the triangular black-topped tail. They are mostly reddish brown in the summer and grayish brown in the winter. Primarily woodlands deer, they are particularly fond of the water and feed on aquatic plants and grasses.

The mule deer is also a white-tailed deer; it inhabits the west coast of the North American continent. It follows regular migration routes, spending the summer in the mountains and moving to the sheltered valleys in the fall. The mule-deer group may be distinguished by their large ears, a large white rump patch and a short black-tipped tail. Deer of this type may stand 42 inches at the shoulder and reach a length of 60 inches, excluding the 8-inch tail. Twin young are born in the summer about six months after mating. There are several dozen species of mule deer, each predominating in a particular area, which may be rather large. Thus the Columbia black-tailed deer ranges from British Columbia to California, the Southern mule deer inhabits the Lower California region, and the Sitka deer migrates from north of Juneau, Alaska, to British Columbia.

The marsh deer is found in the Amazon Basin, from the Orinoco River to the Guiana watersheds. This animal has enormous ears and large antlers. It is lustrous red in color, white on its undersides, with

Both photos, N. Y. Zool. Soc.

The axis deer (*Cervus axis*) is found in India and also in the East Indies.

black legs. The pampas deer dwells in the pampas and savannas of South America. It is a delicate animal with simple antlers; it is red, but the underside is white. When running, this little animal may jump as high as ten feet. The guemals or tarugas occur in the high Andean regions from Ecuador to Chile. They are heavy-bodied with coarse hair and a peculiar salt-and-pepper coloration. The antlers are rather simple, forming a **Y** as seen from the side, with a long main stem and a short tine over the brow.

The taruga (*Hippocamelus antisensis*) is a graceful South American deer.

Photos, N. Y. Zool. Soc.

Chinese water deer; adult female (above) and fawn (left).

DEER OF
SEVERAL LANDS

The South American marsh deer. It occurs in swampy areas from Brazil to northern Argentina.

The fallow deer has palmate, or hand-shaped, antlers. It is native to Mediterranean countries.

Philip Gendreau

Virginia deer (*Odocoileus virginianus*). This is a fawn a month old.

The brockets also occur in the Amazon Basin. They are wholly nocturnal, as opposed to other deer which feed at dawn and at dusk. The brockets are small and delicate and are reddish brown in color. The antlers are one-tined and straight. The tufted head is an identifying characteristic of this animal. The tiniest of all deer is the pudu, which stands only fifteen inches high at the shoulder and is about the size of a small terrier. It lives in the temperate forest zone of the Andes and ranges all the way from Colombia to southernmost Chile. It is colored reddish brown to pale gray and its horns consist of small spikes.

About one hundred years ago caribou abounded in the northern United States and Canada. Today they are numerous only in the Canadian Rockies and the Barren Grounds of northern Canada. In the caribou both sexes are provided with antlers. The main stems of the antlers sweep back and up from the head, spreading as they rise, then bending forward and ending in a flattened palm. The animal is large; it may stand as much as sixty inches high at the shoulder. The nose is generally covered with hair and there is a pronounced dewlap at the throat. The hoofs are broad and the two cloven sections are spread wide apart; dewclaws (hoof vestiges) are attached to them. The caribou subsists mainly on mosses. The fast, steady trot of the animal protects it from one of its enemies, the wolves. Man is its greatest enemy, for the flesh of the caribou forms the main staple of the diets of Eskimos and northern Indians. The partially digested contents of the stomach are valued as a delicacy by Eskimos.

There are various other caribou species. The Greenland caribou is a small, light-colored animal with slender antlers. The dwarf caribou is darker and heavier than the Barren Grounds caribou; its antlers are stouter and less flattened. It roams the evergreen coniferous forests of northeastern United States into Canada from the Great Slave Lake to Newfoundland. In the Eurasian arctic region, a species of caribou known as the reindeer has been domesticated. (See Index, under Reindeer.)

Among the most widespread deer groups in Eurasia is that of the roe deer. Their species extend from the British Isles to eastern Siberia and Korea. These animals stand about twenty-five inches high at the shoulder. The antlers are often knobby and start to fork near the main stem. The tail is hardly visible. In summer the roe deer are reddish brown, with an orange tone on the flanks and white on the underside. In winter they are brown, with a large white patch on the rump.

A close cousin of the deer is the mouse deer or the chevrotain. This little animal may be traced back to a basic stock from which sprang the majority of the even-toed hoofed animals. It thrives on aquatic plants and fallen fruits. The oriental species is about the size of a hare; the male has small tusks in the upper jaw but no upper front teeth. The central African species — the water chevrotain — is a shy animal. When running it does not bend its legs; this stilted gait enables it to outdistance its many enemies. It is colored olive brown with white streaks, and it blends well with its marshy home. Both sexes remain apart except for mating, which occurs in the summer. All species are hornless and must rely on their shy habits for survival. They feed only at dusk and rest in holes during the day.

See also Vol. 10, p. 275: "Mammals."

CIVETS, MONGOOSES AND HYENAS

Carnivores of the Old World Tropics

The African civet (*Civettictis civetta*) — a small catlike mammal.

IN AFRICA and tropical Asia we find a numerous family of carnivores, or flesh eaters — the Viverridae, or viverrids, which range from small to fair-sized. These animals, whose name comes from the Latin word *viverra,* meaning "ferret," include the civets and mongooses and their allies. They lead much the same kind of life in the Old World tropics as do the weasels and their kin in the more temperate regions of the world. Some dwell on the ground or in trees; others burrow into the ground; several species are aquatic. They are generally secretive, nocturnal creatures; comparatively little is known, therefore, about their way of life.

Like the rest of the modern carnivores, the viverrids evolved from a group of small, weasellike, forest-dwelling flesh eaters known as miacids (see Index). At the end of the Eocene epoch, about 40,000,-000 years ago, the miacid stock branched out, giving rise to weasels and dogs, on the one hand, and to the viverrids, on the other. The cats, in turn, evolved from some of the early viverrids.

The viverrids are comparatively primitive and unspecialized animals; some of them have changed little, apparently, from their miacid ancestors. The skull is long

The ring-tailed genet (*Genetta genetta numanni*).

and narrow, and the face suggests that of a fox. The teeth are typical of carnivores; the carnassials (see Index) are extremely sharp and form efficient shearing blades.

There are over a hundred species of **viverrids**, divided into six main groups: the civets, genets and linsangs; the palm civets; the hemigales; the galidines; the mongooses; the fossa.

The animals of the first group — the true civets, genets and linsangs — belong to the subfamily Viverrinae. They are more closely related to the cats than are any of the other viverrids. Their bodies are more streamlined, however, than those of cats and their heads are foxlike. Though some of them climb trees, their feet are better adapted for walking and running on the ground. The curved claws can be more or less retracted, or drawn back, into sheaths, much as in the case of cats. The Viverrinae inhabit Africa, India and Malaysia.

The African civet (genus *Civettictis*) and the Oriental civet (*Viverra*) resemble each other closely. They are heavy-bodied creatures about as large as domestic cats. They look as if they had been compressed vertically; they are exceptionally narrow and stand high off the ground. The hind legs are longer than those in front, causing the body to slope downward from the rear. The fur shows dark streaks and blotches; dark hair can be erected along the back to form a crest. The bushy tail of these civets is shorter, comparatively

speaking, than that of other viverrids. Beneath the base of the tail are scent glands that secrete a thick, dark yellow substance called civet, or musk. It is composed of a complex mixture of ammonia, fats and volatile oils, and has a strong, pungent odor. To extract musk for commercial use, civets are kept in captivity and "milked" every two or three weeks. The musk is used as a fixative in the manufacture of perfume and also as a drug.

The rasse (*Viverricula*) is a smaller Oriental civet, lacking a crest on the back but possessing well-developed musk glands. It is an expert tree climber. The water civet (*Osbornictis*), first discovered in 1916, is an otter-shaped animal with beautiful reddish brown fur, dark and light markings on the face and a bushy tail. Its feet are webbed and the soles are naked; presumably it can travel easily on mud flats. The water civet is apparently a fish eater; its sharp teeth enable it to hold onto its slippery prey.

Another beautiful civetlike animal is the genet (*Genetta*) of Africa, southern Europe and southwestern Asia. Its yellow or grayish fur is darkly spotted, and dark bands ring the very long, tapering tail. A crest of dark stiff hairs can be elevated along the back. The lithe and graceful animal haunts open forests or rain forests, where it seeks its prey; it is equally proficient as a hunter on the ground or in the trees. Genets, with their pointed faces, prominent ears and agile movements, make

engaging pets.   Peoples of the Mediterranean regions have trained them to hunt
mice and rats since the days of the ancient
Egyptians.   When pursuing prey, a genet
elongates and flattens its body as much as
possible and glides along the ground or on
a branch in an almost snakelike fashion.

Like the genet, the Asiatic linsang
(*Prionodon*) stalks its prey by crawling
with its belly to the ground.   Since it is so
slender, it is often mistaken for a heavy-
bodied poisonous serpent.   The linsang
looks like a small genet.   Its tail is exceedingly long; its coat is light in color and
marked with bands or spots.   The animal
does not possess the musk glands characteristic of civets.   The African linsang

slowly and deliberately through the dense
forest canopy.   It has a long, shaggy coat
of coarse, black hairs; long tufts of hair
project from the top of its short ears.   Its
teeth are comparatively small.

Other palm civets of the East Indies
and southeastern Asia include the small-
toothed palm civet (*Arctogalidia*), which
has tiny teeth except for the large curved
canines; the musang (*Paradoxurus*), a
gray-brown animal, somewhat resembling
a raccoon but with a much longer tail; the
fairly large and powerful masked palm
civet (*Paguma*); and the Celebesian palm
civet (*Macrogalidia*).   The west African
palm civet (*Nandinia*) looks somewhat
like a miniature binturong.

The binturong, or bear cat (*Arctictis binturong*).

(*Poiana*) is a spotted, genetlike carnivore
of western Africa.

The animals of the next viverrid
group — the palm civets (subfamily Para-
doxurinae) — are less agile and catlike
than the true civets; their feet are better
adapted for climbing.   The palm civets are
almost completely arboreal and they possess musk glands.

Among the most striking of the palm
civets is the binturong, or bear cat (*Arctictis*), which dwells in the East Indies and
southeastern Asia.   The binturong, the
largest of the viverrids, is the only mammal
in the Old World, other than the marsupials (see Index), to have a prehensile,
or grasping, tail.   Apparently the binturong
uses its tail as an extra hand as it moves

The viverrids known as hemigales
(subfamily Hemigalinae) include several
little-known carnivores of Malaysia and
Madagascar.   The banded palm civet, or
hemigale (*Hemigalus*), resembles the linsang.   It is prettily marked with black
bands across the back; its face and neck
are reddish.   It has climbing-type feet; its
scent glands are not well-developed.   The
Indo-Chinese civet (*Chrotogale*), which
has an elongated muzzle and small teeth,
and the *Diplogale* of Borneo are rare close
relatives.   The otter civet (*Cynogale*) is
equally at home in the trees and water.
Like an otter, it has broad, webbed hind
feet, a swollen upper lip, a broad muzzle,
a flat head, small ears, a short cylindrical
tail and thick, dense, soft fur.   Hemigales

of Madagascar include the Malagasy civet (*Fossa*), which looks like a genet, and the falanouc (*Eupleres*), which is a long-snouted animal with tiny peglike teeth, weak jaws and a short tail like a bottle brush.

The Galidines, or Malagasy mongooses (subfamily Galidiinae), seem to be a link between the civets and true mongooses. They are small, graceful animals, some no bigger than rats. Galidines are common in the Madagascan forests of giant ferns, where the ground is densely covered with vegetation. One type (*Galidia*) is bushy tailed and scampers during the day through the trees like a squirrel. Madagascar, apparently, became isolated from Africa fairly early in the Cenozoic era. Consequently, quite a few unique mammals are found there; for example, the lemurs, which are primates, and the tenrec, an insectivore. Of all the great order of carnivores, only the viverrids are native to this island.

Great numbers of true mongooses (subfamily Herpestinae) live in Africa, southern Asia and the East Indies. These animals lack scent glands. In general, they are slender and long-bodied creatures, with pointed snouts, very short, rounded ears and fairly long, bushy tails. Some are weasellike, and short-legged; others are like dogs, standing high off the ground on rather long legs. The fur is long and coarse, and is grizzled or striped. Mongooses can climb trees but rarely do so. Their long, slightly curved claws, which are not retractile, are more or less suited for digging. Mongooses are small and active carnivores, moving about by day much more than do the civets. They are easily tamed, if caught young, and are often kept to catch mice and rats.

Common mongooses (*Herpestes*) inhabit Spain, Africa and Asia. The ichneumon, or Egyptian mongoose, was sacred to the ancient Egyptians; the crab-eating mongoose is a semiaquatic animal of Asia; the Indian mongoose, made famous in Kipling's Jungle Book, has been introduced into the West Indies. There are different kinds of mongooses in Africa. Among these are the meerkat (*Suricata*), a plains-dwelling animal that likes to bask in the sun and often sits up on its hind legs like a ground squirrel; the dwarf mongoose (*Helogale*), only about ten inches long from nose to rump; the marsh mongoose (*Atilax*), which dives and swims expertly; the cusimanse (*Crossarchus*), which spends much time underground in burrows; the white-tailed mongoose (*Ichneumia*) and the dog mongoose (*Bdeogale*), doglike, forest-living animals; the bushy-tailed mongoose (*Cynictis*), a long-legged, plains-dwelling animal; and the *Xenogale*, which is shaped like an African civet.

The fossa (*Cryptoprocta*), belonging to the last of the viverrid groups, is the largest carnivore of Madagascar; it is the sole representative of its subfamily (Cryptoproctinae). The animal may be a link between cats and civets, or it may have evolved from common ancestors of the cats, true civets and mongooses. The teeth are catlike; the needle-sharp claws are strongly curved and retractile. About twice the size of a house cat, the fossa has somewhat the same shape as the genet. The short fur is reddish or grayish brown. The animal walks flat on the soles of its feet, unlike a cat, which walks on its toes. The fossa climbs trees readily and some-

The cobra-killing mongoose (*Herpestes mungo*).

times seeks its prey there; it also hunts on the ground.

The viverrids are active and alert hunters; they carefully and silently stalk their prey and then pounce upon it swiftly. The white-tailed mongoose, which has quite long legs, may actually run down its prey, however, instead of stalking it. Small animals of all kinds are game to the viverrids: birds, lizards, snakes and insects, as well as rats, mice and other small mammals. Birds' eggs are also eaten. The dwarf mongoose seems to be extremely fond of hens' eggs. It breaks each egg by throwing it, with its forepaws, between its hind legs against a stone or wall. The fossa hunts lemurs and may occasionally kill sheep. The powerful masked palm civet can kill animals twice its size. Certain viverrids, including the Indo-Chinese civet and the falanouc, feast on earthworms. The semiaquatic otter civet, marsh mongoose and crab-eating mongoose, feed on fish, frogs, crabs and other crustaceans and mollusks. The cusimanse uses its elongated snout to probe for insects in soft ground; it also eats carrion.

Mongooses have won fame as killers of poisonous serpents, especially cobras. Actually the mongoose does not go out of its way to prey on snakes, but when it is hungry, it never hesitates to attack one. Bristling out its fur (thus presenting a larger but indefinite target) the little carnivore provokes the snake to strike. The mongoose agilely dances away, and if quick enough, pounces on the snake before

it has a chance to strike again, biting the serpent through the head. Mongooses may be somewhat immune to snake venom, but if they are bitten and receive a sufficient dose, they succumb to it.

The Indian mongoose was imported into the West Indies to rid plantations there of rats. Unfortunately the animal became a pest, killing off great numbers of the native birds and small reptiles and markedly altering the animal population of those islands. Ironically, the mongoose has not eliminated the fer-de-lance, which is a swift-striking viper, not so vulnerable as is the cobra to the mongoose's attack.

With the exception of the fossa, linsangs and Malagasy civet, which are almost exclusively meat eaters, the viverrids supplement their flesh diet with a considerable quantity of vegetable matter, including berries and other fruits and also roots. The binturong is extremely partial to bamboo shoots. The musang (sometimes called toddy cat) drinks the juice, or toddy, of palm trees from collecting vessels that natives have attached to the trees. Musangs also raid native kitchens and storerooms for whatever they can find. Civets help plant trees (unwittingly of course). The seeds of the fruit they eat pass undigested through the digestive tract and are spread over considerable distances, by way of the feces, as the animals wander over their home range.

Viverrids hunt mostly at night; during the daytime they seclude themselves in some protected spot. A hideaway or den may be a hollow tree, an abandoned burrow made by some other animal or a crevice among rocks. Palm civets hide amidst tree branches or in the tops of palms; but thatched roofs of houses and drains and outbuildings may serve as well. Many mongooses dig their own burrows. The cusimanse spends most of its time underground, even finding some of its insect food below the surface. Like most mammals, viverrids confine their prowlings to a home range, which they do not abandon unless shortage of food forces them to hunt elsewhere. After a night's activity, the genet returns to the same hideaway;

The aardwolf (*Proteles cristatus*).

this is probably the case with other viverrids too.

The civet smears tree trunks, rocks and the ground with musk and also leaves heaps of excrement at definite places, thus indicating the location of its home range to other members of the species. Musk plays a part in bringing together the male and female; it is also used as a means of defense, for the civet can discharge the secretion into the face of an attacker.

Most viverrids are solitary, living alone except when male and female pair for breeding. Some mongooses, however, show a social tendency. Ichneumons, for example, may pair up for a fairly long time; bands of as many as fourteen of the animals have been seen hunting together. Dwarf mongooses associate in groups of about ten; they sometimes combine to attack snakes. Two or more families of striped mongooses (*Mungos*), including the young, may be found living together in a network of burrows. Meerkats and bushy-tailed mongooses, both burrowers, live in fairly large colonies.

Little is known about viverrid reproduction. The Indian civet seems to breed in May or June, and three or four young are born. Two young are produced by the Asiatic linsang, which breeds in February and August. The rasse has four or five young; the masked palm civet and the Indian mongoose, three or four. The genet, which bears from one to three young, has no fixed breeding season; the male seems to take part in rearing the young. We know the life span of a few viverrids that have been kept in captivity. The fossa lives seventeen years; the Oriental civet, fourteen; the genet and the mongoose, twelve.

Hyenas and aardwolves, derived from civet stock

During the Miocene epoch, about 20,000,000 years ago, a line leading to the hyenas evolved from the central civet stock. The hyenas (family Hyaenidae) became fairly large and powerful scavengers resembling ungainly dogs. The skull of the hyena is heavy; the teeth and jaws are large; the jaw muscles are exceedingly strong. The animal stands high on long, stout legs. The front legs are longer than the hind ones, causing the back to slope downward from the neck and shoulders to the rump. The claws are blunt.

One of the modern hyenas does not quite fit the above description. This is the aardwolf (genus *Proteles*; subfamily Protelinae), which is somewhat similar to the striped hyena but smaller and more slender. The aardwolf has weak jaws; except for the large canines, there are only small pegs for teeth. Its fur is striped across the back, a mane runs down the center of the back and the tail is bushy. Aardwolves live in brush, plains and rocky country of eastern and southern Africa.

The hyenas proper (subfamily Hyaeninae) include the striped hyena (*Hy-*

The doglike spotted hyena (*Crocuta crocuta*).

aena) and the spotted hyena (*Crocuta*). Striped hyenas range from India through southwest Asia to north and east Africa. They are dirty-gray animals with black stripes on the flanks and legs, a mane on the nape of the neck and along the back and a short and bushy tail. Along the coastal areas of southern Africa lives the rare brown hyena, a dark brownish animal with long mane. It may be a separate species closely related to, or, merely a race of, the striped hyena.

The spotted hyena is somewhat larger than the others, standing about thirty-two inches at the shoulder. It is yellowish buff with dark blotches, has shorter, more rounded ears than other hyenas and lacks a mane. A common African carnivore, the spotted hyena is found over much of that continent south of the Sahara.

The hyena runs down its prey or feeds on the remains of kills made by the big cats and wild dogs. It has a keen sense of smell; it can detect the whereabouts of a carcass several miles away. Its vision is so acute that it sights vultures at a great distance; it makes for any place where these large, carrion-eating birds are gathering. The animal also keeps a sharp lookout, from a distance, for any stragglers wandering away from herds of zebra and antelope. The sense of hearing is good, too; cries of several hyenas fighting over a carcass will attract many more from miles around.

The hyena's stout teeth crack buffalo leg bones with ease, and even demolish the largest bones of the hippopotamus and elephant. Many times all that is left of a carcass, after vultures and jackals have finished with it, is a pile of bones and perhaps a few scraps of meat and hide. The hyenas entirely consume these remains. Brown hyenas eat marine refuse.

Hyenas often make their own kills, usually preying on sick, injured, aged and young animals. They have also been known to bring down stronger animals, such as young rhinoceroses and the larger antelopes. In some areas, spotted hyenas attack donkeys, goats and sometimes cattle; they will on occasion bite sleeping natives in the face or on the buttocks. The aardwolf, not having the powerful teeth of its larger relatives, feeds exclusively on termites and other insects, newly born animals and rotten meat.

Hyenas and aardwolves are more or less nocturnal, spending the day in natural cavities among rocks or in abandoned burrows. Where such dens are not available, the animals dig their own. Usually the spotted hyena does not have any one home range but temporarily confines its activities to an area where zebra and antelope graze. When the herds move to a new region, the hyenas follow in their train. Hyenas regularly deposit their droppings in a certain place; at a distance such an area looks like a fall of snow because of the chalky whiteness of the feces. Fossilized droppings of hyenas have been found in European caves tenanted by these animals thousands of years ago.

Aardwolves and hyenas have no organized social behavior. However, two or three female aardwolves with young may occupy the same den; half a dozen or so sometimes travel together in search of food. Several spotted hyenas will combine in running down an antelope or zebra; as many as twenty of the animals will gather around a carcass. They utter cries that closely suggest human laughter — a low-pitched hysterical chuckling that frequently rises to a higher key.

The spotted hyena has no fixed breeding season; after a gestation period of 110 days, one or two young are born. After the pups are weaned, they are said to be fed on regurgitated food that has been brought up from the parent's stomach. The gestation period of the striped hyena is about ninety days, and usually four young are born. Aardwolves have from two to four young. The natural life span of the spotted hyena is probably about ten years, but in captivity it may reach the age of twenty-five years. We do not know how long the striped hyena and the aardwolf live in the wilds; in captivity the striped hyena may live to be twenty-four years and the aardwolf thirteen years.

See also Vol. 10, p. 275: "Mammals."

SCIENCE AND PROGRESS
(1815-95) VIII

BY JUSTUS SCHIFFERES

STUDENTS OF THE WORLD'S FAUNA AND FLORA

FROM time immemorial man has been compelled to take an interest in the flora and fauna — the plants and animals — that share the earth with him. Some are his food; others, his enemies. As hunter, herder and farmer, he has been forced to understand their ways. Primitive and savage tribes must and do have a practical working knowledge of some of the subject matter that we should place in the category of botany, agronomy or animal husbandry.

However, as we have observed in earlier chapters of THE BOOK OF POPULAR SCIENCE, almost from the first days of recorded history some men have taken a more comprehensive and scientific viewpoint toward the fauna and flora of the globe. Remembering that science itself was once called "natural philosophy," we designate as naturalists the men who accurately observe and describe plants and animals in their natural habitats. Among the famous naturalists of earlier centuries we can list the Greek philosopher Aristotle (384–22 B.C.), the Greek botanist Theophrastus (372?–287? B.C.), the Roman scholar-soldier-diplomat Pliny the Elder (23–79 A.D.), the Holy Roman Emperor Frederick II (1194–1250), the Swiss editor Konrad von Gesner (1516–65), the

Dutch microscopist Anton van Leeuwenhoek (1632–1723), the French physicist René-Antoine Ferchault de Réaumur (1683–1757), the Swedish botanist Carolus Linnaeus (1707–78), the English clergyman the Reverend Gilbert White (1720–93), an American president — Thomas Jefferson (1743–1826) — and a horde of others.

The nineteenth century had its full quota of naturalists, who added immeasurably to man's knowledge of nature. Some of them roamed far and wide to seek out the mysteries of plant and animal life. For the first time the sharp distinctions between species in different areas of the world was brought vividly to the attention of scientists.

The romantic islands of the South Pacific attracted numbers of eager scientific adventurers. Brawny, cheerful Joseph Dalton Hooker (1817–1911), English botanist and physician, served as assistant surgeon on Sir James Ross's expedition to the Antarctic regions; he studied the flora of New Zealand and Tasmania and of the Far South, too. Thomas Henry Huxley (1825–95), staunch ally of Charles Darwin, was a surgeon on H.M.S. Rattlesnake, which did surveying work in Torres Strait, between New Guinea and Australia. Huxley studied the surface life of the South Pacific Ocean; among other things, he made some important discoveries concerning the family of the Medusae. The naturalist Alfred Russel Wallace (1823–1913) spent eight years (1854–62) in the Malay Archipelago. Here he collected the materials for his absorbing narrative THE MALAY ARCHIPELAGO, which appeared in 1869, and he built up the vast insect collection that passed, after his death, to the University of Oxford and the British Museum.

The notable contributions of Von Humboldt to the knowledge of the fauna and flora of South America were discussed in an earlier chapter. His traveling companion, the French naturalist Aimé-Jacques-Alexandre Bonpland (1773–1858) discovered many new species of plants on the South American continent and Mexico.

Charles Darwin, whom we shall discuss later, also did a vast amount of observing and collecting in South America while serving as a naturalist on H.M.S. Beagle. Alfred Russel Wallace joined forces with Henry W. Bates (1825–92) in a collecting expedition in the Amazon Basin. Bates carried on after Wallace's departure in 1850; in 1859 he returned to England with eight thousand hitherto unknown species of insects.

Some of our most fascinating pictures of South America — its birds of many colors, its lush foliage, its sprawling plains (pampas), its lofty mountains — have come from the inspired pen of William Henry Hudson (1841–1922). Born in Argentina of American parents, he wandered about Argentina and Uruguay in his youth, studying the wildlife of these regions. He had chosen to be, so he said, "a naturalist in the old, original sense of the word, one who is mainly concerned with the 'life and conversation of animals.'" Later he settled in England where he became a prolific writer. His ARGENTINE ORNITHOLOGY, written in collaboration with P. L. Sclater and published in 1888–89, is an admirable work on bird lore. Hudson's best-known work, perhaps, is GREEN MANSIONS (1904), a novel whose scene is laid in a forest of South America.

North America also attracted many a roving naturalist. Rafinesque and Audubon, both of foreign birth, were misunderstood pioneers of scientific curiosity in America's backwoods. Constantine Samuel Rafinesque (1783–1840), born in Constantinople (now Istanbul) of French parents, settled in the United States in 1815 and traveled throughout his adopted country on collecting trips. He described many new species of plants, including a number that were medically useful; he also discovered many new species of fishes, particularly in the Ohio River and its tributaries.

John James Audubon (1785–1851) was probably born in Haiti, and was the son of a retired French naval officer. He was educated in France, studying drawing for some time under the great French painter Jacques-Louis David. In

1803 he came to America in order to take possession of a farm owned by his father not far from Philadelphia. After marrying, he tried to settle down as a small businessman in Ohio and Kentucky but failed miserably; to make a living he was forced to paint portraits and to teach dancing and fencing.

From his boyhood, he had spent much time in sketching birds and studying their habits. He now took up seriously the task of watching and painting birds in their natural habitats. He traveled through the wilderness and up and down the Missis-

Am. Mus. of Nat. Hist.

John James Audubon, great American naturalist; two drawings from his *Birds of America*.

sippi River, plying his brush. In 1826 he took his sketches to England, where he worked them up into the great collection of hand-colored plates that won him fame and fortune — THE BIRDS OF AMERICA (1827–38). These plates, with descriptive material supplied by William McGillivray, also appeared in Audubon's ORNITHOLOGICAL BIOGRAPHY (1831–39). He was collecting material for a similar work on AMERICAN QUADRUPEDS when he died; this work was completed after his death by his two sons. Audubon was one of the earliest American conservationists. His name is perpetuated in the famous National Audubon Society, which seeks to protect the birds and other wildlife of America.

The catalogue of later North American naturalists is a long one. It includes John Burroughs (1837–1921), a New Englander who loved song sparrows, bluebirds and juncoes; Scottish-born John Muir (1838–1914), who grew up to love the woods of Wisconsin and the mountains of California; English-born Ernest Thompson Seton (1860–1946), who lived in the backwoods of Canada and the western plains of the United States and who wrote animal tales and travel stories; and Donald Culross Peattie (born in 1898), who combines a profound knowledge of nature with a delightful style.

Switzerland has produced a number of brilliant naturalists. Prominent among them is gallant François Huber (1750–1831). He began to lose his sight at fifteen and later became totally blind. With his faithful servant and, later, his wife and son acting as his eyes, he studied the beautifully complicated life of the bee. He was the first to observe the mating flight of the queen bee, the killing of the drones by the workers, the ventilating system of the beehive and a thousand other intimate details of the community life of the bee.

Augustin-Pyrame de Candolle (1778–1841) was born in Geneva, became famous in Paris and later returned to his native land to become professor of natural history at Geneva. He set forth the first complete natural system of classification for the vegetable kingdom. His work of classification was carried forward by his son Alphonse-Louis-Pierre-Pyrame (1806–93), and by his grandson Anne-Casimir-Pyrame (1836–1925).

The Forel family of Switzerland is also important in the annals of natural history. Auguste-Henri Forel (1848–1931) is chiefly known, perhaps, as a psychiatrist and a pioneer in sex hygiene. But he was also a first-rate entomologist (authority on insects). He was particularly interested in the psychology of ants; his book called THE PSYCHIC CAPABILITIES OF ANTS AND OTHER INSECTS (1901–04) is a classic in the field. His cousin François-Alphonse Forel (1841–1912) was a physician whose real interests lay in natural history and geology. He studied the freshwater creatures of Lake Geneva and Lake Constance, as well as the glaciers of the Alps.

In France, Cuvier and Lamarck were the great names in natural history in the early years of the nineteenth century; we discuss them in an earlier chapter. Pierre-André Latreille (1762–1833) was another

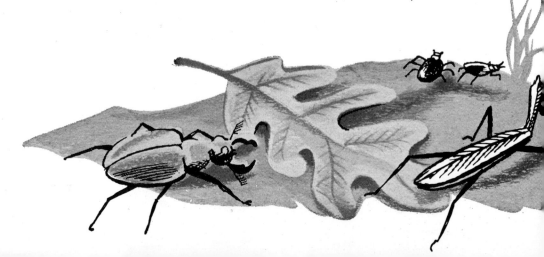

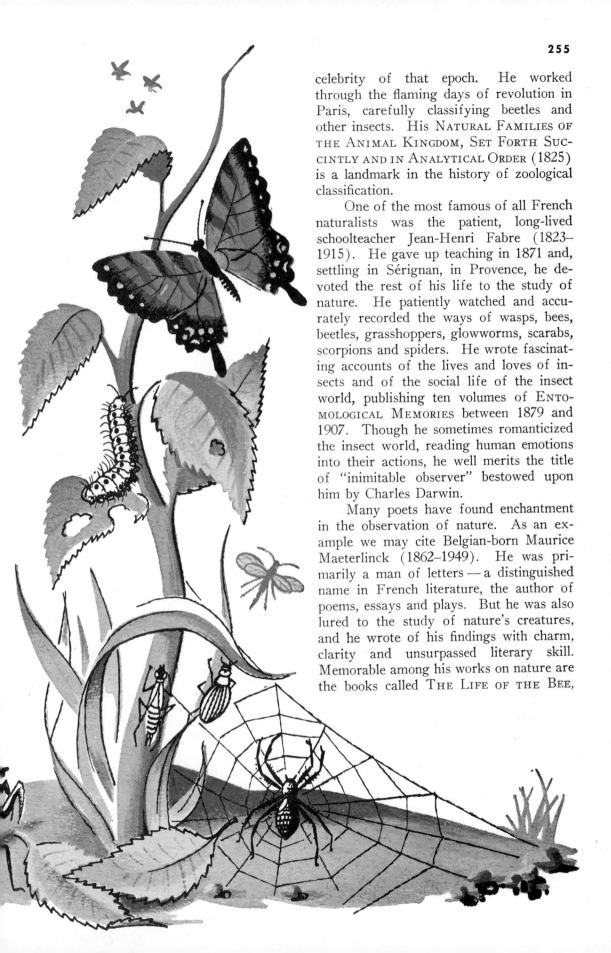

celebrity of that epoch. He worked through the flaming days of revolution in Paris, carefully classifying beetles and other insects. His NATURAL FAMILIES OF THE ANIMAL KINGDOM, SET FORTH SUCCINTLY AND IN ANALYTICAL ORDER (1825) is a landmark in the history of zoological classification.

One of the most famous of all French naturalists was the patient, long-lived schoolteacher Jean-Henri Fabre (1823–1915). He gave up teaching in 1871 and, settling in Sérignan, in Provence, he devoted the rest of his life to the study of nature. He patiently watched and accurately recorded the ways of wasps, bees, beetles, grasshoppers, glowworms, scarabs, scorpions and spiders. He wrote fascinating accounts of the lives and loves of insects and of the social life of the insect world, publishing ten volumes of ENTOMOLOGICAL MEMORIES between 1879 and 1907. Though he sometimes romanticized the insect world, reading human emotions into their actions, he well merits the title of "inimitable observer" bestowed upon him by Charles Darwin.

Many poets have found enchantment in the observation of nature. As an example we may cite Belgian-born Maurice Maeterlinck (1862–1949). He was primarily a man of letters — a distinguished name in French literature, the author of poems, essays and plays. But he was also lured to the study of nature's creatures, and he wrote of his findings with charm, clarity and unsurpassed literary skill. Memorable among his works on nature are the books called THE LIFE OF THE BEE,

THE LIFE OF THE WHITE ANT and PIGEONS AND SPIDERS, among others.

Generally speaking, the naturalists of the world do not create much stir. They go their own way, content to "open a few windows upon the world that is unexplored," to quote Fabre. In the nineteenth century, however, a patient English naturalist, pondering over the data that he and others had collected, developed a theory that was to set the world by the ears. In the following pages we shall discuss this naturalist — Charles Darwin — and his theory of organic evolution.

THE BOMBSHELL — EVOLUTION

In the year 1860, the British Association for the Advancement of Science was holding its annual meeting in the cloistered halls of Oxford University. On the third day of the meeting, June 30, there was to be a debate on a revolutionary book published by Charles Darwin the year before. This work, entitled THE ORIGIN OF SPECIES BY NATURAL SELECTION, maintained that all species of living things had reached their present state through a process of evolution. Among the debaters was to be the Reverend Samuel Wilberforce, Bishop of Oxford and a sworn foe of the evolutionary theory, and the biologist Thomas Henry Huxley, one of Darwin's most ardent supporters.

After some preliminary skirmishing the Bishop of Oxford rose heavily to his feet and launched into an impassioned attack on Darwin's theory. Toward the end of his speech, he turned to Huxley and said smilingly, "I should like to ask Professor Huxley, who is sitting by me and is about to tear me to pieces when I have sat down, as to his belief in being descended from an ape. Is it on his grandfather's or his grandmother's side that the ancestry comes in?"

The good Bishop evidently was under the impression, as so many people have been from that time to this, that the central idea of the ORIGIN OF SPECIES is that man is descended from the apes. Actually, the doctrine of organic evolution, as expounded in the ORIGIN OF SPECIES and in other works of Darwin, stresses particularly the idea that all living things — men, monkeys, amoebas, starfish, birds and plants, for example — have attained their present structure and function through a slow eon-long series of changes in living tissue, beginning with the simplest "primordial protoplasm." The doctrine states further that the direction of changes is guided (though not necessarily altogether) by the process known

Left: Charles Darwin. Below: giant iguana, found in the Galapagos Islands.

George Stone, from Ewing Galloway

either as natural selection or the survival of the fittest. This doctrine of organic evolution, greatly modified since it was first presented, has profoundly influenced the thinking of mankind.

The author of THE ORIGIN OF SPECIES — Charles Darwin — was a mild, modest and persevering gentleman, whose personal life was not nearly so exciting as his ideas. He was born at Shrewsbury, England, in 1809, the grandson of the physician-poet Erasmus Darwin (of whom more later) and the famous potter Josiah Wedgwood. Darwin received his higher education first at the University of Edinburgh, where he had the idea of working for a medical degree, and later at Christ's College, Cambridge, where he planned to study for the ministry.

The turning point in his career came in 1831 when he signed on H.M.S. Beagle to serve without pay as a naturalist. (We described the Beagle expedition in a previous chapter.) For five years Darwin roamed the seven seas; he carefully noted the fauna, flora and geology of many little-known lands. The Galapagos Archipelago, off the western coast of South America, gave him most food for thought. On almost every one of the tiny islands making up the archipelago, he discovered different kinds of living creatures — species unknown to science. That, among other things, set him thinking about the origin of all species — "that mystery of mysteries," as he called it. Upon his return to Eng-

land, the young naturalist became the secretary of the Geological Society. In 1839 he married his cousin Emma Wedgwood; three years later the young couple moved to Downe House, near the village of Downe, in Kent. Here Darwin lived the rest of his life.

In 1837 he had begun to collect facts about the formation of different breeds of domestic animals and plants. He observed in his notebook that selection was the keystone of man's success in bringing about variation in different breeds. "But," he added, "how selection could be applied to organisms living in a state of nature remained for some time a mystery to me." The clue to the solution of this mystery was offered to him by the ESSAY ON THE PRINCIPLE OF POPULATION, by Thomas Robert Malthus (see page 403, Volume 4).

"In October, 1838," he observes, "I happened to read for amusement Malthus on Population, and being well prepared to appreciate the struggle for existence which everywhere goes on from long continued observation of the habits of animals and plants, it at once struck me that under these circumstances favorable variations would tend to be preserved and unfavorable ones to be destroyed. The result of this would be the formation of a new species. Here then I had a theory by which to work."

Darwin now set about to put this theory to the test. Some of the data that he used came from the breeding of domestic animals and birds. (He himself was a pigeon fancier.) He also studied the "record in the rocks" — the geological story and the regular sequence of fossils; "the story in the egg" — the evidence of unfolding layers in the growing organism; the classification of animals and plants according to their characteristics and their geographical distribution. The Darwin and Wedgwood family fortunes made it possible for the young naturalist to devote himself exclusively to his studies. He did not let himself be turned from his self-imposed task by the severe headaches that tortured him with increasing frequency. As the years went on, his views gradually crystallized and a reasonably complete the-ory of organic evolution emerged in his mind. It was one of the most revolution-ary generalizations in the history of sci-ence.

The central idea of evolution goes back to antiquity. The ancient Greek phi-losopher Aristotle, as we noted in our dis-cussion of ancient science, classified living things according to the amount of soul stuff they possessed. He set up a "ladder of nature," in which man is represented as the highest point of one long and continu-ous ascent from the very lowest forms to be found in nature. The seventeenth-century German philosopher Leibniz pointed out that the different classes of animals are so connected by in-between forms that it is practically impossible to determine where a class begins or ends.

In the eighteenth century, the French naturalist Count de Buffon (see Index) had pointed out the close similarities be-tween kindred species such as the wolf and the dog and implied that one species may have developed from the other. He be-lieved that environment has a direct in-fluence on the development of species. "How many species," he wrote, "being per-fected or degenerated by the great changes in land and sea, by the favors or disfavors of nature, by food, by the prolonged in-fluence of climate, are no longer what they formerly were?"

Erasmus Darwin (1731–1802), the grandfather of Charles, was greatly in-fluenced by Buffon's views. In his treatise called ZOONOMIA, OR THE LAWS OF OR-GANIC NATURE, he introduced certain evo-lutionary doctrines of his own. He was impressed by the natural changes produced in animals after their birth (thus the cater-pillar gives rise to the butterfly); the changes produced by man (as in the case of dogs bred for strength, courage, acute-ness of smell or swiftness); the changes produced by climate (as in the case of hares whose fur becomes white in the winter).

He held that all organisms are ulti-mately derived from one and the same kind of "living filament," and that species are

transformed one into the other. "All ani-mals," he said, "undergo transformations which are in part produced by their own exertions . . . and many of these acquired forms or propensities are transmitted to posterity." This statement is only partly true. Animals undergo transformations, but acquired characteristics are not trans-mitted.

The French naturalist Lamarck (see Index) gave his views on evolution in his ZOOLOGICAL PHILOSOPHY (1809). He thought of nature as a creative force con-tinually fashioning living creatures, which he arranged in a broad "staircase" leading upward to the higher apes and man. He held that changes in species had come about in order to meet the challenge of the en-vironment. The giraffe, he argued, got its long neck because it reached for leaves on high trees; moles lost their eyes because they had lived underground for several generations. Lamarck also held (wrongly) that the new characteristics that one gen-eration of animals acquired is passed on directly to the following generation.

Charles Darwin's views on the transformation of species

We have mentioned above only a few important names in this brief account of the predecessors of Darwin. The number of naturalists and others who had referred, more or less vaguely, to the idea of the transformation of species is legion. But Darwin was the first to bolster theory by abundant data. Furthermore, nobody had anticipated the central idea of THE ORIGIN OF SPECIES — that new species may arise as a result of the action of external condi-tions on variations from a specific type. This concept was, as Huxley pointed out, "as wholly unknown to the historian of sci-entific ideas as it was to biological special-ists before 1858."

A thorough, painstaking man, Darwin would probably not have presented his ideas to the world as soon as he did had it not been for an unusual circumstance. On June 18, 1858, he received a manuscript in the mail from the naturalist Alfred Russel Wallace, who was at that time in the Malay

ALFRED RUSSEL WALLACE

Archipelago. Wallace asked Darwin to give him his opinion of the manuscript, which he was to forward to the famous geologist Sir Charles Lyell.

To Darwin's dismay, his own theory, formulated many years earlier, was set forth in Wallace's manuscript. He sat down that very day and wrote to Lyell about this amazing development. "I never saw a more striking coincidence," he wrote. "If Wallace had my manuscript sketch written out in 1842, he could not have made a better short abstract. Even his terms now stand as heads of my chapters."

At the suggestion of Lyell and the botanist Hooker, Darwin sent Wallace's essay to the Linnaean Society, together with an abstract of his own theory. These two essays were read to the Linnaean Society on July 1, 1858, as a joint paper, called On the Tendency of Species to Form Varieties and on the Perpetuation of Varieties and Species by Natural Means of Selection.

Darwin now realized that it was high time to present his theory in its more or less complete form to the general public; and so, on November 24, 1859, he published ON THE ORIGIN OF SPECIES BY MEANS OF NATURAL SELECTION. This work was, next to Newton's PRINCIPIA, perhaps the most important single book in the history of science. Perhaps the best way to give you some idea of its content is to set down what the author himself says in his introduction.

"In considering the origin of species," he writes, "it is quite conceivable that a naturalist . . . might come to the conclusion that species had not been independently created but had descended, like varieties, from other species . . . Naturalists continually refer to external conditions, such as climate, food et cetera, as the only possible cause of variation . . . But it is preposterous to attribute to mere external conditions the structure, for instance, of the woodpecker with its feet, tail, beak and tongue so admirably adapted to catch insects under the bark of trees.

"It is, therefore, of the highest importance to gain a clear insight into the means of modification and co-adaptation. At the commencement of my observations it seemed to me probable that a careful study of domesticated animals and of cultivated plants would offer the best chance of making out this obscure problem. Nor have I been disappointed.

"We shall thus see that a large amount of hereditary modification is at least possible. What is equally or more important, we shall see how great is the power of man in accumulating by his selection successive slight variations. I will then pass on to the variability of species in a state of nature.

The struggle for existence among all organic beings

"We shall be enabled to discuss what circumstances are most favorable to variation. Next the struggle for existence among all organic beings throughout the world, which inevitably follows from the high geometrical ratio of their increase, will be considered. This is the doctrine of Malthus . . .

"As many more individuals of each species are born than can possibly survive; and as, consequently, there is a frequently recurring struggle for existence, it follows that any being, if it vary however slightly in any manner profitable to itself, under the complex and sometimes varying conditions of life, will have a better chance of surviving, and thus be naturally selected. From the strong principle of inheritance, any selected variety will tend to propagate its new and modified form.

"We shall see how natural selection almost inevitably causes much extinction of the less improved forms of life, and leads to what I have called divergence of character . . . I am fully convinced that species are not immutable . . . [and] that natural selection has been the most important, but not the exclusive, means of modification.

"We shall best understand the probable course of natural selection by taking the . . . [imaginary] case of a wolf, which preys on various animals, securing some by craft, some by strength and some by fleetness.   Let us suppose that the fleetest prey, a deer for instance, had from any change in the country increased in numbers, during that season of the year when the wolf was hardest pressed for food. Under such circumstances the swiftest and slimmest deer would have the best chance of surviving and so, of being preserved or selected.

Natural selection, or
the survival of the fittest

"This principle of preservation, or the survival of the fittest, I have called natural selection.   It leads to the improvement of each creature in relation to its organic and inorganic conditions of life; and consequently, in most cases, to what must be regarded as an advance in organization. Nevertheless, low and simple forms will long endure if well fitted for their simple conditions of life."

In the course of the years that followed, Darwin made other notable contributions to his theory of organic evolution — particularly THE VARIATION OF ANIMALS AND PLANTS UNDER DOMESTICATION (1868), THE DESCENT OF MAN (1871) and THE EXPRESSION OF THE EMOTIONS (1872).   He died on April 19, 1882; and seven days later he was buried in Westminster Abbey, where lie many of England's most famous sons.

The most effective tribute, perhaps, to the man and his work was paid by Alfred Russel Wallace, the co-discoverer of the theory of natural selection.   He wrote of Darwin in 1870: "I have felt all my life

and I still feel the most sincere satisfaction that Mr. Darwin had been at work long before me, and that it was not left for me to write the ORIGIN OF SPECIES . . . Far abler men than myself may confess that they have not that untiring patience in accumulating and that wonderful skill in using large masses of facts of the most varied kind, that wide and accurate physiological knowledge, that acuteness in devising and skill in carrying out experiments, and that admirable style of composition, at once clear, persuasive and judicial — qualities which in their harmonious combination mark out Mr. Darwin as the man, perhaps of all men now living, best fitted for the great work he has undertaken and accomplished."

Darwin had remained aloof from the violent controversy aroused by the appearance of the ORIGIN OF SPECIES.   The theory of organic evolution had burst like a bombshell upon the England of the late 1850's and early 1860's.   The theory was violently attacked by certain clergymen

Brown Bros.

Ernst Heinrich Haeckel, above, drew up a genealogical tree, displaying the different groups of living things in a series that ranged from the ameba to man himself. Lopped-off branches represented "missing links." These were sometimes found in fossil forms, such as *Archaeopteryx* (half bird and half lizard), shown at the right. It was found at Solenhofen, in Bavaria.

(including, as we have seen, the Bishop of Oxford) as contrary to the teachings of the Bible. Certain scientists, too, including the comparative anatomist Sir Richard Owen and the geologist Adam Sedgwick, objected to Darwin's findings. Other scientists rallied to Darwin's support: his generous rival Wallace, the botanist Hooker and above all the redoubtable Thomas Henry Huxley.

Why did Darwinism create such a furore? One reason was undoubtedly that the ORIGIN OF SPECIES lumped man together with other living things as a product of organic evolution, thus apparently contradicting the Biblical story of creation. The political and social conditions in England at the halfway mark of the nineteenth century also offer some explanation of the heat of the Darwinian controversy in that country.

Politically speaking, the battle over Darwin represented a fight between liberals — for him — and conservatives — against him. England was feeling the full effects of the Industrial Revolution, which was breaking down old forms, old customs, old families, old fortunes. Those who had benefited by the change — first and foremost, the industrialists who were creating new fortunes and who wanted a louder voice in public affairs — were likely to view all changes as beneficial. They saw a new world of iron and steel, of fast transportation and easier communication growing up around them. Hence they thought of change as progress, and they used the word evolution as a synonym for progress. Again, the fierce competition for markets and for business survival impressed men — that is, those who were successful in the competition — with the validity of the doctrine of the survival of the fittest. (They rarely asked the question: "Fittest for what?") Opposed to these apostles of change were those who mourned the passing of the "good old days."

Evolution became something more than a scientific theory, whose significance

Am. Mus. of Nat. Hist.

few could truly understand; it became a battle cry and even a philosophy. In England Herbert Spencer (1820–1903) built up an entire philosophical system around the concept of evolution. In 1852 — seven years before the publication of THE ORIGIN OF SPECIES — he had used the word "evolution" to describe the production of higher forms from lower forms. He became an eager disciple of Darwin and fitted the latter's theories into his own philosophy. He saw evolution at work in nature, in the human mind, in human society. He carried the philosophy of evolution into political affairs, too. If nature provides that the fittest shall survive, he reasoned, it follows that the freest are the fittest. Personal liberty became, for him, the highest goal. Governments, he argued, must not interfere with "rugged individualists" or with free enterprise, since these represent natural forces at work.

Germany's political liberals and Darwinism

In Germany *Darwinismus* (Darwinism) became closely tied up with the political liberals or radicals of the 1860's and 1870's. Among the scientific radicals was Ernst Heinrich Haeckel (1834–1919), biologist and philosopher. He was the leader of scientific expeditions to the Canary Islands, the Red Sea, Ceylon and Java; from 1865 to 1908 he was the provocative professor of zoology at the University of Jena. He was also the author of many books, including THE RIDDLE OF THE UNIVERSE (1899), in which he dogmatically claimed that he had "solved" the great problems of life and death.

As a young man Haeckel was an enthusiastic and powerful champion of Darwin. He was an important figure in the development of evolutionary theory. For example, he was the first to draw up the now familiar genealogical tree, relating the various orders of living creatures in a series ranging from amoeba to man. Lopped-off branches represent the "missing links." These are sometimes found in fossil forms, like the birds with teeth discovered by O. C. Marsh and E. D. Cope in the Badlands

of South Dakota, or the famous archaeopteryx — half bird, half lizard — found preserved in a piece of shale picked up at Solenhofen, in Bavaria.

Haeckel was the first to present the so-called biogenetic law with his jawbreaking statement that "Ontogeny recapitulates phylogeny." This means simply that in the life history of each individual animal (ontogeny) there are repeated, and in the same order, the stages of growth through which the whole species has passed (phylogeny) in its historic evolution. Or, to put the matter still more simply, every individual is a "historic document." The biogenetic law, so tersely stated by Haeckel, was based on the first-hand research of a number of men, including Thomas Henry Huxley, the German geologist Fritz Mueller and the Russian zoologist Aleksandr Onufrievich Kovalevski.

In France, where the influence of Cuvier (see the Index) was still strong, Darwinism made slow headway at first; in fact, some important French men of science, including the great physiologist Claude Bernard, were never completely won over. However, the evolutionary theory, known in France as transformism, at last prevailed. This led to a renewed interest in the evolutionary ideas of Lamarck; in fact, French Darwinism was strongly tinged with Lamarckism.

Flaws in Darwin's theory of evolution

By the end of the nineteenth century, the evolutionary theory had won wide acceptance. However, even the firmest supporters of the doctrine of organic evolution had to concede that there were glaring inaccuracies and gaps in Darwin's original theory. For one thing, they found flaws in his complicated explanation of the reproduction process in the germ cell. Darwin held (wrongly) that every cell in the body at every stage of the body's growth is represented in the germ cell (sperm or ovum) by a tiny unit called a gemmule. The reproduction power, he thought, really lies with all the body cells, acting through their representatives, the gemmules. The germ

Sir Francis Galton helped introduce fingerprint identification.

cell itself is important only as a convenient meeting place for the gemmules. Darwin called this theory, appropriately enough, pangenesis ("all-reproduction": that is, reproduction by all the cells).

This arbitrary doctrine has been discarded. Modern scientists now believe that certain paired bodies in the germ cells — microscopically small units called chromosomes — are the true bearers of hereditary characteristics. Among the Darwinians who fought most eagerly for this concept was the German biologist August Weismann (1834–1914), professor at the University of Freiburg from 1866 to 1912. Relying on his amazing powers of intuition rather than on the facts actually known to him, Weismann located the physical basis of inheritance in the chromosomes and insisted on the continuity of the germ plasm.

He maintained that only hereditary characteristics could be transmitted by means of the chromosomes; he denied that characteristics acquired in one generation could be carried over to the next. To prove his point he snipped off the tails of

a number of mice at birth; then he raised and bred the little animals. He dealt with their offspring in the same way, and so with many following generations of mice. In all these generations there was not a single case of a mouse born without a tail.

Darwin gave no adequate explanation of the mechanism of heredity and the way in which "spontaneous variations" are carried on from one generation to the next. An Austrian monk, Gregor Johann Mendel (1822–84), was working on this vital problem at the very time that THE ORIGIN OF SPECIES appeared. In 1865 he published a memorable paper called Researches on Hybrid Plants, in which the mechanism of heredity was accurately set forth. But this paper remained unknown to or disregarded by practically all of Mendel's contemporaries; as we shall see, it was not rediscovered until the year 1900.

Though much of Darwin's original theory has been revamped or discarded, its influence upon almost every field of human activity has been very great. History, archaeology and ethnology have undergone profound changes because of the theory; in these fields, as the Swedish biologist Erik Nordenskjöld has pointed out, "the development from earlier to later stages has been the one clue for research." The influence of the doctrine of organic evolution has been particularly strong upon the old biological sciences — anatomy, physiology, morphology, parasitology, embryology, cytology, genetics and so on. It has brought into being at least two new biological sciences — eugenics and biometry.

Sir Francis Galton,
a man of many talents

Versatile Sir Francis Galton (1822–1911) had a great deal to do with the development of these two sciences. Galton, a first cousin of Charles Darwin, was a man of many talents. He was a scientist, linguist, explorer (he sought the source of the Nile), mathematician, meteorologist (he devised the modern weather map), novelist (at the age of eighty!) and the inventor of the ticker tape and the slide whistle.

The name "eugenics" was invented by Galton in 1885; he derived it from the Greek *eugenes:* well-born. According to his own definition, eugenics "is the study of the agencies under special control which may improve or impair the racial qualities of future generations either physically or mentally." He believed that society should seek to increase, by a conscious selective process, the number of individuals possessing desirable hereditary qualities. He admitted that a statesman might consider tolerance highly desirable and pugnacity just as undesirable, while a soldier might take the opposite view. But Galton pointed out that few people would deny that such qualities as health and energy and ability should be fostered.

Galton was one of the pioneers in biometry, the science of statistics applied to biological observation. Beginning with a study of the stud books of basset hounds, Galton deduced a law of ancestral heredity that says, in effect, that the average contribution of each parent to the heredity of his child is one fourth; of each grand-parent, one sixteenth; of each great-grandparent, one sixty-fourth and so on.

Galton set up the first anthropometric laboratory for the measurement of the physical characteristics of man. Among other things, he helped introduce the modern method of fingerprint identification used in crime investigation. Some of the measuring machines he devised for the International Exhibition of 1884, in London — like the punching bag and the test-your-grip machines — are still to be found in penny amusement arcades.

The earlier controversies caused by Darwin's theory of evolution eventually petered out. New controversies followed in their wake. Eugenics became a battleground; the issue was nature (heredity) versus nurture (environment). Is heredity or environment more important in the development of an eminent citizen? Or a genius? Or a moron? We now realize that all this is rather pointless; for a child not only inherits his physical and psychic characteristics but also becomes the product of his physical and social environment. Another violent conflict arose between the mechanists and the vitalists. The mechanists held that all the processes of life can be explained on the basis of chemical and physical laws; the vitalists vehemently denied that this is the case. This controversy no longer excites scientists. It seems to have hinged on the interpretation to be given to certain words, rather than on genuine differences of belief.

The vitalists point out that mechanism is simply a point of view. Human beings and other living creatures act like machines only if you choose to describe them that way. Life is a chemical process only if you insist on analyzing all things chemically. But such a view is narrow and cannot explain everything. As a distinguished American chemist, Jerome Alexander, recently put it: "Many if not all the basic material facts of life are understandable on catalytic [chemical] principles . . . but mental and spiritual phenomena which emerge and which are just as real as the material ones are as inscrutable as ever."

Continued on page 400.

DEEP-SEA EXPLORATION

Venturing into the Earth's Watery Spaces

BY PAUL J. FOX

US Navy

United States naval oceanographic vessel, USNS *Mizar*, used by the Naval Research Laboratory; it is 266 feet in length.

THROUGH the centuries, man has conquered, inhabited and studied some of the harshest environments on the earth's continents. The oceans, however, which cover about 70 per cent of the surface of our world, have not yielded so easily to man's curiosity. In many ways, we know more about outer space than about the great ocean deeps.

Deep-sea exploration, as the name indicates, is concerned with the ocean and its floor and life-forms at great depths. More specifically, it deals with the sea beyond the edges of the continental shelves (see Index) to depths of 600 feet and more. The entire mass of water in this region, however, from the surface down to the greatest known depth of 7 miles, is also included in the subject. (For a series of articles on the sea and the tides, see Volume 6 of THE BOOK OF POPULAR SCIENCE.)

A tremendous variety of methods and instruments is used to investigate the sea and the ocean basins. The one device that is virtually indispensable for most forms of marine exploration today is the oceanographic ship. The first part of this article deals with the oceanographic ship, its equipment and some of the scientific results of its use. The second part of the article takes up special oceanographic research that requires far more than the oceanographic vessel and its apparatus.

Oceanographic ships and their equipment

The oceanographic ship provides a platform upon which scientists collect data from the ocean. It contains laboratory space for biological, chemical, geological, geophysical and meteorological (weather) investigations. There is ample deck room for oceanographic equipment and instruments. All ships have large winches capable of holding three or more miles of strong cable, which is used to lower oceanographic devices to the floor of the ocean.

A typical oceanographic ship is generally not large, averaging a few thousand tons displacement, and is about the size of a small freighter. There is a small crew, and also a number of scientists.

An oceanographic vessel at sea operates in two *modes:* the *underway mode* and the *on-station mode*. During the underway mode the ship moves slowly at 7–12 knots and measures various properties of the ocean and its floor continuously with various instruments. These properties include sea depths, gravity and magnetic patterns and the nature of the bottom layers as determined by seismic (quake) recordings.

During the on-station mode the ship may not move at all. Properties of the sea that cannot be measured continuously are determined by means of coring devices, bottom cameras, water-current meters, heat-flow meters, dredges, biologic trawls, water samplers, and salinity and temperature devices. We will now consider a number of oceanographic instruments in detail and some results of their use. We start with the types used during the underway mode.

Operations of the underway mode

Depth recording. Before the twentieth century, the only known method of determining the depth of the sea at any point was by means of a cable. The ship stopped and a hemp line or wire with a weight at the end was lowered to the bottom. This was a slow, laborious task and not very accurate.

In 1911 an engineer devised a method of determining depths from a moving ship by means of sound pulses. The time required for a sound pulse to travel from the ship to the bottom, from which it was reflected, and back to the ship was measured. Since the velocity of sound waves in seawater is known, the depth of the ocean at that point can be easily determined. This system is known as echo sounding or echo ranging. Later, a continuous echo-ranging method was developed.

The echo sounder uses a pinger mounted aboard the ship below the water-line; it releases the sound pulse. The echo is picked up later by the ship through a sensitive underwater microphone called a hydrophone. The hydrophone converts the sound pulse into an electrical signal, which is then recorded on a continuously moving chart. As the ship sails along, a topographic profile of the bottom is traced on the chart through a series of signals.

Echo sounding has shown that the bottom of the sea is not predominantly flat, as it was once believed to be. On the contrary, its topography may be as rugged and as varied as that of the continents. Jagged mountain ranges, steep valleys and flat plains are found in all the major ocean basins of the world.

A continuous series of mountainous ridges (and associated rifts), the so-called Mid-Oceanic Ridge, extends through the Atlantic, Indian, Arctic, Antarctic and South Pacific oceans and the Norwegian Sea for a total distance of over 35,000 miles. The deepest parts of the ocean floor (down to 7 miles) occur as linear or arc-shaped trenches not far off certain coasts of the continents and island chains. (See Index under Trenches, oceanic.)

The method described above simply gives water depths. It says nothing about the kinds of rocks and other deposits, their thickness and structure, underlying the base of the water itself. One technique that reveals the layers in the ocean floor is seismic, or earthquake, surveying.

Seismic (earthquake) surveying of the ocean floor. An underwater explosive charge is set off; the disturbance produces earthquake-like waves in the layers of the ocean bottom. These waves travel at different velocities through the layers and into the overlying water, where they are picked up by hydrophones. The shooting ship explodes the charge; the hydrophone ship receives the waves. The three bottom layers are described in the text on page 267.

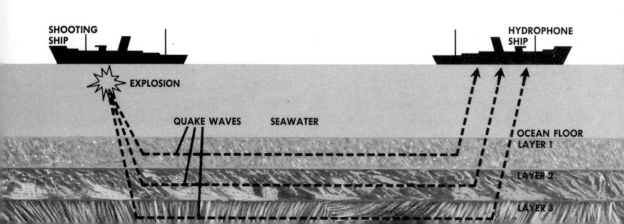

SHOOTING SHIP   HYDROPHONE SHIP

EXPLOSION

QUAKE WAVES   SEAWATER

OCEAN FLOOR LAYER 1

LAYER 2

LAYER 3

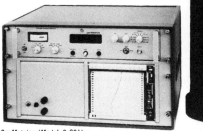

A marine proton magnetometer, used to measure variations in the earth's magnetic field at sea. The tall dark element at the right is the sensor, which is trailed in the water behind the ship. The main instrument with its panel, left, is kept on board and is connected to the sensor.

GeoMetrics (Model G-801)

Seismic surveys. The quakes that are studied here are not natural ones, but relatively weak vibrations induced in the ocean floor by explosions or other intense man-made sounds propagated in the seawater or in the bottom. These sound waves, as they pass through the rocks of the sea floor, are reflected off certain layers or are refracted, or bent, by others. These waves are received back aboard ship through hydrophones as electric signals that are recorded. Refraction surveying uses refracted quake waves, reflection surveying, reflected earthquake waves.

The oceanographer measures the total times it takes for the sound waves to travel from their origin to the hydrophones. From this and with ordinary geometric reasoning the paths of the sound waves can be traced. By recording either a line of explosions at one fixed receiving position or a single shot along a line of receiving positions, a relationship of sound-travel times to distances from the explosion site(s) can be established. This information enables oceanographers to determine the number and thickness of layers in the ocean bottom.

Also, the velocity of sound propagation through a given layer can be determined. This figure enables the oceanographer or seismologist to infer the kinds of material making up the layer in question.

One of the most important discoveries of seismic surveying has been that the earth's crust underlying the oceans is much thinner than that portion underlying the continents. The average thickness of oceanic crust is about 10 kilometers (1 kilometer = about 0.6 mile); that of the continental crust is about 30 kilometers.

Moreover, the oceanic crust is remarkably similar in different oceans. Three crustal layers are almost always present underneath the sea. The uppermost, *Layer 1*, often consists of loose deposits called *sediments*; it is 0.1 to 4 kilometers thick. The velocity of sound waves in Layer 1 is from 1 to 2 kilometers per second. This layer is thicker near the margins of the continents because the continents supply sediments in enormous volumes to the adjoining deep ocean basins.

On the crests and flanks of the mid-oceanic ridges, Layer 1 is unusually thin, only 0.1 to 0.5 kilometer. This fact has led many oceanographers to propose that the ocean basins are growing steadily wider around the ridges and that the crests of the ridges are the youngest portions of the basins. (For a discussion of sea-floor spreading, see the article Landmasses in Motion, in Volume 2.)

Layer 2 ranges in thickness from 1 to 4 kilometers; the velocity of sound waves in it is 4 to 5.5 kilometers per second. It appears to be composed of sediments and solid rocks, consisting of any combination of the following: limestone, sandstone, basalt, metamorphic rock (here, altered sediments) or silica (see Index under these names). Most investigators believe that basalt, a dark, dense rock of volcanic origin, is the most abundant.

Layer 3, the lowermost and often called the oceanic layer because it is identified in all ocean basins, ranges in thickness from 4.5 to 5.5 kilometers; the velocity of sound waves in it is 6.7 kilometers per second. The composition of this layer is not really known, but it may well be gabbro or diabase, rocks of igneous (molten) origin related to basalt (see Index, under the entry Igneous rocks).

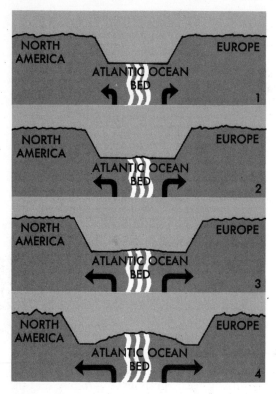

The widening of the Atlantic Ocean and the drifting-apart of North America and Europe through time. Stage 1 (top) represents a period, about 140,000,000 years ago, when the two continents were much closer. Stage 4 shows positions of Europe and North America nearly 20,000,000 years ago.

Underneath Layer 3 is the earth's mantle, which is the part of the interior of our planet that lies between the crust above and the heavy core of the earth below. The upper part of the mantle has a wave velocity of 8.2 kilometers per second.

Thus far, we have measured such physical properties of the ocean and of the underlying crust as depth and the velocities of sound waves in these mediums. In both cases the oceanographers had to create a disturbance in the waters and the earth to determine these properties. Other instruments are more passive in that they simply measure some energy or force in the earth and the sea that is already there. These forces are magnetism and gravity.

Magnetic surveying. The earth's magnetic field converges at the north and south magnetic poles (see the article Magnets Large and Small, in Volume 7). Sensitive instruments have been devised that not only measure the different properties of this field but that also detect very small changes and variations in it. This procedure not only gives information about the magnetic field itself, but also about the magnetism of local rock masses. Rocks and minerals are often affected by magnetism, becoming polarized like the earth itself or like an ordinary permanent magnet.

Magnetometers measure the intensity of the earth's magnetic field at a given spot. The intensity not only represents the strength of the general field there, but also the effect of any magnetized rock bodies present. When the effect of the general field is mathematically removed, there may still be a magnetic pattern left, which must be due to local rock magnetism in the crust itself.

This local effect is called a *magnetic anomaly*. An anomaly may indicate that the rock is strongly or weakly magnetic or nonmagnetic. Geologists know the magnetic properties of numerous rocks and minerals. From a pattern of magnetic anomalies in a small locality or in a larger region, they can guess the compositions of the rocks and the kinds of layers and structures present in these rocks.

Magnetic studies of the earth and the sea have revealed another startling fact about the earth's field. This is that the entire field can suddenly change its polarity — that is, the north magnetic pole becomes a south magnetic pole and vice versa.

If a molten rock mass is solidifying, it acquires the polarity of the existing magnetic field of the earth. It retains that polarity ever afterward, no matter how often the earth's field changes its polarity, or direction. In other words, a rock may preserve the record of a former magnetic field of the earth. This condition is called *paleomagnetism* ("ancient magnetism") and is frequently studied by scientists for clues to the past history of our globe.

From magnetic studies at sea, geophysicists (specialists in the physics of the earth) have discovered that the magnetic anomalies of the ocean basins tend to occur in lines, or narrow belts. Moreover, they are symmetrical with respect to the mid-oceanic ridges:

Space Science & Engineering Center, Univ. of Wisconsin

"In throwing wide the horizons of space, we have discovered new horizons on earth."

Richard M. Nixon

DEEP-SEA EXPLORATION

The ocean, brilliantly blue, teeming with life and covering nearly three fourths of our planet, is mankind's last frontier on earth. In a variety of projects, we are beginning to delve into the mysteries of this vast "horizon."

G.E. Ocean Systems Projects

Perhaps the most significant survey of the sea bottom is the Deep Sea Drilling Project. Cores of sediment recovered by the drilling ship, Glomar Challenger, are providing exciting data on the age and structure of the ocean floor.

At a drilling site (right), a sonar beacon is dropped to the ocean floor where it emits signals used in positioning the ship. The drilling assembly is rigged up and lowered into the ocean. When it touches the ocean floor, power is turned on, and the drill bit starts to penetrate the sediment. The expedition has drilled to depths of more than 3,300 feet below the ocean floor. Below: day's end.

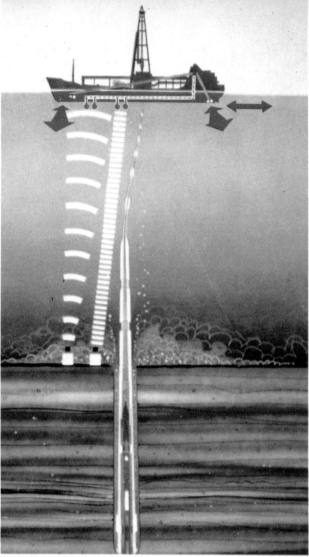

Photos this page, Scripps Institution of Oceanography

The drilling crew assembles the drilling tools. Sediments have been recovered from deposits three miles underneath the ocean surface. They have provided confirmation of sea-floor spreading and, in the words of Dr. Maurice Ewing of Columbia University, opened "a new era in the science of geology."

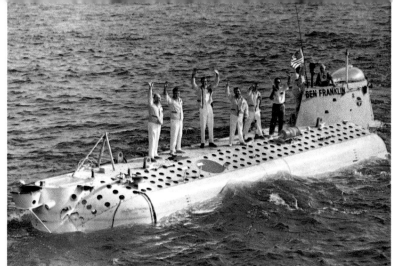

Exploring the Gulf Stream. On July 14, 1969, the research submarine Ben Franklin (right) and her 6-man crew slipped beneath the ocean's surface off the Florida coast. One month and 1,444 nautical miles later, she surfaced, having silently drifted along the mysterious Gulf Stream.

Grumman Aerospace Corp.

Mining the riches of the seas. A number of operations are under way to recover minerals from the oceans. One company has begun to mine manganese nodules (below). It is estimated that there are at least 90,000,000,000 tons of these nodules on the ocean bottoms. Here, the nodules are brought aboard the research vessel R. V. Prospector (left) in wire dredge baskets (below, left).

Three photos Deepsea Ventures, Inc.

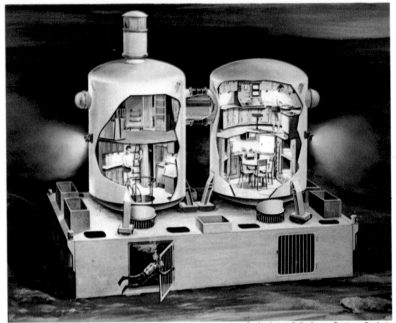

On February 15, 1969, 4 aquanauts began a 60-day underwater mission entitled Project Tektite. Living in a double-domed habitat (left) on the ocean floor in Virgin Islands National Park, the men moved freely from their dry "home" into the surrounding waters.

The coming years. Engineers have designed a number of deep-sea habitats and vehicles which will be tested in the early 1970's. Below: Bottom-Fix, a planned project fur housing men and equipment at 12,000 feet or deeper.

268-d

Both photos G.E. Ocean Systems Projects

lines of anomalies occur parallel to a ridge on each side of it. This suggested to some geologists that thin vertical layers of once molten rock, pushed up through crustal breaks along the rifts, could be the cause of these striplike anomalies.

From their understanding of past magnetic fields, two British oceanographers in the early 1960's proposed that new crustal material, originating in the upper mantle, is injected continuously along the axis of a mid-ocean ridge through the rift associated with the latter. The molten rock, as it solidifies, acquires the polarity of the earth's magnetic field that was present at the time of injection. Since the earth's field reverses polarity periodically, a sequence of rock strips alternately magnetized in opposite directions is found lying parallel to the ridge axis and on each side of it.

The youngest rocks are closest to the ridge, on each side. From the dating of the rocks and the sequence of magnetic reversals, geophysicists have been able to determine that new ocean floor is being formed at each ridge and moving away from the latter on

Life at nearly 16,000 feet down in the oceanic abyss: *Umbellula*, an animal, photographed on the Atlantic Ocean floor.

US Navy

both sides. This means also that the continents are being carried away from each other as the oceans widen. (See the article Landmasses in Motion, in Volume 2.)

Gravimetric surveying. The methods of gravimetric surveying, or measurement of the earth's gravitational field, are similar to those used in magnetic surveys. There are local variations in the strength of the gravitational field owing to differences in the densities of the various rocks that make up the earth. The effects of the earth's general field are subtracted, and the remaining readings indicate the anomalies due to local or regional rock masses. Dense rocks have high anomalies, less-dense ones, low anomalies. Even the absence of any considerable rock formations gives low or negative gravimetric readings.

The typical gravimeter in use today aboard ships consists basically of a spring supporting a weight. Changes in the extension of the spring indicate changes in gravitational attraction at that location. Since the earth's dense core surrounds the center of gravity in our planet, increase in distance from this center weakens the pull of gravity on any body.

Gravimetric surveys of the oceans have shown that the ocean basins are generally in balance with the continental masses. What this actually means is that the mass of a given column of continental crust balances the mass of a given column of oceanic crust plus the overlying column of water. General crustal balance is called *isostasy* (see the Index).

Oceanographers have found, however, that there are striking departures from isostatic balance in some areas of the ocean basins. Strong negative gravity anomalies, pointing to deficiencies in the mass of the rocks, have been detected in surveys over the deep ocean trenches mentioned earlier in this article. These trenches border most of the Pacific coastlines and the island arcs (consult the Index).

Most scientists believe that the trenches represent long, linear zones of compression, where the crust is being buckled down into the underlying mantle. The forces causing this are thought to be related to those widening the ocean basins and thinning the crust at the mid-ocean ridges. In other words, the ridges are places where crust is being formed, while the trenches are places where crust is being destroyed.

The methods of oceanographic surveying described above are used during the underway mode, while the ship is moving. The following discussion of oceanographic techniques and some of their results centers on those employed during the on-station mode, when the ship is often stopped.

Operations of the on-station mode

Oceanographers must find out about the ocean water itself and also about the rocks and sediments on the sea bottom in contact with the water. The underway-mode type of operations gives relatively little information about these factors.

Since ocean depths range from 100 to 11,000 meters (1 meter = about 39 inches), oceanographers have designed specialized items of equipment capable of collecting data to extreme depths. They are lowered by cable to any point under the sea surface. Most oceanographic ships stop once or several times a day to "take a station," that is, take measurements and samples while stationary. These measurements and samples are of various kinds.

Dredging and trawling. Among the first tools developed by oceanographers were trawling and dredging devices to take specimens of marine life and also sediments and rocks from the sea floor. Trawls are for biological work and consist of two basic types.

One kind of trawl has a sturdy iron rectangular frame to which a ridged mesh bag of iron is attached. The trawl is lowered by wire cable to the bottom. Then the ship is moved ahead slowly for a short distance, enough to catch some marine organisms in the bag as it drags along the bottom.

The second type of trawl has a bag of woven nylon attached to the frame. This trawl is pulled only through the water at a selected depth by the ship, and nets specimens of sea life.

Trawl sampling has revealed the presence of life in most depths and regions of the ocean. Marine organisms, both animal and vegetable, are usually concentrated in the uppermost several hundred meters of the ocean, where most of the sunlight and food materials are. Nevertheless, some animals have been adapted to living in the extreme pressures and dark cold of the deepest oceanic abysses.

Dredges are used to obtain samples of rock and loose sediments from the bottom of the ocean. To collect large volumes of sediment, oceanographers use a dredge called a *grab sampler*. This device resembles a large, box-shaped clam. When lowered to the bottom and triggered, the jaws of the sampler close suddenly and scoop up a sample of bottom material.

Dredges for sampling hard rocks make use of a large iron-link bag attached to a steel rectangular "mouth." The dredge is lowered by a strong cable to the sea floor, but only where hard rocks are not covered by deep layers of sediment, as on steep slopes where sediment cannot accumulate. The dredge is dragged by the ship along the bottom; it breaks off protruding edges of rocks and catches them.

Work with dredges shows that much of the ocean bottom the world over is covered by layers of mud and ooze, some of it very deep. This sediment is composed of land-derived material, the minute shells of sea organisms and meteoritic dust. The solid rock is commonly basalt or related igneous rocks, as well as metamorphic rocks and hardened sedimentary rocks (see Index under these rock names).

The mid-oceanic ridges along their crests are composed of basalts and metamorphosed igneous rocks. Many of these have been dated by radiometric methods (radioactivity measurement) and have been found to be young by geologic standards — 1 million to 10 million years.

Igneous rocks dredged from the flanks of the crest are markedly older — 50 million to 100 million years. This situation confirms the theory that the ridges are the sites of new crust formation and that the crust moves away very slowly from both sides of a ridge. In fact, the greatest known age of any oceanic rock is about 170 million years; some continental rocks are well over 3,000,000,000 years old. The present ocean basins, then, are much younger than the continents.

Dredging is a rather crude method of sampling the ocean bottom. It disturbs the rocks and sediments so much that oceanographers may not be able to get an exact idea of what they have obtained and where it came from. They would much rather get small cross sections of as much of the bottom as possible. In other words, a device such as a tube that goes down vertically into the mud and rock and extracts a long *core* preserving all the essential features of the material is far more preferable to a dredge.

Coring. Short coring devices were first developed in the late 1800's. Improvements in later years, such as explosive charges, triggering mechanisms and pistons inside the tubes, have produced very long cores 10 to 30 meters in length.

After a core is taken on deck, it is extracted from the tube intact; it may be studied aboard ship or stored in a container for transportation to a land-based laboratory for future examination.

A core is packed with a tremendous amount of information for the specialists who examine it. They probe it microscopically, chemically, and physically. Any layering in the core reveals the history of the sediments and rocks of the sea bottom. Age analysis dates the core and also gives the rate at which the sediments composing it have accumulated. It may have taken millions of years for a few inches of mud to collect, for example. If possible, the history of the ocean bottom and of the sea itself exposed in a core is tied in (correlated) with geological events in the history of the continents.

Geologists can tell about former current patterns and the temperature and salinity changes the oceans underwent in the past by studying the chemical constituents of the rocks and sediments in a core. The shells of small long-dead plants and animals in the core also tell a similar story. For example, from these core contents historical geologists (specialists in earth history) have worked out the temperature changes in the ocean that took place as the glaciers advanced and retreated on the continents during the last Ice Age, up to 2 million years ago.

The temperatures in the ocean are caused not only by climate and weather and by the ocean's own internal conditions, but also by the flow of heat from the earth's interior. Heat flow also reveals the state of the earth itself at any point. Instruments to measure this heat flow have been devised and are used in oceanographic work also.

A deep-sea bottom-coring device. Suspended on a wire cable, the corer is lowered to the floor of the ocean. When the counterweights hit the bottom, they activate the release mechanism through the control cables. The piston is forcefully driven, as a result, into the sea bed and extracts a core of the rocks and minerals there. The coring device, with its core, is then raised.

DEEP-SEA CABLE

RELEASE MECHANISM

WEIGHT

CORING TUBE

WIRE

CONTROL CABLES

PISTON

COUNTER-WEIGHTS

Heat-flow measurement. Much of the earth's heat energy originates in the interior. It is generated by the extreme pressures in the interior and also by the energy of radioactive mineral concentrations at various spots. Much of this heat energy eventually escapes through the crust and the surface. In general, heat flow from the earth is uniform for land and sea, but there are significant variations.

Information about heat flow enables geologists and oceanographers to draw conclusions about the amount and distribution of heat sources in the earth and about the thermal history of our planet.

Two measurements must be taken to calculate heat flow: the range of temperatures (temperature gradient) and the heat conductivity. These measurements are taken on cores extracted from the ocean floor. Electrical instruments mounted on the coring device give the temperature gradient at different positions on the core. When the core is extracted on deck, heat-conductivity determinations are made. Temperature gradient and heat conductivity, when multiplied together, give the heat flow through a given standard area (unit area) of the ocean floor.

From numerous heat-flow measurements it has been found that in general the heat flow per unit area of the ocean floor is nearly equal to that through each unit area of the continents. High values of heat flow, however, exist along the axes of the mid-ocean ridges. This probably means that hot molten rock matter is indeed being pushed up from inside the earth along the crustal ridges.

All the methods described above do not give an actual picture of the sea bottom, but only certain of its properties. But scientists would like to see what is down there, if they cannot actually go themselves. Photography of the sea floor is one answer to this problem.

Sea-floor photography. Extensive photography of the ocean bottom has been perfected in recent years. Strong, watertight cameras capable of operating almost continuously at great depths have provided oceanographers with numerous photographs of the bed of the ocean. Now scientists can study the distribution and kinds of life far below the surface of the sea, as well as the rocks, sediments, forms of erosion and the patterns of current-ripple marks and other kinds of water markings on the bottom. Today's deep-sea cameras can take many pictures in a single lowering to the sea floor, because of electronic flash units that can be operated over and over again.

Many photographs of the sea bottom have shown that animal life exists at the greatest depths; some of these organisms represent newly-discovered species. In many

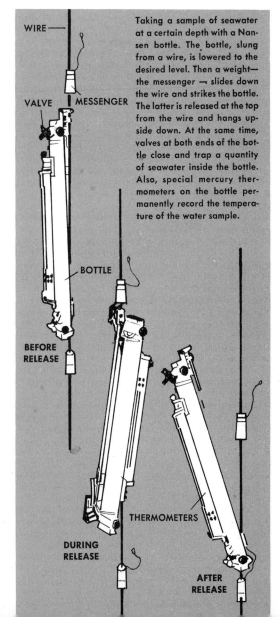

WIRE

VALVE   MESSENGER

BOTTLE

BEFORE RELEASE

DURING RELEASE

THERMOMETERS

AFTER RELEASE

Taking a sample of seawater at a certain depth with a Nansen bottle. The bottle, slung from a wire, is lowered to the desired level. Then a weight—the messenger—slides down the wire and strikes the bottle. The latter is released at the top from the wire and hangs upside down. At the same time, valves at both ends of the bottle close and trap a quantity of seawater inside the bottle. Also, special mercury thermometers on the bottle permanently record the temperature of the water sample.

The marine research submersible craft *Deepstar-4000*. It is capable of descents to 4,000 feet for periods of up to 24 hours. Submarine vessels of this kind have opened up the ocean deeps to man's curiosity and exploration. A great variety of them are in operation at present.

Westinghouse

areas the deep-ocean floor is marked by the action of powerful currents whose existence was hardly suspected before.

Many of the oceanographic methods thus far described have dealt with the bottom, rocks, sediments and life of the ocean. The water itself and its properties, of course, are highly important in oceanography. These properties include its clarity (transparency), its content of mineral and organic matter, temperatures, salinities, pressures, currents, waves and so on. We will be able to touch on only a few of the methods in use here.

Water sampling. Oceanographers take samples of seawater at many localities and depths in order to measure many of the properties mentioned above. They use special containers called *water samplers* for this task. The latest types are sometimes very large and are made of rubberized nylon; most are much smaller, however. There is a valve at each end of the container.

Several samplers, with valves open, are attached at different levels to a wire cable. They are then lowered into the sea until each sampler reaches the desired depth. Next a series of weights is automatically slid down the wire, closing the valves of the samplers in succession and thus trapping the water that has entered the containers. As this happens, special thermometers attached to the samplers record the temperatures of the water samples.

Water-sampling stations determine the temperatures and salinities of the seawater at a given locality. Because of the spacing, however, it is not possible to get detailed readings at all depths in an entire region. Devices that monitor temperature and salinity continuously at all depths are being developed at present.

Current measurement. There are numerous methods of tracking currents in the sea, both on its surface and below it. These involve the use of floating radioactive substances, bottles and other markers, as well as highly visible dyes.

More-sophisticated methods involve the use of special current meters to determine the rates and directions of oceanic currents at the surface and below it. These meters may be used aboard ships or with special buoys, or floats, moored at various spots in the ocean. Neutral-buoyancy floats are set to sink to any desired depth. As they move along, they emit sound signals that are detected by a research ship's hydrophones. The paths of the floats are thus followed, so that current directions and velocities can be charted.

The result of all this work shows that the ocean is full of complex currents moving in many directions and at various depths.

For example, it has been found that a large surface current often has a countercurrent moving along below it, but in the opposite direction.

We have given an idea here of the more-or-less-routine work of an average oceanographic ship. There are other kinds of marine research that require special methods and equipment, such as deep-sea drilling and manned descents to the depths.

Specialized ocean research

Manned descents. Man has long desired to see the ocean deeps for himself. No matter how ingenious his instruments, they are no substitute for going himself for a look. Diving suits allowed men to go down only a few hundred feet at best. Recent scuba or skin-diving methods may eventually allow a human being to descend directly for several thousand feet.

But machines have thus far been the best means for men to enter the abyssal realms of the ocean. In the 1940's and 1950's, the Swiss scientist Auguste Piccard developed the bathyscaphe, a deep-diving vehicle which carried its own electric power and air supply. Ballast and buoyancy chambers allowed the vehicle to control its vertical motions. In the first bathyscaphe Piccard and his crew descended nearly 36,000 feet to the bottom of the Marianas Trench in the Pacific, the deepest known ocean trench. (For a picture story in color of the bathyscaphe, see Volume 6, pages 74-a–74-d.)

Since that time, a great variety of submersible craft have been designed and built to explore underneath the ocean's surface. Their range is from 200 to 8,000 feet down. Many of them travel around and are fully maneuverable at these depths; they are equipped with all kinds of devices — lights, cameras, mechanical arms, nets, water samplers, corers and so on — to explore the sea and its bottom.

Men have also been placed in undersea stations from which they sally forth to study the world underneath the ocean surface. Such stations have been sunken thus far in relatively shallow waters only, but scientists see the day when entire communities of investigators may be settled on the deep-ocean floor. In the recent Sea Lab experiments of the United States Navy, 1968–70, men and women were sent to live at depths of 600 feet in special chambers on the ocean bottom. From such experiments scientists hope to train human beings to live and do research at great depths.

Deep-sea drilling. In the 1950's the idea of actually *drilling* into the deep-ocean bed from the surface of the sea was born. Formidable technical problems were later solved, primarily with the knowledge gained by oil companies drilling for petroleum in offshore wells.

In 1966 the United States Joint Oceanographic Institution for Deep Earth Sampling (JOIDES) was organized. This is a plan for drilling shallow holes into the ocean bed at many places in the world in order to extract cores for scientific study. In 1968 a ship named the *Glomar Challenger* was built basically as a deep-sea drilling platform for this JOIDES project. A dynamic positioning system keeps the ship exactly on the desired location where drilling goes on. The drill can go through as much as twenty thousand feet of water at the end of a series of drill pipes to reach the sea floor. It makes a hole in the bottom and extracts a core.

Since her launching, the *Glomar Challenger* has collected many thousands of feet of cores at locations in the Atlantic and Pacific oceans, the Gulf of Mexico and in the Caribbean Sea. This work has uncovered or confirmed many of the facts about the ocean floor which we have already discussed, particularly in connection with the mid-oceanic ridges. It has also discovered oil and gas deposits at depths of 12,000 feet in the Gulf of Mexico.

As time goes by, new and undreamed-of means of marine exploration will be developed. Entire new chapters will be added to man's knowledge of the deep sea and its history. And there will be increased practical benefits as man learns to use the resources that the sea offers him.

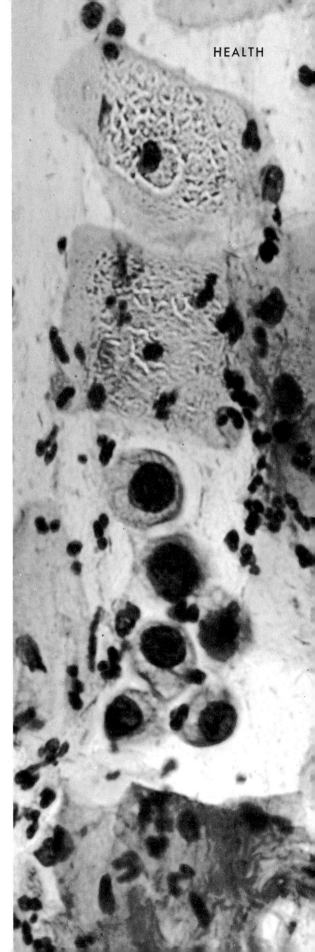

THE CANCERS

A Survey of Malignant Growths
in Man *

BY W. W. BAUER

A CANCER is a tumor — a growth or enlargement of the tissues; but not all tumors
by any means are cancerous. Actually, they
are divided into two general classes: malignant and benign (nonmalignant). A *malignant tumor* grows rapidly or slowly, and
spreads from the place in which it started to
other parts of the body, nearby or remote.
A cancer is a malignant tumor. A *benign
tumor* is confined to the area in which it
originated; its growth is usually slow. The
term "benign" is somewhat misleading, since
a benign tumor may do harm. It would be
better to refer to tumors of this type as nonmalignant.

People sometimes use the word "cancer" in the singular to refer to malignant
growths in general, as if a single specific
disease were involved. It is more accurate
to refer to the cancers — a group of diseases
that have certain features in common but
that may differ widely in other respects.
However, the use of "cancer" in the sense of
"cancers" still persists.

There has been a steady increase in the
number of known cancer cases in the last
generation. The cancers now cause the
second-largest number of deaths in the
United States and in most countries where
statistical records are well kept.** It is estimated that one out of every four Americans
will eventually have a malignant tumor. In

* Article revised by Marie A. Hinrichs, M.D., Ph.D.,
medical consultant, Department of Health Education, the
American Medical Association, Chicago, Illinois.
** Diseases of the circulatory system (including diseases
of the heart, which is the pump of this system) account for
the greatest number of deaths.

Malignant cells (near center of photo) have enlarged, darkened nuclei and shrunken cytoplasm. Above these are two
normal cells with large masses of cytoplasm and small nuclei.

American Cancer Society

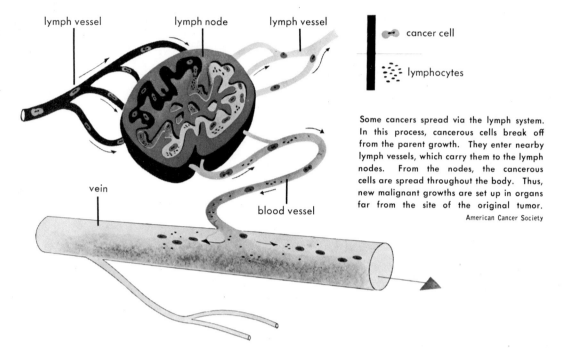

lymph vessel   lymph node   lymph vessel

cancer cell

lymphocytes

Some cancers spread via the lymph system. In this process, cancerous cells break off from the parent growth. They enter nearby lymph vessels, which carry them to the lymph nodes. From the nodes, the cancerous cells are spread throughout the body. Thus, new malignant growths are set up in organs far from the site of the original tumor.

American Cancer Society

vein

blood vessel

children, the cancers are the leading cause of death, not including accidents.

There are various reasons why the cancers appear to be on the increase. For one thing, though their victims are to be found in all age groups, they are particularly apt to attack persons in their middle and advanced years. Naturally, since people live longer these days, there are more cancer victims than formerly. Again, with better methods of diagnosis, doctors can more readily recognize cancerous growths that would formerly have passed unnoticed or that would have been wrongly diagnosed. It is believed, too, that certain habits and conditions of modern living, including heavy smoking and the pollution of the air, may expose people to more cancer-causing circumstances than before.

Cancerous growths
in the human body

Growth is a normal characteristic of the human body, as it is of living things in general. The body develops from a single cell that has been fertilized. This cell divides into 2; the 2 cells divide into 4 and so on, until a little ball is formed. The differentiation of cells into various tissues then begins. The normal human being develops

until he is mature, and then he stops growing. Only when there is an injury is growth resumed and then only to the point of normal healing. New cells are constantly being formed in the skin and other places, it is true, but these cells simply replace those that have died.

Cancers represent wild and uncontrolled growths of cells, which form a highly erratic pattern. These cells resemble somewhat the noncancerous cells in a given tissue, but there are marked differences. Cancerous cells have a more primitive appearance; they are quite irregular in shape and have large and irregular nuclei. They carry out few or none of the functions of normal adult cells. They are most likely to develop, in the case of men, in the skin, lungs, prostate gland, stomach and intestines. In women, the commonest sites are the breast, uterus, intestines, skin and stomach.

Normal cells may be completely replaced by tumor cells. Cancerous cells tend to invade surrounding tissues. They may even break off from the parent growth and migrate through the blood or lymph circulation to other parts of the body, perhaps far from the original site. Here they start new malignant growths. This process of cancer spread within the body is called *metastasis*.

As they continue to grow in uncontrolled fashion, malignant tumors become a menace to life. They interfere with bodily function; the pressures that they create produce severe pain when nerves are affected. Malnutrition goes hand in hand with cancerous growth. The cancerous tissue seems to draw to itself the nutrients taken into the body; normal tissues are deprived of their fair share of these nutrients.

Various cancers cause various signs and symptoms, or changes in function, according to the kind of tissue or body part involved. For example, a cancer of the pituitary gland, or hypophysis, may cause abnormal growth, and the victim may become a dwarf or a giant. When the pancreas is affected, there are often digestive disturbances and diabetic symptoms. In the brain and spinal cord, a malignant growth sometimes causes paralysis. Some cancers can erode blood vessels and cause sudden death by hemorrhage.

Physicians classify cancers according to the tissues in which they originate. We can list only a few representative types here. The *carcinomas* arise from the cells of the skin, the lining membranes of internal organs and the glandular organs. The *sarcomas* develop in muscles, bones, cartilages and connective tissues. Cancers originating in the network of supporting connective tissues in the brain and central nervous system are called *gliomas*. A particularly dangerous type of cancerous growth is the *melanoma*. This is a rapidly growing pigmented tumor; it originates in certain types of pigmented moles found on the skin. If these moles are irritated, they may become malignant, and the resulting cancers metastasize rapidly.

The lymph nodes (sometimes called lymph glands) and other tissues of the lymphatic system (see Index) give rise to cancerous growths called *lymphomas*. The condition known as Hodgkin's disease is now recognized as a malignant lymphoma; it was formerly regarded as an infectious disease. Another lymphoma is lymphatic leukemia. It is marked by enlargement of the lymph nodes (see Index); the white-cell-forming centers in these nodes become unduly active,

FALLACIES AND FACTS ABOUT CANCERS

Will eating hot foods cause a cancer? **No**

Will using aluminum cooking utensils cause a cancer to develop? **No**

Is there any evidence that irregularity of eating or drinking causes cancer of the stomach? **No**

Can a cancer be caused by a bruise or injury?
Note: Not from a single injury. **No?**

Is there any evidence that the use of alcohol bears any relationship to cancer of the stomach? **No**

Could one get a cancer from physical contact (such as kissing) with a person who has cancer? **No**

Could an animal with a cancer pass it on to a person who touched it? **No**

Will eating meat from an animal who had a cancer cause a cancer to develop in a human being? **No**

Do hemorrhoids ever turn into a cancer? **No**
Note: They may hide a cancer and should always be brought to the attention of the physician.

Do corns or calluses ever turn into a cancer? **No**

Can one's mental condition have any influence on the development of a cancer? **No?**
Note: Anyone can develop a cancer.

Is it true that cancers usually develop in persons with poor health? **No**

Once cured of a cancer, can you develop another cancer? **Yes**

In the same place? **Yes**

In some other part of the body? **Yes**

Can a person with tuberculosis develop a cancer?
Note: It has been suggested, but not proved. **Yes**

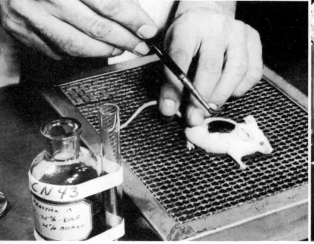

Both photos: American Cancer Society

Tars found in cigarettes will produce skin cancer in mice. The "smoking robot" (above, right) is an automatic device that simulates the smoking manner of human beings. It traps tobacco tars, which are then painted on the backs of white mice (above, left). Cancerous growths result.

and the number of white cells becomes abnormally great. The liver and spleen are enlarged; an acute form of anemia sets in.

The causes
of cancer

In the present state of our knowledge, it is still not clear just what causes a cancer. However, there are certain more-or-less well-established contributing factors.

It is quite certain that *chronic irritation* predisposes to cancers. There are various examples. Cancers of the lip may occur when tobacco pipes are smoked. Irritation due to broken or malformed teeth has often led to cancer of the tongue or cheek. People who have been exposed for many years to sun and wind have developed cancers of the skin. We should point out, however, that a great many people experience chronic irritations of various kinds without developing malignant growths. It would appear therefore that irritation by itself would not suffice to cause cancer. There must also be something in the tissues that increases the likelihood of malignancy.

The theory of chemical irritation through hormones or other chemicals has been advanced. It seems clear that there is a relationship between hormones and cancer. Cancer of the prostate in the male can be checked by means of female hormones;

breast cancers in the female are controllable to a certain extent by the male hormone.

As yet, there is no conclusive proof that viruses cause all cancers. We know that tumors containing viruses have been transplanted successfully from one animal to another. So far, however, there is no real evidence that viruses or any other infectious agents can cause cancers to be transmitted from one human being to another.

There is much controversy over the role of *atmospheric pollution* as a cause of cancers. It is true that these diseases are more apt to strike city dwellers than country dwellers. It is also true that the atmosphere over cities is often seriously polluted with the products of combustion of coal and oil; with the gases arising from various chemical processes; with motor-exhaust gases; with a heavy blanket of soot. The increase in industrialization runs virtually parallel with the increase in lung-cancer cases. Yet we cannot say that there is necessarily a cause-and-effect relationship here, since many other factors are involved.

Statistical studies in the United States and elsewhere leave not the slightest doubt that heavy smoking of cigarettes is associated with the increase in the number of lung-cancer cases.* There is as yet no positive

* *Editor's note*: The relationship between smoking and lung cancer is discussed in the article The Tobacco Problem, in Volume 5.

evidence that cigarette smoking alone causes lung cancer. However, the statistics are disturbing, to say the least.

It is well known that excessive *radiation* will cause cancerous growths. Early investigators who used X rays without appropriate shielding and other protective devices often fell victim to the radiation, and some of them developed cancers. Within recent years, more and more attention has been drawn to radiation as a cause of malignant growths, what with weapons based on nuclear fission and the increasing use of radioactive substances in medicine and in industry.

Careful studies have been made of the amount of radiation that an individual can endure without running into danger. The safety margin is not so great as was formerly believed. The average person can withstand only a limited amount of radiation in a lifetime. Cosmic rays (see Index), to which we are all exposed, account for some of it; a certain amount stems from medical and dental X rays. More may come from radioactive fallout resulting from the setting off of atomic bombs or from the escape of radioactive materials into the community from industrial establishments or wastes.

Chemical substances added to food may have some bearing on the development of cancers. Federal law in the United States requires that any substance that is shown to cause malignancy in animals must be excluded from human foodstuffs and cosmetics. With this in mind, researchers are now making extensive studies on many substances used in the preservation, coloring, flavoring, conditioning and preparation of foodstuffs.

What of *heredity*? Is a person more likely to develop a cancer if other members of his family have been cancer victims? The bulk of the evidence seems to indicate that the answer is "No." The statistics showing the occurrence of cancers in certain families can be accounted for on a non-hereditary basis, with a few conspicuous exceptions. One thing seems more or less certain: that a cancer is not inherited. What may be inherited (and this is by no means certain) is a tendency to develop a cancer.

The diagnosis and treatment of the cancers

Attempts to develop an anticancer vaccine have not been successful; we cannot at present prevent the formation of malignant growths. It is extremely important therefore to detect a malignant condition at the earliest possible stage. We should be on the lookout for precancerous growths — growths that may ultimately develop into cancers. For example, there may be a change in the mucous membrane covering the tongue or lining the mouth cavity, causing a patchy appearance — a condition known as leukoplakia, or white patches. It is very definitely a precancerous condition; the person in whom it appears should not only consult a doctor but should at once give up the use of tobacco. Persistence of sore places in the mouth or on the lips is potentially precancerous. In older persons, a small sore may develop on the skin, heal and then come back. Such sores should not be neglected, because they may be and often are precancerous. In some cases, precancerous growths rapidly develop into cancers; in others, as much as ten years may elapse before the growths become malignant. Sometimes they never become cancerous.

The American Cancer Society has given wide publicity to the symptoms that may indicate precancerous or cancerous growths. Here is a list of these symptoms:

1. Unusual bleeding or discharge.
2. A lump or thickening in the breast or elsewhere.
3. A sore that does not heal.
4. Change in bowel or bladder habits.
5. Hoarseness or cough.
6. Indigestion or difficulty in swallowing.
7. Change in size or color of a wart or mole.

If a person has any of these symptoms once and only once, there is no reason to be alarmed. However, if they persist or if they keep coming back, it is time to consult a competent physician.

It will be easy for him to recognize certain types of cancers that can be seen or

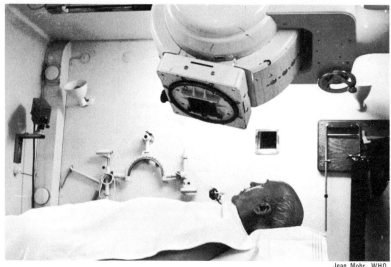

Applying radiation treatment for cancer at the Karolinska Hospital in Stockholm, Sweden. A beam of radiation is focused on a malignant tumor. Such therapy is effective in eradicating or retarding some types of localized cancer.

Jean Mohr, WHO

felt. In other cases, he will use various tested methods of diagnosis. Diagnostic methods in the detection of cancer will vary according to type and location of suspected cancer tissues. For example, transilluminating (shining light through) breast tissues may reveal opaque (dark) masses that can later be further explored by palpation and biopsy. The most reliable is the technique known as biopsy, which consists of a microscopic examination of a small portion of the

Dr. George Papanicolaou, developer of the "Pap" test, a simple painless method of detecting early cancer of the cervix.

American Cancer Society

suspected cancer. In some cases, such as those involving growths on the breast, the biopsy can be performed quickly. After an incision is made, a small bit of tissue is removed from the suspected area. It is frozen and examined under the microscope while the patient is in the operating room. If the report indicates that the growth is cancerous, the surgeon proceeds with the necessary extensive operation; if the report is negative, he merely removes the tumor.

The so-called Papanicolaou technique ("Pap" test) is used in examining secretions from the lung, stomach, mouth, intestines, bladder and uterus. It is based on the fact that malignant growths on the surface of internal organs shed their cells into the secretions from these organs; the cells can then be identified under the microscope. X rays are also a valuable means of diagnosis, particularly in cases involving the digestive tract, kidneys and chest.

There was a time when a diagnosis of cancer was equivalent to a sentence of death after prolonged suffering. Today one third of all persons with cancers can be cured. This number could be increased to one half the total number of cases if more people paid attention to the warning signs and procured medical treatment promptly.

All cancers start at some definite location in the body, and few start in more than one place at the same time. It is vitally important to root out a cancer completely at its starting point before it invades nearby tissues or sends malignant cells to other parts of the body.

There may be a choice of therapy in the treatment of a given cancer. The physician will select the most promising method. For example, some malignant tumors are preferably treated with various kinds of radiotherapy. Others respond to drug treatment, although, at present, there is seldom any permanent cure to be obtained with drugs. Surgery is preferred if it is at all possible. The surgery, which may be extensive, involves completely removing the cancerous growth. In the radiation treatment, X rays or radioactive substances are employed. Beams of radiation are focused on the tumor, killing the cancer cells. In some cases, minute particles of radioactive material are inserted directly into the cancerous tumor, so that the radiation reaches the tumor cells without having to pass through normal tissues. Whether surgery alone will be recommended or radiation alone or a combination of surgery and radiation will depend on the location of the tumor, the stage of its development and the condition of the patient. In some cases, radiation is used before the operation; in others, after the operation; in still others, both before

and after. Chemotherapy is used in some cases to supplement other treatments. Many different chemicals have been used singly or in combination with other chemicals, as in the treatment of certain forms of leukemia.

After a person has been treated for cancer with apparent success, a five-year waiting period must follow. If no cancerous growths develop within this period, he or she may be presumed to be cured, although this is not absolutely certain. It is important that he be under medical supervision during this five-year wait.

Even when there is no hope of saving the patient's life, radiation or surgery or both may still be useful in relieving pain and increasing the patient's comfort. In addition, sedatives, pain-killing drugs or, in some cases, selected hormones can help to keep pain under control.

Many quacks have offered to provide relief from the pain of cancer or even a complete cure without being able to make good on their promises. In some cases, a sincere but misguided person may be convinced that he has stumbled upon a "cure," which he then presents to the public. Whether a false cure is offered in ignorance or with fraudulent intent, the result is the same. The cure

The effect of radiation on cells. Left: this microphotograph shows a normal plant cell about to divide. The chromosomes of its nucleus have split into two similar groups. Right: this plant cell has been exposed to a damaging dose of X rays. The chromosomes are tangled, frayed and broken. Such a cell may die or, if it lives, may pass along its altered heredity to future generations. It may even produce a cancer.

Brookhaven National Laboratory

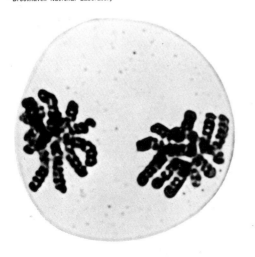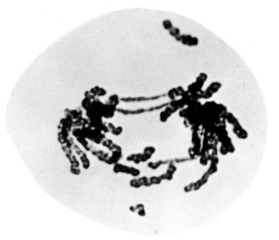

can do no good, and it may do harm. If a given case is curable and time slips by while the patient is taking the bogus "cure," he may resort to tried-and-true methods too late to save his life.

Modern cancer research

Research in cancer is being conducted today in many laboratories and clinics. Physicians and biologists have noted many peculiarities about the cancer cell, but they do not yet fully understand the significance of all their discoveries. Among the abnormalities of cancerous cells are the following. There are more chromosomes, or units of heredity, in a cancer cell than in a normal cell. Fragmentation of chromosomes appears in some cancer cells. Also, malignant cells are interconnected by very thin strands of cell substance, or protoplasm. Another phenomenon displayed by cancer cells is that they can be most dangerous when they are disintegrating. When they do so, they release harmful substances that damage the organs and tissues of the body.

Cancerous tissues are also covered with antigens (see Index) which are somewhat like the antigens produced by infectious germs in the body. The body secretes antibodies to combat invading antigens, as part of its immunity defense. Certain authorities believe that some people develop cancers without knowing that they do so; the cancers, however, are rapidly destroyed by antibodies. Drugs and vaccines that stimulate a person's immune responses to cancer are being sought.

The body also secretes a biochemical known as *interferon*, which interferes with the development of viruses invading the body. It has been found that interferon also causes some tumors to regress or even vanish in animals. Drugs that stimulate the body's production of interferon may prove to be highly effective against certain cancers. One such drug, Poly-I-C, has combatted tumors in animals. Some authorities, however, do not think that interferon production is the real solution to the cancer problem in humans.

Another class of body secretions is the so-called *chalones* (from the Greek *chalan*, "to slacken"), which are normally present and prevent many kinds of regular body cells from multiplying much of the time. Tumor cells have been found to be low in chalones, while otherwise-normal cells near the site of the malignancy are high in chalones. Experimental injection of chalones into cancerous laboratory animals causes certain tumors to decrease or even disappear.

Cancer cells are actually more susceptible to drug action when they are growing and multiplying. Certain drugs may be very potent in attacking cancer cells during growth. For example, the drug L-asparaginase stops the spread of leukemia (blood-cell cancer) by destroying certain chemical foodstuffs needed by growing leukemia cells. L-asparaginase is undergoing clinical tests on human patients at present.

Evidence linking certain cancers and viruses is increasing. Recent refined analytical techniques have disclosed the presence of viruses in certain cancerous tissues long thought to be virus-free. One theory holds that such viruses are weak and defective, lying inactive for long periods, even years. But they become active and cause cancers when body conditions become right for them. Experiments with animals have shown that leukemia-causing viruses can activate other types of cancers (such as sarcomas) when they are injected. The virus that causes herpes (a nerve and skin disease) is suspected of also causing leukemia and similar blood diseases in human beings.

The early detection of cancer is vital for its cure. Simpler and surer tests for cancer are being sought, especially those based on blood and urine sampling. A recent blood test has revealed the presence of intestinal cancer very precisely in humans.

There is no reason to believe that researchers will not eventually be successful in finding cures for cancers. In the meantime the terrible toll that cancers take would drop significantly if people learned to recognize the symptoms and consulted a doctor as soon as possible.

See also Vol. 10, p. 278: "Medicine, Progress of."

THE AMATEUR WEATHERMAN

How to Set up Your Own Weather Station

BY ROBERT MOORE FISHER

TWO thousand years ago, the Roman poet Ovid wrote that "wet weather seldom hurts the most unwise; so plain the signs, such prophets are the skies." For the amateur weatherman today, the signs are still plain to read. They can be found far aloft in the clouds as well as close at hand in the weatherglass. By learning to interpret them, you can follow the weather and predict its major variations.

A continuous record of daily observations will reveal weather trends, averages and extremes for your locality. Take at least one daily observation at a regular hour, preferably toward evening. If convenient, the hour should coincide with the hour of the nation-wide weather observations published in your local newspaper. An early morning "ob" (observation) is also useful.

A daily observation should include a description of the sky, instrument readings of air temperature and pressure, a notation of wind direction and speed, measurements of relative humidity and precipitation and an estimate of visibility. You can make your own record sheets, using the column headings and symbols as shown in the table on the following page. In this article, we shall show how to record various kinds of weather phenomena on a record sheet.

Judging sky cover and types of clouds

The first items on the sheet, you will notice, are "sky cover" and "cloud type." The phrase "sky cover" refers to the extent to which the sky is filled with clouds; the proportion of cloud cover is expressed in tenths. You can best estimate the total

THE AMATEUR WEATHERMAN'S DAILY RECORD

Date	Sky		Temperature					Pres-	Wind		Humid-	Rain-	Visi-	Remarks
Sept.	sky cover	cloud type	cur- rent	maxi- mum	mini- mum	mean	degree day	sure (in.)	direc- tion	speed (mph)	ity (%)	fall (in.)	bility (mi.)	Remarks
1	⊕R	Ns	55	61	42	52	13	29.81 −	SE	20	91	.47	2	Low overcast
2	◑	Ac	68	76	50	63	2	29.75+	SW	10	72	.12	7	Further clearing
3														
4														
5														

Average												
Extremes												
Normal												

SYMBOLS
O = clear, 0/10−1/10 cloud coverage.
◐ = scattered, 2/10−5/10 cloud coverage.
◑ = broken, 6/10−9/10 cloud coverage.
⊕ = overcast, over 9/10 cloud coverage.

T = thunder-storm.
F = fog.
R = rain.
S = snow.
L = drizzle.

READ Temperature to nearest degree. 79.5°: read 80°; 82.5°: read 83°. If two daily observations are taken, extra columns can be provided. Pressure to nearest 1/100 inch. + means rising during past 3 hours; − means falling. Humidity to nearest 1 per cent. Wind speed to nearest 1 mile per hour. Rainfall to nearest 1/100 inch. T = trace (less than 1/100 inch). If none occurs, leave column blank. Observations concerning thunder, lightning, smoke, haze or other special phenomena are entered in the "remarks" column.

cloud coverage by viewing the four quarters of the heavens successively and then averaging the separate amounts of coverage. For example, if the northern quarter of the sky is 7/10 covered by clouds, if the southern quarter is 5/10 covered and the other two quarters are clear, the sky as a whole is 3/10 cloud-covered (7 + 5 + 0 + 0 = 12, divided by 4 = 3).

"Cloud type" represents the particular cloud formation visible at a given time; each formation is indicated by a symbol. In the table on page 222, Volume 2, you will find a list of familiar cloud formations as well as a brief description of each cloud and the symbol used to identify it. Consult this table and use the appropriate symbols when making entries on the record sheet.

As weather heralds, the clouds foreshadow future developments by their direction of movement and order of appearance. In the Northern Hemisphere, you can generally predict warmer temperatures as long as low clouds travel from the south. Cooler temperatures will prevail as long as they move from the north. Rainfall is likely if high clouds thicken and are replaced by formations of middle and then low clouds. The faster this sequence occurs, the sooner you can expect rain or snow to begin, and the sooner it may clear afterward. The ground will be wet within the hour when fluffy heaped-up clouds boil upward, the sky darkens and static crackles on AM radios.*

Editor's note: For the difference between AM (amplitude modulation) and FM (frequency modulation), see the article An Introduction to Radio, in Volume 3.

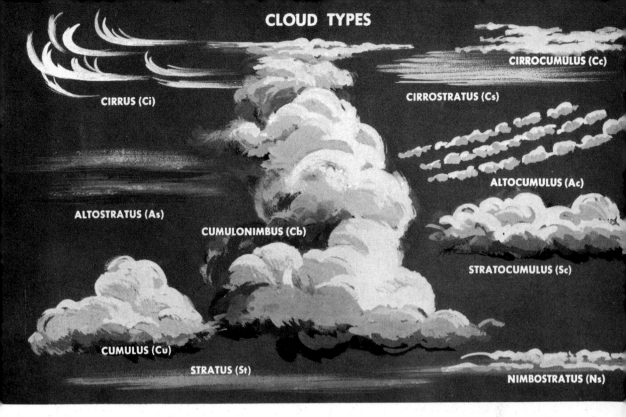

CLOUD TYPES

CIRRUS (Ci)

CIRROCUMULUS (Cc)

CIRROSTRATUS (Cs)

ALTOCUMULUS (Ac)

ALTOSTRATUS (As)

CUMULONIMBUS (Cb)

STRATOCUMULUS (Sc)

CUMULUS (Cu)

STRATUS (St)

NIMBOSTRATUS (Ns)

How to measure air temperature

The most common instrument for measuring temperatures is a liquid-in-glass thermometer containing mercury or alcohol. Mercury is better than alcohol except where temperatures fall below the freezing point of mercury — about 38° F. below zero. The liquid is sealed in a slender glass tube with a bulb at the bottom; above the liquid in the tube is a partial vacuum. The mercury or alcohol expands and rises in the tube as the temperature goes up; it contracts and goes down as the temperature falls. A scale showing degrees of temperature should be etched on the glass tube.

For a dollar or so, you can buy an inexpensive thermometer large enough to be read easily and accurate enough for your purpose. You should test, or calibrate, it by comparing it with an instrument of known accuracy. To check the 32° F. mark, immerse the bulb of the thermometer in a pan filled with a mixture of water and ice; keep stirring this mixture. The temperature should read 32° F. — the melting point of ice. Test for the boiling point at sea level — 212° F. — by bringing water in a teakettle to a boil and holding the bulb at the kettle's spout, from which steam is escaping. If the readings on your thermometer deviate from 32° F. and 212° F. respectively, make the necessary adjustments in all future readings.

Mount your thermometer a few inches away from a wall surface that faces north. Never expose it to the direct or reflected rays of the sun; if you do so, the thermometer will be warmer than the air itself and its readings will be inaccurate. In reading a liquid-in-glass thermometer, keep your eye exactly level with the top of the column of mercury or alcohol.

Perhaps you will want to make a shelter for your thermometer and other weather instruments. We show you a typical shelter in Figure 1. It should be securely mounted, preferably over grass at a height of about five feet above ground in an unshaded site; the door should be on the northern side. The shelter should permit free circulation of air around the instruments and should protect them from rain, snow and direct or reflected sunlight. It should have louvered or slatted sides, a ventilated floor and a double roof (with an air space between the two sections), sloping away from the door side. The shelter should be painted white to reflect the sun's rays.

1. A shelter for weather instruments is illustrated in A. In B, we see details of the shelter's construction.

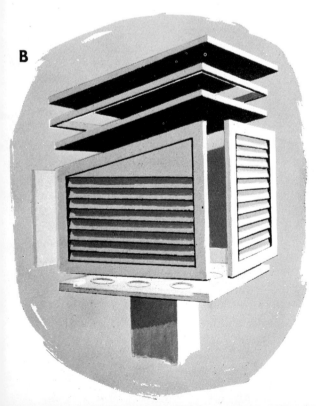

The day's maximum and minimum temperatures are usually published in your local newspaper. The twenty-four-hour mean temperature is calculated by adding the maximum and minimum readings and dividing by two. A degree-day occurs whenever the daily mean temperature drops at least one degree below 65° F. For instance, a day with a mean temperature of 60° F. is a five degree-day (that is, five degrees lower than 65° F.). A knowledge of the degree-days to be expected during a month is useful in estimating the amount of heat needed for homes, offices, buildings or factories.

Normally, the temperature reaches its minimum at daybreak and its peak about three o'clock in the afternoon. The greatest daily range comes most often with clear skies and light winds; the least with a low overcast and a steady breeze. The daily mean temperature varies with the kind of air mass that is present. An air mass from the north ordinarily brings much lower mean temperatures than one traveling from the south. A change from one air mass to another — caused by the passage of a weather front (see Index, under Fronts) — results in a marked rise or fall in daily mean temperatures.

How to measure changes in atmospheric pressure

One of the most helpful weather instruments is the barometer, which measures the weight (*baros*, in Greek), or pressure, of the air overhead. If you plan to become a dyed-in-the-wool weather hobbyist, you should invest in a good barometer.

The easiest instrument to use is a box-like aneroid (without liquid) barometer. It is composed of a disk-shaped metal cell, or box, with a flexible top. (In some models, the bottom is flexible too.) The box is sealed after nearly all the air has been pumped out of it; a strong spring keeps it from collapsing under the pressure of the outer air. When the atmospheric pressure outside the box increases, the top is pushed slightly downward against the pressure of the spring; when the pressure decreases, the spring pulls the top slightly upward.

The movement of the top is communicated by a system of levers to a pointer on a dial, from which the pressure can be read.

The mercury barometer is more accurate than the aneroid type, and is the kind used for many official readings. It consists of a glass tube about three feet long, sealed at one end and filled with mercury. The tube is inverted, with its open end immersed in a glass container of mercury. An increase or decrease in air pressure causes the mercury column in the tube to rise or fall. Hence, the length of the mercury column becomes a measure of air pressure. That is why barometer readings are given in inches or millimeters.

You can make your own mercury barometer, as shown in Figure 2. Provide yourself with a pound of mercury (which you can buy from a druggist) and a slender but strong glass tube of uniform bore about three feet long. Close one end of the tube by melting it over a hot flame until it seals itself. Fill a small glass, such as a wineglass, half full of mercury. Next pour mercury into the open end of the glass tube. Work slowly and tap the tube sharply now and then to keep out bubbles of air. When the tube is full, hold a finger firmly over the opening, turn the tube upside down and lower it into the glass. Do not remove your finger until the mouth of the tube is completely immersed in the mercury of the glass. When you take away your finger, the mercury column will drop down in the tube.

You will have to provide some permanent support for your barometer in order to keep the tube in a vertical position on a plumb line. If you wish, you can clamp it to a ring stand; a pair of wall clamps will also serve your purpose. The tube should be adjusted so that it does not rest on the bottom of the glass.

On a card, measure and mark a scale from 28 to 32 inches, graduated in tenths and twentieths. Now fasten the scale to the tube in such a way that the 28-inch mark is exactly twenty-eight inches above the mercury level in the small glass. From now on you will be able to note changes in air pressure fairly accurately.

One of the simplest kinds of barometers is the old-fashioned Cape Cod or Dutch weatherglass, provided with an S-shaped spout (Figure 3). To fill the weatherglass, pour water through the spout until the level of the liquid remains part way up the spout. This level will dip or mount as the air pressure rises or falls.

Your barometer should be kept away from drafts or sunlight; it should be set up in a spot where the temperature remains as constant as possible. Since the pressure inside or outside a house is the same, a barometer can be kept inside in a convenient place. You can test it by comparing the readings with those given in radio, television or newspaper weather reports. If a Weather Bureau office is nearby, you can take the instrument there so that it may be properly adjusted.

When reading an aneroid barometer, tap it gently with your finger or a pencil to make sure that the pointer is not sticking. Do not put too much faith in a single observation; watch, instead, for any definite trend. During the day, there is a normal fluctuation in pressure, because of the action of the so-called atmospheric tides. The highest pressures tend to occur about 10 A.M. and 10 P.M., the lowest, about 4 P.M. and 4 A.M. If the barometer rises rapidly, you can usually anticipate clearing skies. But "when the glass falls low, prepare for a blow."

How to determine wind direction and speed

Wind direction corresponds to the sector from which the air moves. A north wind, for example, blows from the north; an east wind, from the east. Direction is ordinarily indicated by a wind vane designed to point into the wind. You can build a serviceable wind vane of wood or sheet metal, as illustrated in Figure 4. The split wings and the weighted pointer, often in the form of an arrow, should be balanced at the point of mounting so that about two-thirds of the length is to the rear. The vane should be suspended at its center of gravity so that it swings easily in a horizontal plane. Set up markers showing directions on the

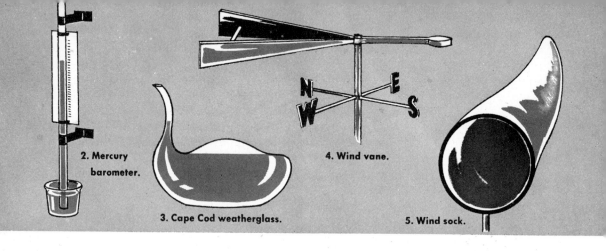

2. Mercury barometer.

3. Cape Cod weatherglass.

4. Wind vane.

5. Wind sock.

shaft of the vane, as shown. You can also find the direction from which the wind is blowing by a wind sock, made out of a basket ring or barrel hoop and water-repellent cloth (Figure 5).

Wind speed can be measured by a deflection anemometer, consisting of a vertically hung wood or metal plate, which swings freely when the wind blows. The anemometer should be combined with a vane, as shown in Figure 6, so that the plate will always face squarely into the wind. The plate should measure 6 by 12 inches and weigh 7 ounces, or 200 grams. (To meet this figure, fix an added weight to the back of the plate if necessary.) As the wind blows, the plate swings at an angle to the vertical, above an attached arc, as shown. The markings on the arc represent angles (from 0° to 90°); you can work them out by using a protractor (see Index). To find the miles per hour of wind speed, note the particular angle to which the plate swings and use the following table:

Angle (degrees)							
0	4	16	31	46	58	72	81
Wind speed (miles per hour)							
0	4-7	8-12	13-18	19-24	25-31	32-38	39-46

You can check the calibration on a calm day by holding the anemometer upright outside the window of an automobile that is driven at varying known speeds.

You can determine the wind direction approximately, without instruments, from the drift of smoke or the motion of waves or flags; or you can keep turning in the general direction of the wind until it blows with equal force on both cheeks. Approximate wind speed can be estimated by means of the Beaufort Scale (see Index).

During the day, there is a normal fluctuation in average wind speed; it is highest in the afternoon and lowest near daybreak. Whenever the wind blows more strongly than usual, you can expect some change in the weather. A shift toward the south or east often spells trouble; a shift toward the west or north means possible clearing.

How to find the relative humidity

Relative humidity is the ratio of the amount of moisture (invisible water vapor) in the air to the maximum amount that the air could contain at a given temperature and pressure. At 50 per cent relative humidity, for example, the air contains one-half as much moisture as it could hold; at 75 per cent, three-fourths as much moisture.

Relative humidity is ordinarily measured by an instrument called a wet-bulb psychrometer. You can make one, as shown in Figure 7, by mounting two matched thermometers side by side. Cover the bulb of one with a tight-fitting muslin wick tied with a thread; the lower end of the wick should be immersed in a small bottle of water. Before reading the thermometers, ventilate them thoroughly by fanning them for a minute or two with a cardboard or electric fan. The water evaporating from the moist wick will lower the temperature of the wet-bulb thermometer. The difference in degrees between the two thermometers is called the depression of the wet-bulb; it is directly proportional to the dryness of the air.

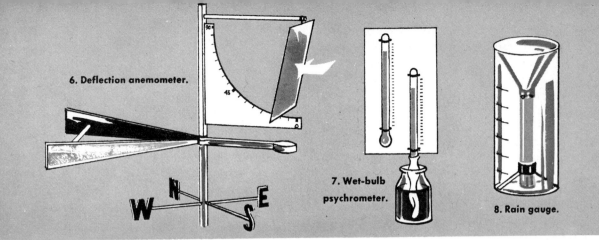

6. Deflection anemometer.

7. Wet-bulb psychrometer.

8. Rain gauge.

Once the wet-bulb depression is known, you can find the relative humidity and also the dew point (the temperature at which the relative humidity would be 100 per cent) by consulting special psychrometric tables. If you wish, you can work out the relative humidity by using the following formula:

$$\text{Approximate relative humidity in percent} = 100 - \left(300 \times \frac{\text{dry-bulb temp.} - \text{wet-bulb temp.}}{\text{dry-bulb temp.}}\right)$$

For example, if the dry-bulb reading is 80° F. and the wet-bulb reading is 70° F., we would substitute these figures for "dry-bulb temperature" and "wet-bulb temperature" and obtain the relative humidity by

$$100 - \left(300 \times \frac{80 - 70}{80}\right) =$$

$$100 - \left(300 \times \frac{1}{8}\right) =$$

62.5 per cent relative humidity

For accurate readings, use a clean muslin wick that has been washed to remove the sizing. (Sizing is material that is used to fill pores in cloth and paper; it keeps water out.) Change it often, since a dirty wick will give incorrect results. Moisten the muslin, if possible, with rain, snow, distilled or soft water. If you have only one thermometer, read the dry-bulb temperature first, after fanning. Then put on a moistened wick, ventilate the instrument and read the wet-bulb temperature.

We can also use the dry-bulb and wet-bulb temperature readings to calculate the temperature-humidity index, or THI, which indicates how bearable the weather is. (See Index, under Temperature-humidity index.)

How to measure the amount of precipitation

Precipitation is a general term for all forms of falling moisture, such as rain, snow, hail or sleet. Rainfall includes rain and the water equivalent of snow, hail and sleet. It does not include frost and dew.

To record the quantity of the rain that falls, you can place a tin can with straight sides on a level site. If you measure the depth of water in the can with a ruler, you will have a rough idea of the amount of precipitation. It is not difficult to prepare a more accurate gauge, shown in Figure 8. It consists of an outer cylinder, or tube, and an inner one, which is held upright in the middle of the larger cylinder by means of a support. Both tubes are to be open at the top and closed at the bottom. The radius of the large cylinder should be as nearly as possible 3.16 times the radius of the small one. Prepare a funnel whose wide end will have a diameter equal to that of the large cylinder and whose small end will fit into the small cylinder, as shown. Both the funnel and the small cylinder are to be freely removable. The latter is to be provided with a flow hole near the top so that when a heavy rain fills it, the excess rain water will pass out of the hole and collect at the bottom of the big cylinder.

The funnel will conduct rain into the small cylinder. When it has stopped rain-ing, remove the inner tube and, with a yardstick, measure the depth of water that has accumulated in it. The ratio between the diameters of the big and little cylinder is such that the depth of rainfall is magnified 10 times. This means that if the water is 10 inches deep in the measuring tube, the actual rainfall has been one inch.

Water may have passed through the flow hole of the small cylinder into the larger one. In that case, first measure the depth of water in the inside cylinder, record the measurement and empty the contents (but *not* into the large cylinder). Pour the excess rain water that has collected at the bottom of the large cylinder into the small tube for measurement. Add together the two sets of figures that you have obtained.

You can use the same apparatus to measure snowfall. The funnel and inner tube of the gauge are removed, so that snow can collect in the outer cylinder. The snow catch can be recorded by measuring some warm water and adding it to the snow in the outer cylinder. Pour the melted snow plus the water into the inner tube for measurement. From the total, subtract the amount of warm water added to the snow. The remainder gives the water equivalent of the snow catch. An average of ten inches of snow equals the water content of one inch of rain. Wide variations from this average figure are possible.

How to determine
average visibility

Visibility, or visual range, represents the average maximum distance (toward more than one-half the horizon) at which an observer with normal eyesight can identify prominent objects, such as buildings, trees or towers. You can estimate the visibility by referring to a homemade visibility chart, showing distances from the observation point to several prominent objects in all directions.

Visibility varies not only with the presence or absence of fog or haze, but also with the amount of air pollution by smoke, dust or other foreign matter. It also depends upon the type of air mass overhead. Air masses that are warmer than the ground over which they pass tend to bring about poor visibility; if they are colder than the ground, the visibility is usually good.

Additional information
for the amateur weatherman

Daily weather observations, combined with information provided by newspapers and radio or television broadcasts, will help you read the weather signs correctly. After you have acquired experience in keeping daily weather records, you may want to become a voluntary observer for the United States Weather Bureau or the Meteorological Service of Canada. Volunteer observers, who ordinarily serve without pay, are furnished with standard equipment and supplies for their back-yard stations. Their reports, published in special bulletins, help to define the climate of North America. When a post as volunteer becomes available, you will be eligible if you will accept custody of instruments, take accurate observations in accordance with instructions and write the required reports.

Special information
for the amateur weatherman

Weather hobbyists in the United States will find it advisable to become associate members of the American Meteorological Society, 45 Beacon Street, Boston, Massachusetts. In Canada, eligible applicants can join the Royal Meteorological Society, Canadian Branch, 315 Bloor Street West, Toronto 2, Ontario, Canada.

A list of useful government weather-and-climate publications, including daily weather maps and psychrometric tables, can be obtained from the Superintendent of Documents, United States Government Printing Office, Washington 25, D.C., or the Controller, Meteorological Division, Department of Transport, 315 Bloor Street West, Toronto 7, Ontario, Canada.

Amateur weathermen will find the following books especially useful:

Robert Moore Fisher, *How about the Weather?*, Harper and Brothers, New York, revised edition, 1958.
———, *How to Know and Predict the Weather*, Mentor Books, New York, 1953.
Charles and Ruth Laird, *Weathercasting*, Prentice-Hall, Inc., Englewood Cliffs, N. J., 1955.
Eric Sloane, *Eric Sloane's Almanac and Weather Forecaster*, Duell, Sloane and Pearce, Inc., New York, 1955.

GEYSERS AND VOLCANOES

How Earth's Internal Heat Bursts Forth

GEYSERS are jets of boiling water issuing from the earth in a few volcanic regions. The term is derived from the Icelandic word, "geysir," meaning "to burst out in violence." The true geyser is a spouting tower of glistening steam, often rising several hundred feet in the air, persisting anywhere from a few moments to an hour or more. It gushes forth in an area of extensive hot springs. There are only three principal regions where geysers are known to exist: Iceland, New Zealand and Yellowstone National Park in North America. Of these three, Yellowstone has by far the most impressive display of geysers.

The steam-jet wonderland of Yellowstone was probably first glimpsed by an official exploring party in 1807, at the time of the Lewis and Clark Expedition to the Northwest Territory. Subsequently trappers and scouts explored the Yellowstone and reports were sent back to Washington from time to time. Despite the accumulating information about geysers, Easterners remained skeptical about their existence. In 1870 Nathaniel Langford published in SCRIBNER'S MONTHLY a narrative of a several weeks' trip through the Yellowstone region. Langford's stirring description of Firehole Valley with its 3,000 boiling springs, its 100 turbulent geysers, its many fountains of dazzling steam, was scornfully dismissed as "balderdash" by armchair experts in the East. Langford himself was denounced as the "champion liar of the Northwest." When David Folsom, who had taken a similar trip in 1869, attempted to publish his experiences, he received a most ironic note from LIPPINCOTT'S MAGAZINE: "Thank you, but we do not publish fiction!"

Nevertheless, interest in the unique natural marvel of Yellowstone continued. The official report of Gustavus C. Doane, 2nd Lieutenant, 2nd U. S. Cavalry, himself a member of the Langford party, proved to be the most comprehensive survey ever made of the Yellowstone. A movement began to incorporate the territory into the national domain. No worse time could have been chosen. The country was in the grip of the scandals of the Grant Administration and there was strong pressure on Congress to throw open the Yellowstone to private exploitation. To the credit of the American nation, the area was saved for the permanent benefit of the entire people.

On March 1, 1872, Congress passed the Act creating Yellowstone National Park as a "public pleasure ground and game preserve." Nathaniel Langford was appointed the first superintendent of this magnificent wildlife sanctuary. At the present time, Yellowstone National Park has a total area of 3,458 square miles. It is almost three times the size of the state of Rhode Island.

In the seventies, a British visitor, the fourth Earl of Dunraven, traveled through Yellowstone following a big game hunt with Buffalo Bill and Texas Jack. Dunraven stood in awe before Firehole Valley and the fury of the geysers, the Giantess, the Castle, Old Faithful and the Grand Geyser. This last flung a column of steam ninety feet high, from the apex of which spouted five torrential jets to a height of 250 feet from the ground.

In THE GREAT DIVIDE, published in 1874, Dunraven described Castle Geyser in the following words: "[We] could hear a great noise . . . Louder and louder grew

Iceland Tourist Bureau

The Great Geyser of Iceland is located about 30 miles northwest of Mount Hekla. In its prime, it was able to hurl boiling water columns up to 150 feet into the air.

the disturbance, till with a sudden qualm he would heave out a few tons of water, and obtain momentary relief. After a few premonitory heaves had warned us to remove to a little distance, the symptoms became rapidly worse; the row and the racket increased in intensity; the monster's throes became more and more violent; the earth trembled at his rage; and finally, with a mighty spasm, he hurled into the air a great column of water. I should say that this column reached at its highest point of elevation, an altitude of 250 feet. The spray and steam were driven through it up to a much greater elevation, and then floated upwards as a dense cloud to any distance. The operation was not continuous, but consisted of a strong, distinct pulsation, occurring at a maximum rate of seventy per minute, having a general tendency to increase gradually in vigor and rapidity of utterance until the greatest development of strength was attained, and then sinking again by degrees. But the increase and subsidence were not uniform or regular; the jets arose, getting stronger

and stronger at every pulsation for ten or twelve strokes, until the effort would culminate in three impulses of unusual power. . . . The volume of water ejected must have been prodigious; the spray descended in heavy rain over a large area, and torrents of hot water, six or eight inches deep, poured down the slopping platform."

In the Middle Basin of the Firehole region is "The Devil's Paint Pot," a caldron of boiling, many-hued mud, which bubbles and steams with a wonderful play of changing colors.

The hot water ejected by geysers is always beautifully clear, but still it is highly charged with mineral matter, mainly dissolved silica. Around the orifices of geysers the incrustations of such materials often assume beautiful or fantastic forms and colors. Thus, the Castle Geyser has made a beautiful white cone for itself, and the Great Fountain Geyser a broad, circular pedestal about two feet high. The tints assumed by the mineral deposits are produced by algae, which grow luxuriantly in the hot water.

Though Yellowstone Park boasts of the most numerous and famous geysers, Iceland, the birthplace of geyser study, must not be passed by without notice. This country is itself one of the wonders of the world — weird, grotesque, and sinister, a wonder of desolation. It is a land of fire and ice, of glacier and geyser, of lava and slush, of avalanche and volcano.

From the top of Odahahraun, a lava field extending over 1,422 square miles, the surface of the earth resembles a gigantic stiffened corpse, petrified, black as the night. All together, no less than 4,650 square miles of the island are black with lava, and 5,170 square miles are white with snow. Rising from this black-and-white desert are great volcanoes, such as Hekla (5,108 feet). Most of the volcanoes are quiescent now, but Askja, whose crater is 34 square miles in extent, and 3,000 feet deep, was in eruption in 1875, and covered about 2,000 square miles with its ashes.

The Great Geyser has built a mound of siliceous material for itself about 40 feet high, and from the saucer-like basin on its top a tube 10 feet in diameter descends about 74 feet. In its prime, it used to eject water to the height of 150 feet, but now the average height of its jet is only 70 or 80 feet. The sound of the geyser in eruption has been compared to the roar of an angry sea, intermingled with the regularly recurring sounds of guns.

The Strokkr has no regular basin, merely a funnel-shaped tube, narrowing from a diameter of 8 feet at the surface to 10 inches at 44 feet down. It can usually be made to erupt by throwing in turf or stones, the amount thrown in regulating the time of eruption. The Strokkr can spout higher than the Great Geyser, but its usual height is only 30 or 40 feet.

A third great geyser wonderland is found in the North Island of New Zealand, in the volcanic country around Lake Taupo and Rotorua. Here we find the greatest geyser in the world. Its tube, 80 feet deep, is situated in the middle of a hot lake lying in a crater formed during the great volcanic eruption of 1886. This tremendous geyser has not the beauty of the Yellowstone Park geysers, since the column of water it flings is muddy and inky-black, but it surpasses the Yellowstone geysers in violence. It often throws its water column to a height of 500 feet, and on one occasion three times that height. At times it is dormant for weeks; at other times, it is active for weeks. Near Rotorua there are a number of geysers, but only the smaller ones are now active. The Wairoa Geyser, which used to spout spontaneously to a height of 200 feet, now plays only when fed with bars of soap.

It has been found that the addition of certain alkaline substances, such as soap or lye, makes the water somewhat viscous, causing it to retain the smaller steam bubbles till the accumulated volume of steam breaks out with increased eruptive power and a sudden lowering of pressure. Before the great eruption of 1886 two geyser lakes, by their overflow, had produced a set of marvelous siliceous incrustations known as the White Terrace and the Pink Terrace. At the eruption, these beautiful formations were destroyed.

In the picture below we see the great Pohutu Geyser in Whakarewarewa, Rotorua, New Zealand. It was created along with six other geysers in the great eruption which took place in 1886.

Nat. Publicity Studios, New Zealand

What causes the eruption of geysers to take place? The famous German chemist Robert Wilhelm Bunsen (1811-99), who invented the Bunsen burner, developed a quite satisfactory theory to account for the phenomenon. The Bunsen theory of geyser eruption is still pretty generally accepted at the present time.

It is based on the fact that the boiling point of water is raised as the pressure is increased and is lowered as the pressure is decreased. Water will boil at a temperature of 212° Fahrenheit at sea level, where the atmospheric pressure is about 14.7 pounds per square inch. The pressure will be twice as great — about 29.4 pounds per square inch — at a depth of 33 feet. At this depth the water will boil at a temperature of 248° Fahrenheit.

If ground water penetrates into a more or less vertical fissure deep within the earth, it will form a liquid column as it accumulates. As the bottom of the column comes into contact with volcanic heat, its temperature will rise quite a bit above 212° Fahrenheit, which, as we have seen, represents the boiling point of water at sea level. However, it will not boil at first because it is subjected to the pressure exerted by the column of water above it and its boiling point is raised.

As the water at the bottom of the column becomes hotter, a part of it will be turned into steam. The pressure that is exerted by the steam will cause the water column to rise, and some of it will overflow at the surface. This will mean that less water will press down upon the water at the bottom; the pressure will be decreased and the boiling point will be lowered. The result will be that more of the water will be converted into steam. At last the accumulating steam will blow the whole column of water out of the fissure; an eruption of the geyser will take place. The phenomenon will be repeated with more or less regularity according to the supply of water and heat. A laboratory model built and operated on the basis of the Bunsen theory will work.

Hot springs in the Minerva Terrace, Yellowstone National Park. According to one theory, hot springs are fed by ground water penetrating to great depths and heated by the steam arising from molten rock.

Jean Speiser — National Park Service

New Zealand Embassy

Scalding hot water and live steam erupt from Papakura Geyser, one of the most spectacular in New Zealand.

Volcanoes and geysers must be considered together, for they are akin. Volcanoes, however, are much more spectacular in their action than geysers; in fact, they represent the most impressive display of natural forces, with the possible exception of earthquakes. All over the world volcanic craters are glowing and steaming and grumbling. Ever and anon, in paroxysms of violence, they devastate the land with clouds of fire and steam, with broadsides of boulders and rocks and with rivers of lava and hot mud. Volcanoes such as Vesuvius, Etna and Stromboli seemed to the men of old forbidding gateways leading to the infernal regions.

A volcano is a vent or fissure in the earth's crust; from it there emerge, at intervals and more or less violently, hot solids, liquids and gases. The materials that issue from the vent sooner or later build up a cone about it. The term volcano is often applied, not only to the vent, but to the mountain itself. Volcanic mountains differ greatly in height and shape; generally they are conical and often very lofty. Orizaba and Popocatepetl, in Mexico, Cotopaxi and Aconcagua, in the Andes, Mount St. Elias, in Alaska, Kilimanjaro and Kenya, in Africa, and Ararat and Demavend, in Asia, are all between 17,000 feet and 23,000 feet high.

Many of the conical peaks of volcanoes form unforgettable landmarks. Among these are the Peak of Tenerife, rising from the sea; Etna, girdled by the Sicilian surf; Chimborazo, in Ecuador; Mayon, in the Philippines; Mount Osorno, in Chile; and the Japanese peak of Fujiyama (or Fuji), whose cone stands out in majestic isolation above the countryside.

The top of a volcano generally shows a more or less cup-shaped or basin-shaped depression: that is, the crater. The characteristic conical shape of a volcano is undoubtedly due to the manner in which the mountain arises. It is an accumulation of solidified lava, rock fragments and cinders ejected from the vent. The ejected material assumes the shape of a cone, with a pipe or funnel through its center leading to a cup-shaped cavity at the top. The angle of the cone depends on the viscosity of the lava (the sluggishness of its flow) and the cohesiveness of the material that has been thrown out. The more viscous the lava and the more cohesive the ejected material, the steeper the cone. The volcanoes in the island of Réunion, in the Indian Ocean, have been formed from very viscous lava, and they are steep. In the Hawaiian Islands, on the contrary, the cones were built up by fluid lava. They have a gradual slope; their bases range up to seventy miles and even more in diameter.

The composition of the cones of various volcanoes differs. Some are made up of cindery and slaglike fragments, others of sheets of lava and still others of a mixture of both. The more massive volcanoes are usually built up out of alternating layers of lava sheets, rock fragments and volcanic dust. Crossing these layers in various directions there are numerous cracks, filled with lava. It may seem hard to believe that volcanic mountains 15,000 or 20,000 feet high have been erected in this way, layer by layer. We must not forget, however, that this process has been going on for many thousands of years.

Mountains that have arisen in this way are apt to be broken down as well as built up, and they may change their shape considerably in the course of their growth. Explosions may tear away part of the summit; new vents and new cones may be formed. In the course of the centuries, Vesuvius has changed its shape many times. In the first century B.C., the summit of the volcano was a great depressed plain nearly three miles across; it was here that the gladiator Spartacus and his followers were besieged by a Roman army. After the great eruption of 79 A.D., which destroyed the cities of Pompei and Herculaneum,

Inside the crater of Mount Vesuvius, the famous volcano located some ten miles southeast of Naples, Italy. The crater is gradually being filled up with materials ejected from small, active cones.

Amer Mus of Nat. Hist.

Vesuvius developed a huge crater, within which a new cone with a smaller crater was formed.

In 1822, a great eruption reduced the height of the volcano by 400 feet and produced a huge crater a mile in diameter and a thousand feet deep. By 1843, three small cones with craters had sprung up within the great crater. The eruptions of 1872 and 1906 again changed the shape of Vesuvius. In 1922, a new cone 230 feet high was built up within the old crater. All active volcanoes are subject to such changes in shape. The general effect of volcanic activity is to increase both the bulk and height.

Almost all great volcanoes develop new vents. These give rise to subsidiary, or "parasitic," cones, which tend to destroy the original conical shape. The volcanoes of the Hawaiian Islands have thousands of subsidiary cones of this type.

The size of craters varies within very wide limits. As far as we know, there is little or no connection between the size of the crater and the height of the volcano. For example, the extinct Mexican volcano of Orizaba is 18,250 feet high; its crater is about a thousand feet in diameter. Popocatepetl is not quite so high (17,887 feet); yet its crater is about twice as large as that

Majestic Fujiyama, or Fuji, an extinct volcano about seventy miles southwest of Tokyo, Japan. Its impressive cone, rising from a base sixty-five miles in circumference, stands in splendid isolation

Japan Air Lines

of Orizaba. Haleakala, a volcano in the Hawaiian Islands, is only 10,000 feet high; its crater is twenty miles in circumference. There does not seem to be any connection, either, between the size of the crater and the explosive potentialities of the volcano in question.

The name "caldera" has been given to very broad and comparatively shallow craters. This word is Spanish and means "caldron." It was first applied to a very large pit in the Canary Islands — a pit more than three miles in diameter and surrounded by cliffs that rise to a height of some 3,000 feet. A caldera results from the explosion or collapse of a former volcanic cone, causing the original crater to become much wider and shallower.

There are many calderas and some of them are huge. That of Mauna Loa, on the island of Hawaii, is three and a half miles long and a mile and three-quarters wide. The caldera of Kilauea, on the same island, is two and a half miles long and a mile and three-quarters in width. Both of these immense calderas were formed by the collapse of the old volcanic cones.

Many calderas become filled with water, forming lakes. Among the most beautiful and picturesque of these is Crater Lake, in what is now the Crater Lake National Park in southwestern Oregon, on the crest of the Cascade Mountains. Crater Lake is over 6,000 feet above the level of the sea. It is nearly circular in form and is surrounded by rock walls that range in height from 900 feet to 2,200 feet. The lake is about five miles in diameter and in some places it is some 2,000 feet deep. The Indians called it the "Sea of Silence." To the American poet Joaquin Miller, it was "a sea of sapphire set around by a compact circle of . . . grizzly rocks."

Volcanoes vary greatly in activity. As a rule, periods of partial or complete inactivity alternate with periods of explosive violence. Until the eruption of 79 A.D., Vesuvius had been in repose for centuries. The eruption of Krakatoa in 1883 took place after an inactive period of two hundred years. Some volcanoes are constantly active. Stromboli, a volcanic island in the Mediterranean, off Italy's Lipari Islands, has been pouring forth lava for more than 2,000 years. Izalco, in El Salvador, began in 1770; it has been in constant eruption ever since. Volcanoes of this type are relatively tame.

Certain volcanoes, as far as we know, are entirely extinct. Among these are the great volcano of Palma, in the Canaries; Mount Shasta, in California; Mount Hood, in Oregon; and Mount Rainier, in Washington. The peaks of certain dead volcanoes are snow-clad; their slopes have been carved by erosion to such an extent that they are fast losing their characteristic shape. In some instances, the forces of erosion have worn down dead volcanoes almost to their foundations.

Sometimes, gas issuing from a vent at the surface of the earth may carry particles of sand and clay with it. In the course of time, a cone may be built up from these particles. If they are moistened by heavy rainfall, they will form mud; this will dry and harden at the surface of the mound. The gas will accumulate under the hardened surface; the pressure will build up until finally the gas will blow off the top of the cone. These mud volcanoes, as they are called, are frequently found in oil and gas fields; some arise in regions where volcanic steam escapes through mud. Certain mud volcanoes rise to considerable heights; that of Bog-Boga, in the Baku region of Russia near the Caspian Sea, is more than a hundred feet high.

The distribution of the world's volcanoes

There are several thousand volcanoes in the world, and about 450 of them are known to be alive. They are distributed in a series of belts. The great belt of the Atlantic Ocean includes a considerable number of volcanic islands; among them are West Spitsbergen, Jan Mayen Island, Iceland, the Azores, Madeira, the Canary Islands and the Cape Verde Islands. Another volcanic belt extends southward from Arabia; it takes in the Red Sea, Ethiopia and the large island of Madagascar, off the coast of eastern Africa.

A line of volcanoes may be traced along the eastern side of the Pacific Ocean from the Aleutian Islands to Cape Horn. There are many active volcanoes in the Aleutian and Coast ranges of Alaska. Katmai Volcano erupted as recently as 1912, when it blanketed Kodiak Island with lava and ash. The Cascade Range, in the United States, has many volcanoes, but only a few are active. Lassen Peak, in northern California, erupted violently in 1914–15. Mount Rainier, in Washington, and Mount Hood, in Oregon, give off vapor. Active volcanoes stretch across Mexico from Colima to Tuxtla. Popocatepetl, in central Mexico, gives off clouds of smoke.

A series of volcanoes in South America begins with the Nevado del Tolima, in Colombia, and continues down to Corcovado Volcano, in southern Chile. Cotopaxi, in Ecuador, is 19,344 feet high; it is probably the highest active volcano.

Another line of volcanoes swings from the Bering Strait to the west side of the Pacific Ocean and then runs down to the Indian Ocean. Kamchatka Peninsula has about fourteen active volcanoes. The islands of Japan are a long volcanic mountain chain lifted above the ocean. The string of volcanoes passes through the Philippines, the Moluccas and the Sunda Islands to New Guinea.

Ewing Galloway

Crater Lake, in Crater Lake National Park, Oregon. It was formerly the pit of an extinct volcano.

The condensation of the enormous amounts of steam rising from even small active volcanoes results in the addition of considerable quantities of water to the earth's supply. It is estimated that during a single eruption of Mount Etna nearly 600,000,000 gallons of water were given off as steam. A small parasitic cone of the same crater yielded about 460,000,000 gallons in a hundred days. The steam that issues from volcanic vents is believed to come partly from ground water that has seeped far into the earth and that has become superheated; partly from the water that was present in the rocks when they were formed.

Volcanoes have pushed out incredibly great masses of solids from the earth. All volcanoes owe their mass to the interior of the earth — extinct and active volcanoes; volcanoes only a few hundred feet high and those that tower to a height of three miles (or six miles, if we measure cones such as Mauna Loa, in Hawaii, from the ocean bed from which they arose). The average annual discharge of sediment by the Missis-sippi River is some 560,000,000 tons; in a few hours, during the eruption of 1902, Mount Pelée belched out five hundred times as much in the form of volcanic dust.

Warning signs of a volcanic eruption

There are usually a number of warning signs when a volcano is about to erupt. There are earthquakes and loud rumblings, like those of thunder. These rumblings are probably caused by the movement of gases and molten rock, held in under great pressure. Hot springs may suddenly appear in the vicinity of a volcano shortly before an eruption takes place. There may be a considerable increase in the activity of the volcano. The level of nearby lakes may rise or fall appreciably; the lakes may be entirely drained of their water.

Then the eruption proper takes place. Out of the crater rush enormous quantities of superheated water vapor and other gases, carrying with them vast quantities of rocks, stones, ashes, lava and other materials.

Mount Etna, in the eastern part of the Sicilian commune of Catania, is one of the most destructive of all volcanoes. More than eighty major eruptions, costing many lives, have been reported.

Italian State Tourist Office

GEYSERS AND VOLCANOES

The interior of the earth is very hot. We are reminded of this fact whenever tall columns of boiling water and steam rise from the hot springs we call geysers, or when volcanoes eject hot rock fragments, gases and lava. Some well-known geysers and volcanoes are shown in the full-color photos on these pages.

Union Pacific R. R.

Old Faithful (above), in Yellowstone National Park. This geyser erupts every sixty-five minutes or so on the average; but the intervals between eruptions may vary considerably. An eruption is preceded by a rumbling sound; then a column of hot water and steam rises to a height of a hundred feet or more.

Izalco (right) is an active volcano, 6,184 feet in height, in the western part of the small Central American republic of El Salvador. This volcano is of comparatively recent formation; it first appeared in the year 1770. Because of its continuous activity, it has been called the "Lighthouse of the Pacific."

Dave Forbert, from Shostal

300-b

The left-hand photograph on page 300-b shows Pohutu, a geyser in New Zealand's Whakarewarewa region. Most of the world's geysers are found in New Zealand, Yellowstone Park and Iceland. In the right-hand photograph on page 300-b we see the recently formed volcano of Parícutin, in the west central part of Mexico. Parícutin arose on February 20, 1943, from a corn field, following a violent earthquake. First a crater appeared and began to belch sand and rock. Then immense streams of lava were emitted. As the volcanic materials piled up around the crater, the volcano increased rapidly in height.

Left-hand photo, New Zealand Consulate General
Right-hand photo, American Airlines

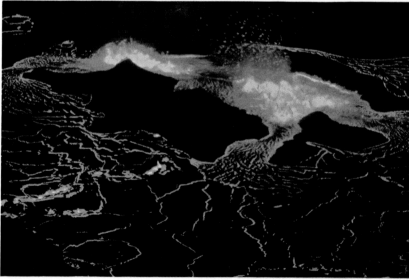

FPG

In the upper photo on this page we see an eruption of the Hawaiian volcano of Puna. At night the hot lava looks like white-gold fireworks as it leaps out of the crater cone. The other photo shows lava gushing from Halemaumau, a pit within the crater of Kilauea, another volcano of the Hawaiian Islands.

Union Pacific R. R.

The wind catches the column of hot water and steam rising from the Daisy Geyser, one of the more than a hundred geysers in the Yellowstone National Park. This scenic region is mostly volcanic in formation. Besides its many geysers, the park has some four thousand hot springs.

Popocatepetl, a dormant volcano 17,887 feet high in central Mexico, about forty-five miles southeast of Mexico City. The name comes from an Aztec word meaning "Smoking Mountain." Popocatepetl last erupted in 1702, but it still emits vast clouds of smoke from time to time.

American Airlines

300-d

The ground shakes violently and there is a tremendous roaring sound. It is said that the roar that accompanied the 1915 eruption of Tamboa, a volcano on Sumbawa Island in Indonesia, could be heard at a distance of a thousand miles.

The column that
rises from a volcano

The column of gases and solid particles issuing from a volcano in eruption goes up vertically to a great height. It assumes the form of a pine tree, to which it was compared by the Roman naturalist Pliny the Elder. The friction of the particles issuing from the vent creates static electricity. Forked lightning flashes to and fro and peals of thunder are heard. The rising column may be lit up by the glowing lava in the crater so that it seems to be aflame. In some cases, the column that arises from the crater is so dense that it hides the sun and creates impenetrable darkness. The condensation of the steam as it rises aloft brings about torrential rain. The downpour is also partly due to the condensation of moisture-laden air, drawn upward to high altitudes by the updraft created as volcanic gases are discharged. As the rain falls in torrents, muddy rivers flow down the slopes of the volcano. They may bury the cities that lie at the base of the mountain.

The composition
of volcanic gases

By far the greatest part of the gases that issue from volcanic vents is made up of steam — superheated water vapor, which may attain a temperature of 1,000 degrees Fahrenheit. Other gases present may include hydrogen chloride, carbon dioxide, carbon monoxide, methane, hydrogen, oxygen and argon, all at a high temperature. Various compounds of sulfur, such as hydrogen sulfide and sulfur dioxide, are given off by some volcanoes. Gases are emitted not only from the volcanic vent itself, but also from lava flows. These flows may continue to emit gases for weeks or even months after being thrust out of the vent at the time of the eruption.

In some ways, lava is the most characteristic product of a volcanic eruption. It is true that in some cases, as in the great eruption of Mount Pelée in 1902, no lava at all is ejected. As a rule, however, after the volcano has "cleared its throat" with the discharge of superheated steam and other gases, lava begins to flow from the crater or even to spout high in the air like a fountain of molten metal. Usually the lava pours over the edges of the crater or from fissures in the sides of the volcano. Where there are a great many fissures, the volcano seems to "sweat" lava from numerous pores.

In some cases, when lava is pushed out from the volcanic vents, it is too viscous to flow, either because of its comparatively low temperature or its low gas content. Hence it is piled up in the form of great domes over the vents. These volcanic edifices are called plug domes. In Lassen Volcanic National Park, there are thirteen domes of this type within an area of approximately fifty square miles.

The spine of
Mount Pelée

An unusual type of plug dome appeared on Mount Pelée after the disastrous eruption of 1902. The lava that filled the vent after the eruption became hardened into rock at the surface. The interior was highly viscous, however, and the whole mass was gradually pushed up until it formed a tower over a thousand feet in height above the summit of the volcano. It was called the spine of Mount Pelée. Gradually it crumbled away, largely as the result of continuous explosions of gases within it. Later, another plug was formed in the course of the new series of eruptions of Mount Pelée from 1929 to 1932.

The rate of flow of the lava that issues from the earth is often rapid at first. It varies with the viscosity of the lava and the slope of the land. When the lava reaches a steep downward slope, it pours tumultuously over it; on gentle slopes, however, it may move very slowly. As the lava cools, it becomes more viscous and its flow becomes much more sluggish.

The surface appearance of the cooled lava depends in part on its chemical composition, which determines whether the resulting rock will be a basalt, rhyolite or andesite. (See Index for these varieties of rock.) The nature of the lava flow is another extremely important factor. Lavas that are particularly viscous (that is, that flow very sluggishly) break up, as they cool, into multitudes of rough blocks. These grate together as the lava continues to flow and they are often piled up in heaps or mounds. When the flow finally stops, the result is an exceedingly irregular formation — an accumulation of sharp-edged blocks and jagged fragments. Rough lava of this kind is called block lava; it is sometimes known by the Hawaiian name *aa*. When more fluid lava hardens, it shows a twisted, ropy structure with a vari-colored and satiny surface. The Hawaiian word *pahoehoe* is used for this corded lava.

Streams of lava
cool very slowly

Lava is red-hot or white-hot when it first issues from a crater. It is a poor conductor of heat and cools very slowly. Consequently, after the surface crust has formed and has become cool enough to be walked on, the lava may be red-hot below. Many examples might be given of the slowness with which lava cools. In 1830, steam was still issuing from lava that had flowed from Etna, in Sicily, forty-three years before. The volcano of Jorullo, in Mexico, discharged streams of lava in 1759; eighty-seven years later, two columns of steam still rose from the lava.

Materials that are
flung out by volcanoes

Besides lava, as we have seen, volcanoes discharge huge quantities of rocks, stones, cinders and ashes. During the eruption of 1779, mighty Vesuvius hurled cinders to a height of 10,000 feet. In 1815, Tambora, in the Indonesian island of Soembawa, or Sumbawa, covered the sea for miles around with such copious quantities of pumice stone that ships could hardly force their way through it. Cotopaxi, a volcano in north central Ecuador, is said to have flung a 200-ton block 9 miles.

The accumulation of lava and rocks and sand in the vicinity of a crater sometimes reaches almost unbelievable proportions. The creation of the volcano known as Parícutin offers a striking example.

How Parícutin
came into being

On February 20, 1943, there was a severe earthquake in the state of Michoacán, Mexico, some 180 miles west of Mexico City. Following the earthquake a crater appeared in a corn field and began to belch forth rocks and sand; later it emitted immense streams of lava. The accumulation of volcanic materials about the crater mounted with fantastic speed; soon a new volcano, Parícutin, was born. Within the space of a year its height was more than a third that of Vesuvius, which had been many thousands of years in the making.

A vast amount of dust is emitted from volcanoes. When Tambora erupted in 1815, dust fell on the island of Borneo, 870 miles away, in vast quantities. In the course of the eruption of Cosigüina, or Consegüina, in Nicaragua, it is estimated that some 6,500,000,000 cubic yards of dust were cast to the winds and were carried at least 800 miles. In 1877, the dust from an eruption of Cotopaxi plunged the city of Quito, Ecuador, into pitch darkness.

Dust from the
eruption of Krakatoa

Dust from the eruption of Krakatoa, in 1883, darkened the sky for 150 miles from the site of the eruption and fell in appreciable quantities 1,000 miles away.

The dust in the vicinity of an erupting volcano produces eerie darkness, penetrated only by flashes of light from the crater and the glow and glare of molten lava. Later the dust produces startlingly beautiful effects in the skies. As it floats aloft, sometimes making several circuits of the globe, the fine volcanic dust refracts or reflects the individual colors of sunlight and thus brings about gorgeous sunrises and sunsets all over the world.

The deluge of water from the condensed steam of the eruption often converts the volcanic dust into a sticky mud that overwhelms and buries everything in its path. The ancient Roman city of Herculaneum was buried during the eruption of Vesuvius in 79 A.D. by such an inundation of mud. The great eruption of Cotopaxi in 1877 covered many villages with a deposit of mud, mixed with lava and various kinds of debris. Whole forests were laid low as a consequence of the eruption.

The submarine volcanoes of the world are among the most spectacular of all. Etna, Stromboli, the Peak of Tenerife and many other volcanoes had their birth under the sea. Nobody knows how many active volcanoes are now at work in the ocean depths. Often they throw up cinders and steam and agitate the surface of the sea. Vast outpourings of lava on the ocean floor accompany these eruptions; sometimes layer upon layer of lava is formed and at last an island appears above the ocean surface. The volcanic chain of the Hawaiian Islands was created in just this way. Occasionally submarine volcanoes betray their presence in a most striking way: through their action a new island is suddenly thrust up or an old island sinks below the surface. This action is quite different from the gradual deposition of lava to which we have already referred.

Graham Island, a short-lived islet in the Mediterranean Sea

The Mediterranean Sea has often witnessed the sudden appearance and disappearance of new volcanic islands. In 1831, a new islet, called Graham Island, suddenly appeared between Sicily and Africa as a result of volcanic action. This island reached a height of 200 feet above water and attained a circumference of something like three miles. Since it was composed of loose materials, it could not withstand the constant pounding of the sea. It was rapidly eroded by the waves and in a few brief months it was reduced to a shoal, called Graham's Reef. New islands, produced by volcanic action, are thrown up quite frequently between Sicily and Greece, but they generally disappear almost as quickly as they arise.

The birth of three volcanic islands in the Bering Sea

Some islands that have been thrown up by submarine volcanoes are more permanent. Three islands of the Bering Sea are good examples. A new body of land, called Bogoslof Island, appeared about 40 miles to the west of Unalaska Island in 1796 after an eruption. In 1883, another eruption threw up a volcanic cone of black sand and ashes, known as New Bogoslof, or Fire Island. Still another volcanic island made its appearance in that vicinity in 1906; it was larger than either of the others.

Wide World

Smoke billows upward as a new volcano erupts in 1952 from the depths of the Pacific, 150 miles south of Tokyo.

The lava of
fissure eruptions

Not all the materials that are ejected from the earth are piled up to form mountains. In many instances, lava flows out of fissures in the earth's crust and spreads over large areas without producing anything suggesting volcanic cones. Such outpourings are called fissure eruptions. The American geologist H. S. Washington suggested that they should be called plateau flows, because the area they cover is as flat as a plateau. The lava that issues from fissure eruptions, or plateau flows, is not very viscous.

Fissure eruptions have taken place on a vast scale. In the Deccan region of western India, an area of 200,000 square miles has been covered with lava, which is 2,000 feet thick in certain places. Parts of Washington, Oregon, Idaho, Montana and California lie under a sea of solidified lava about 225,000 square miles in area and with an average depth of 500 feet. Similar flows have covered large regions in the northern part of Wisconsin and Michigan and along the northern shores of Lake Superior. Some 40,000 square miles in England were were at one time under a sea of fissure lava. Large tracts in Ethiopia are covered with such lava.

Fissure eruptions have been common in Iceland, an island that has been built up to a great extent by volcanic activity. In 1783, two streams of lava issued from a fissure, and they flowed forty and twenty-eight miles respectively. The great lava desert of Odahahraun, which covers an area of more than 1,400 square miles, is also a product of fissure eruptions.

What causes materials to
be ejected from the earth?

A volcanic eruption or an outpouring of lava in a fissure eruption is due to the heat within the earth. We do not fully understand the origin of this heat. Some geologists maintain that it is a residue of the original molten state of the entire earth during the early period of its formation. The earth's internal heat has been retained for so many millions of years, they say, because rocks are such poor conductors of heat. Other geologists maintain that heat is produced because of the breaking down of radioactive elements such as uranium and thorium. Both of these elements are present in all rocks, though generally in very small quantities.

The formation of
magma in the earth

Whatever the source of the earth's heat, it is great enough to liquefy vast masses of rock. This molten material, called magma, consists mostly of solutions of silicates with oxides and sulfides; it also contains steam and other gases, held in solution by pressure. Where this magma forms, the liquid rock weighs less than the adjacent solid rock, and the gases dissolved in it make it even lighter.

The magma is subjected to the pressure of the heavier rock that surrounds it and it is forced upward. Some of it reaches the outer part of the earth's crust and ultimately makes its way to the surface through fissures or vents. The molten materials are then hurled aloft by the explosive forces within the earth, or else they flow out from craters or fissures in the form of more or less viscous lava. (Lava is simply magma that has reached the surface of the earth — magna from which most of the gases have escaped.)

Utilizing the energy
released in volcanic areas

Man has made a beginning in controlling for his needs the tremendous energy released by subterranean heat in volcanic areas. In Tuscany, Italy, steam jets issuing from boreholes sunk in the earth have been used to run turbines which generate electricity. At Larderello, in that area, considerable power has been generated in this way, while quantities of boric acid, ammonium carbonate and sodium carbonate have been recovered from the steam that runs the turbines. Some day, perhaps, vast stores of power will be made available in the far-flung volcanic belts of the earth.

See also Vol. 10, p. 271: "Volcanoes and Geysers."

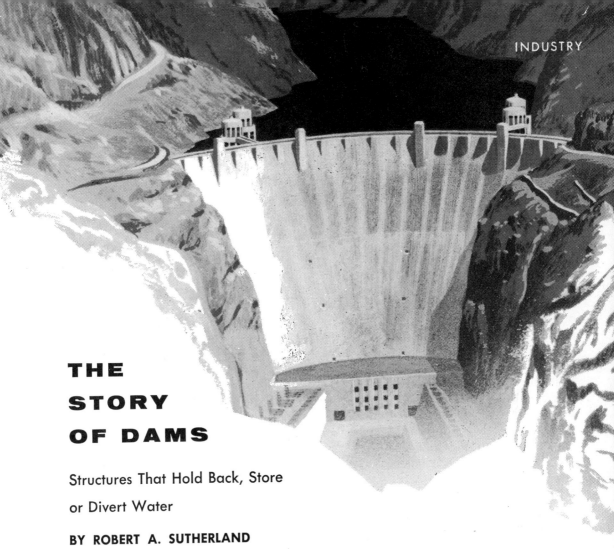

THE STORY OF DAMS

Structures That Hold Back, Store or Divert Water

BY ROBERT A. SUTHERLAND

STRICTLY speaking, the name "dam" can be applied to any structure that is built or formed to hold back a liquid. This structure may be only a fraction of an inch in height, as in the case of the "tinker's dam" — a strip of clay used by a tinker, or traveling tinsmith, to hold molten solder or lead. A dam may be a fairly large temporary structure, such as a cofferdam, which serves to keep water out of an excavation. Dikes and levees are also dams. In this article we are going to use the word "dam" in the sense of a more or less massive structure used to store water on a large scale, or to divert water from a stream.

A dam of this kind is generally built across a river or other stream. In many cases, however, it serves to form a water storage area at some distance from a stream. Such a dam may completely surround the storage area. Sometimes it is very simple. In dry areas of the United States, Australia and various other countries, farmers' reservoirs are formed by digging a large hole in the ground and using the dirt thrown up in this way to form a bank around the hole.

There are many different kinds of dams. So-called fill dams are made of local materials, such as earth, gravel and rock; there are earth-fill and rock-fill varieties. Masonry dams are built of concrete or of stone laid in mortar. Gravity dams, arch dams and buttress dams are all of this type. A gravity dam looks like a retaining wall, set across a river. The typical arch dam suggests a section of concrete pipe, split lengthwise and turned up on end. A buttress dam is supported by a

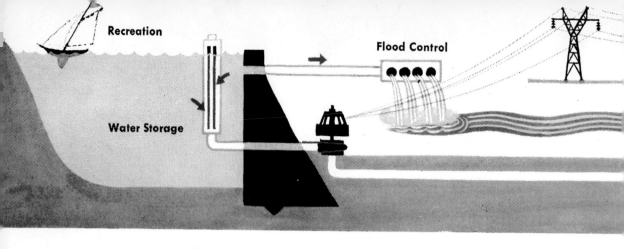

Recreation

Flood Control

Water Storage

series of buttress walls, set at right angles
to the dam on the downstream side. All
these types are more fully described later.

Some dams are low-lying structures;
others tower hundreds of feet into the air.
Certain types are collapsible and disappear
from view in time of flood. Most dams,
however, form a permanent obstruction to
a stream. They are almost always provided with a spillway, which allows flood
waters to pass by in such a way that the
dam will not be overtopped by the stream.
The Miena multiple-arch dam, which forms
the Great Lake in Tasmania, has no spillway. Its waterspread area is so great in
relation to the water inflow that it can
meet any flood situation that could possibly arise.

Purposes for
which dams serve

The function of a dam is to store
water. It is designed to make the most
effective use, at reasonable cost, of the
available supply of the water in a stream.
Nearly all rivers have a quite variable
flow. Those that have their source in snow-
covered mountains are swollen during the
spring and early summer, while the snow
on the mountains is melting. In the fall
and winter, however, the water level drops
sharply. Dams serve to regulate the flow.
In warm climates there is often a rainy
season and a dry season. It is important
to retain as much water as possible from
the rainy season so that it may be available
during the dry season. The total flow of
certain streams may vary greatly from year
to year. In that case, water that accumulates in wet years must be stored up for
use in dry years.

The water stored by a dam may serve
in various ways. Probably the most important use, as measured by human needs
rather than by the volume of water stored,
is for the water supply of towns and cities.
Most cities built near hilly country have
one or more water-supply reservoirs, in
which water is impounded, or collected,
by dams. The reservoirs are usually surrounded by wooded areas and carefully
fenced in. As a result the danger of contamination is lessened, and human beings
and certain animals are kept away. Reservoirs that are at a distance from the city
that they serve are called supply reservoirs;
those near the city are known as service
reservoirs. The latter supply water to a
community as required.

Another important use of stored water
is for irrigation. Perhaps the most famous
dam ever built for this purpose is the Aswan Dam on the Nile River in Egypt, about
700 miles from the Mediterranean. This
dam was constructed between 1898 and
1903 and has since been twice raised in
height. It discharges through sluices, or
water gates, the whole flow of the Nile
during the period when the water is heavily
laden with silt. Later, when the water runs
clear, the water gates are closed and the
water is stored; still later, this water is released. Many other irrigation storage dams
have been built in the United States, India,
Australia and Latin America.

Water stored in dams also serves to
generate hydroelectric power. Nearly all
countries have dams of this type. In some
countries that are dependent mainly on
hydroelectric power for their needs, such
as France, Italy, Switzerland, Austria and
Norway, large dams have been built in the

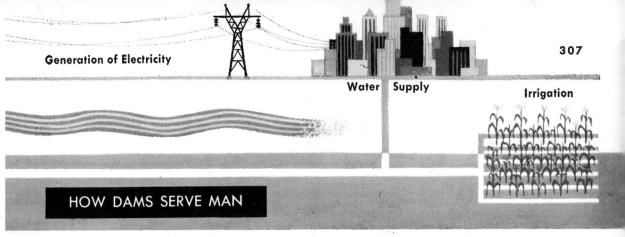

HOW DAMS SERVE MAN

mountains to store the snow melt for use during the fall and winter.

Water storage helps control floods. The city of Dayton, Ohio, built several dams for this purpose. The structures have permanently open sluices, so designed as to pass a reasonable amount of flood water and to retain the rest until the flood has abated. The stored water is then gradually released. Such flood-control dams are often known as retarding basins.

Some storage dams have been built for recreational purposes. The outboard motor has given a tremendous impetus to water sports, which can be carried on in artificial lakes created by dams. In most cases recreation is a secondary benefit derived from lakes that were created for other purposes. For example Lake Mead, which is now a favorite center for water sports, was formed for water storage and flood control when the Hoover Dam was constructed. This vast lake, extending along the Arizona-Nevada border, has a shore line of 550 miles.

In California and certain other places some dams have been built primarily to hold back debris. They protect fertile and populous low-lying areas from gravel and boulders brought down from the mountains by sudden floods.

Many dams divert water from a river. An outstanding example is the Parker Dam, on the California-Arizona border. It diverts water from the Colorado River to an aqueduct that supplies water to Los Angeles and other communities. Diversion dams are generally rather low structures. In some cases, however, high diversion dams have been built, particularly to carry water from one watershed to another. Dams of this sort generally serve for power purposes.

The storage of flood waters may serve to improve navigation on various rivers. At certain times of the year the water level on such streams would be lowered so greatly as to interfere with navigation if it were not for dams. They hold back enough water from the total flow so that the low flow can be increased by an adequate amount for navigation. The Fort Peck Dam on the Missouri River in Montana was constructed partly for this purpose. There may be a series of such navigation dams, provided with locks. These make it possible for barges and other craft to pass through stretches that would otherwise be impassable. Many rivers in Europe have been made completely navigable through the use of navigation dams. In the United States the Ohio River is the best example of navigation control through the use of dams and locks. In many cases navigation locks are limited in height of lift to ten or twenty feet; but several much higher lifts have been used on some rivers.

Seldom does a storage dam serve only a single purpose. Power is frequently generated at dams built primarily for irrigation, navigation or flood control and sometimes also at water-supply dams. Most dams have some recreational value; many serve for silt control. One of the most striking examples of a multipurpose dam is the Hoover Dam. This vast structure develops electric power, supplies water for domestic purposes, controls the flood waters of the Colorado River, provides irrigation for farm lands in Arizona, New Mexico and California and controls silt carried down by the river.

Horse-Drawn Scraper

The history of dam construction

We do not know when man first began to store the water of a stream for irrigation or other purposes. The very earliest dams were probably built of logs and boulders, and were made reasonably tight by spreading gravel, sand and clay on the upstream side. As civilization developed, more ambitious structures were undertaken by rulers using the abundant manpower supplied by slaves. Some of these ancient dams were very large.

There still exist the ruins of a masonry dam built by the Pharaoh Menes I about 4000 B.C. on the Nile River, at a point about twelve miles south of Memphis. This dam is mentioned by the historian Herodotus; it was about fifty feet high and fifteen hundred feet long. A large dam about two miles in length was built by an Arabian king called Lokman about 1700 B.C.; the flood caused by its collapse is recorded in Arabian history. Many thousands of dams have been built in India from the earliest days to the present time. Captain R. Baird Smith reported in his book IRRIGATION IN THE MADRAS PROVINCES that there were 4,300 earth dams, called tanks, still in repair in the Madras Provinces and about 10,000 out of repair. The tanks totaled about 30,000 miles in length.

It may seem surprising that there are so few remains of dams built by the Romans, since many of their aqueducts are still standing and some are still in use. The reason seems to be that the Romans were more proficient in building massive structures than they were in designing dams that would hold back for many centuries the raging waters of floods.

The oldest existing dams in Europe are the Almanza and Alicante dams in Spain; they were built some time before 1586. There is a complete record of the Ponte Alto Dam, in Italy. It was built in 1611 and has been raised several times since then. In Mexico several buttress dams still in existence were built by the Spaniards over two hundred years ago.

Many gravity dams were constructed in France, Spain and other European countries before the middle of the nineteenth century, but they were not based on scientific principles. In 1853, a French engineer, De Sazilly, developed a logical theory for the design of gravity dams. This type of dam has enjoyed steady favor from that time to this.

The first large arch dam, the Zola Dam, was built in France in 1843; it was 123 feet in height. The Australian engineer L. A. B. Wade constructed many notable dams of this type in Australia in the latter part of the nineteenth century. In 1912, the American engineer Lars Jorgensen developed an improved design for arch dams. In the years that followed, they became increasingly popular. In 1910, there were only six arch dams over a hundred feet high. In 1920 the number had risen to twenty-four and in 1930 to seventy-five. At the present time there are well over two hundred.

Until recent times earth dams were designed and built largely by rule of thumb. The Indian tanks we mentioned before were constructed in a primitive manner. Workmen dumped basket loads of earth at the designated spot and then kept treading upon the newly deposited earth until it was firmly in place. Centuries of experience had shown the appropriate proportions that the dam should have and the precautions that should be taken. Only since about 1920 have scientific principles been observed in the design of earth dams.

The other types of dams — rock-fill, buttress, multiple-arch — have had rather checkered careers. Many rock-fill dams were built in California by gold miners in the nineteenth century, since rock was abundant and cement was extremely expensive. However, not many large dams of this kind were built until the period from 1920 to 1940. It is believed that the large rock-fill dam has a promising future. If it is well built, it should be about as permanent as the natural dams that have been formed by the debris transported by glaciers. Many lakes in northern latitudes have been dammed behind such natural barriers, which have endured for thousands of years.

Buttress dams were built in large numbers in the first three decades of this century, but few high dams of this type were constructed. Design changes — particularly the development of more massive buttresses — have made high buttress dams more popular in recent years.

Mechanization and the development of metals such as cast iron and steel have played an all-important part in the construction of large dams. One may well ask why a ruler who could build the Great Pyramid, which contains about 3,000,000 cubic yards of masonry, could not also build a dam like the Aswan Dam. The answer is undoubtedly that he had no means of controlling the water flow. Only with the use of hoisting mechanisms, applied to gates of cast iron and later of steel, was it possible to control water flow adequately.

Another reason why large dams could not be built until quite recently was that no explosive stronger than gunpowder was available. In constructing many (if not most) large dams, rock must be excavated. Not until the development of dynamite and other powerful explosives could dam-builders blast rock quickly and economically. The introduction of new tunneling techniques has also played an important part in the construction of large dams. In such construction, tunnels must often be dug in rock for the temporary or permanent diversion of water. With modern techniques tunnels can now be bored through rock in a fraction of the time formerly required. (See Index, under Tunnels.)

The rock used in early masonry dams, say two hundred years ago, was sometimes obtained by gathering stones from stream beds or from debris slopes. Sometimes it was blasted out of quarries by means of gunpowder. The invention of dynamite made it much easier to obtain great quantities of rock for masonry and other dams.

Formerly rocks were transported by men or horses; they were lifted and placed in the dam by derricks also operated by horse power or man power. When steam power was introduced it became possible to organize all this work in a more systematic way. Of course the development of the internal-combustion engine has revolutionized construction work, as indeed it has revolutionized our entire way of life. The derricks and other equipment used to move rock are now run by internal-combustion engines or electric motors.

Mechanization has played an important part in the building of earth-fill dams. Within the memory of many people still living, the moving of earth for highways, canals and large earth dams was done by horse-drawn scrapers and to some extent even by wheelbarrows — particularly in tight corners. This antiquated equipment has been replaced by huge bulldozers and other mechanized earth-moving equipment.

In calculating the real cost of dams in terms of human effort, the man-hours * of labor per cubic yard of masonry supply us with a convenient measure. Using this yardstick, we have striking evidence of the great progress that has been made in dam construction. For example, it took about 400 man-hours per cubic yard to set in place the 3,000,000 cubic yards of masonry in the Great Pyramid of Egypt. There are 10,500,000 cubic yards of concrete in the Grand Coulee Dam, in the state of Washington; only 3 man-hours per cubic yard were required to erect this structure, and the figure includes the task

* A man-hour is a work unit. It represents the work performed by one man in an hour.

Modern Bulldozer

310

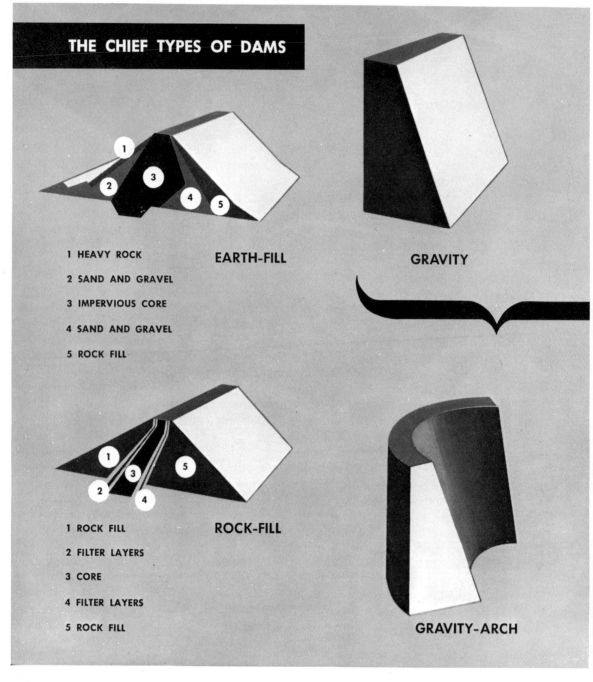

THE CHIEF TYPES OF DAMS

EARTH-FILL

1 HEAVY ROCK
2 SAND AND GRAVEL
3 IMPERVIOUS CORE
4 SAND AND GRAVEL
5 ROCK FILL

GRAVITY

ROCK-FILL

1 ROCK FILL
2 FILTER LAYERS
3 CORE
4 FILTER LAYERS
5 ROCK FILL

GRAVITY-ARCH

of making cement for the dam. To give another example, the labor required for a tunnel made by the Romans to drain Lake Fucino in Italy in the first century A.D. came to 300 man-hours per cubic yard of excavation. By way of contrast, large rock tunnels built in recent years in Sweden have required as little as 1¼ man-hours per cubic yard.

A survey of the different kinds of dams

There are various ways of classifying dams. From a practical viewpoint they can be divided into the following types: (1) earth-fill; (2) rock-fill; (3) gravity; (4) buttress; (5) arch; (6) multiple-arch and (7) miscellaneous. Gravity and arch

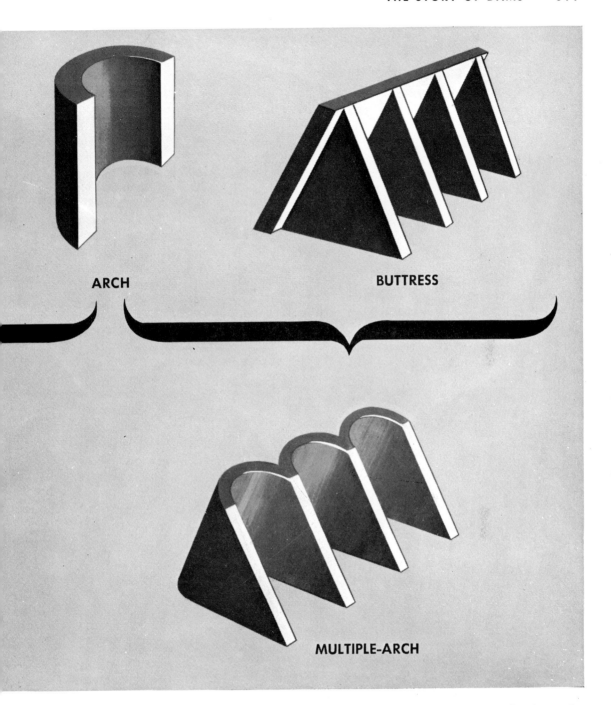

ARCH   BUTTRESS

MULTIPLE-ARCH

dams generally receive more publicity than the other types. As a matter of fact, however, there are more earth-fill dams in the world than all other kinds combined.

Earth-fill dams. An earth-fill dam is made up partly or entirely of impervious material, consisting of fine particles — usually clay, or a mixture of clay and silt or of clay, silt and gravel. The dam is built up with rather flat slopes, as is shown in the above diagram. It is rather difficult to illustrate a "typical" earth-fill dam, because the arrangement of materials varies greatly.

When the fine, impervious material of an earth-fill dam occupies a relatively small part of the structure, it is known as the core. The core is located either in a cen-

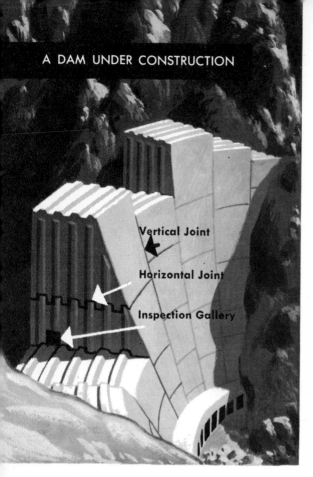

Vertical Joint

Horizontal Joint

Inspection Gallery

tral position or in a sloping position upstream of the center. If the remaining materials consist of coarse particles, there is a gradation in fineness from the core to the coarse outer materials. Some earth dams have a large proportion of rock in the outer zones; this makes for stability.

A vast amount of material must be set in place in the construction of an earth dam. There are several ways of doing this. In rolled-fill construction, the material for the dam is spread out and compacted (pressed together) by means of mechanical equipment, including rollers. In the semihydraulic method, the material is excavated by machines and is hauled to the site where it is to be used. It is then sluiced into place on the dam.

Water is used in the hydraulic-fill method to transport excavated materials to the site of the dam and to set them in place. For example, in building the Fort Peck Dam, on the Missouri River in Montana, materials were dredged from the bed of the river. They were then transported in a current of water pumped through steel pipes to the site of the dam. Finally the materials were washed into place. The Fort Peck Dam, with a volume of over 100,000,000 cubic yards, was the largest dam ever built by this method. It is still the largest earth dam, in point of volume, in the world. The Kingsley Dam, on the North Platte River in Nebraska, with about a quarter the volume of the Fort Peck Dam, was the last large dam built by the hydraulic method. This method is becoming obsolete in earth-dam construction because of the development of enormously powerful and economical earth-movers, such as bulldozers.

Rock-fill dams. Rock-fill dams have attained their greatest development in the western part of the United States. The early dams were sheathed on the upstream side with timber planking as a waterproof facing. Later, a concrete slab was laid on a carefully prepared bed of rock, which was set in position by derricks or by hand.

The main body of a rock-fill dam consists of a mass of dumped rock, which is allowed to take its own angle of repose — that is, to settle naturally. This results in a slope of about 36 degrees. When the body of rock has settled, a blanket, or outer shell, of derrick-placed and hand-placed rock is laid carefully on the upstream face. A concrete facing slab is then put in position. Several well-known dams have been built in this way.

About 1940 a new development was introduced by James Growdon. He made use of an impervious clay core, called a blanket, for watertightness, as shown in the diagram on page 310. The clay core is separated from the rock on both sides of it by layers of graduated material, known as filter layers, to prevent the clay from moving into the rock.

Gravity dams. The gravity type of dam depends upon its weight (and in some cases the weight of part of the water it impounds) to stay in place. It consists of a wall with a roughly triangular cross section (see diagram). The upstream face (the face, or side, facing upstream) is vertical or nearly vertical; the downstream face has a slope of about 50 degrees from the horizontal. In the case of dams of considerable height the lower part of the upstream face may deviate considerably from the vertical. In this way the base of the dam is broadened, and there is not so much pressure at the heel, as the upstream side of the base is called.

As viewed from above, a gravity dam may be straight or curved. It may be composed of two or more straight sections meeting at an angle. This shape is used in order to set the dam on the highest or best rock available.

The wall of a gravity dam is generally made of concrete at the present time. However, in countries where labor is cheap and cement is expensive, it is sometimes made of stone masonry. A recent example of a large gravity dam of this type is the Tungabhadra Dam in India. This structure contains 1,200,000 cubic yards of masonry, exclusive of mortar. It is 160 feet high and 6,000 feet long. It took about 20,000 workers, toiling two shifts a day, a number of years to erect the structure. The project was begun in 1946; most of the masonry was placed between 1951 and 1956.

Even if the body of the dam is of concrete, a facing of stone masonry may be used for the sake of appearance, just as stone or brick veneer is used to cover the wood or stucco of certain small homes.

While concrete is one of the most valuable building materials known to man, it has some regrettable shortcomings. Shrinkage is the most troublesome of these. It is due to the fact that the material is first warmed by the chemical action that takes place in setting and later cools considerably. As a result of the shrinking process, long retaining walls sometimes develop vertical cracks. These may be at intervals of about fifteen or twenty feet.

Great care is taken to avoid such cracks in large gravity dams built of concrete. In many cases cooling pipes are imbedded in each "pour" of concrete; cold water is kept circulating through the pipes in order to keep the temperature down during the setting process. Sometimes a special type of low-heat cement is used for the concrete mixture.* As a further precaution vertical construction joints are used to divide the dam into a number of blocks, called monoliths. The construction joints are sealed at the upstream end, generally by a number of copper strips imbedded in the concrete.

It is also necessary to avoid longitudinal, or lengthwise, cracks in each monolith. One method is to divide each of the monoliths into smaller blocks by means of construction joints running parallel to the length of the dam. These joints are spaced about as far apart as the vertical joints. They have a saw tooth shape so that each section will fit more or less snugly in the section above or below it. The joints are grouted: that is, they are injected with cement under pressure, in order to make them fit together more tightly.

Most gravity dams have at least one inspection gallery running lengthwise through them near the upstream face. This gallery serves, as the name indicates, for inspection purposes; it is generally from six to eight feet wide and from seven to ten feet high. The inspection gallery (or the lowest gallery, if there is more than one) is set about ten or fifteen feet above the foundation rock. During the construction period holes are drilled in the floor of the gallery and into the foundation. Cement under pressure is then injected in the holes in order to strengthen the foundation rock. The inspection gallery also serves as a drain to carry off water that has seeped into the dam.

As we have pointed out, the great majority of gravity dams have a vertical or at least very steep upstream face and a comparatively steep downstream face. However, certain dams show different cross-section shapes, depending upon the nature of the terrain. For example, the spillway dam of the C. J. Strike project in Idaho rests on consolidated silt (silt that has become solidified), which makes a poor foundation. It was necessary to give both the upstream and downstream faces a much

---

* Ordinary concrete consists of a mixture of portland cement, sand, gravel or crushed stone and water.

gentler slope than is customary, to provide a broader base and thus spread the load.

Buttress dams. The use of buttresses to support a wall is of very ancient origin; it is not surprising, therefore, that this idea was applied to dams. The principle involved is simple enough. Suppose we put a number of triangular bricks side by side and a few inches apart. If we set a plank in place so that it rests on one set of sloping edges, we would have a crude model of a buttress dam.

A certain number of dams in Mexico built of stone masonry by the Spaniards over two hundred years ago used buttresses to support a relatively slender vertical masonry wall. As we have seen, some of the dams are still standing. In the early part of the twentieth century, this type of design was made more scientific by the use of reinforced concrete buttresses supporting a sloping reinforced concrete slab. Dams of this type were often called Ambursen dams, after the name of the inventor; they were far more stable than the previously mentioned buttress dams with vertical walls. Many hundreds of Ambursen-type dams were erected from about 1910 to 1940. Ambursen dams are of rather limited height. They are well adapted to poor foundations, such as shale, because the load is light, well spread and well centered. Only a few big dams of this type have been built.

About 1930 a new type of buttress dam came into use. It had massive, unreinforced buttresses, enlarged at the upstream edges. This type has proved very successful. It combines most of the advantages of a massive gravity dam, and at the same time requires less concrete than a gravity dam. There are various other advantages, including better cooling of the concrete and easy inspection and maintenance.

Arch dams. An arch dam is generally built in a canyon with solid walls. As seen from above, the dam has the form of an arch, with the bulging part facing upstream. The dam resists by its own weight part of the pressure of the dammed-up water. It transmits the rest of the water pressure, or load, by arch action * to the walls of the enclosing canyon. The arch dam depends primarily on the strength of its concrete (or occasionally stone masonry) for its stability. Its cross section is generally thinner, and sometimes notably thinner, than that of a gravity dam of the same height. The arch dam is always stronger than a gravity dam, and is generally more economical to boot. Some arch dams have a vertical upstream face; others are curved not only horizontally (that is, as seen from above) but also vertically. In some cases dams of this type may assume almost a dome shape. We show several types of arch dams, in cross section, on this page.

One of the most remarkable advances in arch-dam design has been the use of wider sites. Some of the older authorities recommended arch dams only for sites where the top width did not exceed a few hundred feet. At the present time, if suitable rock is available as a foundation, there

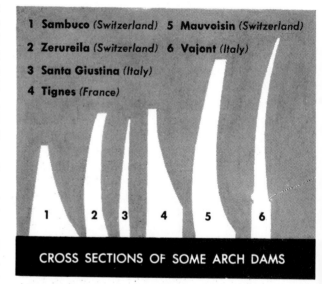

1 **Sambuco** (*Switzerland*)   5 **Mauvoisin** (*Switzerland*)

2 **Zerureila** (*Switzerland*)   6 **Vajont** (*Italy*)

3 **Santa Giustina** (*Italy*)

4 **Tignes** (*France*)

1 2 3 4 5 6

CROSS SECTIONS OF SOME ARCH DAMS

is little hesitation in considering a site with a top width of up to 1,500 feet.

As in the case of gravity dams, an arch dam is divided into monoliths by vertical joints. These joints are often grouted, to allow for concrete shrinkage. Instead of grouting, spaces of three to five feet may be left between the adjacent blocks.

---

* Arch action refers to the transfer of the water load to the ends of the arch.

These gaps are later filled with concrete after the monoliths have cooled.

The design of an arch dam is somewhat complicated, since many factors have to be taken into consideration. It has been found helpful in working out an efficient design to build a small-scale model of the dam and to substitute mercury for water. European engineers, who have built many notable arch dams, have been particularly skillful in working with such models.

Certain dams have a cross section comparable to that of a gravity dam and yet are curved in the form of an arch. Such dams are commonly known as arch-gravity or gravity-arch dams.*

Multiple-arch dams. If we replace the flat slab of a reinforced concrete buttress dam by a series of arches, as shown in the diagram on page 311, we have a multiple-arch dam. This dam consists of a number of different elements that can be precisely designed. Like the Ambursen type of dam, the multiple-arch dam is adapted to poor foundations. It has proved very useful in many places where local timber for formwork is available and concrete materials are rather scarce. However, the great amount of timber formwork required in its construction adds greatly to labor costs. As a result the multiple-arch dam has lost favor in recent years.

Miscellaneous dams. There are a number of dam types that do not fall into any of the classes that we describe above.

* An arch-gravity dam is an arch dam with a particularly thick wall. A gravity-arch dam is a gravity dam with a pronounced arch, as seen from above. The Hoover Dam is an example of a gravity-arch dam; the Sambuco Dam (Switzerland) is an arch-gravity dam.

In a bridge dam, openings between the piers of a permanent bridge are closed by gates pivoting from the piers. The shutter, or wicket, dam consists of a series of shutters, or wickets, set upright side by side. The lower edge of each shutter rests against a sill and is supported by a prop set behind it. This type of dam is used for rivers whose waters rise high in time of flood. The shutters may tip automatically or may be released and fall to the river bed when the waters of the river reach a certain height above the top of the dam. When the flood waters have subsided, the shutters are raised again by means of a windlass operated from a boat or service bridge. Several dams of this sort are found on the Ohio River.

A roller dam is formed by a massive cylinder set across a river; the diameter of the cylinder is about equal to the desired height. To permit water to pass, the cylinder is rolled up inclined tracks set at both ends of the dam. In the roller dam across the Yakima River in Washington, the cylinder has a diameter of 14 feet; the width of the river at the site of the dam is 110 feet. There are also steel-faced rock-fill dams, timber dams, steel dams and various other kinds.

The accessory works of dams

Many dams have accessory works that add greatly to their operating efficiency. The most important of these works are the spillway and the controls for the release of stored water.

FREE OVERFALL SPILLWAY   SPILLWAY WITH CONTROL GATES

The spillway is a kind of safety valve, which serves to pass floods that otherwise would endanger the dam. Many water-supply dams are provided with a free over-fall spillway; water is permitted to flow freely over one section of the dam. In the case of other dams, spillway gates of different kinds are generally used. The gates are opened by means of a hoist apparatus in order to let water pass through when the level of the river becomes threateningly high. Spillway gates are generally operated by electric power; but an alternative source of power is often provided.

Water-release controls have become very important as dams have increased in height and size, and as the quantity of water they store has become greater. Controls are generally provided by gates set in tunnels within the dam structure. Early experience brought home to engineers the devastating power of water when it is released under high pressure in enclosed spaces. It is now usual to design controls so as to offer the least possible disturbance to the flow of water in the control tunnel. In many cases means must be provided for dissipating the energy of the water after it is released from the outlets of the tunnel. This can be done by spraying the water into the air, or by letting it flow into a widened basin called a stilling pool.

An important accessory structure of some dams is a fish ladder, for the use of salmon and other fish that make their way upstream in order to breed. The fish ladder consists of a water stairway, with a series of steps up which the fish can jump. Fish ladders and other devices for handling fish may become a costly part of a dam project, as in the case of the McNary Dam, across the Columbia River.

The forces that act on dams

Dams are subject to various forces. The most obvious one is that exerted by the water that presses upon the upstream face of the structure. The sheer pressure of this water is very impressive, for a cubic foot weighs very nearly 62½ pounds. The total pressure on a given length of wall is proportional to the square of the depth. For example, the horizontal water load acting on a vertical strip of dam one foot wide would be 156 tons if the dam were 100 feet high and the water reached the crest. If the height were 200 feet, the total load on the one-foot-wide strip would be 624 tons. When silt builds up against the lower part of the dam, it acts as a liquid that is denser than water. Engineers must take this factor into account in designing a dam.

Ice is another factor that must be considered. In cold climates a thick sheet of ice may form on the reservoir surface. Such a sheet of ice may be warmed by the sun from a temperature of, say, 20° F. below freezing to, say, 10° F. below freezing. The tendency to expand may then cause a tremendous force near the top of the dam. Hence this part of the structure must be made thick enough to withstand the pressure. If pressure due to ice were allowed to build up at the spillway gates, it could easily damage them. To prevent this from happening, means are provided to protect the gates. Sometimes they are heated; sometimes air is bubbled through the water in front of the gates to prevent ice from forming.

The weight of the dam itself is another force that acts on dam structures. This factor is important chiefly in the case of gravity dams and very high arch dams. The vertical downward pressure due to the concrete of a typical gravity dam is about one pound per square inch for each foot of height. This means that the pressure at the bottom of a gravity dam 500 feet high would be about 500 pounds per square inch. Concrete can withstand such pressures. So can most rock foundations, but not foundations of shale, soft sandstone and certain other rocks. To reduce the stress on such weak foundations, a limit must be set to the height of the dam, and the upstream face must be sloped so as to spread the load. The weight of the water pressing down on the slope will act as a stabilizing factor. A buttress type of dam may be used, with the base of each buttress widened.

DAMS
a pictorial survey

The Shasta Dam, shown on this page, is on the Sacramento River, in California. It is a fine example of the curved gravity-type concrete dam. The Shasta Dam is 602 feet high and its length at the crest is 3,500 feet. The water it stores up is used for irrigation and generates hydroelectric power. The big dam also serves for flood control.

U. S. Dept. of the Interior

316-a

The more or less massive structures we call dams are used to store up water for various purposes or to divert water from streams. There are many kinds of dams, built from a variety of materials, such as concrete, rock, gravel and earth. In these four pages of full color we give close-up and long-range views of some important types of dams.

Right: the Kamishiiba Dam, a high-arch type built for a 90,000-kilowatt power plant on the uppermost reaches of the Mimi River, in Kyushu, Japan. This photo shows the dam in the construction stage; it was completed in 1955.

Below: the Hoover Dam, 726 feet high, on the Colorado River, on the border between Arizona and Nevada. The waters dammed up by the structure have formed a large artificial lake, Lake Mead, which is shown in the background.

Kyushu Electric Power Co.

U. S. Dept. of the Interior

316-b

U.S. Army Corps of Engineers

Above: Dam No. 8 on the Ohio River. This is a movable dam, which was built by engineers of the United States Army in order to improve navigation on the Ohio. Below is shown a typical earth-fill dam — the large Yale Dam, on the Lewis River, north of Portland, Oregon. The earth-fill type is the workhorse among dams; there are more earth-fill dams in the world than all other kinds combined.

Ebasco Services, Inc.

Ebasco Services, Inc.

The Bliss hydroelectric development (above), on the Snake River near Gooding, Idaho, has a generating capacity of about 80,000 kilowatts. The spillway is shown at the extreme right.

Below: the Ross Dam, on the Skagit River in Washington. This beautiful arch-type dam is 545 feet high and 1,275 feet long. It serves for power production and also for flood control.

Dept. of Lighting, Seattle, Wash.

316-d

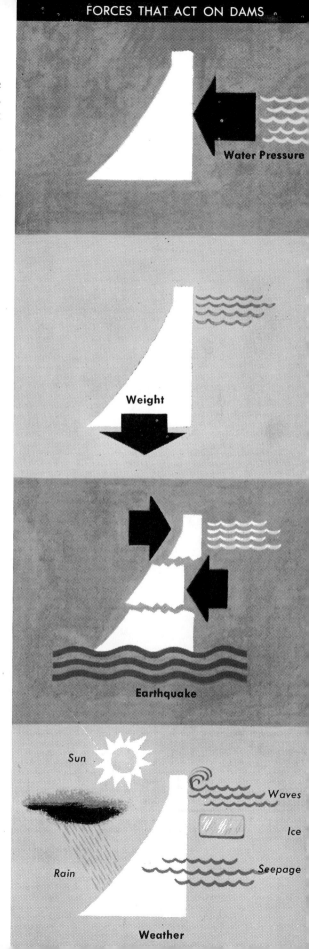

Water Pressure

Weight

Earthquake

Sun

Rain

Waves

Ice

Seepage

Weather

In the case of earth-fill dams, the weight of the material is about two thirds that of concrete. When such dams rest upon the ground itself, as is often the case, and not on rock, the weight of the dam must be carefully considered. There must not be too great a load on the ground.

Earthquakes may exert considerable pressure on dams. An earthquake consists of a rapid to-and-fro motion of the ground, which may cause tall structures, such as smokestacks, to topple. The action is like that of pulling a rug from under a person who is standing on it. The horizontal force exerted by an earthquake may be equal to as much as a tenth of the weight of the dam. Few places are entirely free from earthquakes; hence earthquake forces are generally taken into account in dam design.

Seasonal and daily changes in temperature may cause internal stresses in dams, particularly thin arch dams. These changes must be carefully analyzed.

In a masonry dam there will be some seepage of water. This water may leach some of the lime out of the concrete in time. Water percolating into the outer surface of certain dams may freeze and expand, and break down the concrete in this area. Water will seep into an earth-fill dam until it meets an impermeable barrier. If the water level of the dammed-up stream is rapidly lowered, the water-soaked material may become unstable. This has to be considered in the design of earth dams.

Waves striking against the face of a gravity or arch dam have little effect on the stability of the structure. In the case of an earth-fill dam, however, the waves would soon erode the surface material if it were not protected by a facing of heavy rock laid on a bed of gravel. Such rock is known as riprap. Sometimes hand-laid rock or concrete slabs are used.

The erosive forces of nature — winds, rain, running water and the like — are always at work, of course, on any structure. Maintenance work is required on all dams from time to time to keep these forces in check. If such maintenance is provided, well-designed and well-built dams

IMPORTANT DAMS OF THE WORLD

ABBREVIATIONS: A = arch; **AB** = buttress supported on arch; **AG** = arch-gravity; **B** = buttress; **DC** = debris control; **E** = earth-fill; **ERF** = earth-fill dam containing large amount of rock; **FC** = flood control; **G** = gravity; **GA** = gravity-arch; **I** = irrigation; **MA** = multiple-arch; **N** = navigation; **P** = power; **RF** = rock-fill; **WS** = water supply.

NAME OF DAM	RIVER	COUNTRY	HEIGHT feet	TYPE	LENGTH feet	VOLUME cubic yards	WATER STORAGE acre-feet	PRINCIPAL PURPOSES	YEAR COMPLETED
L'Aigle	Dordogne	France	312	AG	951	320,000	160,000	P	1948
Abitibi Canyon	Abitibi	Canada (Ont.)	290	G	400	299,000	11,200	P	1933
Alder	Nisqually	U. S. (Wash.)	330	A	178	440,000	200,000	P	1944
Ambuklao	Agno	Phil. I. (Luzon)	420	ERF	1,400	8,000,000		P	1955
Ancipa	Troina	Italy	305	B	830	392,000	24,600	P	1952
Anderson Ranch	Boise (Southern fork)	U. S. (Ida.)	456	E	1,350	9,080,000	500,000	I, P	1948
Ariel	Lewis	U. S. (Wash.)	313	A	1,300	300,000	222,000	P	1931
Arrowrock	Boise	U. S. (Ida.)	348	G	1,100	585,000	286,000	I	1915
Aswan	Nile	Egypt	174	G	6,900	1,730,000	4,040,000	I, N	*1934
Balze di S. Lucia	Salto	Italy	341	G	605	470,000	225,000	P	1939
Bartlett	Verde	U. S. (Ariz.)	287	MA	800	182,300	182,000	I	1939
Beauregard	Grisanche	Italy	434	AG	1,280	550,000	56,800	P	1957
Bekhme	Greater Zab	Iraq	550	G					1957
Bimont	Infernet	France	290	A	850	120,000	32,400	P	1952
Bin el Ouidane	Oued el Abid	Morocco	434	A	950	340,000	1,060,000	P, I	1953
Bort	Dordogne	France	393	A	1,279	872,500	324,400	P	1952
Cabril	Zezere	Portugal	432	A			500,000	P	1955
Cap de Long (Pragnères)	Ise	France	295	AG	1,863	334,000	54,300	P	1954
Cardenas (El Palmito)	Nazas	Mexico	302	E	787	7,000,000	3,263,000		1948
Castelo do Bode	Zezere	Portugal	377	A	1,150	563,300	893,200	P	1950
Castillon	Verdon	France	328	A	656	163,700	121,600	P	1949
Chambon	Romanche	France	367	G	1,080	392,000	46,500	P	1935
Chastang	Dordogne	France	288	AG	984	340,600	145,900	P	1951
Cherry Valley	Cherry	U. S. (Cal.)	315	ERF	2,600	6,500,000	270,000		1954
Cleveland	Capilano	Canada (B.C.)	325	G	640	150,000			1954
Cohila	Nansa	Spain	381	A	680		9,000		1950
Detroit	N. Santiam	U. S. (Ore.)	463	G	1,580	1,500,000	455,000	P, FC	1953
Diablo	Skagit	U. S. (Wash.)	389	A	1,142	350,000	90,000	P	1930
Dique la Vina		Argentina	344	A		223,000	185,000		1943
Dixence	Dixence	Switzerland	285	B	1,500	552,000	40,500	P	1935
Doiras	Navia	Spain	290	G			75,400		1935
Elephant Butte	Rio Grande	U. S. (N.M.)	306	G	1,155	605,000	2,219,000	I	1916
El Fuerte	Fuerte	Mexico	300	G			25,000,000		1940
Exchequer (Lake McClure)	Merced	U. S. (Cal.)	326	G	955	350,000	289,000	P, I	1926
Fiastrone	Fiastrone	Italy	286	AG		200,000	16,500	P	1953
Folsom	American	U. S. (Cal.)	280	EG		20,300,000	1,000,000	I, FC	1957
Fontana	Little Tennessee	U. S. (Tenn.)	470	G	1,600	2,800,000	1,600,000	P, FC, N	1944
Forte Buso	Travignola	Italy	361	AG	1,053	340,000	26,000	P	1951
Fort Peck	Missouri	U. S. (Mont.)	270	E	9,000	12,400,000	19,000,000	P, FC, N	1940
Friant	San Joaquin	U. S. (Cal.)	320	G	3,480	2,134,000	520,000	I, P	1942
Fujiwara	Tone	Japan	315	G	1,070	1,000,000	28,000	P, FC	1957
Generalisimo	Turia	Spain	340	G	672	504,500	206,700		1949
Genissiat	Rhone	France	341	G	656	576,000	42,900	P	1948
Granby	Colorado	U. S. (Col.)	295	E		2,901,000	546,000	P, I	1950
Grandas de Salime	Navia	Spain	441	G	826	850,000	227,000		1953
Grand Coulee	Columbia	U. S. (Wash.)	550	G	4,173	10,493,000	9,517,000	I, P, FC	1942
Grande Dixence	Dixence-Rhone	Switzerland	932	G	2,296	7,792,000	324,000	P	1962
Grimsel-Spitallamm	Aar	Switzerland	374	AG	848	445,000	81,000	P	1932
Hiwassee	Hiwassee	U. S. (N.C.)	307	G	1,287	807,200	438,000	P, FC	1940
Hoover	Colorado	U. S. (Ariz., Nev.)	726	GA	1,180	3,250,000	30,500,000	P, I, FC	1935
Horse Mesa	Salt	U. S. (Ariz.)	305	A	810	147,000	245,000	P, I	1927

*Dam heightened and lengthened in 1934.

NAME OF DAM	RIVER	COUNTRY	HEIGHT feet	TYPE	LENGTH feet	VOLUME cubic yards	WATER STORAGE acre-feet	PRINCIPAL PURPOSES	YEAR COMPLETED
Hungry Horse	S. F. Flathead	U. S. (Mont.)	546	A	2,130	3,087,000	3,500,000	P, FC	1952
Ikari	Ozika	Japan	354	G	1,050	655,000	40,000	P	1956
Ikawa	Oi	Japan	325	G	700	1,570,000	120,000	P	1956
Jandula	Jandula	Spain	295	G	830	412,500	283,800		1930
Jubilee (Gorge)	Shing Mun	Hong Kong	285	RF		660,000	10,800	WS	1936
Kajakai	Helmand	Afghanistan	321	ERF	900	4,300,000		I, P	1953
Kamiitsuka	Kuma	Japan	311	A	750	262,000	11,600	P	1956
Kamishiiba	Mimi	Japan	366	A	1,150	407,000	75,000	P	1955
Kenney	Nechako	Canada (B.C.)	317	RF		3,700,000		P	1952
Kuromata No. 1	Shinano	Japan	299	G	790	418,000	32,000	P	1956
La Joie	Bridge	Canada (B.C.)	285	RF	1,400	3,620,000	572,000	P	1950
La Vinia	Los Sauces	Argentina	331	A	1,040	240,700	186,500		1943
Lucky Peak	Boise	U. S. (Ida.)	328	E	1,700	5,775,000	306,000	I, FC	1955
Lumiei	Lumiei	Italy	441	A	454	131,000	58,000	P	1947
Madupatty	Muthirapuzha	India	280	G	805	210,000	45,000	I, P	1953
Marèges	Dordogne	France	295	A	810	242,300	38,000	P	1935
Margaritzen (Moell)	Moell	Austria	302	A	557	46,000	3,200	P	1953
Maruyama	Kiso	Japan	288	G		615,000	46,000	P	1953
Meishan	Hwai	China	290	MA	268			FC, I, P	1956
Monte Surei	Mulargia	Italy	329	GA	847	300,000	246,000	I, P	1957
Monticello	Putah Creek	U. S. (Cal.)	304	A	1,017		1,600,000	I, WS	1956
Mooserboden (Drossen)	Kaprun	Austria	368	A	1,180	440,000		P	1955
Mooserboden (Mooser)	Kaprun	Austria	335	G	1,540	836,000		P.	1955
Mud Mountain (Stevens)	White	U. S. (Wash.)	425	E	700	2,230,000	130,000		1948
Norris	Clinch	U. S. (Tenn.)	240	G	1,570	1,195,000	2,710,000	FC, P	1935
Oberaar	Aar	Switzerland	328	G	1,722	614,000	47,000		1953
O'Shaugnessy	Tuolumne	U. S. (Cal.)	430	G	900	660,000	350,000	WS	1925
Owyhee	Owyhee	U. S. (Ore.)	417	AG	833	521,000	1,120,000	I	1932
Pacoima	Pacoima	U. S. (Cal.)	380	A	640	226,100	6,000	FC	1929
Paradela	Cavado	Portugal	367	RF	1,900	3,537,000	128,000	P, I	1956
Parker	Colorado	U. S. (Cal., Ariz.)	320	A	856	312,000	717,000	WS, P	1938
Peares	Mino	Spain	295	G		523,000	145,900	P	1953
Picote	Douro	Portugal	328	A	390	301,300	12,000	P	1957
Pieve di Cadore	Pieve	Italy	368	A	1,345	494,000	55,000	P	1950
Pine Flat	Kings	U. S. (Cal.)	440	G	1,820	2,400,000	1,110,000	FC	1953
Raeterishboden	Aar	Switzerland	302	G	1,500	366,000	21,900	P	1950
Ricobayo	Esla	Spain	326	G	791	497,000	960,300	P, I	1935
Roosevelt	Salt	U. S. (Ariz.)	280	G	723	342,000	1,420,000	I	1911
Ross	Skagit	U. S. (Wash.)	545	A	1,275		1,400,000	P, FC	1949
Sakarya	Sakarya	Turkey	360	G	1,000	700,000	7,000,000	P, I	1954
Sakuma	Tenryu	Japan	504	G	965	1,280,000	265,000	P	1956
Salime	Navia	Spain	390	G			226,000		1953
Salt Springs	North Fork Mokelumne	U. S. (Cal.)	328	RF	1,300	3,000,000	130,000	P	1931
Sambuco	Maggia	Switzerland	426	AG	1,116	1,050,000	50,200	P	1957
San Esteban	Sil	Spain	380	A	945	599,000	172,700	P	1955
San Gabriel No. 1	San Gabriel	U. S. (Cal.)	310	RF	1,500	10,600,000	53,300	FC	1939
Santa Giustina	Noce	Italy	500	A	408	147,000	147,000	P	1951
Sariyar	Sakarya	Turkey	354	G	825	730,000	1,540,000	P, FC	1955
Sarrans	Truyère	France	371	G	738	589,500	235,000	P	1933
Sautet	Drac	France	416	A	272	131,000	105,400	P	1934
Schraeh	Wagitaler-AA	Switzerland	362	G	510	310,000	119,000	P	1925
Shasta	Sacramento	U. S. (Cal.)	602	G	3,500	6,230,000	4,390,000	I, P, FC	1943
Shima	Nahari	Japan	325	G	920	450,000	52,000	P	1957
Shimoitsuki	Kuma	Japan	299	G				P	1957
Shimokotori	Kotori	Japan	420	G	1,080	1,050,000	· 96,000	P	1957
Shoshone	Shoshone	U. S. (Wyo.)	328	A	200	79,000	456,000	I, P	1910
South Boulder Creek (Gross)	South Boulder Creek	U. S. (Col.)	340	G	1,022	580,800	42,000	WS	1954
South Holston	Holston	U. S. (Tenn.)	285	E	1,550	5,900,000	783,000	P	1951
Sudagai	Tone	Japan	290	G	980	297,000	21,000	P	1955
Suiho	Talu	Korea	350	G	3,280				1944
Sultan No. 1	Sultan	U. S. (Wash.)	310	G			234,000		1952
Suviana	Limentra	Italy	293	G	721	366,000	37,800	P	1933
Tu Fung Man	Sungari	Manchuria	300	G	3,800				1945
Tagokura	Tadami	Japan	477	G	1,410	2,480,000	400,000	P	1957
Val Gallina	Val Gallina	Italy	290	A	738	128,000	4,800	P	1952
Venda Nova	Rubagao	Portugal	315	A	790	288,200	74,700	P	1951
Warragamba	Warragamba	Australia	420	G	1,100	1,400,000	1,680,000	WS	1957
Watauga	Watauga	U. S. (Tenn.)	320	ERF	900	3,500,000	627,000	FC	1948
Yale	Lewis	U. S. (Wash.)	323	E	1,500	4,200,000		P	1952
Zerureila	Valserrhein	Switzerland	554	AG	1,560	850,000	81,000	P	1956
Zeuzier	Lienne	Switzerland	524	A	870	418,000	40,500	P	1957

should be able to defy the elements for centuries.

Methods of building dams

Except in the few cases in which dams are built away from a river, the first step in the building of a dam is to drain off water from the river whose waters are to be dammed up. One or the other of two methods is generally used.

In the first of these methods, the river is totally diverted from its normal course, at least for a part of the year during the construction period. The waters of the river may be carried off by a flume, or by a channel cut in the rock at the side of the river or through one or more tunnels. Flood flows that are greater than the tunnel capacity are allowed to pass over or through the uncompleted work. Sometimes a part of the dam is built "in the dry" near the river. A channel is then cut to divert the river through the work that has already been done.

The second method, which is generally used when a wide river is to be dammed, is to close off one part of the river at a time by means of a cofferdam. Water is then pumped from the cofferdam and a part of the dam is erected on the river bed. Another part of the river is then closed off by another cofferdam, and the flow of the river is directed through the completed part of the structure. Water is pumped from the second cofferdam, and the second part of the dam is erected beside the first.

It may not be feasible to divert the waters of the river in this way. In such case, the side portions of the dam are built "in the dry" at the two sides of the river. A large concrete block is then prepared and is toppled into the river to close the gap between the side portions. This procedure is possible only with rather low dams.

The layman often thinks of a dam as a large concrete wall built directly upon the surface of the river bed. In many cases, however, the river valley is partly filled with alluvial deposits, which would offer a most shaky foundation for a huge structure. The foundations of dams must be sunk through such deposits down to the solid rock beneath. Cut-off walls may have to be sunk below the bottom of the dam, on the upstream side, to prevent water from seeping into the dam from beneath.

The choice of the method to be used in building a dam is one of the most important decisions that must be made by the engineer charged with the work. The electronic computer will help him make the abstruse calculations on which his decision will be based. He must adopt one method or another before much work has been put into the design of the dam, since the method of construction will affect the design. If he has chosen well, work on the dam will proceed quickly and economically.

Outstanding dams of the world

In 1955 I prepared a list of dams over a hundred feet high throughout the world for the Proceedings of the American Society of Civil Engineers. This list contained 1,763 dams, including those already constructed or under construction (1,558) and those proposed (205). Of the 1,763 dams I listed, 569 were in the United States (including Alaska), 222 in Japan, 164 in Italy, 141 in Spain and 96 in France. In other words, the United States had more high dams than any other country, followed by Japan, Italy, Spain and France. The order of precedence was quite different, however, if one considered the number of such high dams per thousand square miles of area. From this point of view Puerto Rico ranked first, followed by Switzerland, Japan, Italy, Portugal and Spain.

In the table on pages 318-19, I provide a shorter list of outstanding dams of the world, giving important data concerning them. I must point out that the iron-curtain countries are not adequately represented in the list, because information concerning their dams is not readily available. It is known, however, that great advances in dam construction have been made in these countries, particularly in the case of the Soviet Union.

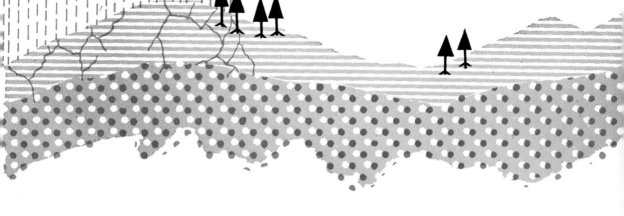

WATER IN THE GROUND

BY OSCAR E. MEINZER

A Vast Reserve Stored Up in Soil and Rock

THERE is more water in the world than we can see in oceans, lakes, ponds and streams. A good deal is contained in the pores and cracks of the soil and rocks under the surface of the earth. This ground water, or subsurface water, as it is called, feeds springs, brooks and rivers and supplies wells. If there were no ground water, some streams would be dry except after a heavy rainstorm or immediately after the melting of snow. In many places the only water supplies would be those obtained by impounding storm waters or catching rain water in cisterns.

What is the source of ground water? This question once puzzled the most learned men. Before the latter part of the seventeenth century, it was generally as-sumed that the water discharged by springs could not be derived from rain. It was believed that there was not enough rainfall for this purpose. Besides, the earth was supposed to be too impervious to permit rain water to penetrate far below the surface.

Various ingenious hypotheses were offered to account for the presence of the water that fed streams and emerged from the earth in the form of springs. The favorite theory was that sea water was conducted through subterranean channels below the mountains. It was then somehow purified and raised, finally penetrating to the surface. This theory was not fully discredited until the Frenchmen Pierre Perrault (1608–80) and Edme Mariotte

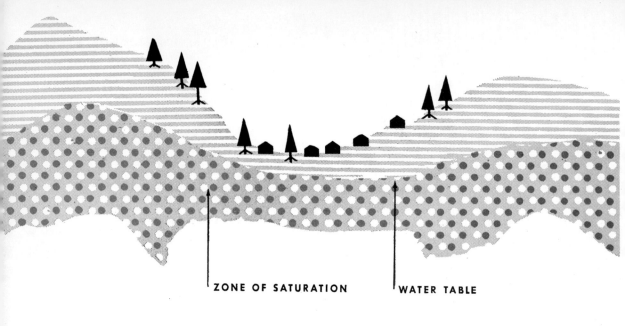

ZONE OF SATURATION   WATER TABLE

This drawing shows the zone of saturation, where all the rock openings are full of water. The water table is the upper boundary of this zone.

(1620–84) made crude measurements of rainfall and stream flow. They showed that the amount of rainfall is ample to supply springs and rivers.

Today we realize that most ground water is derived from rain and snow and other forms of precipitation. There are several other possible sources, it is true. At the time when sediments were laid down on ocean or lake bottoms, water filled the spaces between the grains of material such as sand and silt. The deposits were transformed into sedimentary rocks in the course of the ages. Water was trapped in these rock formations and some of it still remains there. Ground water may originate, too, from the steam rising from magmas — molten rock materials — deep within the earth. But the water trapped in rock formations and that derived from magmas make up a very small part of the total quantity of ground water.

The amount of precipitated water that will seep into the ground from the surface will depend on various factors. Of course an important one is the total amount of precipitation. The rate of precipitation is also important. When rainfall is heavy, the surface quickly becomes saturated and water flows along the surface instead of

making its way into the ground. The slope of the land on which rain falls is another determining factor. The steeper the slope, the greater the surface runoff will be.

The porosity of the rock through which water must pass also helps determine the water content underground. Porous rocks have various openings into which water can penetrate. These openings may be spaces between pebbles and grains in deposits of gravel and sand or between grains in indurated, or hardened, rocks. Cracks and fissures may be produced by the fracturing of hard, brittle rocks, such as sandstone, quartzite, slate and granite. Crevices may form in rocks such as limestone, as percolating water slowly dissolves them. Gaps may be produced by the weathering of rocks near the surface. The porosity of a given formation represents the ratio of its empty spaces to its total volume. It ranges from less than 1 per cent in the case of certain igneous rocks, such as granite, to more than 40 per cent in certain sands and gravels.

The rate at which water will sink into the ground will also depend upon the permeability of the formation through which the water passes. When we say that rock or soil is permeable, we mean that it per-

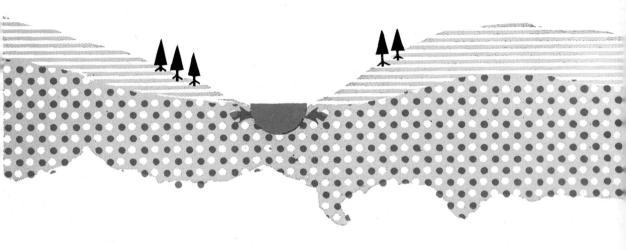

Usually the water table slopes downward toward a stream (shown above in cross section). The zone of saturation supplies water to the stream.

mits water to pass through it. Permeability is not the same thing as porosity. Two rock formations may be equally porous. Yet one will permit water to pass through rather easily, while the other will be almost impermeable. The arrangement of coarse-grained and fine-grained particles within the formation may be the deciding factor. Coarse-grained gravel mixed with fine-grained particles may be less permeable than gravel not so coarsely grained but with fewer fine-grained particles. A clay formation may be very porous; yet when it is saturated, it will be impermeable. Since the pores in clay are tiny, the water in the pores will be held by the molecular attraction of the clay particles.

The nature of the rock strata is another factor in the descent of water. If a given rock stratum is flat, it will be harder for water to make its way downward than if the stratum is inclined. The descent of the water will also be influenced by the presence of vegetation at the surface. Forests and grassy tracts hold back runoff following heavy rains. The result is that more water sinks into the ground. Finally, the amount of water vapor in the atmosphere will help determine how far downward water will penetrate. If the humidity

is low following a rainfall, much of the rain water will evaporate before it can sink below the surface. In desert areas, most of the precipitation that has been yielded up by the atmosphere is returned to it again in a comparatively short time as the water evaporates.

The water making its way downward through porous and permeable rock formations finally reaches a zone of saturation. This is a region in which all the rock openings are full of water. The upper boundary of the saturation zone is called the water table, or the ground-water level. Unlike the surface of rivers or lakes, the water table is not level. It is higher in some places than in others, depending upon the general contour of the land surface. For example, the water table is higher under a mountaintop than it is in the valley below. However, it would lie deeper *below the surface* at the mountaintop than it would at the valley, as the diagram on this page shows.

In most cases the water table slopes downward toward a stream. As the upper limit of the zone of saturation lies above the stream, the zone supplies the latter with water through seepage and springs. In an arid region, the water table slopes

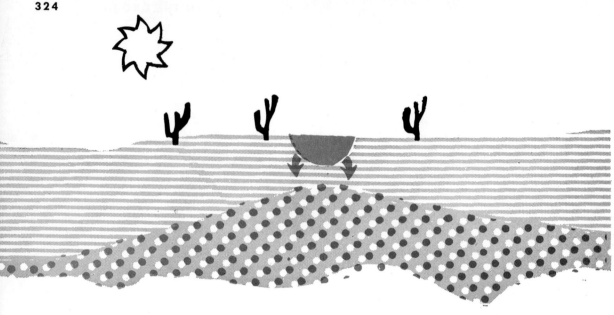

In an arid region the water table slopes upward toward a stream. Water seeps from the stream, in this case, toward the zone of saturation.

upward toward the stream; the saturation zone lies below the stream. Water seeping from the stream replenishes the ground water beneath it.

A body of water trapped in porous and permeable materials may be suspended above the main water table. This happens when ground water is prevented from sinking any farther by a mass of impervious rock or by a basin-shaped bed of clay. The water trapped in such a formation is called a perched water body. Its upper level is known as the perched water table. Water that has been trapped in this way is usually found in arid or semiarid country. It often furnishes a valuable water source.

The character of the rock formations will determine how deeply surface water will penetrate into the ground. This represents the bottom limit of the zone of saturation. In some places it is several hundred feet below the surface; in other places, several thousand feet. It is believed that at a depth of several miles the pressure of the overlying rocks is so great that there are no open spaces in the rocks. Therefore, their porosity and permeability are nil.

The ground water above the zone of saturation descends because it is drawn by the force of gravity. However, it rarely goes straight downward; it tends to follow the path of least resistance. If some of the rock formations it encounters are very permeable and others not nearly so permeable, its course is apt to be very erratic. The downward movement of water in the area above the zone of saturation is called the shallow, or vadose, circulation. (*Vadum* means "ford" in Latin. The adjective *vadosus* came to mean "shallow" because a ford is always at a shallow place in a river.)

The water in the zone of saturation also circulates. Its movement, however, is generally much slower than that of the water above it. One reason is that there are far fewer openings in the rock. The movement of water in the saturation zone is called deep circulation. It depends on various factors. Among these are the character of the openings in the rocks, the way in which rock layers slant and the head, or water pressure, of the water in the saturation zone.

The temperature of ground water very near the surface fluctuates somewhat with the seasons. At depths of as much as twenty-five feet in the temperate zones, water temperature is nearly constant. It

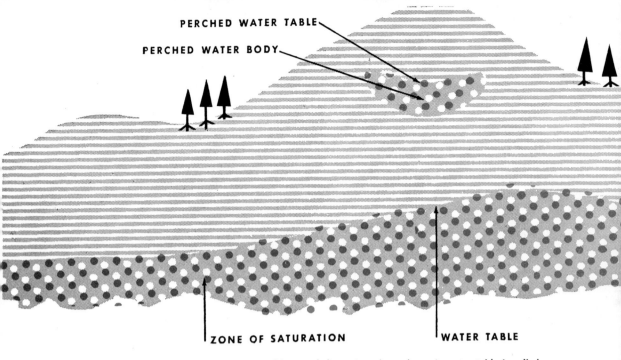

PERCHED WATER TABLE

PERCHED WATER BODY

ZONE OF SATURATION

WATER TABLE

Water trapped in a rock formation above the main water table is called
a perched water body. Its upper boundary is the perched water table.

is generally about the same as the mean
annual temperature of the air, or a little
higher. Ground water temperature rises
with increasing depth. The rate of increase is generally about 1° F. for each
fifty feet of increase in depth. In parts of
Alaska and other arctic regions, the ground
water is perennially frozen to depths of as
much as a few hundred feet. It may thaw
out near the surface, however, in the
summer.

Ground water
emerging as springs

When ground water makes its way to
the surface through natural openings in
the ground, springs are formed. In many
low-lying places spring water seeps im-
perceptibly out of the ground. It flows
from a hundred different sources and
forms so-called seepage springs. In other
places, the water may be seen rising from
the soil or flowing out of crevices in rock.
Certain springs are said to "boil," even
though the water is cold, because the rising
water stirs the clear sand in the spring
basins. Large springs generally issue from
deep underground pools, or from tunnels
or alcoves in the rock. Occasionally
springs arising from deep-seated strata

may discharge fresh water on the ocean
floor. A spring of this sort is found in
the Gulf of Argolis, in Greece. The fresh
water issues with such force that the surface of the ocean above the spring is somewhat convex.

Thermal springs, or hot springs, discharge water that has a higher temperature
than that of ordinary springs. The water
of some hot springs is ordinary ground
water that has sunk to a considerable
depth. As we saw, the temperature of
such water rises as the depth increases. In
time the heated water will return to the
surface. It will make its way through up-
turned strata, faults and other features that
produce a natural inverted siphon. (See
Index, under Siphons.) Hot springs in
volcanic regions may be derived, in part
at least, from so-called juvenile water.
This is formed as steam rising from molten
rock condenses.

In certain instances the water of hot
springs spurts upward from the ground at
intervals. Such springs are known as geysers. They are found m·stly in three parts
of the world — Iceland, New Zealand and
Yellowstone National Park, Wyoming. In
most cases the eruptions take place at irregular intervals; sometimes, however, the

intervals are quite regular. Generally a series of rumblings precedes an eruption. Water begins to flow over the vent of the geyser; then spouts of hot water shoot from the surface. These spouts may reach a height of several hundred feet; in some cases, however, they may be only a few feet high. We describe various famous geysers elsewhere in this set. (See Index, under Geysers.)

How Bunsen explained
the eruption of geysers

Many years ago the German chemist Robert Wilhelm Bunsen investigated the geysers of Iceland and proposed a theory to account for their eruption. This theory is still generally accepted. It is based on the fact that the greater the pressure, the higher the temperature at which water will boil. At the surface of the earth, at sea level, the atmosphere presses down on the water with a pressure of about 14.7 pounds per square inch. Water at this level will boil at a temperature of 212° F. At a depth of 33 feet, the pressure is just twice as much — about 29.4 pounds per square inch. Here water will boil at 248° F. At 295 feet below sea level, the pressure is 147 pounds per square inch. Water will not boil until the temperature of 357° F. is reached.

Suppose ground water makes its way into a more or less vertical fissure deep in the earth. It will form a liquid column as it accumulates. The water at the bottom of the column gradually becomes very hot because of contact with volcanic heat. Its temperature rises considerably above 212° F., the boiling point of water at sea level. But it will not boil at first because of the pressure exerted upon it by the column of water above. At last, however, it becomes so hot that part of it is converted into steam. The steam pressure causes the column to rise. Part of the water then overflows at the surface. This will lessen the pressure of the water column; more of the water is converted into steam. At last the steam blows the whole column of water out of the fissure in the earth. An eruption has taken place.

Ebbing and flowing springs, also called periodic springs, differ from geysers in that their water has the temperature of ordinary ground water. During periods of flow there is a considerable discharge of water. The flow of water is sharply reduced or stops entirely during ebb periods. The interval between flows may be a few minutes, or hours or even days. Nearly all the springs of this type issue from limestone. Modern geologists are inclined to believe that ebbing and flowing springs result from natural siphons in rock. They are quite few in number.

Before men learned how to purify flowing water and to build sanitary wells, springs furnished the cleanest and most attractive water supplies. They still serve this purpose to a limited extent in rural communities. Some manufacturers use quantities of spring water for carbonated beverages.

The curative properties of certain springs have long been celebrated. Some persons seek health by bathing in hot springs Others drink spring water containing various minerals. Many famous resorts have been established at the sites of medicinal springs. Among them are Bad Ems and Baden-Baden, in Germany; Sedlcany, in Czechoslovakia; Luchon and Barèges, in France; Cheltenham, in England; Hot Springs (Arkansas), Saratoga Springs and Palm Springs, in the United States. There are some famous hot springs in Canada. They include the hot sulfur springs at Banff, in Alberta, and Radium Hot Springs, in British Columbia.

The waters of some springs, containing sulfates of magnesium and sodium, act as purgatives and diuretics. (A diuretic tends to promote the secretion and discharge of urine.) Muriated water is used in the treatment of rheumatism and gout. It contains chloride of sodium (ordinary table salt) and is rich in carbon dioxide. Chronic diseases of the skin have been treated by sulfurous waters. Mineral water derived from springs is bottled in great quantities and is widely distributed. Many people prefer it to ordinary drinking water for various reasons.

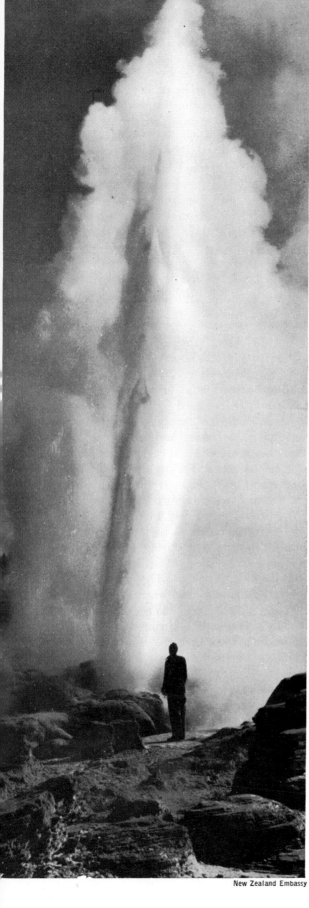

New Zealand Embassy

U. S. Geological Survey

Above: Roaring Spring, at the headwaters of the Pease River, in Texas. Springs are formed when ground water makes its way to the surface through natural openings.

Left: the Pohutu Geyser, near Rotorua, New Zealand. In this area we find hot springs, bubbling mud, colored pools and cliffs delicately tinted by the action of steam.

Below: Champagne Cauldron, hot spring in Wairakei, in New Zealand. Hot springs are derived either from ground water or from the condensation of steam from molten rock.

New Zealand Consulate General

Geologic changes brought about by ground water

Ground water brings about certain important changes in the earth's crust. Water charged with carbon dioxide, derived from the atmosphere, can dissolve carbonate rocks such as limestone and dolomite. In humid regions great quantities of rock may be dissolved in this way and huge caverns may be formed. We discuss this process in the article In the Depths of the Earth, in Volume 5.

Ground water often dissolves the substances that bind together the grains of the sedimentary rock called sandstone. In time it may reduce part of a sandstone formation to a heap of loose sand. The water also attacks igneous rocks, such as granite. It decomposes the iron-bearing minerals and feldspars they contain. Various minerals dissolved by ground water are transported to springs. Later these minerals are carried to rivers, and the rivers transport them to the sea. Here they frequently remain in solution, adding to the salt content of the ocean.

Often changes are brought about in underground formations by the precipitation of the materials that ground water carries in solution. When calcite, iron compounds and silica are precipitated, they may cement together particles of sand and other loose materials. They may convert these materials into sedimentary rock. Sometimes the substances carried in solution in water are precipitated within the pores of solid rock. They greatly reduce the porosity of such rock.

Cavities formed in rock by the dissolving activity of water may be filled at some later time by precipitated mineral matter. If deposits of amorphous silica (silica that is not in the form of crystals) are laid down in this way, they later crystallize and become agate. A mineral, such as quartz or calcite, may be precipitated in a cavity in the form of crystals. If the cavity is only partly filled, the crystals will point inward toward the center of the cavity. Formations of this kind are known as geodes.

When ground water dissolves limestone, calcium bicarbonate is formed and is carried off by the water in solution. Later calcium carbonate is deposited, generally in the form of calcite, on the roofs, floors and walls of caves as the water evaporates. This ultimately results in the creation of stalactites, stalagmites, fluted columns and other structures found in caves. We discuss elsewhere these deposits, which are called speleothems. (See the article In the Depths of the Earth, in Volume 5.)

In some cases ground water dissolves certain kinds of matter and replaces it with other kinds, which it is carrying in solution. This process is called replacement. Suppose a tree trunk lies buried below the water table. Its woody matter will gradually be dissolved away by water, and will be replaced by silica carried in solution in the water. Eventually the tree trunk will be replaced by a perfect replica in stone. The water table may later be lowered in this area, and erosion may lay bare the tree trunk. The Petrified Forest of Arizona contains many fine examples of such materials. Some of the petrified logs found there are a hundred feet long.

Mineral substances carried in solution in ground water are often precipitated at or near the site of a spring as evaporation takes place. Various types of deposits are laid down in this way. They are generally very compact and hard when slow evaporation has taken place. When calcium carbonate is precipitated through rapid evaporation in hot springs, the resulting deposit has a spongelike texture. It is called calcareous sinter. Sometimes it forms impressive-looking mounds and terraces. Various other substances, including gypsum, limonite (an ore of iron), sulfur and silica, are deposited by springs.

Making ground water available through wells

A water well is a pit or hole dug or drilled into the ground to reach a supply of water. The pit will remain empty or practically empty until it reaches the water table. If the hole is sunk into the zone of saturation, water will enter it and will

The name "artesian well" is given to a well drilled into a permeable bed between two impermeable layers, forming a species of natural basin.

come to rest at the water table. As water is pumped from the well, the water level will be lowered. However, a fresh supply of water will percolate into the hole until the level of the water table is reached again. Wells sunk in very permeable rock formations will naturally be replenished more quickly than wells in less permeable rock. If a well does not reach the water table at all times, it will become dry as the table falls. Wells range in depth from less than a hundred feet to several thousand feet.

The construction of durable and dependable wells is an art that requires specialized knowledge and skill. It is necessary to sink the hole to the proper depth to reach a productive aquifer, or water-bearing formation. It is then necessary to provide proper casing for the well in order to prevent cave-ins and also surface pollution.

Most domestic wells are pumped intermittently at the rate of only a few gallons a minute. But many wells that furnish water for public water works, industrial plants, or irrigation yield several thousand gallons a minute. One well on the Hawaiian island of Maui, is pumped at the rate of about 28,000 gallons a minute, or 40,000,000 gallons a day. It consists of a shaft and radiating tunnels below the water table in lava rock.

In many places a bed of clay, shale or other relatively impermeable material overlies a permeable bed, perhaps of sand or sandstone. Under this second bed there is an impermeable or almost impermeable stratum. Suppose these alternating layers of rock form a sort of natural basin, whose surfaces slope toward a central point. There may be an outcropping of the permeable stratum at the surface some distance above the central point of the basin. In that case water will penetrate into the permeable bed. As it descends it will be imprisoned between the two impermeable strata.

If a well is drilled at this point through the upper layer, water from the water-bearing stratum will rise in the well above the level of the stratum where it was first encountered. It may come up to the surface and overflow. A water well of this kind is known as an artesian well. The name is derived from Artois (Latin *Artesium*), a former province in France where such wells were first dug in Europe.

The digging of wells goes back to very ancient times. Primitive men, who depended chiefly on hunting and fishing, generally lived near natural watering places. Hence they progressed very little in the art of digging wells. When men began raising large flocks and herds, they faced a different problem. The pasturage within reach of natural watering places did not suffice for their livestock. Therefore, they had to provide for the needs of their animals by building wells. As the twenty-sixth chapter of Genesis shows, the Biblical patriarchs Abraham and Isaac were remarkably expert in locating and digging wells for the use of their stock.

Many rural communities and a considerable number of large cities obtain all or part of their water supply from wells. The largest city in the world supplied mostly from wells is Berlin, Germany. An extensive system of wells on Long Island supplements the water supply of New York City. Wells supply great quantities of water for processing and cooling purposes in industrial plants. They are used extensively, too, for irrigation. Millions of acres in India are supplied with water from wells. Some of these are very ancient, with primitive devices for lifting the water. Others are of modern construction, having efficient electrically driven pumps. Wells are used extensively for irrigation in the United States. The water obtained from this source is used to irrigate fruit and other crops in California and other western states. They serve for rice irrigation in Arkansas, Louisiana and Texas and for raising fruit and vegetables in Florida and other states of the east.

To seek out promising sources of ground water for wells, the terrain is studied by geologists. Then a series of borings are made on the basis of their observations. Deposits of sand and gravel supply most of the wells of large yields. Certain properly constructed sand-and-gravel wells may supply thousands of gal-

Unations

Above: an artesian well dug by the French in Libya. This well supplies water for crops in the desert. Below: a reverse rotary drilling rig, drilling a large well that will be used for irrigation. Wells go back to ancient times; they still furnish abundant water for various purposes to communities all over the world.

U. S. Geological Survey

lons a minute. Sandstones, derived from sand particles that have been compressed and cemented, will provide ample quantities of water if enough of the original open spaces between the grains remain. If these spaces have been filled by precipitated minerals, the sandstone may yield only moderate supplies of water from the various cracks and fissures that develop in the course of time. In some cases the sandstone may be entirely unproductive.

Many kinds of rocks, including granite, quartzite and slate, are very compact and contain almost no open spaces except those formed by cracks and fissures. Some wells drilled into such rocks are entirely dry. In many cases, however, hard rocks are loosened up near the surface by various weathering processes. As a consequence, they become porous and permeable enough to yield small water supplies to wells. Certain clay formations are the most unproductive of all. They are too fine-textured to yield water and too soft to have any water-bearing cracks.

In newly formed limestone there may be abundant open spaces between the fragments of which the stone is composed. However, the original open spaces tend to close off or become filled. The older limestones, therefore, may be compact and impervious. In some places, however, old limestone formations have been dissolved to a great extent by percolating water. They become extremely porous and permeable. The amount of water, therefore, that one can obtain from limestone formations varies greatly. In some limestone regions, it is difficult or impossible to obtain enough water from wells for domestic and livestock needs. Elsewhere limestone may yield great quantities of water.

Certain persons without scientific training claim they can detect the presence of underground water by means of a divining rod, or dowser, which generally consists of a forked stick. These people are known as water witches, or dowsers. They maintain that a divining rod will twitch in the hands of a water witch when a bountiful water supply is near. It will point to the place where a boring should be made.

Many books and articles have been written on the subject of water witching. The practice still persists in certain communities.

Borings at the sites indicated by water witches have sometimes yielded successful water wells. It is hard to determine, however, to what extent such successful ventures have resulted from lucky guesses. It is hard, too, to find out the ratio between successes and failures in the operations of water witches.

Plants that bring up ground water

Ground water is brought to the surface not only in springs and wells but also by certain plants. These put their roots down to the water table or to the moisture just above the water table. They are found particularly in places where the table is relatively near to the surface. Plants of this sort are known as phreatophytes (from two Greek words meaning "well plants"). In the arid regions of the world more ground water is absorbed and transpired by the phreatophytes than is discharged by springs. (For an explanation of transpiration, see Index, under Transpiration.) These plants serve to indicate the presence of underground water.

Among the best known of the phreatophytes are salt grass (*Distichlis*), greasewood (*Sarcobatus vermiculatus*), mesquite (*Prosopis*) and various species of palms. Salt grass is widely distributed along the eastern and western coastal areas of the United States and Mexico, and along the Pacific coast of South America. It is generally confined to areas where the depth of the water table is not much over ten feet. Greasewood is found west of the Great Plains area of the United States and in southwestern Canada. The water supplies that it taps may be thirty feet or more below the surface. The mesquite is a much-branched shrub or small tree, thriving in the arid regions of southwestern North America. Its big taproot sometimes reaches down sixty feet below the surface to obtain water. The palm trees found in the Sahara and Arabian deserts and in other arid areas of the world indicate the presence of ground-water accumulations not far from the surface.

The sanitation of ground-water supplies

When water from rain or snow runs off over the land surface it picks up particles of soil and filth. These particles contain great numbers of bacteria — some of them disease-producing bacteria derived from human beings. Some of this water may seep downward into the soil, reaching the water table sooner or later. The particles held in suspension in this ground water (including the bacteria) are largely removed by filtration. Any bacteria that have been disposed of in this way generally do not survive. For one thing, there will probably not be enough nutrients for them in ground water. This is particularly true if this water remains in the zone of saturation for a long time.

However, not all water supplies obtained from wells and springs are free from bacterial contamination. The water may be polluted for various reasons. It may reach the aquifer through crevices in limestone, fissures in other rocks, the open spaces in coarse gravel or other large openings. Privies, cesspools, or the outflow of sewers may be located near the wells or springs. Surface water may find its way into springs or into wells that are not tightly cased.

Wells or springs can be kept reasonably free from contamination by removing all possible sources of contamination in the vicinity. These sources are particularly dangerous if they are in an upgrade position (that is, if they occur on land higher than that in which the water source is located). It is sometimes advisable to abandon the old well and build a new one. This should be in an upgrade direction from sources of contamination. To combat pollution, a well should have a durable, watertight casing. It should be built far enough above the surface of the land to keep out surface water. If possible, the casing should extend downward beyond the water table to the nearest solid stratum. In any event it should go well below the water table. Wells sunk in limestone, lava rock and other formations that may have large openings present special problems. The only sure way of protecting their water from pollution is to purify it by chlorination or some other method.

See also Vol. 10, p. 269: "General Works."

Well water may be seriously contaminated if a cesspool is in an upgrade position from it: that is, if it is located on higher land than the well.

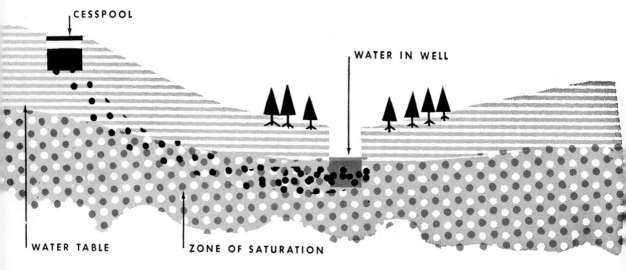

CESSPOOL

WATER IN WELL

WATER TABLE   ZONE OF SATURATION

HEREDITY IN MAN

How Successive Generations Are Molded

BY THEODOSIUS DOBZHANSKY

Union Pacific Railroad

The girl and the giant sequoia at which she is looking each began as a tiny fertilized egg and grew into adult-hood in accordance with a definite hereditary pattern.

AN APPARENTLY unimportant event took place some three thousands years ago on the slopes of the Sierra Nevada Mountains in California. A tiny seed of a tree called the giant sequoia, *Sequoia gigantea,* fell to the ground and found conditions favorable for growth. As years and centuries passed, the tree that developed from the small seed grew larger and larger; today it is more than two hundred feet high and more than twenty feet in diameter.

A human lifetime is a mere episode compared to the lifetime of a sequoia tree; yet man and the sequoia have more in common than we might think. The seed of the sequoia was a product of fertilization. The human body, too, begins its existence with the act of fertilization, when a microscopically small male cell unites with a female egg cell, which is barely visible to the unaided eye. In the course of time, as we have seen, the seed of the sequoia grows to be a huge tree. The fertilized human egg develops in the mother's womb into an embryo, the embryo becomes a fetus and about nine months after fertilization, a baby is born. In due course the baby becomes an adult, increasing its body weight about twentyfold.

In order for a seed or a fertilized egg to become a sequoia tree or a man, much material has to be taken from the outside world and incorporated in the living body. In the case of the sequoia, this material consists of water and mineral salts, taken up by the roots from the soil, and carbon dioxide, coming in through the leaves from the air. The materials from which the human body is constructed are to be found, of course, in the foods that we eat — foods derived from the fields, gardens and meadows and even from the depths of the sea.

Before such foods can become a part of a living body, they must undergo a complete chemical change. Moreover — and this point is vitally important — each organism builds itself up from these transformed substances in a particular way. As a result, the mature product of the building process resembles, in composition as well as in structure, the bodies of the parents and other ancestors of the organism.

333

The process of transformation of food materials by a living body into a likeness of itself and of its ancestors is called heredity. An organism may reproduce rapidly (as bacteria often do) or slowly (as in the case of man and the higher animals and plants). In either case, the urge to multiply according to a definite hereditary pattern exercises an unrelenting pressure upon the environment in which the organism lives. This "pressure" of heredity reveals itself in the gnarled and twisted trees growing out of small fissures in apparently solid rock on lofty mountain tops, in the riot of vegetation in a tropical jungle, in the teeming life in a drop of water from a stagnant swamp, as seen under the microscope. Various factors, however, tend to oppose the "pressure" of heredity. Only certain substances can be used by organisms as food; others are useless or even poisonous. Only certain localities, offering favorable conditions, can serve as dwelling places. Even when the environment is favorable, multiplication is checked by the exhaustion of the food supply, space limitation and competition with natural enemies, parasites and the agents of disease. Except, perhaps, for man in a highly organized society, living beings are continually exposed to countless dangers that make survival a lucky accident and premature death the rule.

The agelong conflict between the pressure of heredity and the deterring factors present in the environment has resulted in the grand process of organic evolution. Instead of a single kind of organism, instead of a single heredity able to use only some few substances as food to be transformed into living bodies, we find in the world about us an immense diversity of organisms and heredities.

In the following pages we shall examine the hereditary process particularly as it applies to man; we shall also consider some of the more important consequences of this vital process.

It is remarkable that the heredity that is able to organize pounds of inert materials into the living bodies of human beings is transmitted from parents to offspring through exceedingly minute quantities of living substance. Only some parts of the female sex cell (the egg) and the male sex cell (the spermatozoon) are directly concerned in the transmission of heredity. Examined under a microscope, an egg cell consists mainly of a sticky substance called cytoplasm, in which are stored certain food materials to be used by the developing embryo. A little blob of a slightly more translucent material — the nucleus — floats in the cytoplasm. The spermatozoon, or male cell, which is to fertilize the egg cell consists of a head, which contains a nucleus, and a contractile tail — one that makes the spermatozoon motile (capable of spontaneous movement).

Rapid strokes of the tail propel the spermatozoon head forward and thus enable it to reach the egg cell. The spermatozoon then enters the egg, and the nuclei of both cells fuse together. This nuclear fusion is the act of fertilization proper. A new individual has now arisen. The fertilized egg and its nucleus divide into two cells, each with a nucleus. The two cells divide into four; the four, into eight. Eventually, the billions of cells and nuclei that compose the human body come into being.

When strongly magnified under a microscope, the cell nucleus reveals some interesting structures. A certain number of minute bodies, called chromosomes, can be seen inside the nucleus, particularly when the latter is in process of division. Most human cells contain forty-six chromosomes; the sex cells, however, have only half that number — that is, twenty-three. When fertilization takes place, the twenty-three maternal chromosomes contained in the egg cell, plus the twenty-three paternal ones brought by the spermatozoon, combine and form the normal complement of forty-six chromosomes in the fertilized egg.

These chromosomes are not all alike. Some of them are larger than others; some are rod-shaped, others hook-shaped, still others V-shaped. After examining many nuclei and chromosomes, one learns to recognize a chromosome by its shape, just as one recognizes the letters of the alphabet or the faces of one's acquaintances. The twenty-three maternal chromosomes are all dif-

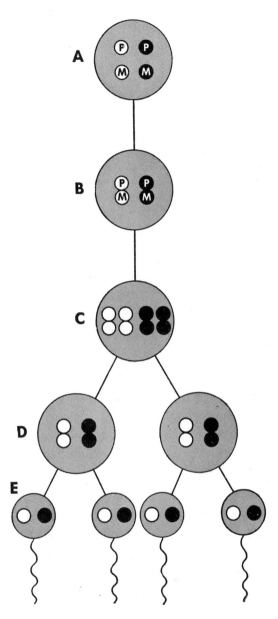

ferent; the same is true of the twenty-three paternal chromosomes. But, with a single exception, which we shall discuss later, every maternal chromosome is matched by a similarly shaped one in the paternal set.

In the formation of most of the body cells, a cell division called mitosis takes place. In this process a cell gives rise to two cells, identical with the original one. A different type of division, called meiosis, takes place in the production of sex cells in the sex glands — the ovaries, in the female; the testes, in the male. First, the matching maternal and paternal chromosomes in a forty-six-chromosome cell of the sex gland are attracted to each other. They approach in the nucleus and become intimately paired. Then they unravel, but still lie side by side, so that the nucleus now shows twenty-three chromosome pairs. Some of these pairs have exchanged sections — a phenomenon called crossing over.

Each chromosome splits lengthwise, so that instead of twenty-three pairs, we have twenty-three tetrads. (A pair is made up of two units; a tetrad, of four units.) No further division of the chromosomes takes place. Next, the cell divides, forming two new cells and then these two also divide. The original cell, therefore, has given rise to four new cells, each having twenty-three chromosomes — one from each of the twenty-three tetrads that were formed when the twenty-three pairs of matched chromosomes divided.

How we inherit from our ancestors

In all his body cells except his sex cells, your father has forty-six chromosomes; he inherited twenty-three of these from his father and twenty-three from his mother. Your mother's cells also have forty-six chromosomes — twenty-three of them from her father and twenty-three from her mother. How many chromosomes have you inherited from each of your grandparents? Have you and your brothers and sisters inherited the same chromosomes?

Let us indicate the chromosomes that your father derived from his father (your paternal grandfather) by letters with primes

Highly simplified diagram showing how a male germ cell gives rise to male sex cells (sperm) by the process of meiosis. We shall assume that the germ cell has two pairs of chromosomes, as in A. The white paternal chromosome (P) matches the white maternal chromosome (M); the black paternal chromosome matches the black maternal chromosome. The matching chromosomes pair together (B). Each chromosome splits, so that we now have four chromosome pairs (C). Next, the cell divides, forming two new cells, each with four chromosomes (D). Finally, the two cells become four as they divide (E). Each of the four cells that have just been formed is a male sex cell, or sperm, containing two chromosomes. It will unite with a female sex cell, also containing two chromosomes, and the fertilized cell, or zygote, that results will have the original number of four chromosomes.

Identical twins, or multiples. They came into being when a single fertilized egg produced two embryos, which developed into two babies. Identical twins have the same chromosome complements and also the same hereditary endowments.

D. W. Corson from A. Devaney, N. Y.

(A', B', C' and so on), and the chromosomes he derived from his mother (your paternal grandmother) by letters with double primes (A'', B'', C'' and so on). Your father's chromosomes would be represented by this combination:

A'B'C'D'E'F'G'H'I'J'K'L'M'N'O'P'Q'
R'S'T'U'V'Y *
A''B''C''D''E''F''G''H''I''J''K''L''M''N''
O''P''Q''R''S''T''U''V''X *

Only one member of each matching pair — A'A'', B'B'', C'C'' and so on — would get into the sex cells of your father; the distribution would be at random. For the sake of simplicity let us consider only three types of chromosomes — A, B and C. In the sex cells formed by meiosis, there would be eight possible combinations of A, B and C chromosomes, as follows:

A'B'C'   A'B''C''   A'B'C''   A'B''C'
A''B''C'   A''B'C'   A''B''C'   A''B'C''

In other words, the three grandpaternal chromosomes, A', B' and C', may all get into your father's sex cells, or one or two of them may get into these cells, together with one or two grandmaternal chromosomes, or else they may not get into the sex cells at all. In the case of three chromosome pairs, there would be 2^3, or eight, different combinations.

Since man has, not three, but twenty-three chromosome pairs, the paternal grandparental chromosomes can be distributed in the sex cells of your father in 2^{23}, or 8,388,608 different ways. The probability, then, that you have inherited a full

* We explain X and Y chromosomes later in this article.

set of chromosomes (twenty-three in all) from your paternal grandfather is exceedingly slight — only 1 in 8,388,608.

Actually your chances of inheriting a complete set of chromosomes intact from a grandparent would be even less than that. We saw that in the process of meiosis some of the paired chromosomes exchange sections; the resulting chromosomes are no longer identical with the original ones. Suppose that chromosome A', received from your grandfather by your father, had exchanged sections in this way in the formation of your father's sex cells. Even if chromosome A' were to get into one of these cells, it would no longer be the same as the A' chromosome in your grandfather's body cells.

Your brothers and sisters are no more likely to obtain exact replicas of all your grandfather's chromosomes than you are. Hence one may safely assert that brothers and sisters always have different hereditary endowments, unless they happen to be identical twins.

It is important to note how identical twins differ from so-called fraternal twins. Fraternal twins arise when two egg cells are simultaneously fertilized by two different spermatozoa. The probability that the simultaneously fertilized eggs have similar combinations of chromosomes is no greater than it is for any two eggs produced by the same mother at different fertilizations. Consequently, the hereditary endowments of fraternal twins are on the average no more similar than the endowments of brothers and sisters not born at the same time.

Identical twins (or multiples) arise when a single fertilized egg gives two or more embryos, which developed into two or more babies. Since the process of cell

The fraternal twins at the left arose when two egg cells were simultaneously fertilized by two different spermatozoa. Their hereditary complements are no more similar than those of brothers and sisters who were born at different times.

Ewing Galloway

division called mitosis involves an accurate splitting of each chromosome and an exact distribution of the halves to the daughter cells, identical twins possess identical chromosome complements and, therefore, have the same heredities.

Apart from identical twins, the possibility that any two human individuals, now living, or having lived in the past, or yet to live in the future, may have identical heredities is remote indeed. It is exceedingly small even in the case of brothers and sisters, with the same maternal and paternal grandparents; it is even smaller for unrelated individuals. Even in the case of identical twins, variations in the conditions of the environment may be expected to produce differences.

What determines the
sex of an individual?

What causes some babies to be girls and others boys; in other words, what determines the sex of an individual? This problem has stirred the curiosity of mankind since the earliest days; it has given rise to a great number of unfounded guesses and unjustified speculations. Certain quacks even claim to have discovered how to control the sex of offspring by means of various tricks and nostrums. Only since the mechanism of chromosome distribution has become clear have scientists been able to offer a valid solution of the intriguing problem of sex determination.

When we made the statement that human body cells contain twenty-three matching pairs of chromosomes, we should have added that, strictly speaking, this is so only for the cells of a feminine body. In the case of a male's body, there are twenty-two matching chromosome pairs, corresponding to those in female cells, plus a pair of chromosomes consisting of unequal members. One of the partners of this twenty-third pair is clearly larger than its mate. The larger partner resembles either member of the twenty-third, or X, pair of the female chromosomes. The smaller partner is not represented in the female chromosome complement at all; it is found in male cells only. Let us call the larger partner the X chromosome and the smaller one the Y chromosome. The female cells, then, contain two X chromosomes; the male cells, one X and one Y chromosome.

When female sex cells are to be formed, the two X chromosomes of the female germ cell come together; they then split lengthwise, forming four X chromosomes. Cell division subsequently yields four nuclei, each containing an X chromosome. In the formation of male sex cells, the single X chromosome pairs with its smaller partner, the Y chromosome. Each of these chromosomes splits lengthwise; the cell then contains two X chromosomes and two Y chromosomes. When this cell gives rise to four new cells — the sex cells — half will contain X chromosomes, the other half Y chromosomes.

If these two kinds of male sex cells, or spermatozoa, are equally efficient in reaching and fertilizing the egg cells, we should expect two kinds of fertilized eggs to be equally numerous. Half of them would have two X chromosomes (XX) and would develop into girls; half would carry one X and one Y chromosome (XY) and would give rise to boys.

As a matter of fact, male births are slightly more frequent than female births. The reason is as yet unknown. Possibly the spermatozoa with a Y chromosome are

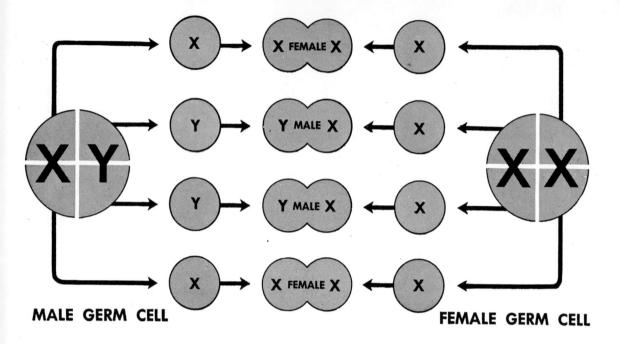

MALE GERM CELL   FEMALE GERM CELL

The male germ cell has one X chromosome and one Y chromosome; the female germ
cell has two X chromosomes. Of the four sex cells formed from the male germ cell,
two contain an X chromosome apiece; the other two, a Y chromosome. Each of the
female germ cells also yields four cells when sex cells are formed; each of these
contains an X chromosome. If the four male sex cells fertilize the four female
sex cells as shown, the result of these unions will be two males and two females.

a bit more efficient swimmers, or penetrate
the egg cell more readily; perhaps some of
the XX fertilized eggs die off at early stages
of their development.

The solution of the age-old riddle of
sex determination, then, has proved to be
surprisingly simple. The sex of a child is
determined at the time of fertilization by
the kind of spermatozoon that happens to
unite with the egg cell. This spermatozoon
may carry an X chromosome, in which case
it will give rise to a girl; if it carries a Y
chromosome, it will produce a boy. It is
all a matter of chance.

Certain facts seem to contradict the
chromosome theory of sex determination.
In a considerable number of families, all
the children are either girls or boys. If
one half of the spermatozoa are always boy-
determining and the other half girl-deter-
mining, why should the Jones family have
six boys and no girls and the Smith family
six girls and no boys? If we examine the
facts in such cases closely, we shall find
that such facts agree surprisingly with the
theory involved.

Suppose that we toss a coin in the air
a number of times, calling the first toss A,
the second B, the third C, the fourth D and
so on. When we analyze tosses A, B, C, D
and the rest, we shall find that in about half
the cases the coin will land heads up and
in the other half heads down. Suppose now
that we consider tosses A and B as forming
part of a single group and that we pair C
and D and all the rest in the same way. We
shall find that in about one quarter of the
cases there will be two heads, that in another quarter there will be two tails and
that in one half the cases heads will be
followed by tails, or tails by heads. That
is the way the law of chance operates. In
the case of families with two children, you
will find that about a quarter of them will
have two girls, a quarter will have two boys,
and a half will have a boy and a girl. If
coins are tossed in groups of three, in about
one-eighth of the groups there will be a
"run" of three heads, and in about one
eighth a "run" of three tails. Of families
with three children, about one in eight has
three boys; one in eight, three girls.

If the letter n stands for the number of children, the probability of having a family of n children all of one sex is approximately $\frac{1}{2}^n$, or $\frac{1}{2}$ to the nth power. (This is an approximate formula, not an exact one since, as we have pointed out, boys are born slightly more frequently than girls; the probability of having n boys is, therefore, slightly greater than that of having n girls.) With the aid of this formula, we can predict that about one in sixty-four families with six children will have six boys, and one in sixty-four will have six girls ($\frac{1}{2}^6 = \frac{1}{64}$).

Genes — the atoms of heredity

Thus far we have been dealing with the chromosomes as if these were the ultimate hereditary units. As a matter of fact, each chromosome contains a number of smaller units, called genes. The genes of all organisms appear to be astonishingly similar in chemical composition — they consist of nucleoproteins (compounds of nucleic acids and proteins). What, then, makes the human genes differ from those of, say, a dog or a sequoia tree? What makes the genes that modify the eye color in man differ from those that modify the hair shape? Recent studies by biologists and chemists indicate that the gene "specificities" — that is, the differences between the genes — reside in their nucleic-acid components. The most important kind of nucleic acid, the so-called DNA nucleic acid, consists of series of four kinds of molecular constituents, or nucleotides, arranged in spirally wound fibers. Now the four kinds of nucleotides can be compared with letters of the alphabet. The twenty-six letters of the English alphabet suffice to build up an infinite number of words, sentences and paragraphs. This can be done also with only the three "letters" of the telegraphic Morse code (dot, dash, gap). Different genes of the same species, such as man, and genes of different organisms differ in the sequences, the arrangements, of the four "letters" of the genetic "alphabet."

We can think of the genes of a given chromosome as grouped together in a string (the chromosome), each gene occupying a definite place in the string. The genes, together with environments, are responsible for the development of a living body from fertilization, through various stages of fetal life (life in the womb), to infancy, youth, maturity, old age and death.

What roles do different genes play in this process? The genes are best compared with separate players in a symphony orchestra. A symphony is the product of all the players working together, and yet each player has a different function. When a gene changes, the result may affect various characters or traits. In man, for example, some gene changes influence the color of the eyes, others the skin color, or the shape of the nose, or the speed of blood-clotting, or the intelligence and so on. Geneticists have been able to draw up chromosomal charts for certain species of animals and plants, showing the position of the genes in the chromosomes. They have made a start, at least, in drawing up similar charts for human chromosomes.

Curiously enough, the laws governing the inheritance of specific factors, or characters, as determined by the genes, were worked out about a century ago by an Austrian abbot long before scientists knew anything about either genes or chromosomes. In the year 1865, Gregor Johann Mendel, abbot of the monastery at Bruenn, read a 20,000-word paper on his experiments in the crossbreeding of common edible garden peas; the paper was published in the proceedings of the society in the following year. In this paper he established certain laws of heredity which have come to be known as Mendel's laws. Contemporary scientists were not particularly impressed, and Mendel's paper was buried in oblivion for years. It was not until 1900 that it was rediscovered independently by three investigators, working in Holland, Germany and Austria. It was found later that Mendel's laws applied not only to peas but to other organisms, including man.

Before discussing Mendel's laws, let us clarify some basic facts about genes. Ordinarily, each inherited trait, such as the color of eyes or hair in a human being, is controlled by a particular pair of genes working together; these genes are always found in the same relative position on a particular pair of matching chromosomes. Some traits, however, are determined by the interaction of many pairs of genes, located on different chromosome pairs.

When a geneticist speaks of an organism that is pure with respect to some trait, that means that both its genes for the trait are identical. As a result, when two pure organisms are crossed, all their offspring will show this trait. A hybrid is an organism that possesses two different genes for the trait in question; if two hybrids are crossed, some of their offspring will show the effects of one gene and some the effects of the other.

Let us suppose that a hybrid individual has one gene C for curly hair and one gene c for straight hair. In this case, the hair will actually be curly, since the gene C is more powerful in determining this trait than the gene c. C is called the dominant gene and c the recessive gene of this pair. In representing the genetic make-up, or genotype, of an individual, capital letters are ordinarily used for dominant genes and lower case (small) letters for recessive ones.

A dominant gene will always show its effects in the physical characteristics, or phenotype, of an organism, even though the recessive is present (as in individuals with either genotype CC or Cc); the recessive gene expresses itself only when the dominant is absent (as in genotype cc). Dominance and recessiveness are met with very often in the heredity of man, as in that of other organisms. Yet, as we shall see later on, there are pairs of genes in which neither gene dominates but each modifies the effects of the other.

Mendel's laws
of heredity

The first of Mendel's laws is that of segregation. When two purebred organisms with two contrasting characters (such as yellow and green seed color, in the case of peas) are crossbred, the offspring will show only one of these characters. In the case of the yellow-green pair in peas, only

The abbot Gregor Johann Mendel, pioneer in genetics.

yellow will appear in the first generation of hybrid offspring. Yellow is the dominant character. The other character — green — will not be evident in the first generation of offspring, but will reveal itself in a later generation; green is the recessive character. Suppose that the hybrids showing yellow seed are mated. Of the resulting generation of peas, 25 per cent will have yellow seeds, 50 per cent will be hybrid but show yellow and 25 per cent will have green seeds. This is known as the 1:2:1 ratio. Thus the dominant and recessive characters that were mixed in the first generation of hybrid offspring have become separated or, as a geneticist would say, segregated.

Mendel's second law is concerned with the independent assortment of genes. It states that in cases in which more than one pair of genes are involved, each pair will segregate independently of the others. In peas, for example, tallness and shortness of stem are contrasted characters that segregate in the same way as do yellow and green seed color. Suppose that a tall race with yellow seeds is crossed with a short race

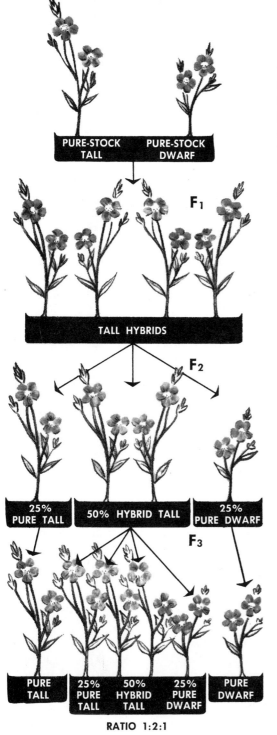

PURE-STOCK TALL   PURE-STOCK DWARF

F₁

TALL HYBRIDS

F₂

25% PURE TALL   50% HYBRID TALL   25% PURE DWARF

F₃

PURE TALL   25% PURE TALL   50% HYBRID TALL   25% PURE DWARF   PURE DWARF

RATIO 1:2:1

with green seeds. In the first generation only the two dominant characters — tallness and yellowness — will appear. If the hybrid forms that result are crossed, out of every sixteen individuals, on the average, nine will show both dominants (tall, yellow), three will show the dominant of one pair and the recessive of the other (tall, green), three will show the other dominant and the other recessive (yellow, short) and one will show both recessives (short, green). This is called the 9:3:3:1 ratio.

We can illustrate the Mendelian inheritance in man by discussing the heredity of a trait that is probably unknown to most people. A weak solution of the chemical known as phenyl-thio-carbamide (PTC, for short) tastes intensely bitter to about 70 per cent of Americans. To the other 30 per cent of the American population, the same solution has no taste at all or only a very faint taste. This is a matter of heredity. To find out whether you are a "taster" or a "nontaster" you will have to try some PTC solution yourself. It will not affect your health in any way.

If a pure "taster" (that is, one whose parents and other ancestors were all "tasters") marries a "nontaster," the children from this marriage will all be "tasters." There will be no perceptible difference in the ability of the pure and hybrid "tasters" to taste PTC solutions. In this case, being a "taster" is a dominant trait; being a "nontaster" is a recessive trait.

Suppose now that two hybrid "tasters" marry. According to Mendel's first law, out of every four offspring, on the average, one will be a pure "taster," one will be a pure "nontaster," and the other two will be hybrids — that is, they will be "tasters" but will carry the recessive, "nontasting" trait,

Gregor Johann Mendel, whose portrait appears on the preceding page, crossbred common edible peas and established the laws of heredity now called Mendel's laws. Above are shown the workings of the laws of dominance and segregation. When a pure-stock tall plant is bred to a pure dwarf, the first generation of offspring—F₁— will all be tall hybrids: that is, the gene for tallness will be dominant in them, but the gene for smallness will also be carried. If the members of the F₁ generation are bred together, giving rise to the second gene-ration—F₂—25 per cent will be pure tall, 50 per cent will be hybrid tall and 25 per cent will be pure dwarf —a 1:2:1 ratio. This means that the dominant and recessive characters, which were mixed in generation F₁, have now become segregated, or separated. What will happen in the third generation—F₃? If the pure tall plants of F₂ mate, all their offspring will be tall; if the tall hybrids mate, their offspring will show the same 1:2:1 ratio that we noted in F₂; if the pure dwarf plants of F₂ mate, all their offspring will be pure dwarf stock.

in latent form. The pure recessives — the "nontasters" — that are born from marriages of two hybrids are just as pure as those found in families in which both parents are also "nontasters." The recessive heredity does not disappear, nor does it become weakened or contaminated by its passage through hybrids, in whom only the dominant trait appears.

This fact is extremely important because it shows how far from the truth are the ideas about heredity held by the general public. According to popular belief, heredity is transmitted from parents to offspring through "blood." The paternal and maternal "bloods" are supposed to mix and to give rise to a kind of solution or alloy. According to this notion, the heredity of children is a compromise between the heredities of the parents, and brothers and sisters have similar heredities. These ideas are reflected in such figures of speech as "blood mixture," "half blood," "blood will tell" and so on. Of course, as we have seen, blood has nothing to do with the transmission of heredity. Mendel's discovery has shown that parental heredities do not mix, but that, on the contrary, they are segregated in the offspring. A child born to parents who are both "tasters" may be a "nontaster." Brothers and sisters may be just as different in this respect as are unrelated individuals.

The sex cells of a pure "taster" contain a gene, T, for tasting PTC; the sex cells of a "nontaster" contain a "nontasting" gene, t. A child of such parents carries both T and t; since T is dominant over t, this child is a "taster." But — and this is the gist of the law of segregation — the genes T and t do not fuse in the hybrid; among the sex cells produced by the hybrid, one half carry T and the other half t.

Among the offspring of two hybrid "taster" (Tt) parents, about one quarter of the children will arise from a union of a T egg with a T spermatozoon and will be pure "tasters" (TT). About half of the children will come from the union of a T egg with a t spermatozoon, or vice versa; they will be hybrid "tasters" (Tt). Finally, one quarter of the children will

result from the union of a t egg with a t spermatozoon and will be "nontasters" (tt).

Sex-linked inheritance

We have seen that one pair of chromosomes is involved in the determination of sex — the XX chromosomes of the female and XY chromosomes of the male. The traits determined by the genes carried in the X chromosomes will show a special kind of inheritance, called sex-linked inheritance. One of the traits in question is color blindness, a condition in which a person is unable to distinguish colors, particularly red and green, easily distinguished by a person with normal vision. Another sex-linked trait is hemophilia. In this condition, the blood fails to clot properly; a slight wound is liable to result in fatal bleeding.

Color blindness is a recessive trait with respect to normal vision, which is the dominant trait. Let us indicate the gene for normal vision as C and the gene for color blindness as c. Consider now a marriage in which the mother is color-blind and the father has normal vision. The color-blind mother carries two X chromosomes, both with c genes, and all the eggs she produces will also carry c. The father, with normal vision, has an X chromosome with a C gene, and a Y chromosome that does not carry a gene for color vision at all. Half of his spermatozoa will have an X chromosome and the other half will have a Y chromosome.

The X-bearing spermatozoa fertilizing the egg cells will give rise to daughters, who will carry one X chromosome with C and another with c gene: that is Cc. Since normal vision is dominant over color blindness, these girls will have normal vision, but will harbor the gene for color blindness, c.

If such a woman is married to a normal man (C), all of the daughters will have normal vision; but half of them will harbor the gene for color blindness. Half of the sons will inherit the X chromosome with the gene for color blindness from their mother, and will consequently be color-blind (c); the other half of the male offspring will inherit the gene for normal vision (C).

A color-blind man (c) married to a normal woman not carrying the gene for color blindness (CC) produces children with normal vision only. But while his sons (C) are free from the gene for color blindness, all his daughters carry this gene in a hidden condition (Cc) and will transmit it to half of their children. The inheritance of hemophilia is like that of color blindness.

Complications
in inheritance

The inheritance of many human traits is not so simple and clear-cut as the ability to taste PTC, or color blindness or hemophilia. One complicating factor is the fact that, according to the second law of Mendel, different genes are inherited independently. Suppose that a pair of chromosomes carry the dominant gene for brown eyes, which we may call B, and its recessive counterpart, b, which causes blue eyes. The genes T and t, for the ability or inability to taste PTC, are carried in the same person in a different chromosome pair. If a person who has brown eyes and is a "taster" (BBTT) is married to a blue-eyed "nontaster" (bbtt), the children will have brown eyes and will be "tasters." But they will carry the recessive genes for blue eyes and for the inability to taste PTC. The genes of these children will be BbTt. At meiosis, the chromosomes with the genes for eye color will be distributed independently from those that carry the genes for tasting ability. A marriage of two BbTt individuals may result in the production of four kinds of children — brown-eyed "tasters," brown-eyed "nontasters," blue-eyed "tasters" and blue-eyed "nontasters." Obviously, the situation will be still further complicated if parents carry more than two such latent recessive genes.

Another complication arises when two or three genes located in different chromosomes produce similar effects on the organism. For example, there are several genes in man that produce a dark color in the skin. Let the genes that darken the skin

be called D^1, D^2, D^3 and so on, and the counterparts of these genes, producing no color, d^1, d^2, d^3 and so on. Suppose that Negroes carry at least six color-producing genes $D^1D^1D^2D^2D^3D^3$ and that whites carry genes that leave the skin pale, $d^1d^1d^2d^2d^3d^3$. A mulatto of the first generation will evidently have the genes $D^1d^1D^2d^2D^3d^3$. It happens that the skin-color genes are neither dominant nor recessive, so that the mulatto has, in effect, three color-producing genes, which make his skin color about intermediate between that of a Negro and that of a white. As we have seen, the sex cells of a mulatto may carry eight different combinations of skin-color genes ($2^3=8$):

$D^1D^2D^3$	$D^1d^2D^3$	$D^1d^2d^3$	$d^1d^2D^3$
$D^1D^2d^3$	$d^1D^2D^3$	$d^1D^2d^3$	$d^1d^2d^3$

A marriage of mulattoes may therefore produce children with as many as six color genes (remember that each set of color genes is paired); other children born of the pair may have no color genes at all. The children, then, may be as dark as their Negro ancestors or as pale as their white ancestors. Most of them, however, will resemble their parents in skin color. This analysis is oversimplified, because more than three pairs of genes are actually involved; besides, some of these genes are more efficient color producers than others.

The hair color, stature, weight, shape of the head and so on are traits that are determined by interaction of several genes each. Their inheritance resembles that of skin color. In the case of eye color, there seems to be one basic color-gene pair that determines the difference between blue and brown eyes and several gene pairs with relatively minor effects. These secondary genes may modify the blue or brown colors by making them darker or lighter, or by restricting these colors to different parts of the iris of the eye.

Mutations —
changes in genes

Cells and chromosomes are continually dividing in the body, and the genes are generally exactly duplicated when the chromosomes divide. In relatively rare cases, however, one of the genes of a sex cell will no longer be an exact copy of the original gene. A change of this sort is called a mutation; the changed gene is known as a mutant. It is known that the rate of mutation in various organisms is increased by radiation with X rays, radium and ultraviolet, and by various other conditions.

A small minority of mutant genes may produce desirable effects. The great majority of mutant characters, however, are harmful; they lessen the chances of survival. In the case of humans, harmful mutations are responsible for conditions such as hemophilia and amaurotic idiocy (mental impairment coupled with loss of vision). One of the dreaded consequences of atom-bomb explosions is that they may build up a dangerous degree of radioactivity and that this may add considerably to the number of unfavorable mutations.

The problem of undesirable genes

Many inheritable defects and diseases in man are caused by recessive genes. This is true, for example, in the case of the condition called albinism. Those who display this condition — albinos — have a pale skin extremely sensitive to sunburn, pink eyes oversensitive to strong light and very light, straw-colored hair. Albinos are pure carriers of a recessive gene, a. Many more people carry not only this recessive gene but also a dominant gene, A, for normal pigmentation. These people are entirely unaware that they are carriers. It has been estimated that they are about 280 times more numerous than the albinos.

Albinism is not a very serious handicap. Other recessive genes, however, are responsible for crippling diseases in man. Like the gene for albinism, they are carried by a number of persons who are quite normal themselves. Marriages between such carriers of defective recessive genes may result, to the utter surprise of the parents, in the birth of afflicted children.

A group of scientists known as eugenists want to improve human populations by freeing them from defective genes, or at least by reducing the frequency of such genes. The eugenists aim, among other things, to prevent the carriers of undesirable genes from having children. This could be accomplished either by persuading carriers to abstain voluntarily from parenthood or by sterilizing them by means of a surgical operation.

If an undesirable condition is caused by a gene that is dominant over the normal or desirable trait, the measures suggested by the eugenists may be effective. If, however, the undesirable gene is a recessive, the eugenical program will yield results very slowly, if at all. This is because the persons who carry both the dominant gene and the undesirable recessive gene are themselves normal people. Therefore they cannot be identified and sterilized. In the case of albinos, for instance, the sterilization of all the albinos would leave nonalbino carriers of albinism, who are about 280 times more numerous than the albinos themselves, to carry on the gene for albinism. The number of albinos in the next generation would not be appreciably diminished, as far as we can see.

Of these twin brothers, born of Mexican parents, the one on the left is a true albino, whose pale coloring is emphasized by the dark hair and eyes of his twin.

United Press

What do we
mean by a race?

If marriages took place regardless of social and economic status, religious persuasion and nationality, gene frequencies would become alike throughout the world. This would not mean that all men would become alike. Since gene differences are not leveled off by hybridization, mankind would show even more varied hereditary endowments than is the case today; however, the same diversity of types would be found everywhere. Actually, marriages are not concluded at random, and certain genes are encountered with greater frequency among the inhabitants of certain lands than others.

The gene for albinism, for example, occurs more frequently in a certain Indian tribe living in the jungles of the Isthmus of Panama than in most other populations. There are fewer cases of color blindness among Indians than among whites. The frequency of the genes for dark skin color is much higher among the natives of central Africa than among people of European origin. The gene for the blue eye color is met with very frequently among the inhabitants of northern Europe, particularly in Norway. It becomes progressively less common as one proceeds southward from Norway, through central Europe, to the Mediterranean countries and to Africa. It must be kept in mind, however, that some brown-eyed individuals occur in Norway, and some blue-eyed ones in the Mediterranean region and even in Africa.

Populations differing in the frequencies of their genes are called races. The modern idea of race follows from the gene theory; it differs radically from the older concept, based on the notion that heredity is transmitted through "blood." According to this outmoded racial doctrine, the hereditary endowments of a human population become more and more similar generation after generation, provided that no interbreeding with other tribes takes place. Eventually, a tribe becomes a "pure race," whose members are more or less uniform genetically.

Some scientists were misled by this notion into believing that, at some remote time in the past, mankind consisted of an unspecified number of such pure races, whose members were all blue-eyed or all brown-eyed, all tall or all short and so on. Distorted versions of such notions became a part of the Nazi race creed; they still poison the thinking of many persons in other lands.

Revulsion from the horrors of Nazi race persecution induced some people to underestimate the differences between members of the human species. The statement was sometimes made, accordingly, that there is no such thing as racial difference. This point of view is also obviously contrary to the facts in the case. It is easy to see, for example, that blond persons are more frequent among people of European extraction than among those descended from Africans, and that they are found in northern Europe more often than in southern Europe.

There is undoubtedly such a thing as race. When we try to classify the different races, however, we run into great difficulty. The trouble is that there are many races and that they are not sharply separated from one another. It is obvious, for instance, that more people have blue eyes in Norway than in Italy. It is possible to say, therefore, that the Norwegian population is racially distinct from the Italian one. Yet we cannot jump to the conclusion that Norwegians represent a Nordic race and Italians a Mediterranean race. There is no clear-cut separation between the Nordic and Mediterranean races, because the populations that live in the countries intervening between Norway and Italy (Denmark, northern and southern Germany, Switzerland) show a chain of intermediate gene frequencies. It does not help matters much to call these intermediate populations separate races, because nowhere can a line separating any of these races from the rest be drawn.

We are faced with a dilemma, therefore. It is an undoubted fact that the populations of Norway, Denmark, Germany, Switzerland and Italy are racially distinct. It is purely arbitrary, however, to recognize

JAPANESE   PERUVIAN   GREEK   RWANDESE

only one, or two, or three, or five or ten races. It is just as arbitrary to discard racial labels altogether.

Heredity and environment

The hereditary endowment — that is, the genes that an individual has inherited from his parents — functions in an environment that does not remain constant. A rainy day is followed by a sunny one; summer, by winter; abundance of food, by scarcity. It would obviously be fatal for life on our planet if each kind of heredity could function in only one environment. In reality, organisms are generally able to exist in a certain variety of environments.

The effect of environment upon man and all other living things is obvious enough. As we have seen, the skin color in man is determined by his hereditary constitution. But everyone knows that skin color is also influenced by exposure of the skin to the sunlight. Some individuals have quite pale skin after a winter spent indoors, but develop a dark brown color after a summer spent on the beach. The genes for skin color remain the same in winter and in summer; yet these genes produce little color in some environments and much color in others.

To say, therefore, that the skin color is determined by the genes is inexact. The skin color is determined by the interaction of heredity (as determined by the genes) and the environment. The heredity of an albino produces no pigment in the skin regardless of sun exposure. Other heredities that occur frequently in what we call rather inaccurately the white race show pale skins

when seldom exposed to the sun but dark skin after prolonged exposure. Still other heredities, mainly in populations of tropical lands, give more or less dark skin regardless of sun exposure.

Some traits are determined rather rigidly by heredity and do not vary much in different environments. Other traits are much more plastic, or liable to be modified; they are easily altered by manipulating the influences of the environment. The ability to taste PTC, or to distinguish between red and green colors, depends upon the possession of the genes T and C respectively and has little to do with the environment. A "taster" will cease to find PTC solutions bitter only if his sense of taste in general is impaired by old age or by some accident; a person with normal color vision will cease to distinguish red from green only when his eyesight deteriorates. Stature is certainly inherited; yet children of Japanese immigrants grow taller in the United States than their parents grew in Japan. Skin color is doubtless a hereditary trait; yet, as we have pointed out, it can be modified to quite a considerable extent.

It is a matter of the greatest importance, from the viewpoint of evolutionary development, whether a trait responds to the influence of the environment or is rigidly determined by heredity. Natural selection favors the hereditary endowments that enable individuals to survive and reproduce in a variety of environments. To our ancestors, for example, the ability of the skin to tan was probably important as a protection against sunburn; it enabled them to gather food in intense sunlight. Such an environment might have proved fatal to

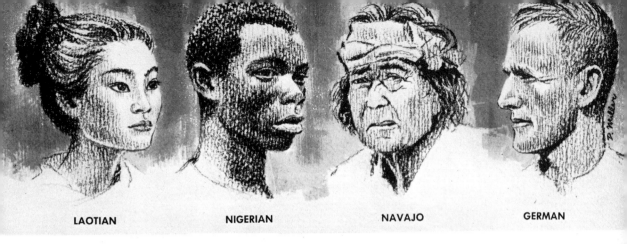

LAOTIAN   NIGERIAN   NAVAJO   GERMAN

organisms not possessing this particular type of adaptation.

Man, more than any other biological species, lives in the social environments that he himself creates. His survival and well-being, therefore, depend principally upon his ability to adapt himself to life with other men. Social environments are extremely changeable not only from place to place but also from time to time. A human individual is confronted in his lifetime with many different environments to which he must adjust himself. He will find himself in turn a boy among his playmates, a pupil in a school, a soldier in an army, a civilian working to gain a living, a father bringing up a family — he will have to adjust to each of these situations.

The number of different hereditary endowments that exist in mankind is almost equal to the number of men living. We cannot change the genes with which each of us has been born. The eugenists may be successful in persuading, or forcing, the carriers of some grossly defective genes to refrain from passing these genes to posterity. Yet this would not solve the basic problem: to live in a world in which every human has a different hereditary environment and differs from every other human.

This problem can be solved. For one thing, we shall have to accept and even admire the differences among men, instead of considering them as a curse and a pretext for conflict. We shall also have to provide each member of society with an environment in which he can realize the full possibilities of his hereditary endowment.

See also Vol. 10, p. 272: "Genetics."

The upper pictures on these two pages show the undoubted differences in race in various areas of the world. However, it is hard to classify the races of mankind since they are not sharply separated from one another.

Human development is determined by the interaction of heredity and the environment. The two youngsters below show a definite resemblance (the effect of heredity) to their parents, in the foreground; however, they are noticeably taller, due to the effect of the environment.

Robert Scott

CHANGING SHORELINES

Beaches, Coasts and Shore Processes

BY FRANCIS P. SHEPARD

EVERY year millions of people make their way to the seashore to enjoy the cool ocean breezes, bathe in the sea and sun themselves on sand beaches. How many of them realize the dramatic processes of change that are going on unceasingly at this meeting place of land and sea? Certainly it would add greatly to the interest of a stay at the beach if one knew of the mighty forces at work carving the land or encroaching upon the sea. Among the most powerful of these forces are the waves and currents of the ocean, eroding here and depositing there.

Waves are present all over the surface of the great oceans, but man's chief contact with them is along the shore. They are complicated formations as they occur in nature. In the diagram on the next page, we show a theoretical water wave, which does not indicate this complexity. Note,

in the diagram, that the high point of any wave is called the crest and the low point the trough. The wave length is the distance between two crests; the wave height, the vertical distance from trough to crest. The name "wave period" is given to the time it takes for two successive crests to pass the same point — say, a pier piling.

A wave may appear to move forward rapidly, but actually the water particles it contains do not keep pace with the apparent movement of the crests. Instead, these particles go around more or less in a circle though there is a very slight forward motion. You can see this for yourself if you throw a floating object in the water from the end of a pier beyond the breakers. You will see it move shoreward when the crest is passing and then back again in the trough of the wave, but usually with a net shoreward movement.

If you watch some wave crests advancing toward the beach, you will note that they move more slowly, get closer together, hump up to a greater height with steeper sides and finally break. This change in the character of the wave is brought about as the water becomes shallower. At the place where the crest collapses, tumbling over like a waterfall, the waves powerfully affect the sea bottom. If you have ever been caught in a breaker, you have become aware of the strong downward push of the water. A large breaker is exceedingly dangerous for this reason. Unless you can jump up clear of the breaker you should always dive under it when you find it overtaking you. In this way you will get through the breaking wave before it can fall on you with much force.

Inside the breakers, the moving mass of water advances up the beach. For every forward motion there is a subsequent backward one. The forward motion, called the uprush, is somewhat more violent than the return motion — the backwash; it is also of shorter duration. As a result of the higher velocity of the approaching waves, larger objects on the sea bottom are likely to move shoreward; because of the longer duration and lessened force of the outflow, small objects have a tendency to move seaward while larger ones are left behind. This accounts for a common (though not invariable) coarsening of material as one advances up a beach. Many beaches have gravel in the upper part and sand in the lower part.

Waves are a good deal more complicated than would appear from the rather simplified explanation we have just given. This is partly because of their origin. They are generally started by winds, largely storm winds, blowing on the ocean surface and producing humps and hollows moving in the direction of the wind. The waves with longer periods move faster and maintain their undulations longer than those with shorter periods.

A great storm will often send these long-period waves several thousand miles beyond the point where it was centered. In their journey from one storm, waves encounter the wave products of other storms, and the motion of one set is superimposed upon that of the other. This is one of the reasons why waves are so irregular in period and height. When the crests from one storm reinforce the crests from another, high waves result; if the two counteract one another, small waves develop.

Waves continue in the same direction until they get into shoaling water. Here they are slowed in their progress. They are not parallel to the shore as they approach, but are diagonal to it. The net forward motion of the water sets up currents that move along the shore in the direction in which the waves are approaching. Such currents may attain relatively high velocities, up to several miles an hour, and may cut troughlike depressions along the shore. The troughs are a source of danger to bathers; they may have deep sides, and the strong currents that create them may carry swimmers beyond their depth.

Because the net movement of the waves is landward, there must be a return current or else the water would keep piling up higher and higher along the shore. The return current is known as a rip current (or popularly, as a "rip tide" or a "sea puss"); it is in many ways even more dangerous than the breakers. The water ordinarily returns toward the sea along certain lanes,

A diagram of a theoretical water wave, showing the wave length, crest, trough and wave height.

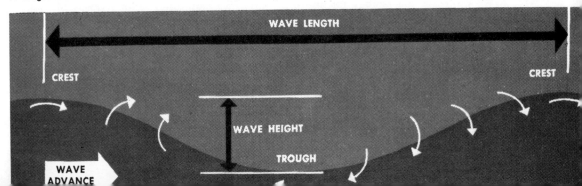

WAVE LENGTH

CREST

CREST

WAVE HEIGHT

TROUGH

WAVE
ADVANCE

U. S. Air Force

Rip currents, indicated by cloudy appearance in water.

F. P. Shepard

A long-period swell, coming in from a distant storm.

which are partly the result of the irregularities of the sea floor and partly due to coastal irregularities. Rip currents move diagonally outward from top to bottom of the water mass, rather than as "undertow."

After going seaward to a point where the water deepens considerably, the currents confine their energy to the surface or near surface. Presently they begin to spread or fan out. The water then moves along the coast and returns shoreward, though at lesser speed than before. Rip currents cut small gullies in the sea floor. Because this causes the water to be deeper, the waves often do not break in this zone, but show an agitated appearance.

Rip currents undoubtedly cause more deaths by drowning among surf bathers than any other factor. Their velocity is often so great that even moderately good swimmers cannot make headway against them. Seized by panic when they find they cannot reach the shore, most people do not realize that all they have to do is swim parallel to the shore until they can escape from the rip. This is the only safe procedure unless one is a good enough swimmer to allow oneself to be carried out beyond the breakers and then return along a different lane.

Occasionally great surges of the sea wreak havoc on the coasts they encounter. They are usually called tidal waves, although they have nothing to do with tides. They are caused principally by sudden crustal movements of the sea floor. These either push up the water surface or suck it down, according to the direction of the motion on the bottom of the sea. Most scientists refer to the resulting waves as tsunamis, after the Japanese word meaning "large waves in harbors."

Tsunamis differ from ordinary wind waves in having very long periods between crests — commonly fifteen minutes or more. Since their crests are miles apart, they produce no noticeable effect out at sea; it is only near shore that they grow to terrifying heights. Often they rise like a tide, only much faster and higher. As a result, they come in over the entire beach, over sea walls and even into harbors.

Another disastrous type of wave is caused by hurricanes or other great storms. The winds drive the water up along the coast; they cause the sea level to rise as much as fifteen feet and occasionally even higher. This has caused the flooding of many low-lying areas. Galveston, Texas, was devastated by such a "storm tide" in 1900. A hurricane, blowing steadily for eighteen hours, heaped up immense waves, which swept over a large part of the city. About 5,000 persons lost their lives; property damage resulting from the "storm tide" was estimated at $17,000,000.

The sudden explosion of a submarine volcano or the sudden engulfment of a volcanic island can bring about even larger waves. In 1883, a large part of the volcanic island of Krakatoa, off the coast of Java, was engulfed after a series of violent eruptions rocked the island. Disastrous waves resulted; thousands of people were drowned in the adjacent islands. The waves spread all over the world, and they caused loss of life and property damage in many widely separated places.*

* There is a detailed discussion of waves in another chapter of *The Book of Popular Science*. See The Waves of the Sea, in Volume 6.

Both photos, F. P. Shepard

Fine sandy beach at La Jolla, California, in summer.   The same beach, stripped of its sand by winter storms.

Beaches: their
origin and characteristics

Waves and currents are the chief agencies in the formation of beaches. Because waves push relatively coarse types of bottom sediment toward the shore under their advancing crests, they generally pile up sand above sea level along vast stretches of coast. This does not happen everywhere, however. On steep, rocky coasts, there is no platform on which sand can be deposited. In certain places, the adjacent sea floor has no sand supply. In other regions, the sea floor drops off so steeply that the waves are not powerful enough to move the sand shoreward against the mighty force of gravity.

Beaches are by no means stable; most of them are constantly growing either wider or narrower. Many disappear completely during stormy periods, only to reappear when the storms are over. These changes appear to be related to the relative power of the uprush and backwash of the waves. When these are small, only the uprush has the power to move the sand, which therefore comes in onto the beach. During large waves, on the other hand, the sand is carried by both the backwash and the uprush. The sand particles that move down with the backwash are often caught in powerful rip currents and swept out into relatively deep water. In such cases, the beach retreats.

In many instances, changes in beaches can be traced directly to the work of man. In order to develop harbors along a straight coast, where no natural protection exists, walls called jetties are built out into the sea. They serve to stop the natural drift of sand along the shore. If the current is moving principally in one direction, the sand piles up on the side from which the current is coming. The beach becomes so wide that the houses originally built near the water are now far away from it.

On the other side of the jetty, the beach becomes "starved." The sand that would normally be supplied to it by the longshore currents and by the shoreward sand migration from the small waves is trapped by the jetty. The result is that the winter storms cut away the beach and it is not replaced.

In many cases, the "starving" of the beach results in the destruction of houses built near the water; it may even bring about the loss of the entire beach. At Redondo, on the south end of Santa Monica Bay, California, the cut in the lee (protected side) of the curving harbor jetty caused the loss of an entire city block before the coast could be stabilized. To the east of the Santa Barbara harbor jetties, the beaches were cut away for many miles, causing great loss to resorts that had been built on that side. To the west, on the other hand, the beach grew so wide that it was too much of a walk from the shore road out to the ocean.

The cutting away of beaches in the lee of a jetty does not go on indefinitely, because after a time the indentation of the coast serves as a natural protection against further cutting. An equilibrium is reached as the beach receives a supply of sand from the other direction during occasional reversals of the longshore currents. Also a certain amount of sand is carried around the end of the jetty to the lee side.

351

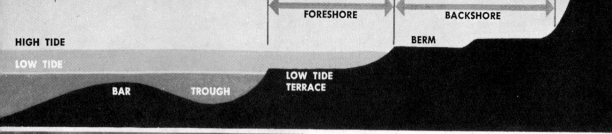

FORESHORE · BACKSHORE

HIGH TIDE · BERM

LOW TIDE

BAR · TROUGH · LOW TIDE TERRACE

Above: the divisions of a typical beach. Upper photo: backwash ripples formed by the retreating waves. The lower photo shows current ripples produced by currents.

Upper photo: D. B. Sayner; lower photo: T. Nichols

Any wide beach has several distinct divisions. At the top, the beach has a more or less horizontal surface called the berm, or backshore. This slopes slightly landward in coarse sand or gravel beaches and gently seaward where the sand is fine-grained. Seaward of the berm the relatively steep slope is called the foreshore. The slope of this section also depends on the size of the sand grains. It may be as much as ten or fifteen degrees in the case of coarse sand beaches. Fine sand beaches slope only from about one to three degrees.

When there are no appreciable waves, the high tide covers most of the foreshore slope; during periods of large waves, the uprush carries the high tide over the berm edge. On many beaches, the low tide ex-

poses either a terrace or a trough and bar (see diagram). Both the terrace and the trough and bar are likely to be cut away during periods of high waves. The trough and bar then move out into deeper water, where they are located well below the lowest tides.

Beach sands along the continents consist mostly of rock minerals, particularly quartz. This mineral is almost entirely lacking in the beaches of oceanic islands in the tropics. Although many of the white sands in these places look rather like quartz to the uninitiated, they really consist of coral and shells, ground up by the action of the waves.

Under ordinary conditions, the foreshore of fine sand beaches is firm enough to support an ordinary car. On the other hand, only vehicles such as tractors or jeeps can be driven on a coarse sand beach. The berm of even a fine sand beach, however, is apt to be treacherous, and even the foreshore is unsafe if the tide has very recently covered it. There is an interesting contrast between beaches on the east coast of Florida. At Daytona, the fine quartz sand beach is so firm that it has been used for auto racing for many years. On the other hand, the shell beaches of Miami, Palm Beach and Fort Lauderdale offer no support for a car. Most of the long beaches of Texas are excellent for cars because they are composed largely of fine quartz sand; but a jeep is required to travel on the sands of Padre Island, to the south, because these sands consist mostly of shells.

Closer examination of a beach reveals many interesting phenomena. On fine sands, the backwash of the retreating waves develops ripple marks, which are low in height and almost always about eighteen inches from crest to crest. In the pools left behind in the ripple troughs, the mica which temporarily floats in the water settles to the bottom. Such troughs often glisten in the sunlight because of the reflection from the flat mica surface. These backwash ripples are not found in coarse sand because the grains fall back in place after the water has retreated.

Another type of ripple appears at low tide, when the troughs of the longshore currents and rip currents are partially exposed. Current ripples are found in both coarse and fine sand beaches. They are formed by the current flowing along the troughs and ordinarily run at a wide angle to the backwash ripples. These current ripples are much more pronounced than the backwash variety and the distances from crest to crest are shorter. Their steep side is away from the current. If you stand in shallow water along the shore in an area with small waves, you can often see two sets of ripple marks on the sandy bottom. One has crests extending at right angles to the shore; the crests of the other are parallel to the shore. The first is due to longshore currents; the second, to the back-and-forth movement of the small waves moving in and out from the shore. If the bottom is covered with fine sand, these wave ripples are closely spaced, about three inches from crest to crest; they are rather widely spaced — commonly from six inches to one or two feet — when the sand is coarse.

Another curious feature of beaches is a symmetrical series of short ridges, each extending down the foreshore and coming to a point. These ridges are called beach cusps. The interval between cusps is al-

most exactly proportional to the size of the waves that produced them. During storms, widely separated cusps, a hundred feet or more apart, are brought about; under small wave conditions, they are much closer together. The smallest cusps are found on the beaches of small lakes, where they may be only a few inches apart.

Most beach sands are stratified: that is, made up of layers. You can often see these layers when a ditch has been cut into the beach. They are generally caused by the variation in the amounts of dark- and light-colored minerals in the sand during the building up of the beach. The larger waves produce a sand with more dark minerals, since they tend to be heavier than light-colored ones and the larger waves can transport them more readily. Other layers are due to a concentration of mica, brought about only when waves are very small. Large waves would carry mica out to sea.

When one digs down to the bottom of a beach, one often exposes a very black sand, particularly along the base of sea cliffs or sea walls. This black layer is the result of especially severe wave conditions, because of which all of the light-colored sand has been carried out to sea, leaving only the heavier mineral concentrates. Black sands often contain valuable minerals such as tungsten and wolframite.

When sand has recently accumulated on a beach, it is likely to contain an excess of air. As water pours over this new sand, the air may be concentrated in pockets, which bulge up like blisters, producing small domes. If one prods these domes, the air escapes and they collapse.

Various other markings are found on the sand. At the top limit of the uprush,

Beach cusps in fine sand at San Onofre, in California.

Accumulation of black sand along the base of a sea cliff.

Both photos, F. P. Shepard

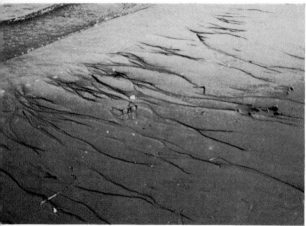

P. H. Kuenen

These rill marks have been cut in the beach at low tide by the run-off of water that had been buried in the sand.

where the water hesitates before returning seaward, fine material is deposited and a thin line is produced. This line, called a swash mark, generally consists of mica or plant fragments. If a new wave comes to a greater height, the swash mark disappears and a new one is formed at the higher level. Such marks, therefore, are left by the highest waves of a tide.

A diamond pattern of dark streaks is produced by the backwash when small pebbles or shells divert the return flow and concentrate it along diverging lines. Sometimes these diamond patterns appear to be due to small sand crabs (*Emerita*) which burrow beneath the surface and push up small antennae to catch organic material drifting past. The antennae have the same diversion effects as the shells and pebbles mentioned above.

At low tide, the water beneath the sand migrates seaward, coming to the surface along the lower beach and developing small streams. These in turn produce miniature valleys, called rill marks. They resemble on a small scale the valleys on land cut by streams — valleys such as one can see when one flies across the arid lands of the southwestern United States.

Other markings on beaches are due to various forms of life. Worms eat their way through the sand in order to extract the organic material it contains. When they are near the surface, they often pro-

duce small ridges. Other organisms crawling over the sand bring about small depressions. Holes are produced by the digging of crabs; the excavated material is built up into cones, resembling miniature volcanoes. Some of the crabs spray out the sand in a pattern of lines, radiating outward from the excavation point. Certain craters produced by crabs in tropical areas are of impressive dimensions.

Types of coasts and shorelines

So far we have been dealing with processes of change involving comparatively small areas — local beaches. Changes on a far grander scale constantly modify the contours of the coasts and shorelines of the world. Before we discuss these changes, it will be helpful to give a few definitions. A shoreline is, literally, the "line" where water and sea meet. The shore is something quite different; it is the zone between low tide and the inner edge of the wave-transported sand. The coast is a broad, rather indeterminate zone landward of the shore; it includes sea cliffs, coastal terraces and the broad lowlands adjacent to the shore. There would be no purpose in taking up shorelines and coasts separately because the two overlap to such an extent. They exhibit unmistakable features which serve to identify them, even in places where the sea that had produced them has long since vanished.

If we examine coastal maps from various parts of the world, we find that there are two principal types of coasts. One has a more or less straight shoreline, such as we find along much of the coast of California and along the outer coast of Texas. The other is deeply indented, or irregular, and shows many bays and promontories.

One of the earlier classifications of coastlines included most of the straight shorelines in one category, labeling them coasts of emergence. Most of the irregular shorelines formed a second category — the coasts of submergence. It was argued that the submergence of valleys and ridges would cause the sea to come up the valleys, making bays and leaving the ridges as promontories. On the other hand, it was maintained, if the coast was elevated, it would raise up the flat sea floor and cause it to be subjected to surf action. Later, the breakers would build up a sand ridge, called a barrier island, above sea level. This would produce a straight coast, because barrier islands are predominantly straight.* If a coast was neither submerging nor emerging, it was classified as neutral. Among the examples given for this kind of coast were deltas, which are built out into the ocean, and volcanoes, which build seaward as they grow in bulk.

The very simple picture suggested by this classification was easy to teach to students and was quite generally accepted by geology professors in the early part of the twentieth century. Then some of us began to test the classification to see if it fitted the geological history of the coasts that had been grouped in this way. As so often happens in science, the theory on which the classification was based proved to have serious flaws.

In the first place, we have come to realize that the sea level has been very unstable during the past million years or so. The most recent event has been a considerable rise in sea level, accompanying the melting of the last great continental glaciers. This rise may be still continuing, although, if this is so, it is proceeding very slowly. It was going on actively up to some 6,000 years ago — that is, up to the time of the building of the early Egyptian pyramids. Such a rise has of course drowned (submerged) all low-level valleys around the coasts of the world and has therefore made them all coasts of submergence except in the rare cases where a recent uplift has counteracted the effect of sea-level rise. It has been shown, too, that many of the straight coasts of the world, such as that of Texas, result from a sinking process combined in its effect with the submergence due to sea level rise. The barrier islands of Texas have been formed as the land sank or the new level rose.

Even many of the so-called neutral shorelines have proved to be badly classi-

* The beaches at Atlantic City, New Jersey, and on the Texas Coast are good examples.

fied. Many of the great deltas of the world are by no means neutral. We know now that they are actually subsiding, so that they would come under the classification of shorelines of submergence. Similarly, volcanic coasts are subsiding in many areas as the result of the great weight of the volcanoes on the earth's crust.

Such difficulties as these have made it necessary to develop a more practical classification of the various types of shores and coasts. I proposed such a classification in the 1940's * and it has been adopted in various new texts. I give a brief outline of it in the following pages. It will undoubtedly have to be modified as we acquire more information about the subject.

First, I divide all coasts into two major groups. In the first, I include the coasts that owe their general shape to nonmarine agencies — that is, processes in which the action of waves and currents plays no part. Only after these coasts have attained their present contours have the waters of the sea penetrated them. My second group includes the coasts that have been shaped largely by the work of waves and currents.

Coasts formed chiefly by nonmarine agencies

Many coasts owe their general shape to the effect of running water on land areas. Through erosive action, streams carved out valleys as they made their way to the sea. The waters of the sea have now penetrated these valleys, but their general conformation remains more or less unchanged. Chesapeake Bay offers a familiar example. Each tributary valley gouged out by running water in ages long past has now become a separate arm of the sea; an oakleaf pattern has resulted.

The running water of rivers carries sediment and builds up deltas at the mouths of rivers. If the delta-building goes on at a faster rate than the sea-level rise, the coast is extended into an arc or a birdfoot,**

* Editor's note: Dr. Shepard's classification is given in detail in his Submarine Geology, Harper and Brothers, New York, 2nd edition, 1963.
** A delta of this type grows outward as a cluster of long, narrow peninsulas, joined together upstream somewhat like the toes of a bird's foot.

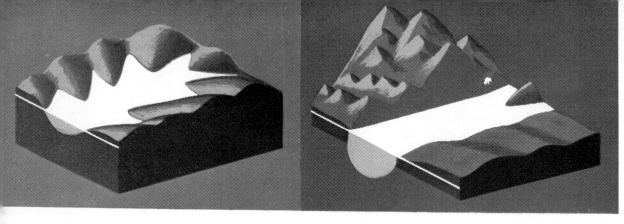

Streams flowing to the sea carve out valleys. Here the ocean has penetrated a number of tributary valleys.

A deep valley (fiord), gouged out by glacier ice and later filled by the waters of the sea as the ice retreated.

356 such as we find at the mouth of the Mississippi. Several deltas may merge and form a continuous plain along the coast. In such cases, a relatively straight river-deposition shoreline may exist. Much of the east coast of New Zealand's South Island has originated in this way.

Glaciers have also helped to shape the coast. If a coast has been glaciated — that is, covered at one time by a great glacier of the Ice Age — its present appearance gives many indications of the action of the ice. Deep valleys, called fiords, were cut out by the ice well below sea level. When the ice retreated, the sea came up into these valleys.

In some localities the glaciers left deposits in the form of elongated ridges and hills, called moraines. The rising sea level brought the shoreline up against these glacial hills. Generally, the sea has so modified the shoreline that the original outline of the moraine no longer exists. There are some exceptions, however; for example, the inner, protected shore of Long Island still shows some of the original glacial-moraine shoreline.

Glacial deposition also resulted in the formation of oval hills, called drumlins; they were elongated in the general direction of the ice movement. The drumlins in what is now Boston harbor formed islands as the sea penetrated the area. These drumlin islands have been so little modified by the subsequent action of the sea that their oval shape constitutes the shoreline.

At least two types of coasts owe their form to volcanic activity. The first, found largely among islands, has a circular or oval outline, representing the base of the volcanic cone. The Hawaiian Islands offer some fine examples of this type of formation. A second type is concave in general outline. It is due to the collapse of old volcanoes or to explosions that have blown away one side of a volcano. Such coasts are quite common in Japan, the East Indies and the Aleutian Islands.

Along the rim of the Pacific, there are many areas where the coastal contour is

Certain coasts may result from volcanic activity. Some of them comprise the bases of volcanoes (left, below).

Others (right, below) are due to the collapse of an old volcano or to an explosion that has blown off part of it.

BASE OF VOLCANO

VOLCANO THAT HAS COLLAPSED OR EXPLODED

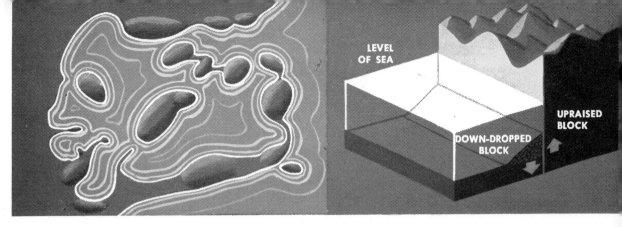

Above we see a series of drumlins (oval hills) which became islands as the sea penetrated the general area.

The border of the upraised block in the above fault has become the coast; the down-dropped block is under water.

due primarily to movements of the earth's crust. Faulting, in which one block of the crust moves against another block, is responsible for many of the straight shorelines in the Pacific area. The border of the upraised block constitutes the coast; the down-dropped block is beneath the sea. Because of the general steepness of fault surfaces, the sea bottom slopes off rapidly along most fault coasts. For example, along the west coast of the Gulf of California, it is possible to cruise along the shore with a thousand feet or more of water under the vessel. Coasts may also be due to folding, or the bending of the earth's crust. When the sea penetrates such areas, a series of islands and intervening sounds are formed. The islands represent the upfolds, or anticlines; the sounds, the downfolds, or synclines.

Coasts formed chiefly
by the action of the sea

Along most coasts of the world, the shorelines have been greatly modified by the effect of waves during the time of the **357** last rise in sea level. Waves may either cut back the coast or build it out by depositing material upon it.

Where the coast has been clearly worn back by the waves, it is straight or has many small indentations. We do not find the deeply penetrating bays characteristic of drowned river valleys and of fiords. The straight wave-cut coast develops where the coastal formations attacked by the waves are relatively uniform and where the rocks are quite soft. The waves attack projecting points and wear them back. The chalk cliffs of Dover are a well-known example of a straight-cut coast.

Where the rocks of a region are of different hardness, the waves cut back the softer portions, developing rounded coves and leaving the harder rocks protruding as headlands. At La Jolla, California, the cliffs that have been cut into the relatively soft Eocene formations to the north are quite straight. To the south, we find a projecting land body called Point La Jolla.

Coast formed by wave erosion. If coastal formations attacked by waves are relatively uniform, a straight wave-cut coast develops (left, below). Irregular contours develop (right, below) if rocks are of different hardness.

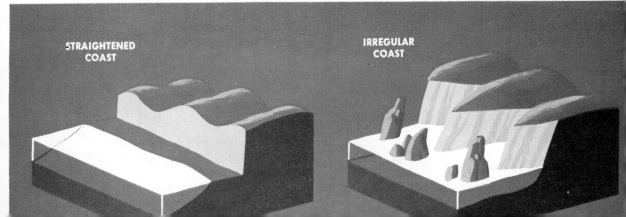

STRAIGHTENED COAST

IRREGULAR COAST

The hard Cretaceous sandstone of this area has withstood more effectively the pounding of the waves.

The laying down of material by the sea has generally resulted in the straightening of the coast line. For example, the long smooth curve of the outer coast of Texas is the result of the deposition of a sand beach, which grew into a barrier island, a mile or more in width. The barrier lies across a highly irregular drowned river valley coast, which extends along most of the barrier length. Similarly, the straight New Jersey coast results from a barrier built across the drowned river valleys; these valleys are recognizable on the inside of the barrier. In other cases, the deposition has simply formed a sand beach across the mouth of a bay, connecting the headlands on either side.

Deposition may develop irregularities, particularly where sand spits are built up from a relatively straight coast. The spits apparently form where there are two adjacent eddying currents and a zone of quiet water in between. If the sediment that is transported by the current is carried into the dead spot, it is deposited.

In some areas, especially in the tropics, the shore is extended as aquatic plants grow out into the water. These are not true oceanic seaweeds, or kelp, but land plants that can live in salt water. The mangroves are the best examples of these plants. They have barbed seed pods which drop from the extended branches into the adjacent water; they plummet to the bottom and there they become rooted. After the plants become established in the shallow waters along a coast, mud is gradually transported into the area and deposited between the roots. Eventually, the region is built up to the sea surface. Such coasts are usually marshy.

The coastal formations called serpulid reefs are due to certain worms, belonging to the family Serpulidae. These worms attach themselves to the shallow bottom near the shore or to coastal rocks. They extract lime from the sea water and form a tubelike frame, which gradually builds up toward the surface. Oysters develop similar reefs in bays of rather low salt content. These reefs may grow to the surface and become a part of the land, partly as the result of wave activity.

Another type of coast, very common in the tropics, is that formed by coral reefs. In many places, they merely produce a rim around a coast of different origin. Reefs of this type are common around volcanic islands, whose volcanoes have become ex-

The laying down of material by the ocean has resulted in the formation of a sandy beach, which has grown into a barrier island. It will be noted that the barrier island lies across highly irregular drowned river valleys.

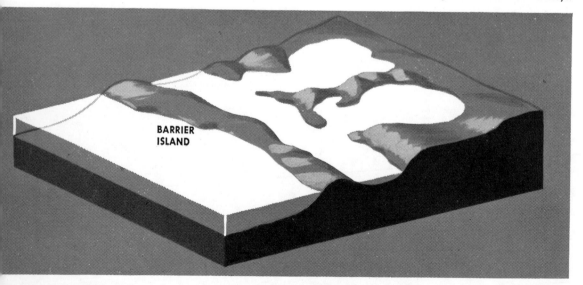

BARRIER ISLAND

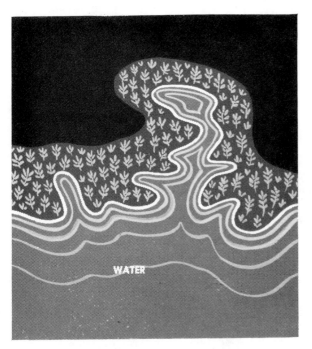

The shore has been built up as mangroves have grown out into the water. A region of this kind is usually marshy.

tinct. The corals grow along the shore; they develop a shoal platform called a fringing reef, which is partly uncovered at low tide. Coral cannot grow above low tide; but storm waves may throw up enough debris on the inner portion of the reef to form a low terrace above the reach of ordinary tides. Elsewhere, these terraces rise somewhat above sea level because of local uplifts of the islands. The chief habitations of South Sea islands such as Tahiti and Moorea are on such terraces.

Barrier reefs form where the coral grows well beyond the coast, leaving a quiet and often deep lagoon on the inside. In some places, these reefs have been built above sea level by the waves or have been sufficiently raised so that human habitations can be found on them. Generally, however, reefs are exposed only at low tide.

The Great Barrier Reef of Australia extends for 1,200 miles along the coast and has a width of up to a hundred miles. In many places it projects above the water, forming islands. Most of these are low and sandy; they are known as faros * or

* A local name derived from the ringlike islets of the Maldive Islands in the Indian Ocean.

cays.* Only a few people inhabit these islands off the Queensland coast. A resort for tourists is located on Great Heron Island, near the southern end of the reef.

Ring-shaped coral islands called atolls are found in profusion in the southwest Pacific. In my article The Depths of the Sea, in Volume 6 of THE BOOK OF POPULAR SCIENCE, I discuss these islands.

The coastal types that we have included in this brief survey are probably the most common; but there are certainly other varieties. For example, the collapse of the roofs of caves along the shore has modified the general shape of the coast in some areas. In other regions, the coastal contour has been altered as wind has blown a sand dune out to sea.

In some instances, a coast may have been formed by a combination of causes. As we have seen, a coast of drowned river valleys may have been straightened because barrier islands, produced by the action of waves and currents, grew up along the outside. A volcanic coast may be built out by coral growth; a fault coast may be so eroded by the action of the sea that it is essentially a wave-cut coast.

* The English version of the American word "keys," found in place names such as Key West.

See also Vol. 10, p. 270: "Oceanography"; "Sculpture of Land."

A coral reef has been raised until it forms an atoll — a ring of low-lying islands around a big central lagoon.

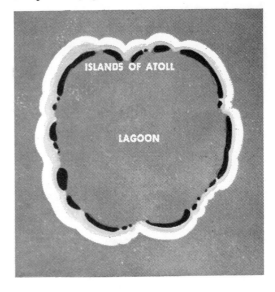

U. S. Public Health Service

On the trail of parasites carried by mosquitoes and other insects: medical entomologists (experts on insects) of the U. S. Public Health Service studying insect specimens recently submitted to them.

THE TRUE PARASITES

Living Things That Dwell within or on Other Living Things

BY LAWRENCE M. LEVIN

ONE of the most common relationships in the world of living things is that between parasite and host. A true parasite is an organism that lives for a longer or shorter period in or on another organism, which is the host. Because of this relationship, the parasite obtains nourishment, supplied by the tissues or waste products of the host. Very often, the parasite helps itself to food that the animal serving as host has digested or partly digested.

The general rule of parasitism is that smaller forms prey upon larger ones. This proposition was stated long ago in the familiar lines:

> Great fleas have little fleas
>> on their backs to bite them.
> Little fleas have lesser fleas,
>> and so ad infinitum.

A great many parasites are exceedingly small. Until the development of the powerful ultramicroscope and the electron microscope, the existence of certain parasites could be determined only by the effects they produced in the host. Among the tiniest are the bacteria, which consist of a single cell. The bacteria called the cocci have a diameter of as little as one micron, the thousandth part of a millimeter. (A millimeter is just a little smaller than a twenty-fifth of an inch.)

There are even tinier parasites. A number of diseases are caused by viruses, particles too small to be seen by the finest optical microscope. By means of the electron microscope, certain viruses have been photographed, and the pictures have been magnified, thus giving us some idea of their appearance. The question that has not yet been answered to the satisfaction of all scientists is: Are the viruses living or nonliving? They act, in many instances, like deadly parasites. If a single drop of juice from a tobacco plant infected with tobacco mosaic, a virus disease, is mixed with a barrel of water, and the liquid is then rubbed on the leaves of healthy tobacco plants, all of these plants will come down with the disease.

At the other end of the scale, in point of size, we find the fish tapeworm *Dibothriocephalus latus*. This parasite of man sometimes reaches a length of sixty feet and perhaps even more.

The true parasite is at least potentially harmful. There are several other kinds of close relationships between living things besides that which we call true parasitism. In some cases, organisms may live within other organisms and not cause them any harm to speak of. For example, man harbors in his own body many species of harmless parasites, among which are various bacteria that live in the intestines. These tiny plants live out their lives in the human body and derive their food from it, but probably neither harm nor help the host in any way. This relationship is called commensalism.

Another type of parasitism — though not true parasitism according to our definition — is found among ants. The queens of parasitic ants are only a little larger than the workers. They get into the nests of other ant species and are tolerated there for some reason or other. Some authorities believe that the parasite queen exudes a pleasant-tasting secretion that is licked by the workers of the invaded nest. Others hold that the invading queen hides until she acquires the nest odor; after that, little attention is paid to her. Eventually, she kills the original queen of the nest, or lets the workers kill that queen.

The workers accept the new queen and care for her offspring. For several years, the ant colony will be mixed, containing workers of both species. However, as the workers of the original queen die off, a pure colony of the invading species will be established in the nest. A number of species of the genus called *Formica* appropriate nests in this way.

Symbiosis is a beneficial relationship

In still another form of relationship, one organism may dwell within another one, to their mutual benefit. The plant called the lichen consists of a fungus, in which algae live. The algae are capable of manufacturing food by the process of photosynthesis (see Index). The fungus lives on the food prepared by the algae; for its part, it aids the latter by absorbing water and mineral salts, which the algae require for food manufacture. The fungus also gives the algae the protection of a "home." This kind of relationship is called symbiosis.

How termites obtain predigested foods

An even more striking example of symbiosis is found in the animal kingdom. Certain termites harbor a vast number of the one-celled animals called protozoans in their intestines. The termites feed on wood, but they cannot digest the cellulose of which the wood chiefly consists. The protozoans feed on the cellulose and predigest it so that it becomes available as food to the termites.

In this article, we shall deal with the true parasites. They are widely distributed in nature; in fact, most plants or animals have been infested with such parasites at one time or another. We have already mentioned the viruses, which have many of the characteristics of living things. They are responsible for many diseases, including yellow fever, infantile paralysis and small pox in man and the foot-and-mouth disease of cattle.

Among the plants, parasitic bacteria and fungi are the most widespread. Certain parasitic bacteria are responsible for human diseases such as tuberculosis, pneumonia, leprosy, diphtheria, tetanus, the plague, syphilis and typhus fever. Parasitic fungi cause ringworm and various other diseases in man; they are responsible for such plant diseases as the blight of potatoes and the mildew of grapes. Certain plants are only partly parasitic. For example, the mistletoe, which attaches itself to various kinds of trees, manufactures its own food through the process of photosynthesis; but it obtains water and minerals from its host.

True parasites that cause disease

The animal kingdom has its share of true parasites. Certain parasitic protozoans are responsible for certain dreaded diseases of man. *Plasmodium malariae* causes malaria not only in man but also in birds, reptiles, frogs, monkeys, apes and other animals. The protozoan called *Entamoeba histolytica* is responsible for dysentery, a particularly disagreeable human disease. Another protozoan, *Trypanosoma gambiense,* is the cause of sleeping sickness. Parasitic worms are numerous. Among them are the beef tapeworm of man, the hookworm, which causes the enervating hookworm disease in man, and the liver fluke of sheep. Another worm that causes disease is the threadlike filaria, which is a parasite of the blood or tissues of vertebrates and causes the condition known as filariasis.

Many insects are parasitic

A great many insects are parasitic. For example, the female of the ichneumon fly thrusts its ovipositor into the larva of another insect species and lays its eggs within the victim's body. When the eggs hatch, the larvae feed on the host and eventually cause its death. True parasites are also found among crustaceans and mollusks. Among the vertebrates, only the lampreys and hagfishes are true parasites.

Some parasitic animals ultimately devour their hosts, thus becoming predators,

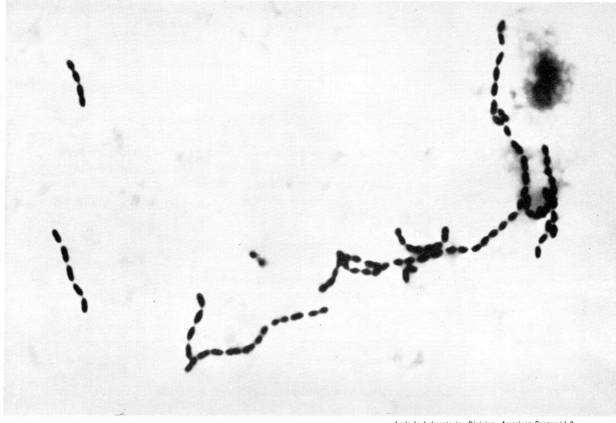

Lederle Laboratories Division, American Cyanamid Co.

A colony of pneumococci, bacteria that cause pneumonia. Four of the many strains of pneumococci are particularly virulent. Pneumococci may penetrate the membranes of the brain and produce meningitis.

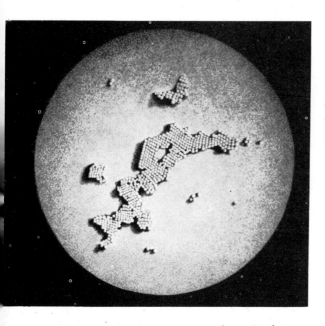

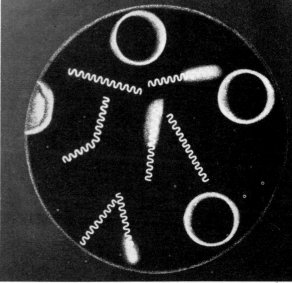

Lower photos, World Health Organization

The viruses shown above are among the strains that produce poliomyelitis. When they attack nerve cells in the spinal cord or brain, they can cause great damage.

Colony of *Treponema pertenue*, the microorganism that causes the contagious disease called yaws. In this, the epidermis thickens and cracks and bleeding is frequent.

or animals of prey. A good example is the hagfish, which is related to the lamprey. The hagfish is a very primitive type, lacking eyes and jaws; it must smell out its food. It wreaks havoc among fishes caught in nets. First it bores a hole through the body wall of its victim; then it crawls inside. Greedily it feasts on the host, devouring intestines and flesh. Finally it leaves its victim a mere hulk of skin and bone. Various insect larvae, including those of various flies, also ultimately devour their hosts, thus becoming predators after having been parasites.

Parasitic and
free-living forms

In certain cases, parasites belonging to a given plant or animal group are similar in appearance to the free-living members of the same group. This is particularly true of the bacteria and the protozoans called amebas. In other cases, parasites differ markedly from their free-living relatives. Where they become attached to the host, their legs and wings tend to atrophy or to disappear entirely. The alimentary canal may undergo great changes or it may be entirely lost, as in the case of the tapeworm. On the other hand, the organs by which parasites become attached to the host become more highly developed than in the free-living varieties.

Two chief classes
of true parasites

True parasites may be divided into two principal classes: the endoparasites, which flourish within the host, and the ectoparasites, which live outside of it. ("Endo" comes from the Greek *endon:* "inside"; "ecto," from the Greek *ektos:* "outside.") Bacteria, tapeworms and liver flukes, which all live within the host for at least a part of the life cycle, are endoparasites. The ectoparasites include such forms as the louse, the cattle tick and the apple scab.

In some cases, it is difficult to classify true parasites as endoparasites or ectoparasites. The organisms living in such areas of the host as the mouth or gills or burrowing completely into the skin might be put in either of the groups.

Endoparasites — true
parasites that live in the host

Many endoparasites select only a limited number of animals as their hosts. Sometimes geographical limits keep a parasite and a prospective host apart. For example, *Trypanosoma gambiense,* which causes sleeping sickness in man, cannot attack humans outside of Africa. The reason is that it is limited to a single insect host, the tsetse fly (genus *Glossina*), which carries the disease; and the tsetse fly is found only in Africa. Hookworm disease is prevalent in certain southern sections of the United States, but not in the northern areas, because the hookworm does not live in these regions.

Some parasites have
a choice of hosts

Other parasites have a wide choice of hosts. For example, the trichina worm (*Trichinella spiralis*) can develop in any meat-eating mammal; the fluke called *Heterophyes heterophyes,* in almost any mammal or bird that feeds on fish. In certain cases, a parasite may be particular about its hosts at one stage of its life cycle and not be at all particular in another stage. Certain types of relapsing fever parasites, for instance, can develop in almost any mammal. However, only one species of the soft tick (*Ornithodorus*) will serve as a combined host and carrier.

The relationship between parasite and host is a complicated one. The parasite tends to invade the host and to maintain itself within the host's body. For its part, the host may have certain defenses which may be inborn or acquired, and these defenses will tend to keep the parasite in check. Depending upon the aggressiveness of the parasite and the effectiveness of the host's defenses, there may be many degrees of parasitism. A prospective host may have such formidable defenses that the parasite will not be able to invade it at all. On the other hand, parasites may invade a host in huge numbers and almost all their offspring may survive. The host will die and so will the parasites, since they will have lost their source of food.

Generally speaking, the relations between parasite and host do not follow either of these extremes. The parasite succeeds in invading the host and in maintaining itself; but it is held reasonably in check. Neither the parasite nor the host succumbs as a result of the relationship. It is believed that this "balance" developed as a consequence of natural selection.

Effective natural
defenses of the host

The natural defenses of the host are often very effective. The skins of the higher animals serve as a barrier; so does the flow of mucus. Even after the invading organisms have penetrated these barriers, there are various other defenses. The digestive juices of the host may prevent the parasites from reproducing their kind. The serum of human blood has the property of killing many of the trypanosomes that infect African animals; that is why men are immune to many of the diseases caused by these protozoans. Unfortunately, human serum does not affect *Trypanosoma gambiense,* which causes sleeping sickness in man.

A host may succeed in warding off the invasion of parasites because it lacks some essential food element that the parasites require. For example, human hosts who are deficient in the vitamin pantothenic acid (one of the B vitamins) resist the malaria germ *Plasmodium malariae* because the latter needs pantothenic acid.

Plants that
secrete antibiotics

In plants, the ability to secrete antibiotic substances, such as penicillin or streptomycin, serves as a natural defense against invading organisms. Man has used these natural defenses for his own purposes; he has developed an entire arsenal of antibiotic drugs to combat disease. (See Index, under Antibiotics.)

In all the cases that we have mentioned above, the ability of the host to resist the invasion of parasitic organisms is determined by its heredity. Plant researchers have put this to good account in developing strains of plants that will resist infection.

How parasites
win a foothold

Parasites may win a foothold within the host in various ways. Some of them, including the bacteria that cause scarlet fever and diphtheria, elaborate poisons, or toxins, which weaken the resistance of the host. Certain bacteria produce substances called leucocidins; these kill the white blood cells that normally attack and devour small invading organisms. It is thought that the parasite often wins passage through animal tissues because it secretes certain enzymes that attack these tissues. In some instances, a mutation (see Index) may enable a parasite to invade a host belonging to a species that has hitherto been immune.

Generally speaking, the endoparasite that has invaded a host is restricted to certain organs, tissues or cells. For example, the protozoan *Plasmodium malariae,* the cause of malaria in man, lives within the red blood cells during most of the period when it is developing within the human body. The parasitic fungus that causes loose smut of wheat is found only in the embryo and the growing points of the plant.

Human beings who serve
as carriers of disease

The invasion of parasites produces a variety of effects in different hosts. In some instances, the host does not seem to be unduly affected in any noticeable manner. Certain human beings can carry the germs of typhoid fever — *Bacillus typhosus* — without showing any of the symptoms of the disease. They may spread it without being aware of the fact that they are responsible. The best-known carrier of typhoid fever was Mary Mallon, who was responsible for fifty-one known cases of the disease and who received the nickname of "Typhoid Mary." She was a cook who worked in various institutions and also in private homes. "Typhoid Mary" was confined during the last years of her life in a New York hospital, where she died in 1938.

Effects of the invasion of parasites

In many cases, the host is affected to a greater or lesser extent by the invading parasites. Blood vessels may be blocked and may then rupture, causing bleeding. The host may be deprived of vitamins or other essential substances. Certain tissues or cells may be destroyed. *Plasmodium malariae* destroys such a great number of the host's red blood cells that the victim suffers from anemia. In extreme cases, so many of the cells are killed that death is the result. Sometimes, the host is seriously affected by the toxins secreted by the parasite. The bacillus called *Clostridium tetani,* which causes lockjaw in man, penetrates the tissues of deep and dirty wounds. It is accompanied in the invasion by the so-called saprophytic bacteria, which feed on dead organic substance. *Clostridium* is able to germinate in the affected tissue. It secretes a toxin that spreads throughout the body and causes the typical lockjaw symptoms. Sometimes invading parasites produce symptoms only in very young hosts. This is true, for example, of the blood parasite *Trypanosoma lewisi,* which affects only very young rats.

How the host reacts to infection

After infection has taken place, the host generally reacts vigorously. Certain substances, called antigens, are released by the parasite and they bring about the formation of antibodies, which counteract the effects produced by the invader. If the parasite secretes a toxin, the antibody (called in this case antitoxin) may neutralize the effects of the toxin. In the case of parasites that do not secrete toxins, antibodies act in various ways. They may alter the surface membranes of the parasite and permit the contents of the cells to ooze through. They may cause the locomotive organs of the parasite to clump together.

In other cases, the antibody may react with soluble materials contained in the parasite and may form precipitates. As all this goes on, an immunity to the invading organism is built up. This acquired immunity is increased as reinfection takes place. That is why, for example, nobody ever has a second attack of typhoid fever.

Defensive role of the white blood cells

A powerful defense against the invasion of parasitic organisms is provided by the hordes of leucocytes, or white blood cells. These attack the small organisms that invade the blood stream and devour them. The white blood cells are mobilized against the invading parasites as inflammation sets in. The more intense the inflammation, the more white blood cells are called into action. Larger parasites, which cannot be devoured by the leucocytes, are often walled off by fibro-blasts. These are primitive cells, which give rise to connective tissue.

Ectoparasites — true parasites that live on the host

There are plant and animal ectoparasites. A good example of a plant that fastens itself to a host and derives its nourishment from the latter's tissues is the dodder. This plant has threadlike, twining stems; its leaves are reduced to tiny scales. The seeds germinate late in the spring. The seedling twines itself around the first living plant that it touches, and it develops a series of adventitious roots, modified into suckers; they are called haustoria. (See Index, under Adventitious roots.)

The haustoria exude, through their thin epidermis, a solvent substance that dissolves the outer cells of the host. The haustoria then make their way to the plant tissue called the phloem, and thereafter the dodder obtains its nourishment from the phloem. The part of the dodder below the point of attachment dies off, and the plant is no longer connected to the ground. The growing point of the dodder finds a new base of attachment upon the host, and new haustoria arise; they make their way

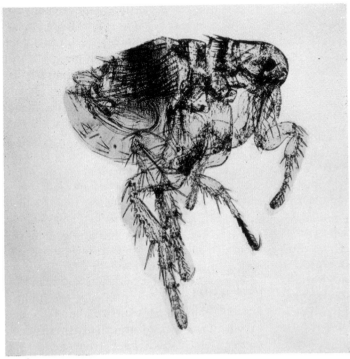

Bausch & Lomb Optical Co

Dog flea *(Ctenocephalus canis)*. **The flea is a cosmopolitan species, infesting dogs, cats, rats and humans and capable of transmitting disease.**

Below: San José scale on Ben Davis apples. This scale insect is a serious pest of fruit trees; it collects in vast numbers on fruit and on bark.

into the tissues of the host in their turn. In this way a skein of dodder "threads" is formed. Yellow or white flowers develop in dense clusters late in the season from these threads.

Special adaptations
of animal parasites

Animal parasites have various adaptations for attaching themselves to the skin of the host or for penetrating it. Ticks fix themselves firmly to the host by inserting the mouth parts into the skin. The common fever tick (*Boophilus annulatus*), which causes tick fever of cattle, remains on the host during most of its life. Adult females that have been swollen with the blood sucked from the host drop off to the ground and lay eggs. The eggs hatch into larvae. These attach themselves to cattle and suck the blood of the host. They change to nymphs (an intermediate stage between the larva and the mature insect) and the nymphs change to sexually mature ticks. Mating takes place; the females become swollen with the blood of the victim and drop to the ground. The cycle is then repeated. If a female has developed on an animal infected with tick fever, the larvae hatching from the eggs of that female will transmit the disease to the cattle to which they attach themselves.

Life cycle of
mussel parasites

The larvae of certain mussels attach themselves to the epidermis of fish, by closing their valves, or shells; they often fasten their hold on the host by means of hooks or spines. The epidermis of the host grows over each mussel parasite and ultimately covers it completely. The enclosed larva is then transformed into a young mussel, which breaks out of its enclosure and falls to the bottom of the sea, where it develops into an adult. This life cycle takes place only if the mussel larva attaches itself to certain specific species of fish. If they become fixed to the wrong species, they are sloughed off from their hosts before the all-important change from larva to young mussel can take place.

Many of the scale insects, belonging to the family Coccidae, fix themselves to plants and feed on their juices. One of the most destructive of these insects is the San José scale, which attacks many fruit, shade and forest trees. Each of these insects is covered with a small, waxy scale. The male develops an elongated scale and can fly; it has no mouth parts for taking food and it dies soon after it becomes full-grown. It is the female that is the parasite. It has no legs or wings and hence cannot move about, once its scale, which is circular, has been formed. It has a long slender beak, which it forces through the outer bark into the tender layers beneath and it sucks up nourishment from the tree.

Development of
scale insects

In the spring, the females produce young, which crawl out from under the mother's scale and go in search of a host. After a young scale finds a host, it settles down on the bark and begins to form the waxy covering over its body. It becomes full grown in about forty-five days.

The female of a certain fly called *Cordylobia anthropophaga,* generally lays its eggs in sand. When the larvae are hatched, they burrow into the skin of man and animals. After reaching a certain stage of development, the larva drops off to the ground, where it enters the pupal stage. (See Index, under Pupae.)

Infected hosts may
acquire immunity

An infected host often reacts vigorously to the ectoparasite. The antigens injected by the ectoparasite trigger the formation of antibodies, and these may bring about acquired immunity. When a fish acquires immunity to the mussel larvae that attack it, it causes the larvae to be cast off from the epidermis before they can attain the required growth. The larvae cannot complete their life cycle and they die. When animals acquire immunity to the larvae of *Cordyloba anthropophaga,* the antibodies they produce attack the guts of the larvae, producing death.

DISEASES OF GARDEN PLANTS

Methods of Prevention and Eradication

EVERY year plant diseases take a heavy toll of cultivated and wild plants alike. Various agents are responsible for these diseases. Bacteria cause blight, wilt, rot and canker in many fruits and vegetables. Fungi produce rusts and smuts in cereals, downy mildew on grapes and other diseases. Flowering parasites such as mistletoe attach themselves to healthy plants and sap their strength. Many kinds of insects attack roots, flowers and leaves. Insects lay eggs in living plant tissue; the larvae that hatch from these eggs are often very harmful to the plant. The mysterious viruses (see Index) are responsible for a great deal of the damage that is done to fruits and vegetables.

The damage from diseases may vary from the malformation or stunting of fruit or flowers to the poisoning of a whole crop of barley, for instance. The crop yield may be cut down to such an extent as to make its cultivation unprofitable, or a product may be rendered so distasteful that the entire crop is useless.

Much can be done to diminish the damage caused by the various plant enemies, but there is still a large element of chance involved over which man has no control. Long periods of rain, for example, will be followed by the rapid growth of plant parasites that thrive on abundant moisture. Winds will carry the spores of fungi from a diseased plant area to a healthy one. Insects and birds are also disease carriers. Even man himself, in transporting plant produce, is often the unwitting transmitter of plant maladies.

While there is no certain way of preventing the diseases that are a threat to healthy crops, some measures are, nevertheless, effective to a degree. Spraying or dusting of plants — applying liquid or powdered chemicals — is probably the most common method. These sprays, however, serve mainly as a protective armor and are of little use if the parasite has already penetrated to the interior of the plant.

Various types of seed treatment constitute another method of disease prevention. The seeds are soaked in a solution designed to kill the spores of the fungi or bacteria that cling to the seed coat.

Removal of the infected plants is still another method of control. Where fungi live alternately on two kinds of plants, called the host plants, it is sometimes sufficient to remove only one type of host plant and thus interrupt the life cycle of the parasitic guest.

Breeding plants that are resistant to disease is an effective preventive measure, though not a foolproof one. Just as individual members of any species differ from one another, so do the individual fungi of any group. Hence a plant that is resistant to one variety of the species may not be resistant to another.

Destruction or control of insect carriers is a highly effective method of checking virus diseases. Plant quarantine is sometimes employed by federal and state governments to prevent the transportation of diseased plants from affected to healthy areas.

Some of the methods we have discussed are simple enough to be applied by even the window-box variety of gardener. No grower, however small or large his planting area, should be without the means for combatting plant diseases at a moment's notice if he hopes to obtain good crops. In spray mixtures, for instance, the owner of a small garden will find it more advantageous to use the prepared mixtures that are put on the market as fungicides than to attempt to mix them himself. Large growers, however, will find it cheaper to buy the ingredients and mix them themselves so that the ideal mixture for what is needed is obtained.

Modern methods of fighting disease

Methods of successfully combatting plant diseases on a small or large scale include sanitation, protection, seed and soil treatment and the breeding of resistant strains. Sanitation refers primarily to the removal and destruction of infected plants so that they will not be able to transmit the disease to healthy ones. In the case of diseases transmitted by contact, sanitation also involves the fumigation or disinfecting of all articles, such as bins, tools and machinery, that might have been in contact with infected plants.

Seed and soil treatment is also a form of sanitation. Sometimes a disease may be carried in a seed. This can be avoided by treating seeds with formaldehyde gas or in other ways. Other diseases persist in the soil over the winter. Soils may be treated with chemicals, such as sulfur, which destroy the spores in the ground. Sometimes, in the case of a disease of a particular plant, the fungus in the soil can be killed by planting a different crop there for a few years, thus starving the fungus. This is crop rotation.

Protection of the growing plants is perhaps the most widespread and effective method of fighting plant disease. It involves the spraying of plants with fungicidal and germicidal agents * so that fungi or bacteria attacking the plant will be destroyed on contact.

The first successful fungicide was discovered accidentally. In the nineteenth century French peasants in the great wine-producing region of Medoc in the Bordeaux River valley used to spray their grapes with a mixture of copper sulfate and lime in order to discourage pilferers. It was

discovered that the vines sprayed in this way were far less susceptible to the attacks of the fungus called *mildiou,* which caused great losses to the vinegrowers of the area. In 1885, P. M. A. Millardet and U. Gayon reported that the copper sulfate and lime mixture was an effective fungicide. The mixture, called *bouille bordelaise,* or Bordeaux mixture, is still used effectively in treating diseases of grapes, potatoes and tomatoes.

Since the invention of Bordeaux mixture, many new fungicides have been developed. Most of them are compounds of the element mercury. They are often used together with Bordeaux mixture.

The most promising method of defeating plant diseases is the development of new plant varieties which are resistant to diseases. There has been a great deal of study and experimentation in this field; many of the plants that are grown today in various parts of the world are products of these experiments.

The diseases of potatoes

Since potatoes constitute such an important article of the diet, it is especially important to control potato diseases. Those attacking the Irish potato fall into three classifications: fungus diseases, virus diseases and nonparasitic diseases.

Fungus diseases, such as late blight, are the most damaging to potato crops. The late blight fungus, known as *Phytophthora infestans,* causes a downy, whitish mildew or flour-like spots, usually on the lower surface of the leaves. From the tops, it is often carried beneath the ground by the rain to infect the tubers. Late blight usually appears soon after the young plants break through the ground: that is, as soon as favorable conditions of moisture and temperature are present. It may cause the young plant to die within as little as two weeks. The late blight fungus usually grows best in the tuber at temperatures of from 36 to 40 degrees. At such temperatures, it develops a dry rot; at higher ones, a soft rot. Humid conditions favor the onset of this disease.

---

* The word fungicide is derived from the two Latin words *fungus,* a microscopic plant, and *caedere,* "to kill." A fungicide is a fungus-killer; a germicide, a germ-killer.

Late blight may be controlled through proper measures of sanitation, protection and development of resistant strains. Inspection and removal of infected potatoes are required. Waste potatoes are usually treated with a solution of trichlorophenoxyacetate to prevent their sprouting. A weekly spraying with Bordeaux mixture discourages infection, although this fungicide is slightly toxic to the foliage. For this reason, milder fungicides such as Dithane and Parzate are used in the early part of the season. Later, when the danger is greatest, Bordeaux mixture is employed. Researchers have bred strains of resistant tubers, which are immune to several different races of the fungus.

Common scab —
a damaging disease

The fungus *Streptomyces scabies* is responsible for common scab, a disease that is quite widespread in the United States. Although the common scab does not ruin the whole potato, it does make it unsightly, lower in grade and less productive. The disease is characterized by raised spots, or lesions, of rough, corky tissue on the tuber. The lesions do not penetrate to the interior of the tuber, but they necessitate extra peeling before the tuber is ready for the dinner table.

The fungus may be harbored in an apparently uninfected tuber. When the plant sprouts in the spring, the fungus is well intrenched. To avoid this, farmers submit all tubers cut for seed to careful inspection. The tubers are then treated with a chemical agent, usually formaldehyde gas, to be sure the fungus is dead.

Treating the seeds does not afford complete protection against the fungus, for it may persist over the winter in the soil. The soil in which an infected crop has been grown is also usually treated with chemicals to destroy the fungus. Scab fungus cannot live in soil whose pH * is less than 5.2 or more than 7.0. The farmer can effectively control fungi, therefore, by lowering or raising the pH of the soil in which they reside.

* See Index for an explanation of pH.

The breeding of scab-resistant strains has helped bring the disease under control in recent years. Certain European varieties, such as Jubel, Hindenburg, Ostragis and Arnica potatoes, resist the disease effectively. Although they do not grow well in this hemisphere, they have served as parent strains for the breeding of scab-resistant varieties which will flourish in this climate. Some of the New World varieties are the Cayuga, Menominee, Ontario and Seneca potatoes.

Verticillium wilt caused by the *Verticillium albo-atrum* fungus occurs most often in the cooler potato-growing areas since its growth is favored by slightly lower temperatures. It is characterized by drooping leaves and discoloration of the stems, roots and tubers. This fungus does not destroy the tissue of the plant at all. It attacks the sap-conducting vessels of the circulatory system. As a result the resistance of the plant against other diseases is seriously weakened. The fungus is harbored in the soils. There are no known chemicals that will destroy the pest. The accepted procedure today is to starve it by rotating crops in areas where the fungus is found. Rotation must cover a three-to-four-year period at least because the fungus has been known to persist for as long as two years.

The potato disease called blackleg, due to the bacterium *Erwinia atroseptica,* produces inky-black lesions at the base of the stems, dwarfing of the plant tops, rigidity in the leaves and soft rot in the tubers. Blackleg is favored by wet weather and will occur most often in areas of fog, extended periods of rain or heavy dew.

A new
disease appears

A rather recently discovered disease that attacks the potato is ring rot, caused by a bacterium, *Corynebacterium sepedonicum.* The disease was first discovered in Canada in 1931, appeared in Maine in 1932 and covered the North American continent by the year 1940.

Ring rot causes yellowing of the leaves; the margins become brown. A soft rot develops in the ring of the sap-conducting vessels just below the skin of the tubers. The disease is not harbored in the soil, but is carried entirely by contact with bins, machinery and other farm apparatus that have already been in contact with infected tubers. *Corynebacterium sepedonicum* is destroyed by removing all infected tubers, washing down bins with a copper sulfate solution and applying formaldehyde gas to machinery. Several varieties of potatoes are resistant to this bacterium.

Potatoes also suffer from a number of virus diseases. The viruses that attack potatoes are divided into two classes: the yellows and the mosaics. The yellows cause such diseases as leaf-roll, witches'-broom and aster yellow. Mosaics include latent mosaic and leaf-banding disease.

The diseases
of tomatoes

Diseases of tomatoes include wilts, leaf spots, fruit spots or rots. Fusarium wilt caused by the *Fusarium lycopersici* fungus is one of the more serious of these diseases. The fungus enters through the roots of the plant and passes into the vessels of the stem. There it produces a toxin which causes the leaves to yellow and die. The fungus then works its way up the plant. Other tomato wilts are caused by soil-borne bacteria and fungi.

Leaf spots may attack the leaf or stem, or both; sometimes they also attack the fruit. Fruit spots attack only the fruit. A leaf spot that often infects tomatoes is caused by the fungus *Septoria lycopersici*. It causes water spots on the leaves, which may be entirely stripped from the plant. Stems and blossoms may be affected also.

More damaging to the tomato is *Phytophthora infestans,* which causes late blight in both the potato and tomato. As in the potato, the fungus attacks both the foliage and the fruit.

The most serious of the fruit rots is caused by the fungus *Colletotrichum phomoides,* or anthracnose. This disease appears only in the ripened fruit; when the tomato ripens, dark depressed spots with concentric rings appear on it. The yield is smaller, as these spots must be removed.

The virus diseases of tomatoes produce symptoms very similar to those of virus diseases attacking potatoes. The most common and most destructive virus disease of tomatoes is tomato mosaic, sometimes known as "curly top." It causes curling, twisting and malformation of the leaves and a streaking of the stem. It is doubly dangerous because it usually brings with it *Virus X* which causes the potato disease called latent mosaic. Plants infected by both viruses develop a disease known as double-virus streak. In this, narrow brown streaks are produced on the leaves and stems of the plants.

A disease
without a germ

Tomatoes also suffer from a rather serious disease for which there is no causative organism. Blossom end rot, which takes a great toll of tomato plants, is caused by great variations in temperature and water supply. In this disease, when the plants are about half grown, brown discolorations appear on the fruit. Blossom-end rot is most likely to occur in plants that have grown rapidly because of high temperature or that have been rather heavily pruned. It is also found in plants that have been sprayed with nitrogenous fertilizers.

The most important defense the farmer has against tomato diseases is sanitation. The large majority of tomato diseases are conducted through the soil. It is most essential that the crop should be planted in clean soil. Crop rotation is effective. The breeding of resistant varieties may offer some help in the future.

Diseases of
the garden bean

Three serious and widespread diseases of garden beans are anthracnose, bean blight and the fungus *Rhizoctonia*. Anthracnose, caused by the fungus *Colletotrichum Lindemuthianum,* may be recognized by the presence on the pods of small brownish or purplish depressed spots. The disease is carried over the winter in the seed. Humid conditions favor the disease; it is far less prevalent in dry areas.

Anthracnose is the most common disease of raspberries. Black raspberries are most severely affected by the fungus. It produces reddish-brown, circular sunken spots on the cane of the plant.

The picture at the right shows the effect of strawberry leaf spot fungus. The leaves of the plants are covered with many small, light-colored spots.

In the picture below, we see cherry brown rot in its advanced stages. The fungus has already rotted the cherry and is now forming spores that will infect other plants.

All photos, Louis Pyenson

The brown rot of cherries is caused by a fungus that often attacks other stone fruits. It is related to the lettuce-drop fungus.

Photos, William G. Smith, Jr., Boyce Thompson Institute for Plant Research, Inc.

In some regions, bean blight, caused by *Xanthomonas phaseoli,* is more destructive than anthracnose. This disease is extremely difficult to control, since the bacterium hibernates in the seed. The use of clean seed from clean pods and the application of three sprays of Bordeaux mixture as for anthracnose will reduce the losses, although no effective remedy is known. Under certain weather conditions, the disease is very destructive, causing large dead brown areas on the leaves. The pods also frequently show spots not unlike those caused by anthracnose.

Another serious disease of the bean is caused by a soil fungus of the genus *Rhizoctonia.* It attacks the plant in the soil and causes the stem to rot at the surface of the ground, forming shriveled reddish brown spots, extensive dry rot in the roots and destruction of the small lateral roots.

Rhizoctonia also attacks sugar beets, carrots, cabbage, cauliflower, cotton, lettuce, potato, radish, sweet potato, pumpkin, watermelon and peas.

Beets are affected by several diseases, of which the leaf spot caused by the fungus *Cercospora beticola* is the most widely distributed and the most destructive. The leaf spots are at first small brown points with reddish purple borders. Later, when they reach a diameter of one eighth of an inch, the center becomes almost white with long, threadlike, many-celled spores. These are blown to other plants or fall to the ground, where they retain their powers of germination for months. In the diseased plants, the crown is more elongated vertically than is normal; the leaves blacken and dry up gradually from tip to base, standing more and more erect. Spraying with Bordeaux mixture controls the disease. Since the spores hibernate on the ground, early spraying is essential.

Lettuce is attacked by damping off caused by fungi of the genus *Rhizoctonia* and by *Pythium debaryanum.* It is also affected by special diseases of which the "downy mildew of lettuce" and "lettuce drop" are the best known. The downy mildew is caused by *Bremia lactucae* and is frequent when lettuce is grown in greenhouses, but it may occur in the fields in cool weather, particularly in the fall. The conidiophores (accessory spore-bearing organs) form a frostlike growth on the underside of the leaf; the affected areas turn paler and wilt. In the greenhouse, the disease is easily controlled by supplying abundant light and ventilation. "Drop" is a more serious disease caused by *Sclerotinia libertiana;* the same fungus attacks other greenhouse and garden plants, cucumbers, tobacco, many forced vegetables and bulbous plants. On lettuce, the disease first produces water-soaked areas on the stem and basal portions of the leaves. The surface of the leaf sometimes shows threads of the fungus on which appear minute black spherical bodies, the sclerotia. The disease spreads rapidly, the plants collapsing and melting to a shapeless mass. This fungus is one of the most serious greenhouse pests. It may be controlled by steam sterilization of the soil to a depth of two inches or more. While the plants are growing, the soil should be kept loose and dry. Another serious disease of lettuce in some regions is the leaf spot caused by *Marssonina panattoniana*.

When onions are grown in quantity, the crop is subject to various diseases. Mildew and smut occasion the greatest loss. The mildew, caused by *Peronospora destructor,* is also known as blight or mold. There have been serious outbreaks of the disease in America. The fungus is a relative of the organism that causes late blight of potatoes and mildew of lettuce. In the onion, the disease appears in June or July. When the plants are covered with dew, the leaves show a peculiar downy appearance of a purplish hue. This is caused by the conidiophores of the fungus. The diseased plants wilt rapidly; a field ravaged by this disorder looks as if it had been sprayed with boiling water. The young plants are less susceptible than the old. The ailment may be stopped if the crop is sprayed frequently with Bordeaux mixture after June 15th, when the disease begins to appear. Carbamate chemicals are effective also. Crop rotation is advisable; infected onion tops should also be destroyed.

Onion smut (*Urocystis cepulae*) is a serious disease. Seedlings are attacked after the first leaf is formed and dark spots soon appear on the leaves. These spots become longitudinal slits filled with black dust; the dust represents a mass of spores and resembles the spore mass of the smut of cereals. To control this disease, it is useful to treat the seed, as the fungus lives in the ground. Onions should be set out in clean soil. Crop rotation should be practiced; sulfur, lime and formaldehyde should be applied to the soil.

Disease is not limited to the herbaceous plants that are cultivated in the garden. The fruit trees and shrubs also have many enemies: insect pests, bacteria, viruses, fungus infections and so on. Today, commercial fruit farmers utilize sprays and dusts of various kinds in order to insure a marketable crop.

Small fruits, such as strawberry, raspberry, currant and blueberry, suffer from various forms of molds, galls and worms. The grape also has a great number of diseases; it was perhaps the first fruit for which spraying was systematically used.

Apples and pears are injured by scab, canker and rot, and also by many insect enemies. Plum is attacked by bladder, silver leaf and various other fungus ailments. Peach often suffers from the yellows and from little peach. Cherry is very susceptible to brown rot, caused by *Sclerotinia fructicola,* in which the fruit shrivels and turns brown, showing on the surface white tufts of mycelium. The fungus passes the winter in the fallen fruit and, the next year, produces a fresh growth of ascospores ready to infect the new crop.

Great progress has been made in the field of horticultural disease control. Constant study of plant disorders has resulted in the discovery of new and exceedingly effective fungicides and insecticides.

See also Vol. 10, p. 272: "Diseases of Plants."

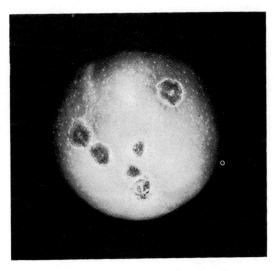

Scab, a disease of the apple, is caused by the fungus *Venturia inaequalis.*

N. Y. Zool. Soc.

Male of the greater kudu. This stately antelope stands almost five feet at the shoulder and its "cork-screw" horns may be over fifty inches long. The animal's range extends from Ethiopia to South Africa.

376

THE ANTELOPE FAMILY

A Varied Group of Hoofed Mammals

THE antelopes are members of the vegetation-eating, hoofed group of mammals known as the Bovidae. These also include the cattle, sheep and goats; as a matter of fact, the name "antelope" is generally given to any species of the family that is obviously not a sheep, or goat or one of the animals of the cattle group. For the most part, the antelopes are fleet creatures with highly developed senses of sight, smell and hearing; they are generally slender and graceful and yet surprisingly sturdy. Many of them, such as bongos and kudus, have beautiful markings. In size these animals range from the dainty little royal antelope, which is only about ten inches in height, to the giant eland, measuring nearly six feet at the shoulder.

Like all the members of the Bovidae family, the antelopes are ruminants, or cud-chewing animals. Like the rest of the Bovidae, too, they have unbranched, hollow horns which, sheathlike, fit over bony cores projecting from the top of the skull. These horns are permanently attached and thus differ from deer's antlers, which are shed annually. Antelope horns are of different sizes and shapes. Some are spiral or lyre-shaped; some curve backward; still others are short and straight.

The early ancestors of our modern antelopes were small, hornless creatures, which lived in Eurasia about 30,000,000 years ago. Their forelegs were much shorter than their hind limbs. The front teeth, or incisors, of the upper jaw were greatly reduced or entirely absent, while the canine teeth were only moderately developed. In the course of time, small and relatively straight horns appeared, the canines were lost, and high-crowned molar teeth were developed; these enabled the animals to grind more effectively the grasses which formed their staple food.

From these early grazers arose many kinds of antelopes, such as the oryx, the klipspringer and the gazelle. About 6,000,000 years ago there appeared the cowlike antelopes — the eland, the nilgai (of India), the gnu and the hartebeest. Our modern cattle developed from types such as these. During the period when ice covered much of the Northern Hemisphere, many of these Bovidae sought refuge in southern Asia and Africa. Today the true antelopes are restricted to those two continents. Of the animals closely related to them, only the bison, mountain sheep, mountain goat and musk ox established themselves in the New World.

There are a number of goatlike antelopes, such as the tahr, goral, serow, takin, Rocky Mountain goat and chamois. These animals, called goat antelopes, are mountain dwellers, extremely rugged and capable of living on the sparse vegetation that grows at high elevations.

The tahr somewhat resembles a goat, but it lacks a beard. It has long and smooth reddish or dark brown hair; in the male, the hair is manelike on the neck and fore part of the body. The tahr dwells in the Himalayas and in the mountains of southern India and Arabia.

The goral is a native of the Himalayas and the mountainous regions of China and Korea. The serow is closely related to the goral; it is quite a bit larger, however, being about the size of a donkey. Its coat is long, coarse and shaggy, and black to reddish brown in color. Serows are to be found in herds in the eastern Himalayas and in the mountain ranges of Burma, Siam and the Malay Peninsula.

Takins are heavily built goat antelopes of secretive habits. They live in the bamboo- and brush-covered mountainous areas of the eastern Himalayas and western China. Keeping to dense thickets by day, they graze on the grassy slopes during the evening and early morning. Takins associate in pairs or in small troops. After mating in July or August, they descend to lower elevations for the winter months. The calves are born in March.

Musk oxen, kin of the takins, live in arctic North America. These ponderous animals have large heads, short legs and shaggy coats of wool with long, coarse guard hairs. Both sexes carry horns, which are enlarged at the base, slope downward on each side of the head behind the eye and then curve upward. Mosses and grasses, dug from under the snow in winter, form the musk oxen's principal fare. Like takins, musk oxen mate in July or August; the calves appear the following spring. When attacked by wolves, musk oxen generally form a ring enclosing the young; the adults stand together facing the foe.

The Rocky Mountain goat is another American goat antelope. It is a sturdy animal and an expert rock climber, spending its life upon treacherous slopes and rocky ledges high above timber line. Even winter's snows do not force it to lower wooded country. It lives in the Rockies and coastal ranges from central Washington and Idaho north to southern Alaska.

This mountain antelope has a woolly underfur and longer hairs of yellowish white. Both sexes have short beards and spikelike horns that curve back slightly. Mating occurs in November; one or two young are born in April or May. Mountain goats are occasionally preyed upon by wolves and bears; eagles sometimes attack the young. The mountain goat is a most courageous fighter.

The chamois is a small, nimble goat antelope inhabiting the rugged mountainous regions of southern Europe and Asia Minor. The structure of the hoofs is such that the animal can "grip" irregularities on rocky surfaces; hence it can travel with amazing speed along narrow ledges and down steep slopes. It can jump agilely from rock to rock and leap wide chasms.

The goatlike tahr, above, is found in the Himalayas.

N. Y. Zool. Soc.

Swiss National Travel Office

The nimble chamois dwells high above the timber line.

Small herds of chamois, composed of females and young males, graze on mountain vegetation above timber line during the summer; in winter they descend to lower altitudes. These herds are joined by adult males during the late-autumn breeding season. In the spring, each female separates from the herd to find a sheltered thicket, rocky retreat or narrow ledge where she gives birth to one or two young. The newborn can immediately rise on its legs and walk, but it does not follow the mother until it is a week or two old.

The chiru of the Tibetan highlands and the saiga of the west Asian steppes are small goat antelopes living in valleys or on open plains. Only males have horns. The chiru has swellings along the sides of its muzzle, while the saiga's nose is large and inflated. It is thought that the many hair-lined channels in the muzzle of these animals serve to filter dust particles from breathed-in air and warm the frosty air before it goes to the lungs.

The numerous true antelopes are divided into several groups. One group includes the handsome, large antelopes having spiral horns, such as the eland, bongo, kudu and bushbuck. A second group of varied antelopes is represented by the addax, oryx, hartebeest and gnu. Duikers are small African antelopes belonging to another division. The graceful gazelle and tiny klipspringer and dik-dik make up a fourth. Antelopes frequent all sorts of localities — open plains and deserts, hilly country, swamps and forests. Some species associate in large herds while others form pairs or lead an almost solitary existence.

The large African eland is similar to an ox in build; however, it is an agile, fleet-footed animal. It has a dewlap, or flap of skin under the neck; its long horns, though spirally twisted, are straight. Elands live in partially wooded, partially open country. To get at leaves and twigs, they sometimes break off small trees by wedging the trunk between the horns and twisting the head. The secretive bongo frequents dense bamboo forests, where it relies on its keen hearing to escape enemies. It is colored a chestnut red, striped with white along the sides of its body. The handsome greater kudu is at home

N. Y. Zool. Soc.

Above: Chinese takin. The takin is a heavily built antelope, found in the mountainous areas of the eastern Himalayas and western China.

Amer. Mus. of Nat. Hist.

Amer. Mus. of Nat. Hist.

Above: the giant eland. Its dewlap (flap of skin under the neck) and its long, spirally curved horns are characteristic of this powerful beast.

Chicago Nat. Hist. Mus.

The bushbuck, shown above, is a medium-sized antelope, with powerful horns and a short, bushy tail. The animal is a bold and strong fighter.

Left: two Rocky Mountain goats in their habitat. These fleet animals are perfectly at home amid the steep slopes and bare ledges of the Rockies.

N. Y. Zool. Soc.

San Diego Zoo

Cape duiker, or duikerbok. Duikers usually associate in pairs. The males and most females have short horns that resemble spikes.

The beisa oryx, of northeastern Africa. When it is seen from the side, this animal presents a silhouette much like that traditionally assigned to the unicorn.

The musk ox, shown below, dwells in the arctic regions of North America. The animal has a large head, short legs and a shaggy coat of wool. It feeds chiefly on mosses and grasses, dug from under the snow.

Amer. Mus. of Nat. Hist.

in dense thickets of the hill country or in tangled scrub and thorn of lower levels. When pursued, these antelopes flee to the highest and roughest part of their range. The kudu may be found from Cape Province in the south to Ethiopia and Somaliland in east Africa. The lesser kudu, short-horned and standing forty inches at the shoulder, is confined to east Africa, from Ethiopia to Tanganyika. It possesses white patches on its throat and chest and lacks the long throat mane that is displayed by the greater kudu.

In the same antelope subfamily are included the bushbucks, or harnessed antelopes. The bushbucks possess spirally twisted horns, and their bodies are crossed by white stripes. The varieties of this antelope consist of the bushbuck proper, the nyala and the sitatunga, or marshbuck. The bushbuck proper is an antelope of me-dium size, standing thirty inches at the shoulder. It is mostly chestnut-colored and has a short, bushy tail. While the range of the species is wide, including all of Africa south of the Sahara, individual bushbucks seldom wander more than a few square miles, and the animal is solitary by nature. The bushbuck is a courageous and strong fighter and has been known to slay a leopard with its powerful horns.

The marshbuck is a large antelope, measuring forty-five inches at the shoulder. Despite its size, the creature moves with ease through the tangled undergrowth of its marshy haunt. It sometimes stands submerged up to its head and horns. The nyala, somewhat larger than the bushbuck, has a coat of smoky grey and wears a long mane. The horns are remarkable for their handsome shape. The nyala frequents the forest regions of southeast Africa.

N. Y. Zool. Soc.

The odd-looking saiga. This small antelope, which is only thirty inches high, is found in the western steppes.

An oriental relative of the African harnessed antelopes is the nilgai of India. This large and rather awkward creature, with stubby crescent-shaped horns, once roamed over a far wider area than it does now. The bull antelope is slate blue in color; that accounts for the name nilgai, which means "blue bull." It may measure up to fifty-four inches in height.

The little chousingha, or four-horned antelope, is another native of India, belonging to the subfamily we have been discussing. This species is unique among antelopes, for the males wear two pairs of short, spiky horns, one behind the other. The fore pair is about two and a half inches long, while the hind pair may measure five inches. The body of the chousingha is about forty inches in length, and the animal stands twenty-five inches at the shoulder. Its coat is red-brown with white-streaked under parts. It frequents the wooded and hilly country of peninsular India, keeping away from the jungles.

Another important antelope subfamily includes the addax, roan and sable antelopes and the oryx. These are large-sized beasts, with long, tufted tails, neck manes and striped faces. The horns, borne by both sexes, are either spirally twisted, curved backward or long and straight. They are all native to Africa except the Arabian oryx, which ranges from southern Arabia to Iraq.

The addax dwells in the desert regions of northern Africa. Its coat is a yellowish white, and it has a patch of black hair on the forehead. The horns of both sexes are ringed, and they form open spirals.

The roan antelope is found south of the Sahara as far as the Orange River. It reaches a shoulder height of five feet or more. The coat is red-brown, and the belly is white. The curving horns of the roan range up to thirty-seven inches in length. Another species, the blue antelope of South Africa, was hunted to extermination by the Boers in the early days of the colony. This antelope, resembling the roan antelope, was a blue-gray in color.

The sable antelope of central and east Africa is a handsome black animal and has

splendid scimitar-shaped horns. At shoulder height the male sable stands about fifty-four inches. The coat is black-brown to glossy black, with a white understreak.

The pale Arabian oryx has long, straight horns, while the white oryx of North Africa bears horns that curve backward in a wide sweep. The gemsbok is another oryx, dwelling in the deserts of southwestern Africa. It is a big antelope with a pair

N. Y. Zool. Soc.

Young nilgai, or blue bull. It is an inhabitant of India.

of straight, slender horns measuring between three and four feet in length. The neck, back and flanks are a pale red-gray, while the belly is white and the head white, marked with black. The tufted tail is black.

The horns of the beisa oryx, of northeastern Africa, are very formidable. Though in medieval times it was the long tusk of the narwhal that was displayed as the horn of the mythical unicorn, it is thought that a glimpse of the beisa may have inspired a poetic conception of the appearance of that one-horned creature.

The body shape of the beisa calls to mind the form traditionally assigned to the unicorn. Though the beisa obviously has two horns, the animal, if seen from the side, would present a silhouette suggesting a beast with one horn jutting from the middle of its forehead.

Antelopes that look like horses

The big antelopes that we have just described — the addax, roan, sable and oryxes — look a great deal like thoroughbred horses. Certainly their gait when walking, running or galloping is quite horselike. The magnificent, high-stepping sable antelope, with its mane flying, neck arched and chin drawn down, resembles nothing so much as a spirited coal-black steed. The hoofs of the sable, roan and oryxes are large, supporting the powerful and well formed legs. The addax's hoofs are especially large, an adaptation that makes it easy for the animal to walk and run on the sand of the desert.

Since all of these antelopes live in fairly open grasslands and deserts, they must depend on fleetness of foot to escape from such predators as lions and the wild, vicious hunting dogs. The roan antelope can attain a speed of thirty to thirty-five miles an hour, and probably the others of this group can run as swiftly. If flight fails and the antelopes are cornered or wounded, they will fight courageously. With quick and powerful thrusts of their sharply pointed horns they can often hold their foes at bay. African natives occasionally come upon the remains of sable and lion lying together, the lion impaled upon the horns of the sable.

Antelopes gather together in small herds

Living in open country as they do, these antelopes congregate in small herds, probably because a number of individuals have a better chance of survival than a single individual or a pair. For one thing, if one or several members of a scattered herd find food, the others of the group will be attracted to the food supply by the feeding

actions of the original discoverers. Approaching predators have a good chance of being spotted by several individuals of the group; their reactions will give warning to the rest. If attack comes, the herd's concerted defensive action may discourage the enemy.

Herds are usually led by large bull antelopes

Usually herds number less than a hundred members; the smallest groups are formed by those species living in deserts where food is not abundant. Often one or several large bulls assume herd leadership. A herd of sable antelopes, containing ten to eighty individuals, for example, is led by a single adult bull, which has to defend its position against all adult males vying for leadership. The unsuccessful males remain separate from the main herd and form a bachelor band of their own. During the breeding season, the herds of some antelope species break up, with males and females forming pairs. After their birth, the young calves and their mothers often remain away from the group for a month or two.

Antelopes are adapted for living in waterless country

The grassland and desert homes of these antelopes are devoid of heavy rainfall during the greater part of the year; consequently, the animals are adapted to survive without drinking water. Occasionally water can be had by lapping up dew from grasses or trickles of rain water from foliage and tree bark. But for the most part the antelopes get the water they need from the grasses and succulent plants upon which they feed. Possibly, the animals also burn up some of their stored fat during the dry season. This fat on being consumed releases not only energy but water, too, which makes up for at least part of the body's water needs. It is truly remarkable that such large herbivores as these can get along without drinking water for great lengths of time. When the water holes are filled, however, these antelopes drink regularly.

Frank Stevens from National Audubon Society

Jackson's hartebeest, of Kenya and Uganda, in Africa. The male shown here kneels in the position of combat.

The mountain reedbuck (*Redunca fulvorufula*) is very shy.

Right: the sing-sing (*Kobus defassa unctuosus*) is a waterbuck found in West Africa.

All photos, except top left, N. Y. Zool. Soc.

FOUR
VARIETIES OF
ANTELOPES

The springbok (*Antidorcas marsupialis*) is a small antelope that lives in South Africa. Both male and female have lyrate (lyre-shaped) horns.

Allied with the oryx are the waterbuck, lechwe, kob, reedbuck and Vaal rhebok. The waterbuck is large, while the others are medium-sized. Only the males have horns, which are moderately slender, ridged and bent in gentle curves. These African antelopes are never found far from water. The lechwe often enters swamps up to its belly to fed on aquatic vegetation, and it swims strongly when pursued. Water-bucks and kobs graze on partially wooded grasslands but take to water if danger threatens. The reedbuck frequents dense reed beds but, when frightened, seeks refuge in a dry, bushy area. The Vaal rhebok prefers hilly country, keeping to open mountain slopes by day and coming down onto the plains to graze and drink at night. Vaal rheboks live in family groups of several females and fawns led by an adult male. Waterbuck males also form harems of females. After the mating season, during which the males fight for mates, several waterbuck families commonly dwell together. Male and female kobs mix in herds, except at breeding time, when each male gathers several females into a temporary harem. Reedbucks are solitary antelopes; only a female and her fawn are bound together for any length of time.

The hartebeests and gnus, or wildebeests, are antelopes built for speed and endurance. They are large animals, standing high at the shoulders; the back slopes downward to the haunches. Hartebeests have long, slender muzzles; the gnu's muzzle is long and broad, and tufts of hair stud the face. A stiff mane fringes the gnu's neck; the tail is long and bushy, similar to a horse's tail. Both male and female hartebeests and gnus carry horns. Those of the hartebeests are ringed and form a **V** or **U** when seen from in front; the upper end of each horn bends rearward. The horns of the gnu are smooth, have a widened base and curve first downward and then upward.

These African antelopes live in large herds on open grasslands. Some species of hartebeests frequent arid regions where they can live almost entirely without water; the gnus depend upon sources of water and migrate considerable distances to find them.

N. Y. Zool. Soc.

The spirited sable antelope arches its neck when it runs.

Relying on speed to escape enemies, the hartebeest can run for some distance at forty miles an hour; the gnu is even faster, making fifty miles an hour. When excited or stampeded, the gnu prances about and leaps in the air; then it tosses its head, kicks its heels and swishes its tail as it breaks into a fast trot, which soon changes to a gallop.

Duikers, or duikerboks, form a subfamily of small or medium-sized antelopes living in Africa south of the Sahara Desert. Some duikers inhabit the tropical rain forests; the others are at home in bushy country, where they take cover in the dense thickets. The horns of the male duikers are short spikes; the females of most species also have horns. Duikers apparently associate in pairs; mating takes place in the month of October, and one or two young are born the following April or May.

Another subfamily includes the oribi, klipspringer, steinbok, grysbok, royal antelope, suni and dik-dik. These African antelopes are, for the most part, small, delicately built creatures standing under two feet; the oribi is a little larger, and the dik-dik, suni and royal antelope are rabbit size. The horns, usually worn only by males, are small and spikelike. Generally these antelopes associate in pairs or are solitary. The klipspringer inhabits rocky regions. It leaps agilely and scales steeply inclined rock faces; its narrow, high hoofs give it a firm grip on the smoothest surfaces. The royal antelope and suni are forest dwellers, while the dik-dik prefers bush country where it dashes through the undergrowth like a frightened hare. Steinboks and grysboks are found in brushy areas and open plains; the oribi lives in the foothills and plains, hiding in tall grasses and dense thickets. Oribi pairs mark their home territory by smearing the secretions of their facial glands on certain trees and bushes, which then serve as "signposts" to any other antelope that may happen to be in the area.

Related to the oribi group are the fleet-footed gazelles and their allies — the gerenuk, dibatag, impala, blackbuck and springbok. They are exceptionally attractive and graceful animals of medium size; their horns are beautifully curved. Most species associate in herds or harems led by an adult male. Many live on semiarid plains or deserts, though the impala prefers openly wooded, grassy country. Gazelles are found widely distributed in Africa and southern Asia. The handsome blackbuck, which has long horns spirally curved like a corkscrew, inhabits India. The gerenuk is a slender gazelle with an exceedingly long neck and elongated legs; it often stands on its hind legs to browse on leaves. The dibatag is another long-legged and long-necked gazelle. Springboks and impalas are noted for their ability to make long and high leaps. The springbok has on its back a long gland lined with white hairs. Whenever it is alarmed, the animal erects these hairs.

See also Vol. 10, p. 275: "Mammals."

The brindled gnu, or blue wildebeest, with its young.

N. Y. Zool. Soc.

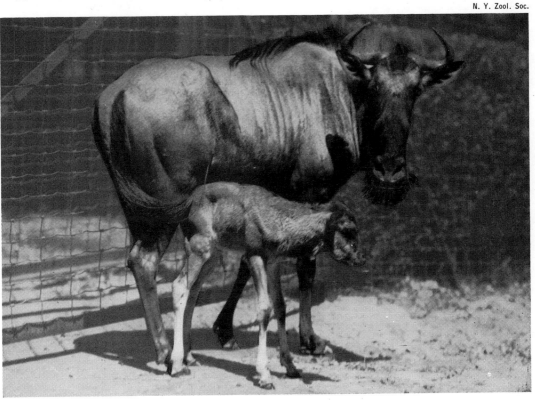

GLASS HOUSES

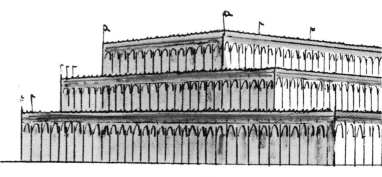

Photo, Alexander von Steiger

THERE was a time when the only glass in a building was to be found in its windows and mirrors. Today architectural glass serves many other purposes. It is used as building material, replacing brick, metal and wood; as decorated or colored tile for walls; and as solar screens regulating the kind and amount of radiation entering the house from the outside and passing back from the house. For these purposes, glass can be made in a great variety of shapes and colors; it can be made proof against breaking, sound, heat, cold or glare. Glass has other advantages; it does not easily corrode, is readily cleaned and requires no painting.

Earlier uses of architectural glass

Even in the past, window glass, or glass panes, met various needs. For example, the ancient Romans, about 2,000 years ago, constructed glass hothouses to protect cultivated flowers that needed continuous warmth to grow. Glass was used because it is able to admit the sun's heat without letting much of it escape back to the outside.

Much of the early glass was not completely transparent. It transmitted diffused daylight into a structure, but it was not easy for a person to see through the glass.

New Windows on the World

Because of the crude methods of manufacture, large panes of clear glass could not readily be made. As a result, sheet glass was rare and costly; only a few small windows of glass existed in a limited number of wealthy homes.

In the Middle Ages, the Gothic cathedrals of Western Europe were built with many tall, arched windows along the walls. Here architectural glass came into its own. Many of these beautiful old churches were literally cages of stone ribs and glass. Stained-glass windows, with their many colored pieces joined together to depict scenes from Scripture, gave the interiors of these cathedrals a mystical, roseate light. Glass was used increasingly for private dwellings, too.

The pictures on these two pages show some effective uses of glass. At the extreme left is the beautiful Lever House (New York City), with its glass facade. Next to it is the Crystal Palace, originally erected in Hyde Park, London, for the 1851 exhibition and later torn down and re-erected at Sydenham, near London. At the extreme right, above, is a stained glass window from Chartres Cathedral, in France. Below is shown the effective use of Thermopane picture windows in a modern suburban home.

Libbey-Owens-Ford Glass Co.

Libbey-Owens-Ford Glass Co.

The roof over-hang or extension in this house is aimed to control the heat of the sun, as is shown in the diagram at the left. In the summer, the over-hang keeps the sun's rays out, but they are admitted in the winter.

In the early years of the nineteenth century, the erection of large factory buildings created new opportunities for glassmakers. To provide adequate light for workers, the size and number of windows were increased. These early plants, however, were still a far cry from the glass buildings of today.

Probably the first large building to be constructed almost entirely of glass was the Crystal Palace, built for the Great Exhibition at London in 1851. It was designed by Joseph Paxton (1801-65), whose profession was raising flowers in glass greenhouses of his own construction. The Crystal Palace was a skeleton of iron girders supporting large panes of glass in wood frames; it covered about eighteen acres. This structure was the marvel of its day and certainly far ahead of its time.*

By the twentieth century, glassmakers had improved the quality of their product and had succeeded in mass-producing it, so that its use on a large scale became possible. As new ideas of architecture were conceived and accepted, glass came to be used extensively in buildings of improved design and great beauty.

Probably the most famous early example of the new architecture of glass was the Bauhaus (House of Architecture) of Walter Gropius (1883-), erected in 1926 at Dessau, Germany. It was intended to house a school of advanced art and architecture, whose principles were applied in the construction of the building. The glass walls of the Bauhaus were margined along top and bottom by bands of stucco, all of which gave an effect of floating transparency. The floors, as seen through the glass, enhanced this impression.

The "International Style" of the Bauhaus was not popular at first, but gradually won acceptance. The American architect Frank Lloyd Wright (1869-1959) helped spread the idea of glass for construction. Two outstanding examples of this type of architecture are the Lever Brothers Building and the United Nations Secretariat Building, both in New York City.

* It finally burned down in 1936.

Modern forms of glass construction

Glass is particularly adapted to the demands of modern architectural design, which stresses a "fluid" and continuous quality in construction. It seeks to create a feeling of unity, with the inner and outer planes of a structure not only in harmony with each other, but with the surroundings of the building. Glass is admirably suited to carry out this ideal of harmonious pattern, whether in the form of blocks, tiles, panels, sheets or tubing. It creates a sense of airy spaciousness, unbroken by the conventional design of window and wall. Broad, flat, gleaming expanses of glass, often enlivened with brilliant colors and bold decorations, conform to the modern taste for simple lines combined with a sense of luxury and gaiety. It has caused the building to be "opened up" and to communicate with the outdoors.

The increased use of glass in a house may simply take the form of larger windows. A so-called daylight wall is in reality a tremendous window, extending from floor to ceiling, or nearly so, and from one corner of a building or room to another. It permits a clear, unobstructed view of the outer world, besides providing ample daylight. The most up-to-date homes of this kind are located and built so as to take full advantage of sunlight throughout the year, to illuminate and warm the interior of the house.

Not only does glass admit solar energy, but it can be made to insulate a house against excessive heat and cold. The glass may be tinted. It may have a double construction: that is, it may actually consist of two panes of glass, held together by a glass or metal bond, and having a uniform and narrow space, partly filled with dry air, between the two panes. The principle is that of the vacuum bottle (see Index). The inner glass is insulated from the outer one, so that it is not affected by extremes of temperature. This protects the home against rapid heating or cooling as a result of temperature changes outside. It prevents the formation of "frost" on the inside of window panes in winter. Double-glass windows also have a soundproofing effect.

Glass panels are also replacing conventional walls between rooms. This not only increases the feeling of spaciousness in even a small house, but allows greater penetration of light to all parts of the home.

The use of glass for the walls of buildings is perhaps the most daring innovation of all. It has become possible because in modern structures, such as office buildings and skyscrapers, the task of supporting the building is no longer allotted to the walls, but to the framework of steel. The walls, freed of this burden of support, can be lightly and quickly built, as little more than a "skin" (curtain wall) on the steel skeleton. A variety of new materials are beginning to be used for modern walls — among them, glass.

The spandrels, or spaces between an upper and a lower row of windows, may be of clear or opaque glass, white or colored.

In the John J. Kane Hospital, near Pittsburgh, glass blocks are used alternately with plate glass in a corridor wall. Below is a close-up of glass blocks used in such construction. They can be assembled by mortaring.

Pittsburgh Corning Corp.

Glass used for such purposes must be very strong and resistant to sudden shocks, such as those caused by solid objects or by the impact of wind buffeting. For example, the *Vitrolux* spandrel glass produced by the Libbey-Owens-Ford Glass Company is only one-fourth of an inch thick, yet it is said to withstand pressures of up to 12,000 pounds per square inch. Moreover, it endures a temperature of as much as 300° F. and will not easily tarnish, crack or warp.

Walls of glass, because of their very thinness, cause more interior space to be available, besides making rooms brighter. Such walls, in combination with broad windows of nondistorting plate glass, give a startlingly mirrorlike appearance to a building. One may wonder whether all this lavish use of glass will not result in loss of privacy. But glass can be processed so as to be as impenetrable to light as the thickest stone wall. Better still, it may be made translucent, admitting daylight effectively, but giving complete privacy to the persons living or working in a building. Compressed glass tubing is often used to make translucent walls covering large areas.

Glass walls can be made to carry many kinds of decorative designs. A kind of photographic process actually prints such designs on glass, so that it may look like anything but glass, perhaps like the finest marble. Walls may be composed of or lined with ceramic glass tiles of unusual beauty. Bits of colored glass may also be put together in mosaics of striking appearance and color.

The most versatile glass-building material is the glass block. Such blocks may be assembled by mortaring to form a wall or partition quickly and easily. Individual blocks have a compressive strength of 400 to 600 pounds per square inch; however, the over-all strength of a glass-block wall is far less than this, and so it is not used to bear loads. A wall of glass blocks can withstand winds of well over 100 miles per hour. Many kinds of blocks are used for different construction and lighting requirements in buildings.

Basic types of glass blocks

There are three basic types of glass block. The ordinary kind has little control of the light passing through it and is used wherever there is not much direct sunlight (such as in a northern exposure), and where exact distribution of light is not necessary. This glass block may be transparent (for viewing) or semitransparent, with raised or wavy patterns on the outer or inner faces, or on both.

The diffusing type of glass block scatters the sunlight passing through it by means of patterns and faces on the surfaces and inside the block as well; it distributes sunlight evenly and with the minimum of heat and glare. It is used to provide light above or below eyelevel and, because it is translucent, insures privacy.

The light-directing glass block has internal faces and prisms, as well as external raised patterns (fluting or ribbing). It will not admit direct low-angle sunlight, but will instead reflect or absorb it. By means of its prisms, this kind of block will transmit the incoming light in a specified direction, usually toward the ceiling, from where the light is reflected down to all parts of the room. It is best adapted for admitting light above eyelevel; it cuts down heat and glare and affords complete privacy.

Glass blocks may be tinted to reduce solar heat further and may have internal filters of silica or glass fiber. They may come in different colors; they can be made so as to admit all wave lengths of light equally. Like twin-paneled glass, glass blocks usually have an internal dead-air space, which serves as insulation and soundproofing. For example, glass blocks transmit only 30 to 75 per cent as much heat as that passing through an ordinary glazed-glass window. They also reduce the amount of water that condenses on the inside of the glass in cold weather. The sound reduction achieved by some glass blocks employed in factories may be so great as to be almost unbelievable.

See also Vol. 10, p. 283: "Glass."

THE COSMIC RAYS

High-Powered Missiles from Outer Space

BY VOLNEY C. WILSON

WHIZZING through our galaxy at all times, in every direction and at fantastic speed, are the charged particles that we call cosmic rays. They are really atomic nuclei —that is, atoms stripped of their electrons. Traveling at nearly the velocity of light, some have energies far greater than those produced with powerful atom smashers.

The cosmic rays that fly through outer space are called primary cosmic rays. A thin rain of these particles constantly strikes the atmosphere of the earth. As they enter the upper part of the atmosphere, they collide with atoms of air. The fragments known as secondary cosmic rays result from the collisions.

About 1 per cent of these secondary rays penetrate the remaining atmosphere and reach sea level. Roughly 600 rays pass through the human body every minute, day and night. This should not alarm us, though; a wristwatch with a radium-painted face exposes us to the same dose of radiation. Doctors estimate that the dose must be 30 to 300 times greater to produce harmful genetic effects on human beings by increasing the rate of mutations.

If we could harness the energy of all cosmic particles striking sea level over the entire world, it would yield little more power than a modern automobile engine. Cosmic rays will never power our machines; but they do provide nuclear physicists with a natural laboratory for the study of super-high-energy phenomena. They may reveal the secrets of the forces that bind together the particles of matter in atomic nuclei. They may tell us about the universe beyond our solar system.

Cosmic rays made their presence felt long before they were identified. About the

Left-hand photo, U. S. Navy; right-hand photo, National Institutes of Health

The left-hand photograph, above, shows a skyhook balloon rising with a rocket that carries instruments for recording cosmic rays. The rocket will be fired off after the balloon has reached an altitude of 70,000 feet or thereabouts. At the right, we see an explosion caused by a cosmic-ray particle striking a photographic emulsion that had been carried aloft by a skyhook balloon. The cosmic-ray particle, consisting of the nucleus of a helium atom, struck a heavier nucleus in the emulsion and gave rise to the jet of mesons shown streaming from the site of the collision. This is a photomicrograph (see Index) of the explosion that had been recorded in the emulsion.

beginning of the twentieth century, physicists discovered that electroscopes and ionization chambers * were affected by certain mysterious rays existing in the atmosphere. These rays were far more penetrating than X rays or any other forms of radiation known at the time. High-energy X rays could be stopped by a lead plate only a sixteenth of an inch thick; but four inches of lead would absorb only 80 per cent of the "new" rays.

The first and most natural assumption was that these penetrating rays issued from some special radioactive material in the earth's crust and atmosphere. Beginning in 1909, however, enterprising experimenters carried ionization chambers aloft in balloons and found the rays grew more intense with

* An electroscope is a device for determining the electric charge on a body. An ionization chamber is used to measure the intensity of radiations. It is filled with a nonconducting gas and contains two oppositely charged electrodes. As radiation passes through the chamber, it transforms many of the gas atoms into ions, or electrically charged particles, which are collected on the electrodes. The ions alter the original charge on the electrodes; when this alteration is measured, it reveals the amount of ionization that has taken place and, consequently, the intensity of the radiation that has passed through the chamber.

Cosmic-ray particles have penetrated a cloud chamber with a series of lead plates across it. As the particles pass through the lead plates, they are slowed down and split again and again, producing a cosmic-ray shower.

Univ. of California Radiation Laboratory

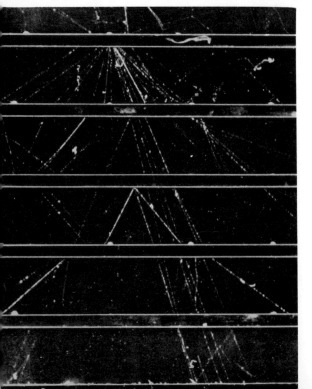

increasing altitude. Between 1910 and 1914, the Austrian Victor F. Hess and the German Werner Kolhörster made the first precise measurements, at altitudes up to 30,000 feet. At this height the rays were ten times more intense than at sea level. Clearly, the penetrating rays traveled downward through the atmosphere. Hess came to the conclusion that the mysterious rays must originate in outer space.

In 1925, a series of experiments by a team of physicists headed by Robert A. Millikan, of the California Institute of Technology, confirmed Hess's hypothesis. Millikan and his colleagues lowered ionization chambers into two snow-fed lakes at high altitudes in California in order to determine the absorption of the penetrating rays in water, as compared with their absorption in air. Measuring the intensity of the rays under different depths of water and at different altitudes, they came to the conclusion that the rays must originate outside the atmosphere. In the report of these experiments, the name "cosmic rays" was used for the first time. Millikan speculated that the original radiation was a very energetic type of gamma ray (see the Index), possibly created by the annihilation of large atoms such as uranium. This idea was later to be discarded.

In a few years, two newly invented instruments revealed important new facts about cosmic rays and later became the cosmic-ray physicist's chief tools. In 1927, the Russian physicist D. V. Skobeltsyn first adapted the Wilson cloud chamber to the study of the rays.*

In 1928 and 1929, Werner Kolhörster and another German, Walther Bothe, devised a research technique using sets of Geiger-Mueller counters. (See the chapter The Geiger-Mueller Counter, in Volume 9.) By arranging the counters in a straight line and providing them with the proper electrical circuits, one could trace the path of a *single* cosmic ray.

The bubble chamber, developed in 1952 by Donald Glaser, also proved helpful in the analysis of the rays. In the bubble

* In the Wilson cloud chamber, the path of an ionizing particle or ray is made visible as a trail of water droplets. See Index, under Cloud chambers.

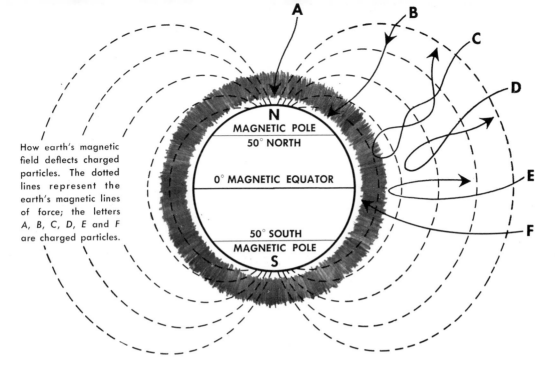

How earth's magnetic field deflects charged particles. The dotted lines represent the earth's magnetic lines of force; the letters A, B, C, D, E and F are charged particles.

chamber, a liquid is kept just below the boiling point. The pressure is suddenly lowered and the liquid becomes superheated. When high-speed charged particles now pass through, their passage will be indicated by a series of bubbles.

By the use of these devices, it has been established that most of the secondary cosmic rays are not high-energy gamma rays but electrically charged particles. This is a significant fact. A beam of gamma rays requires much less energy to penetrate matter than a stream of charged particles. If cosmic rays are particles, their energies must be incredibly high.

Cosmic rays and the earth's magnetic field

In 1927, the Dutch scientist Jacob Clay discovered that primary cosmic rays are affected by the earth's magnetic field. During a voyage between Amsterdam and Java, he observed that the intensity of cosmic rays drops as one approaches the magnetic equator from higher latitudes. The existence of this "latitude effect" seemed to show that primary cosmic rays are electrically charged particles.

A neutral (uncharged) particle moving in a straight line through a magnetic field is not influenced by the field. However, a charged particle in the same situation will have its path bent into a curve, if

it moves approximately at right angles to the magnetic lines of force.

The earth's magnetic field is very weak, but it extends for thousands of miles into space. Any charged particle approaching the earth must travel immense distances through the field and will be appreciably deflected by the curving force. This force is greatest on particles that approach exactly at right angles to the lines of force, and weakest on those that travel parallel with the lines. The bending is more pronounced for slow-moving particles (low momentum) than for faster ones (high momentum).

This is illustrated in the diagram on this page; here the dotted lines represent the earth's magnetic lines of force. A, B, C, D and E are all charged particles approaching the earth with the same momentum or energy. A is only slightly deflected by the field because it approaches near the pole and thus moves almost parallel with the lines of force. On the other hand, E approaches near the equator, moving exactly at right angles to the lines. The strong curving force on this particle eventually turns it back into the direction from which it came. The paths taken by B, C and D will depend upon the angle formed by their direction as they approach the earth and the magnetic lines of force. For a particle to penetrate the magnetic field and the atmosphere at the equator (particle F), it

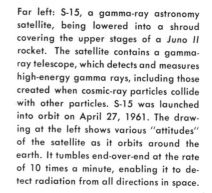

Photo, NASA; drawing based
on material supplied by NASA

Far left: S-15, a gamma-ray astronomy satellite, being lowered into a shroud covering the upper stages of a *Juno II* rocket. The satellite contains a gamma-ray telescope, which detects and measures high-energy gamma rays, including those created when cosmic-ray particles collide with other particles. S-15 was launched into orbit on April 27, 1961. The drawing at the left shows various "attitudes" of the satellite as it orbits around the earth. It tumbles end-over-end at the rate of 10 times a minute, enabling it to detect radiation from all directions in space.

would have to start out with many times more momentum than the other particles.

If the particles are indeed electrically charged, we should expect many fewer rays to reach the earth's surface at the equator than at higher altitudes. This was soon to be conclusively shown. In 1930, a world-wide survey was begun, under the direction of Arthur H. Compton, of the University of Chicago, to find out cosmic-ray intensities at different latitudes and altitudes all over the globe. The report of the survey in 1933 established that, from the geomagnetic latitudes of 50° north (or south) to the equator, cosmic-ray intensity at sea level drops about 10 per cent. It showed that primary cosmic particles are electrically charged.

Is the charge positive or negative? Manuel S. Vallarta, a Mexican mathematician, calculated in 1933 that a positive particle with low momentum could reach the earth more easily when approaching from the west than from the east; the directions would be reversed for a negative particle. By 1938, experiments proved that the rays falling from the west were definitely more intense than those from the east; this led to the conclusion that the primaries are positively charged.

The American physicist Scott E. Forbush discovered that cosmic-ray intensities decrease during periods of high sunspot activity. The sun is continuously throwing out tremendous quantities of protons moving at a speed of about 1,000 miles per second — a phenomenon sometimes referred to as the solar wind. During periods of sunspot activity, when a solar flare (see Index) happens to be at a particular place on the sun's surface, the solar wind in the direction of the earth is greatly increased. This in turn increases the strength of the magnetic field in the vicinity of the earth to such an extent as to shield out some of the low-energy cosmic rays. This is called the Forbush effect.

Rockets and artificial satellites have been used increasingly to measure cosmic-ray intensity above the earth. They have also served to analyze the effect of the solar wind upon the earth's magnetic field.

Cosmic-ray showers; electrons and positrons

The Italian-born physicist Bruno Rossi showed in 1932 that if three or more Geiger-Mueller counters were spread out horizontally in an irregular pattern, they would sometimes be tripped simultaneously, indicating that showers of particles were traveling together. These showers could be produced as cosmic-ray particles passed through lead or other substances containing heavy atoms. Rossi concluded that each shower was produced from a single cosmic-ray particle as it passed close to the nucleus of a nearby atom. Showers could also be observed in a Wilson cloud chamber that had a lead plate across it. The track of a single cosmic-ray particle is seen entering the chamber. As it passes through the lead plate, the track splits in two; these tracks

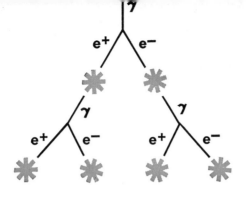

The pair-production theory of cosmic-ray showers. A gamma ray (indicated by γ) yields a pair of electrons, positive and negative, indicated by e+ and e−, respectively. Each of these electrons may disappear, producing another gamma ray as it does so. Each of the gamma rays in turn may yield two electrons. As the process is continued in this way, a cosmic-ray shower is produced.

split again and again, producing a shower of particles.

In the same year in which Rossi revealed the existence of cosmic-ray showers, the American Carl D. Anderson discovered a new fundamental particle in cosmic radiation — the positive electron, or positron. It appeared in a photograph of a cloud chamber containing a lead plate in a strong magnetic field. A particle with the same mass as an electron was seen passing through the plate, but its track was curved in a direction that indicated a unit positive charge instead of a unit negative charge. Anderson's historic achievement won him a share of the 1936 Nobel Prize in physics.

Cloud-chamber photographs soon showed that each time a shower track split, the two branches were oppositely curved; an electron and a positron were created. A pair-production theory of cosmic-ray showers was now formulated. It was maintained that the process of shower formation represented the conversion of energy into charged matter and vice versa.

This is what takes place according to the pair-production theory. As a gamma-ray photon (parcel of radiant energy) penetrates the nuclear field of an atom, a portion of its energy is suddenly transformed into a pair of electrons, positive and negative. The remaining energy provides the velocities of the electron and the positron. Each of the daughter electrons may disappear in its turn, producing another gamma ray, if it is suddenly slowed down in the field of another nucleus — a process described as bremsstrahlung, the German word for "braking radiation." The two gamma rays may go on to produce four electrons; four gamma rays will be produced next, then eight electrons and so on. This multiplication creates the shower.

In 1938, giant cosmic-ray showers were discovered in the atmosphere. They were sometimes a fifth of a mile in diameter and contained millions of particles. Presuming that a single particle initiates the shower, its energy must reach the staggering figure of 100 million billion electron volts.* The fission of a uranium atom releases only 200 Mev. Apparently some rare primary particles possess 500 million times more energy!

The primary cosmic rays

Primary cosmic rays are made up principally of protons — nuclei of the lightest element, hydrogen. Next in abundance are the alpha particles — nuclei of the second lightest element, helium. Experiments in which some specially sensitized photographic emulsions were carried to extremely high altitudes by balloons proved that much heavier nuclei, such as oxygen and iron, are also represented. The following table shows the approximate composition of primary cosmic rays:

- 91 per cent protons (nuclei of hydrogen atoms)
- 8 per cent alpha particles (nuclei of helium atoms)
- 1 per cent nuclei of heavier elements

These figures compare closely with the distribution of elements in the stars, nebulae and interstellar dust that form the bulk of the universe.

The nuclei that make up primary cosmic rays seem to be evenly distributed in space. Their average energy is about 6 Bev; a few nuclei, as we have seen, may have energies up to 100 million Bev.

* An electron volt is the energy acquired by an electron as it is accelerated by a potential difference of 1 volt. In nuclear and cosmic-ray physics, energies are often expressed in Mev (million electron volts) and Bev (billion electron volts).

The two components of
secondary cosmic rays

Secondary cosmic rays are composed of two very different classes of particles, which can be separated by a piece of lead about five inches thick. One component is completely absorbed in this thickness; this soft component consists mainly of electrons, positrons and gamma-ray photons — the typical particles of a cosmic-ray shower. The intensity of this soft component *increases* from the top of the atmosphere down to an altitude of about 55,000 feet, where it makes up roughly four fifths of the total radiation. From this altitude downward, the intensity *decreases* until it makes up only one quarter of the total at sea level.

The other component passes through five inches of lead almost unobstructed; it is called the hard, or penetrating, component. It decreases in intensity continuously from the top of the atmosphere down to sea level; approximately one half of these rays at sea level can still penetrate fifteen inches of lead. The difference in the absorption pattern and penetrating power of the two components represented a puzzling problem to physicists for many years.

In 1937, the riddle was finally solved by the teams of C. D. Anderson and S. H. Neddermeyer, at the California Institute of Technology, and J. C. Street and E. C. Stevenson, at Harvard University. Working independently, these teams showed that all the evidence pointed to a new type of fundamental particle, intermediate in mass between the proton and the electron.* This was the mesotron, now called the meson. In the meantime, the author of this article had been measuring the penetration of cosmic rays below the ground in a copper mine in northern Michigan. He reported that some rays were able to penetrate as much as 1,600 feet of rock. Further investigation showed that these rays were corpuscular (made up of tiny particles) and that they ionized the rock and lost energy all the way down. Calculations showed that they must

* *Editor's note*: The mass of the proton is 1,836 m_e. m_e represents the mass of the electron, used as a basic unit for giving the masses of other particles; it equals .91 × 10^{-27} grams. (For an explanation of the notation 10^{-27}, see Index, under Exponents.)

be the newly discovered mesons. It is now agreed that mesons make up the bulk of the hard component of secondary cosmic rays. The high energy of mesons accounts for the penetrating power of this component.

The existence of the meson had been predicted by the Japanese nuclear physicist Hideki Yukawa in 1935. He also predicted that the meson would be unstable outside the nucleus, decaying with the emission of an electron. In 1940, this decay was actually photographed for the first time in a Wilson cloud chamber.

In 1948, researchers at Berkeley, California, first succeeded in producing mesons artificially, by bombarding carbon atoms with alpha particles accelerated to 380 Mev. Both positive and negative mesons were produced, with a mass of about 276 m_e. These are known as pi mesons, which decay in about two hundred millionths of a second into a lighter variety called mu mesons (mass approximately 209 m_e). Mu mesons live for roughly two millionths of a second before they disintegrate into electrons or positrons. The heavier pi mesons of secondary cosmic rays are created by nuclear explosions in the upper atmosphere as primary cosmic rays collide with the nuclei of atoms — the atoms of the gases that make up the atmosphere. The pi mesons usually decay in flight and produce mu mesons, which live long enough to travel great distances. It is the mu mesons that form the hard component of secondary cosmic rays at sea level.

The origin of
cosmic rays

What is the source of cosmic rays and how do they acquire their fantastic energies? Some astronomers have long held that the universe was created in a single primordial explosion, and that cosmic rays are tiny remnants of this colossal event. Millikan also believed in some explosive process, involving the complete annihilation of matter. But the presence of heavy nuclei in the primary rays shows that their energies must be acquired gradually; in an explosive process, the larger nuclei would be completely shattered.

SUN AND
ITS SATELLITES

SUPERNOVA

Edge-on view of the Milky Way galaxy, showing the position of the sun and its family in the system. According to widely accepted theory, the cosmic rays originate in this galaxy. They are supposed to represent the nuclei of atoms that were thrown out by the exploding stars called supernovae.

Present evidence suggests that cosmic rays must originate and remain largely within our own galaxy. Let us suppose that the distribution of rays observed in our galaxy extends far beyond its borders throughout the universe. We know the average energy of the rays in our region; from this we can calculate the total energy carried by cosmic rays everywhere. This calculation has been performed; the figure reached would equal the entire mass of the universe if this were converted into energy. Scientists consider such an idea impossible. Moreover, the solar system is constantly rotating along with the rest of our galaxy.* If rays are entering the galaxy from outside, certain complex effects due to the rotation of the galaxy should be visible. None have ever been observed.

The sun was once thought to be the source of cosmic rays, but scientists now consider this unlikely. For one thing, the intensity of the rays shows only minor variations as the earth changes its position with respect to the sun. Cosmic rays contain elements known to be absent in the sun's surface. Finally, the magnetic fields of the sun are not strong enough to account for the higher cosmic-ray energies.

The nuclear physicist Enrico Fermi proposed a theory that would account for the vast energies built up by cosmic rays. Large clouds of charged dust are known to travel through interstellar space, creating slowly varying magnetic fields. A charged

particle moving into or near one of these clouds will have its path bent sharply by the field; in effect, it collides with the cloud and bounces off. Mathematical computations show that, given a particle with a certain minimum kinetic energy (200 Mev for protons, which make up the majority of the primary cosmic rays), a gain of energy is more probable in these collisions than a loss; the gain becomes greater as the particle moves faster. The particle will continue to accelerate indefinitely since it gains more energy in the magnetic collisions than it loses by ionizing the interstellar matter it encounters. Thus, the fabulous cosmic energies would be built up.

What would be the source of the cosmic-ray particles caught up in the magnetic fields of interstellar space? It has been suggested that the particles are emitted by supernovae. A supernova is a variable star whose light increases amazingly in intensity as the result of an inner explosion and then dims again. In the last nine hundred years, there have been three supernovae, each of which threw out an amount of matter equal to the mass of the sun. If only one hundred thousandth of this matter were ejected with energies of 200 Mev or more, it would account for the cosmic rays in our galaxy.

Scientists today are inclined to accept the idea that cosmic rays are the nuclei of atoms thrown out by supernovae and accelerated as they come in contact with magnetic fields in our galaxy. But thus far, this is only a working hypothesis.

See also Vol. 10, p. 280: "Cosmic Rays."

* *Editor's note:* The stars that make up a galaxy rotate around the nucleus, or center of the system. See Index, under Rotation of galaxies.

1895 & 1905

Planes, television, use of X rays, electron microscopes, knowledge of atomic structure, vacuum tubes and many other developments stem from discoveries made in the period 1895-1905.

THE TWENTIETH CENTURY (1895-   )[1]

BY JUSTUS SCHIFFERES

THE ACHIEVEMENTS OF THE MIRACLE DECADE

SCIENCE dominates the twentieth century, carrying the world on its shoulders like the mythological giant Atlas. We are still debating whether it is to be our master or our slave — a Frankenstein monster or a good genie, bringing us a longer and a fuller life.

When did the ideas that rule our twentieth-century science come into being? It is difficult, of course, to set exact dates for scientific events. The birth of an idea and its proof or acceptance are often separated by years or even, as in the case of Copernicus, by centuries. However, we may with confidence place the birth of present-day science in the miracle decade that bracketed the turn of the century: the ten years from 1895, when Roentgen announced the discovery of X rays, to 1905, when Einstein published his special theory of relativity.

If all the new scientific facts that were discovered and the new scientific theories that were developed between 1895 and 1905 were wiped out, science in its present form simply would not exist. Twentieth-century scientific achievements are firmly rooted in that fateful decade. The developments of that period were the culmination of centuries of thought. But they came upon the scene with dramatic suddenness. Here are some of the developments of the miracle decade — theories, techniques, discoveries and inventions.

Physical sciences

Aerodynamics — the building of the airplane.

Astrophysics — the composition of the heavens.

Atomic physics and nuclear chemistry — the new world of the atom.

Electronics and "wireless telegraphy," or radio.

Quantum theory — how "bullets of energy" act.

Radioactivity; the discovery of radium.

Relativity.

X rays and vacuum tubes.

Biological sciences

Biometrics (measurement of living things), biochemistry and biophysics.

Bacteriological triumphs — the discovery of the infecting organisms of plague, syphilis and many other diseases.

Genetics — reborn with the discovery of the "lost paper" of Mendel.

Insect transmission of disease — "mosquitoes carry yellow jack."

Psychoanalysis and Freudian psychology.

Ultramicroscope.

Viruses — infecting agents so small that they can pass through the finest filter.

Vitamins.

The miracle decade saw the establishment of philanthropic foundations, like the Carnegie Foundation, and of industrial research laboratories, like that set up by General Electric at Schenectady, New York — developments that were to have a powerful effect on scientific research. In this decade, also, Nobel prizes in physics, chemistry and medicine and physiology were established. These awards, which were first granted in 1901, proved to be a stimulus to research men the world over; they also served to call the attention of the world at large to the achievements of science.

Just before the miracle decade, scientific and industrial progress seemed to have come to something of a standstill. The nineteenth century was resting on its laurels — it was viewing its wonderful accomplishments with smug pride or with gloomy forebodings. Scientists had become cautious; the resounding failure of Robert Koch's "cure" for tuberculosis, announced in 1890, served as a brake upon them. In biology Darwinism had spent its initial force. In the physical sciences, the end point of classical Newtonian physics seemed to be reached. "The future of physics is in the fifth decimal place," said the leading physicists. They meant that from then on physics would offer nothing more exciting than the tedious task of calculating physical constants (like specific heat) to greater degrees of accuracy. New developments were not expected in industry, either. The United States Commissioner of Labor reported that the "era of rapid industrial advance has ended for the civilized world, and the future of the great industrial countries offers no such opportunities as have the fifty preceding years for the creation of new tools and profitable employment." But all this was the lull before the storm.

To give some idea of the crowded happenings in the miracle decade, we are going to present them year by year. We are going to assume that the importance of each new discovery was *immediately apparent* so that it would make newspaper headlines. The headlines and the news items under them might then have read about as follows (our own comments are put in brackets):

1895

X RAYS DISCOVERED

New Era in Physics Opens

Wuerzburg, Germany. — Wilhelm Konrad Roentgen announces the discovery of X rays. These rays — a hitherto unknown phenomenon — are produced in a vacuum tube and have the astounding property of passing through solid matter. Bones of the human body can be clearly revealed on photographic plates by sending X rays through the body. X rays will probably prove valuable in medicine and surgery.

On May 6, 1896, Samuel P. Langley launched a steam-driven model plane from a houseboat on the Potomac River. This was the first sustained flight made by a mechanically propelled heavier-than-air craft.

French Embassy—
Information Division

ANTOINE-HENRI BECQUEREL

SUBCONSCIOUS MIND REVEALED
Sensational Development
in Psychology

Vienna, Austria. — In their STUDIES IN HYSTERIA Sigmund Freud and Josef Breuer make the astonishing revelation that emotional reactions in human beings are largely controlled by "forgotten memories" repressed in the subconscious mind. People reveal their true opinions by slips of the tongue and in dreams. The authors tell how a young, well-educated girl, suffering from hysterical paralysis, was cured when she was made to realize her abnormal relationship to her father. [This was the beginning of psychoanalysis.]

1896

A MACHINE THAT FLIES

Steam-Driven Model
Makes History

Quantico, Virginia, May 6. — Today marked a dramatic *first* in the history of aviation: a mechanically propelled heavier-than-air flying machine achieved sustained flight in the air. This machine was a nine-pound steam-driven model, constructed by Samuel P. Langley, secretary of the Smithsonian Institution of Washington. Lang-

ley launched the model from a houseboat on the Potomac River; it took the air at 3:05 P.M. and flew for half a mile under its own power. Photographs of the first flight were made by Alexander Graham Bell, inventor of the telephone. Mr. Langley, who is well known for his work on astronomy, has founded a new science of aerodynamics, which has its roots in the studies of birds in flight made by Leonardo da Vinci. Langley began his work in Pittsburgh, where he invented a "whirling table" to study the effects of streams of air against an airfoil.

TELEGRAPH WITHOUT WIRES
Messages Sent Two Miles

London, England. — Guglielmo Marconi, young Italian inventor, has sent wireless-telegraph messages for a distance of two miles through space. Marconi has set up his instruments in the London Post Office Building, with the permission and help of Sir William Preece, chief electrician to the Post Office. The brilliant young inventor expects ultimately to send communications by wireless across the Atlantic Ocean.

AN UNUSUAL PHOTOGRAPH
Peculiar Behavior of Uranium

Paris, France. — Antoine-Henri Becquerel, professor of physics at the Polytechnic School in Paris, has called attention to the peculiar behavior of the chemical element uranium. He had left a photographic plate in a desk drawer with a key and a lump of uranium salt (the double sulfate of uranium and potassium). When he developed the plate, Becquerel found an image of the key on the photograph. Apparently uranium gives off something that acts like rays of light or X rays. [The peculiar behavior of uranium was due to radioactivity.]

1897

"CORPUSCLES OF ELECTRICITY"
Something New Inside the Atom

London, England, April 29. — Joseph John Thomson, professor of physics at Cambridge University, announced tonight, in a meeting at the Royal Institution, that

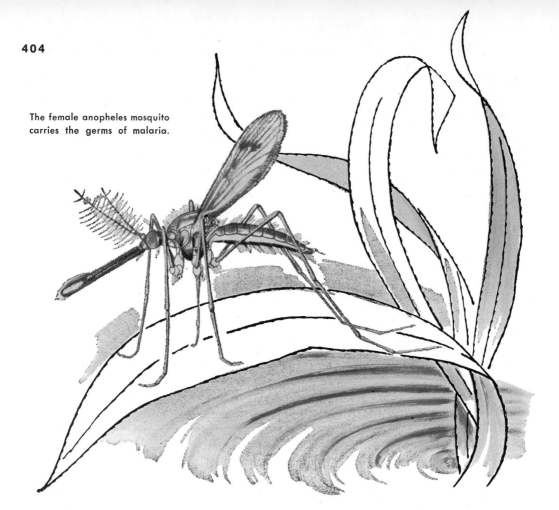

The female anopheles mosquito carries the germs of malaria.

he had discovered a subatomic particle that he calls a "corpuscle of electricity." [This corpuscle corresponded to what is now called the electron; Thomson's discovery represented the starting point of the science of electronics.] Dr. Thomson has been working with electrical discharges through gases in vacuum tubes. Other physicists attending the meeting were openly skeptical.

A NEW FOOD FACTOR
A Cure for the Oriental Disease Beriberi

Batavia, Netherlands East Indies [now the Republic of Indonesia]. — Christiaan Eijkman, a Dutch army surgeon, reports a striking experiment. A crippling nervous disease resembling beriberi — a common ailment in the Orient — was produced in chickens and pigeons by feeding them only polished rice and was then cured by putting the birds on a diet of rice bran. The disease is apparently caused by the lack of an essential food factor [later called the antineuritic vitamin]. Here is the greatest new clue for the study and improvement of human nutrition since Beaumont's researches on the physiology of digestion. [A whole alphabet of vitamins was later to be discovered.]

1898

NEW ELEMENT DISCOVERED
Radioactivity within the Atom

Paris, France. — Pierre and Marie Curie announce the discovery of a new element, which they call radium because of its powerful radioactivity. It is said to be worth 100,000 times as much as gold. This young husband-and-wife scientific team has been following up Becquerel's work on the behavior of uranium; his findings gave the clue to the discovery of radium.

Radium is constantly disintegrating, giving off rays. This process, called radioactivity, is due to changes within the atom.

A NEW "LAZY GAS"
Striking Electric Effects
Produced with Neon

London, England. — William Ramsay and Morris Travers announce the discovery of another inert, or "lazy," gas, called neon. [The previously known inert gases were helium, argon, krypton and xenon.] When an electric charge is passed through a glass tube filled with neon, the tube glows with a brilliant colored light, usually red. [This effect was later put to use in street signs.]

FILTERABLE VIRUSES
A Newly Discovered Cause of Disease

Berlin, Germany. — Friedrich Loeffler and Paul Frosch, of the staff of the Kaiser Friedrich Wilhelm Institute, have discovered disease-causing agents smaller than bacteria. On the borderline between the living and the dead worlds, they can pass through the finest porcelain laboratory filters; hence they have been called filterable viruses. They cannot be seen through an ordinary microscope. Loeffler and Frosch have demonstrated that certain viruses are responsible for hoof-and-mouth disease in cattle. They suspect that other viruses are the infective agents in a good many human diseases. [We now know that they cause yellow fever, rabies, dengue, infantile paralysis, measles and typhus fever in man. Certain plant diseases, like the mosaic disease of the tobacco plant, are also due to viruses. Iwanowsky called attention to these plant viruses in 1892.]

1899

INSECTS THAT CARRY DISEASE
Mosquitoes as Vectors of
Malaria and Yellow Fever

Calcutta, India. — Major Ronald Ross, a surgeon in the British Army, who has been studying the life history of the malarial parasite, has proved that malaria can be transmitted to birds by the bite of an infected mosquito.

Rome, Italy. — Giovanni Battista Grassi, professor of comparative anatomy at the University of Rome, has demonstrated that the anopheles mosquito carries the germs of human malaria. Professor Grassi has been conducting experiments in various malaria-ridden spots in Italy.

1900

NEW INDICTMENT of the MOSQUITO
Reed Shows That One Species
Carries Yellow Fever

Havana, Cuba. — Major Walter Reed, head of the United States Army Yellow Fever Commission in Cuba, reports that "the essential factor in infection . . . with yellow fever is the presence of mosquitoes that have bitten cases of yellow fever." Dr. Reed reached this conclusion following experiments "in the interests of science and for humanity" with United States Army volunteers. He was assisted in these experiments by Drs. James Carroll and Jesse William Lazear of the United States Army and the Cuban physicians Drs. Aristides Agramonte and Carlos Juan Finlay. Dr. Lazear died on September 25 of the effects of the disease that he was studying.

THE FOUR BLOOD TYPES
Blood Transfusion Made Safer

Vienna, Austria. — Karl Landsteiner, a young Austrian pathologist, reveals that there are four main types of human blood, which he labels A, B, AB and O. He finds that, in giving blood transfusions, if the blood of the donor belongs to the same type as the blood of the recipient, transfusion is safe and practical. Landsteiner's discovery will undoubtedly save many lives.

A MILESTONE IN GENETICS
The Rediscovery of
Mendel's Lost Paper

Bruenn, Austria. — Gregor Mendel (1822–84), for many years Abbot of the monastery of Koeniginkloster, has at last come into his own. A 20,000-word paper by him, read before the Natural History Society of Bruenn in February and March 1865 and published in the proceedings of the society in 1866, has been rediscovered after being neglected for some thirty-five years. This paper represented a report on

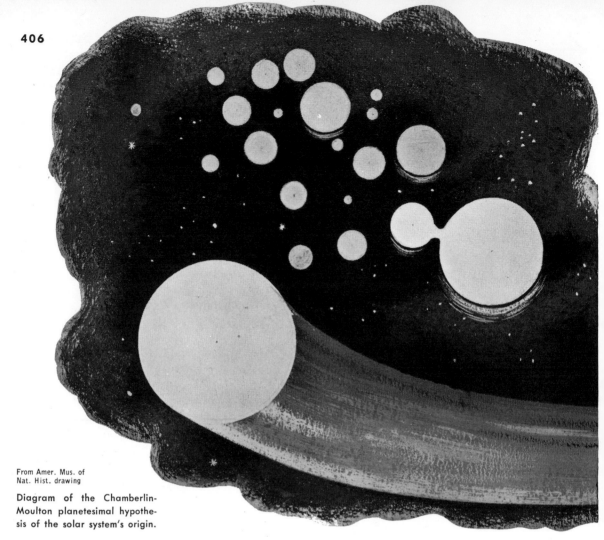

From Amer. Mus. of
Nat. Hist. drawing

Diagram of the Chamberlin-
Moulton planetesimal hypothesis of the solar system's origin.

the Abbot's experiments, extending over eight years, in the crossbreeding of common edible garden peas. The Austrian cleric had found the key to the riddle of heredity.

He introduced the ideas of dominant and recessive traits to help explain the subtly complicated course of heredity. He analyzed it on a mathematical basis, in terms of ratios. One of the fundamental Mendelian ratios is three to one — that is, in the generations bred from once-crossed hybrids, three out of four will display the dominant and one the recessive characteristic, whether it be length of stalk or color of hair.

Mendel's lost paper was rediscovered at about the same time in Holland, Germany, Austria and England. Among those who have revealed its importance are Hugo de Vries, the Dutch botanist who works with primroses and has observed mutations

in the hereditary factors; the German biologist Karl Correns; the Austrian scientist Erich Tschermak; and the pioneer English geneticist William Bateson, who has translated Mendel's paper into English. Mendel's experiments are being repeated and verified all over the world. It is quite generally agreed that his system represents the greatest idea in biology since Darwin's theory of organic evolution.

THE NEW QUANTUM THEORY
Does Energy Travel in "Bundles"?

Berlin, Germany, Dec. 14. — Max Planck, professor of physics at the University of Berlin, told the German Physical Society today that "radiant heat is not a continuous flow and indefinitely divisible." He said that "it must be defined as a discontinuous mass made up of units all of which are similar to one another."

Dr. Planck has been studying a beam

of radiation escaping through holes in a hollow body heated up to incandescence. He asserts that radiant energy does not flow out of the holes in the hot body continuously like a stream of water; instead, it is emitted, he says, in tiny "parcels" or "bundles" [now known as quanta], like bullets from a machine gun. These units are the smallest coin of energy that nature mints. To measure the energy contained in one quantum, Dr. Planck has worked out a constant of proportionality between

planets were formed by a number of little pieces of planet material ("planetesimals"), attracted to each other under the normal operation of the law of gravitation and growing in size like "snowballs in space."

MOTOR WAGONS
Another Use for the Horseless Carriage

St. Louis, Missouri. — The St. Louis POST-DISPATCH has ordered fourteen motor delivery wagons to speed its newspa-

energy and frequency, called Planck's constant. [It later turned out that this constant holds good in the analysis of a great many different kinds of natural phenomena — for example, the photoelectric effects of light, the orbits of electrons in the atom, the wave lengths of lines in the spectrum, the frequency of X rays and the velocity with which gas molecules move.]

THE PLANETESIMAL HYPOTHESIS
A New Explanation of the Earth's Origin

Chicago, Illinois. — Two University of Chicago professors have a new theory of how our earth originated; they call it the planetesimal or spiral nebula hypothesis. These professors are the geologist Thomas C. Chamberlin, formerly President of the University of Wisconsin, and the astronomer-mathematician Forest Ray Moulton. They maintain that the earth and the other

pers to all parts of the city. The first motor wagon ever built appeared this year on the streets of Denver, Colorado. This new version of the horseless carriage, hitherto used only for passenger travel, may replace the horse in time. Motor wagons come equipped with pneumatic tires, which the Goodrich Rubber Company first made for Alexander Winton's automobile in 1896. An automotive industry is rapidly growing up in Detroit.

THE NEW GENERAL ELECTRIC LABORATORY
A Commercial Firm Turns to Fundamental Research

Schenectady, New York, September 15. — For the first time a commercial firm — the General Electric Company — has opened a research laboratory devoted to fundamental scientific research without regard for practical results. The new laboratory is

CHARLES P. STEINMETZ

General Electric

located in a barn behind the home of the inventor Charles P. Steinmetz. Steinmetz, a mathematical wizard, who received his scientific training in Germany, has done a good deal of research in alternating electric currents. He has been a consulting engineer at General Electric since 1893.

1903

A MEMORABLE FLIGHT

A Man-Carrying Airplane
Takes to the Air

Kitty Hawk, North Carolina, Dec. 17. — The brothers Wilbur and Orville Wright made several successful flights today in their man-carrying airplane. It marked the first time that a heavier-than-air flying machine carrying a human being took to the air, remained aloft for an appreciable period of time and then descended without mishap. The Wright brothers, who are in the bicycle business in Dayton, have done a good deal of experimenting with gliders.

NEW ULTRAMICROSCOPE

An Optical Instrument That
Magnifies 2,500 Diameters

Jena, Germany. — Richard Zsigmondy, a colloid chemist, and Henry F. W. Siedentopf, chief of the microscopic division at the Zeiss glass works and professor of microscopy at the University of Jena, have invented a new microscope that will magnify objects up to 2,500 diameters. By using ultraviolet light, which has a shorter wave length than white light, and by immersing the microscope lenses in oil, they have made it possible to see worlds within worlds. It is believed that the new microscope, which is called the ultramicroscope, will be helpful in the study of genetics.

SLEEPING PILL AND PAIN-KILLER

New Triumphs for
Organic Chemistry

Berlin, Germany. — Emil Fischer, one of Germany's foremost chemists, has created a new sleeping medicine and pain-killer, called veronal, or phenobarbital. [This was the first of a long line of pain-killing drugs — barbiturates — which can be made in the chemical laboratory.] Dr. Fischer has already produced synthetically a great many drugs and other substances that were formerly manufactured only by natural processes. He has synthesized simple sugars, caffeine — the active ingredient in coffee — and the purine derivatives of uric-acid compounds. His work is leading the way toward a whole host of substitute products of every type and description.

1904

A PRACTICAL USE FOR ELECTRONS

An Electron Tube for Wireless Telegraphy

London, England. — Sir John Ambrose Fleming, professor of electrical engineering at London's University College, has found a practical use for electrons. He has built an electron tube for use in wireless telegraphy. [This was the first step in the development of the vacuum tube used in electronic devices.]

Fleming's discovery is based on the so-called Edison effect, first observed and reported in 1883 by the distinguished American inventor Thomas Alva Edison. Experimenting with his newly invented electric light, Edison observed a glow inside the bulb as the carbon filament rapidly disintegrated. He then sealed a metal plate inside the tube. When the plate and the positive side of the supply circuit were connected, an electric current flowed through the vacuum tube *across space* from filament to plate.

Dr. Fleming has found a way to control the flow of electrons across space. His electron tube, or Fleming valve, will serve as a detector in wireless telegraphy.

1905

DISEASE GERMS ISOLATED

New Drugs Likely

Hamburg, Germany. — Fritz Schaudinn, zoologist at the Institute for Tropical Diseases in this city, has discovered the organism that causes syphilis (lues). This organism, called a spirochete, resembles a twisted thread and differs from ordinary bacteria. It is seen by reflected light in a dark-field microscope. This discovery rivals that of the trypanosomes, organisms that are the cause of African sleeping sickness and that are carried by the tsetse fly. Dr. Schaudinn has hunted out disease germs in the Arctic as well as the tropics; he went on an expedition to the Arctic Ocean in 1898.

The discovery of the syphilis spirochete caps a decade of discoveries of infecting organisms, including the causative agent of the bubonic plague and the germ that causes dysentery. Now that the organisms that cause these diseases are known, perhaps scientists will discover new drugs that will cure them. Is a new science of chemotherapy [curing by chemistry] in the offing?

THE NEW RELATIVITY THEORY

Einstein Solves the Riddle of the "Ether"

Zurich, Switzerland. — A clerk in the Swiss patent office, Albert Einstein, has published a paper in which he offers an explanation of the Michelson-Morley experiments. [These experiments, described previously, had cast in doubt the existence of the ether.] Einstein calls his explanation a "special theory of relativity." This theory makes it unnecessary to assume the existence of the ether or of all absolute notions of time and space; it calls for a fourth dimension, called space-time. If young Einstein is right, many of the doctrines of Newton must be modified.

These "news items" give some idea of the startling scientific developments of the miracle decade. In it was born a "higher physics" — more subtle than that of Galileo, Newton, Laplace, Faraday, Helmholtz and Maxwell. Today's physicists often deal with an absolutely unseen world, existing only in their imaginations. Facts concerning the invisible atoms, for example, can be accurately expressed only in mathematical equations. Pictures, models and diagrams of the atom are fanciful; they are useful enough for teaching purposes, but they correspond to nothing real. But, although the atom itself remains invisible, its predictable effects have been brilliantly analyzed.

New atomic concepts radically changed the nature of chemistry. In the past, chemical researchers kept collecting vast amounts of new data, which was duly stored in appropriate pigeonholes. But there were so many of these pigeonholes and, all too often, so little connection between them that chemists wallowed in a mass of detail. Now, however, atomic theory has cast a flood of light on chemical relationships.

Other sciences, too, have made striking advances. In biology, new knowledge of the cell has led to many important discoveries in physiology and genetics; the use of radioactive tracer elements has proved a priceless tool in biological research. Anthropology and archaeology have thrown light on man's beginnings; psychology has traced the workings of the human mind. Medicine has continued the marvelous progress of the nineteenth century. It has provided new surgical techniques and lifesaving drugs; it has stressed, as never before, the effect of mind on body as a source of mankind's ills. Other sciences — astronomy, geology, meteorology, oceanography, among others — have also registered spectacular gains.

Not only in the field of theoretical science has the twentieth century distinguished itself. It has brought a host of technological improvements, which enable men to eat better, travel farther and faster, communicate more quickly and live longer and better than any previous generation in the long history of mankind. Unfortunately, advances in technology have not always been equally distributed throughout the world. Nor can all of them be classed as directly beneficial to mankind. New scientific weapons — poison gases, bombing planes, guided missiles, atom bombs and the like — have made war even more horrible to contemplate than before.

In the following chapters we shall have more to say about the fateful period extending from 1895 to 1905. We shall also trace the consequences of the discoveries made in the miracle decade.

THE ROAD TO MODERN ALCHEMY

During the Middle Ages and long afterward, too, alchemists were eagerly trying to transmute, or transform, one metal into another — particularly, to change a base metal, like iron or lead, into a precious metal, like gold or silver. (See the chapter on the Middle Ages, in Volume 2.) Occasionally an alchemist would announce that he had succeeded in the quest; he would even display the "gold" that he had produced in his laboratory. But such claims always proved to be fraudulent.

By the nineteenth century, chemists and physicists had come to believe that it was impossible to transform one chemical element, like iron, into another chemical element, like gold. Each element, in its pure form, they held, was unchangeable. Its basic unit, the atom, was defined as the smallest possible particle existing in matter — a particle that could not be subdivided or altered in any way.

These concepts, which were held up to almost the end of the nineteenth century, have been abandoned by scientists. We now know that, far from being the smallest possible unit in nature, the atom is divided up into a number of subatomic (less-than-atom) particles. Furthermore, the atoms of one element can be changed into the atoms of another element by adding or subtracting subatomic particles. In the course of such transformations vast stores of energy are released. We have already succeeded in utilizing this energy in war (the atom bomb), and we are now seeking to apply it efficiently to the peacetime work of the world. Modern alchemy — the transmutation of the elements by man — represents an entirely new departure in science.

The history of modern alchemy goes back to the 1890's. It really begins with a complaint by an eminent man of science, Sir William Crookes (1832–1919). In the course of his career, Sir William discovered the chemical element thallium, studied the rare earths, distinguished different forms of uranium and took up the investigation of psychic phenomena. He accomplished much; yet he just missed making many discoveries that made other men famous.

Crookes was greatly interested in the study of electric discharges through gases, a field that had already attracted numerous

SIR
WILLIAM
CROOKES

tricity passed through such a tube caused a green glow near the cathode, or negative electrode. Later investigators had held that the glow was caused by rays originating at the cathode — cathode rays, as the German physicist Eugen Goldstein (1850–1930) called them. Crookes developed an improved vacuum discharge tube, which was called the Crookes tube after his name. He performed many experiments with this apparatus and made a number of important discoveries about cathode rays.

Now we come to the noted scientist's complaint, which must have seemed quite unimportant at the time. The photographic plates that he was using in his laboratory in connection with his work were giving him a good deal of trouble; they often became fogged and did not give clear negatives. Crookes wrote to the manufacturers of the plates, pointing out their defects. The manufacturers apologized profusely, but they were not able to do anything about the plates, which continued to be fogged. Other scientists working with Crookes tubes noticed that their photographic plates were also being spoiled. For quite a while, however, nobody thought it worth while to investigate the matter.

The German physicist Wilhelm Konrad Roentgen (1845–1923) was the first to explain the phenomenon. While conducting a series of experiments with a Crookes tube in 1895, he made a momentous discovery. He noted that even when the tube was covered with black paper, through which no light could pass, a nearby sheet of paper coated with barium platinocyanide began to glow when electricity was passing through the tube. Evidently something that affected the coated paper was penetrating the black paper wrapped around the Crookes tube. Roentgen came to the conclusion that the phenomenon was due to some form of "penetrating rays." According to his own account, he called these rays X rays "for the sake of brevity." It is generally believed that he chose the letter X because it represents an unknown quantity in algebra.

Roentgen found that his X rays caused photographic plates to be fogged even when

investigators. A skillful German glass blower, Heinrich Geissler (1814–79), had succeeded in sealing wires, attached to metal electrodes, in glass tubes from which almost all the air had been evacuated — that is, vacuum tubes. A German physicist, J. Plücker, had found in 1859 that elec-

the plates were wrapped in black paper. Apparently this paper, opaque to ordinary light rays, was transparent to X rays. That is why Crookes' photographic plates had been spoiled; they had been subjected to the X rays emitted by the discharge tubes in his laboratory.

Roentgen had a man place his hand between a Crookes tube, in operation, and a photographic plate, and then he developed the plate. The internal structure of the hand was revealed, because the different parts of the hand were not equally penetrated by the rays.

Soon afterward a doctor in the United States used X rays to locate a bullet imbedded in a human body. Other physicians employed the rays to diagnose disease in the bones and to watch the growth of new bone after a fracture. In time X rays were used for a great many other purposes (see the article The Wonder-Working X Rays, in Volume 7). We now know that the rays are electromagnetic radiations, like light and radio waves.

The French physicist Antoine-Henri Becquerel (1852–1908) became interested in X rays after listening to a lecture on the subject given in 1896 by the mathematician Jules-Henri Poincaré (1854–1912) at the Academy of Sciences in Paris. In response to a question, Poincaré had remarked that X rays seemed to originate in the luminescent, or glowing, spot produced when cathode rays struck the glass of a discharge tube. Becquerel had long been interested in the luminescence of substances that glowed for a time in the dark after being exposed to a strong light. Among these substances was the uranium salt called potassium uranyl sulfate. When it glowed, was it giving off rays like the X rays discovered by Roentgen?

To find the answer to this question,

PIERRE AND MARIE CURIE

Becquerel wrapped a photographic plate in black paper, put a crystal of the uranium salt on top of the paper and then exposed the salt and paper to sunlight. When he developed the plate, he found that it had been darkened; this showed that the uranium salt emitted radiations that could penetrate paper. He later showed that these radiations could also pass through thin sheets of aluminum and copper.

Becquerel thought at first that the uranium salt had emitted the rays because it had been acted on by sunlight. But he soon found that even when the salt was not exposed to light, it could affect a photographic plate. This unusual kind of radiation, which apparently arose spontaneously within a substance, represented something quite new in the experience of scientists. The name "Becquerel rays" was given to this phenomenon; later, as we shall see, it received the name "radioactivity."

The new radiations were next studied by perhaps the most remarkable husband-and-wife team in all the history of science —Pierre Curie (1859–1906) and Marie Curie (1867–1934). Pierre Curie, the son of a French physician, had distinguished himself by discovering the so-called piezoelectric effect produced by pressing upon a crystallized substance such as quartz. In 1895 he had been appointed to a professorship in the School of Physics and Chemistry at Paris.

A Polish girl, Marie (Marja, in Polish) Sklodowska was carrying on a series of experiments in the school. She had studied at the Sorbonne, living alone in an unheated garret, and had graduated with honors in physics and mathematics. After her graduation, she had begun to work on an industrial research job — a study of the magnetic properties of certain types of steel. She had begun this work at the Sorbonne, but the laboratory there was too crowded for comfort. That is why she had transferred her activities to the School of Physics and Chemistry.

The shy but distinguished professor and the brilliant young Polish girl fell in love and were married in the year 1895. Pierre continued with his own research

National Film Board

Radium is found in the mineral pitchblende.

projects, while Marie began to study the radiations produced by compounds of uranium. She reported that all the uranium compounds she had examined were active; so were the compounds of thorium. She suggested the use of the word "radioactivity" to describe this particular kind of radiation.

While at work on various uranium ores, Marie noticed that the minerals pitchblende and chalcolite gave off more radiation than pure uranium itself. She decided that these minerals must contain an element even more radioactive than uranium. Obviously the next step was to try to isolate the element. This project promised to be so fascinating that Pierre Curie gave up his own work in order to take part in research with his wife.

The Curies set out to isolate the radioactive agent in the pitchblende ore that was available to them. When they began to work on this ore, it was about two and a half times as active as uranium. After they had eliminated nonactive substances from the pitchblende one by one, they finally obtained a product whose radioactivity was something like four hundred times as great as that of uranium! They reached the obvious conclusion that the substance contained a hitherto unknown chemical element — a metal that they called polonium, after Marie Curie's native land (July 1898).

Continuing their investigation, the

Curies found that pitchblende contained still another radioactive substance, different from polonium in its chemical properties. They sought to isolate this substance and at last they obtained a product that was nine hundred times as active as uranium. This high radiation, they decided, was caused by still another new element, which they called radium, from the Latin *radius,* or ray (December 1898).

The radium-containing product that the Curies had prepared was not pure enough so that they could determine the atomic weight of radium or find out other essential facts about it. The next step was to extract from pitchblende a compound that would have more radium and less of other chemical elements. Through the cooperation of the Academy of Sciences in Vienna and the Austrian Government, the Curies obtained a ton of pitchblende ore, from which much of the uranium had been extracted. They now set to work to obtain, in as pure a form as possible, the tiny fraction of radium that it contained.

They worked under great difficulties in the old shed that served as their laboratory. Finally, after much back-breaking

Brown Brothers   FREDERICK SODDY

ERNEST RUTHERFORD

toil, they succeeded in extracting from the ton of pitchblende a tenth of a gram (less than four thousandths of an ounce!) of radium chloride, pure enough so that the atomic weight of radium could be determined (1902). The scientific world, which had been rather skeptical at first, now freely accepted the Curies' findings. In 1903 they received the Nobel Prize in physics (together with Becquerel) for their outstanding work.

The Curies had led an idyllic married life. A daughter, Irène (who was also destined to be a Nobel Prize winner) had been born in 1897; another daughter, Eve, in 1904. Everything was going well now. The Curies had a well-equipped laboratory to work in; Pierre had been named to a professorship at the Sorbonne. Then tragedy struck; on April 19, 1906, Pierre was run down by a heavy cart and instantly killed.

Bowed down by grief at first, Marie Curie carried on. She succeeded her husband as professor of general physics at the Sorbonne — the first woman to be so honored. She continued her scientific research, working particularly with radium and its compounds. Honors were showered upon her; among other things, she received a second Nobel Prize in 1911, the only person ever to win this award twice. She died in 1934 of pernicious anemia; the doctors who attended her said that she was "the eventual victim of the radioactive bodies that she and her husband discovered." She was only one of the long series of martyrs to science who died as a result of exposure to X rays and radium before it was recognized how dangerous these substances are when they are not shielded.

Marie Curie had insisted that radioactivity takes place within the atom itself. The New Zealand-born physicist Ernest Rutherford (1871–1937) and the English chemist Frederick Soddy (born in 1877) followed up this line of investigation. They worked together on the problem in the early 1900's at McGill University in Montreal, Canada, where Rutherford was professor of physics and Soddy a demonstrator in chemistry. In 1902 they pro-

Above is the inner structure of the carbon atom, with six electrons revolving around the nucleus.

Science Illustrated

posed a theory of radioactive disintegration, or decay. They stated that the atoms of radioactive elements undergo "spontaneous disintegration" and that as they do so they emit two kinds of particles — alpha particles and beta particles. Ultimately the atoms are transformed into the atoms of a new element.

Rutherford and Soddy pointed out that it is possible to measure the rate of decay for each radioactive element. This rate was determined in 1904 by Rutherford when he introduced the constant called the half life of radioactive elements. The half life of a radioactive element represents the time it takes for half its atoms to disintegrate — to be transformed into something else. If a given radioactive element has a half life of one minute, one half of its atoms will disintegrate in one minute; one half of the remaining atoms will have disintegrated by the end of the second minute and so on.

We now know that most radioactive elements found in nature radiate either alpha particles or beta particles; in a few cases both are emitted. In certain instances the alpha or beta rays are accompanied by a third kind of rays, called gamma rays. Alpha particles are really the nuclei of helium atoms; beta particles are electrons; gamma rays are electromagnetic radiations, like X rays.

Both Soddy and Rutherford went on to further scientific triumphs after their epoch-making report of 1902. Soddy received the Nobel Prize (in 1921) for his contributions to the discovery of isotopes; it was he who invented the word in 1913. Isotopes represent two or more forms of the same element; these forms occupy the same place (*isos* means "same" and *topos* means "place" in Greek) in the periodic table of the elements. Chemically, one isotope of a given element cannot be distinguished from the other isotopes of the same element. But isotopes differ from one another in atomic weight and sometimes also in radioactivity. By isolating the radioactive isotopes from the stable forms of an element, scientists have succeeded in producing atomic bombs. They

have also produced tracer elements, which are used in medical, geological and botanical research. More than three hundred isotopes are found in nature.

Rutherford won a Nobel Prize (in 1908) for his work on the disintegration of elements. Later he became the first man to transform one chemical element into another. In 1919 he began bombarding nitrogen gas with alpha particles produced from radium. Now an alpha particle, as we have seen, is really the nucleus of a helium atom; its atomic weight is approximately 4. Rutherford found that a nitrogen nucleus, with atomic weight 14, would trap an alpha particle and would temporarily become an isotope of fluorine. Then the fluorine atom would break up into heavy oxygen (atomic weight 17) and hydrogen (atomic weight 1). In other words, Rutherford had made oxygen and hydrogen out of helium and nitrogen; he had changed one chemical element into another.

Rutherford was the first man, too, to try to picture the inner structure of the atom. It was clear by now that the atom was not the single and indivisible particle that nineteenth-century chemists had imagined. Certain atoms were known to emit various particles and to be changed thereafter into something else — that is, atoms of a different kind. In 1897 the English physicist Joseph John Thomson (1856–1940) had made the first effort to identify the actual composition of atoms. After experimenting with cathode rays, he came to the conclusion that atoms are made up of still smaller particles, negatively charged, which he called "corpuscles of electricity." These "corpuscles," which were later called electrons, were, according to Thomson, the ultimate particles of which all matter was composed. Later, scientists found that this idea was wrong, for other particles, positively charged, were also found in the atom. These positive particles (now known as protons) carry positive charges, equal in magnitude to the negative charges carried by the electrons.

At first, investigators had only the vaguest notion of how these negative and

HENRY G.-J. MOSELEY

positive particles are arranged in the atom. In 1911 Rutherford developed a new and revolutionary theory about atomic structure. He thought of each atom as a kind of miniature solar system. There was a tiny core, or nucleus, made up of positively charged particles, at the center of the atom; this nucleus corresponded to the sun. Around the nucleus revolved the negatively charged electrons, like planets around the sun. According to this concept the atoms — even the atoms of heavy, normally solid elements like iron and copper and gold — consist chiefly of empty space, in which electrical particles ceaselessly whirl.

This picture of the inside of the atom — a miniature solar system of massed protons and orbital electrons — was modified somewhat by the Danish physicist Niels Bohr (1885–1962). Bohr maintained that the electrons whirling around the nucleus of the atom are to be found in shells, suggesting somewhat the successive layers of the skin of an onion. Bohr also pointed out that the electrons give off energy when they jump from one shell to another. He explained many of the obscure events taking place inside the atom in terms of the accepted "classical" system of mechanics worked out by Kepler, Galileo and Newton.

In time it developed that the electron-shell theory could be used to explain the chemical reactions between different elements. It accounted for the firmness with which they grip each other, the speed with which they react with one another (as in explosions) and also their failure to react (as in the case of some of the so-called inert, or noble, gases). The chemical properties of the atom depend upon the number of electrons in its outermost shell. Different kinds of atoms are held together by bonds known as valence bonds. The electron theory of valence was worked out about 1916 by the American chemists Gilbert Newton Lewis (1875–1946) and Irving Langmuir (1881–1957).

How many electrons are whirling around the nucleus of the atom of a given element? This question was answered by the brilliant young English physicist Henry Gwyn-Jeffreys Moseley (1887–1915). Moseley, a pupil of Rutherford, measured the changes in the wave lengths of X rays as they were passed through the crystals of different chemical elements. He came to the conclusion that "there is in the atom a fundamental quantity which increases by regular steps as we pass from one element to the next. This quantity can only be the charge on the atomic nucleus . . . The number of charges increases from atom to atom by a single electronic unit . . . We are led by experiment to the view that . . . the number of charges is the same as the number of the place occupied by the element in the periodic system. This *atomic number* is, then, for hydrogen 1 [No. 1 in the periodic table], for helium 2 [No. 2 in the periodic table], for lithium 3 . . . for zinc 30, etc." The atomic number of a given element gives the number of electrons spinning around the nucleus. Thus hydrogen has one electron, helium 2, lithium 3 and zinc 30.

Moseley enlisted in the British Army in World War I and fell in battle in the disastrous Gallipoli campaign in 1915. Warned by his tragic fate, most of the warring countries in World War II did not allow their promising scientists to enlist for front-line war service.

The large electrostatic generator at the East Pittsburgh Research Laboratories of Westinghouse.

Westinghouse

A colleague of Rutherford, Francis W. Aston (1877–1945), turned his attention to the study of isotopes. How could one separate the different isotopes of a given element from one another? Aston suggested that this might be done by (1) evaporation, since the lighter atoms would escape from a liquid faster than the heavier ones; (2) by the diffusion of a gas through a porous clay barrier; (3) by electromagnetic separation. He used all three methods in separating isotopes.

To bring about the electromagnetic separation of isotopes, Aston was forced to invent a new and, as it turned out, a very important, scientific instrument, called the mass spectrograph. This consists of a tube that produces beams of positive ions (electrified particles) and a magnet that curves the beams. The lighter isotopes are curved more than the heavier ones; hence, each isotope will form its own pile. The piles can be made to fall on photographic plates, where they can be studied.

In 1932 Professor Harold Clayton Urey (born in 1893) of Columbia University made an important discovery by using the mass spectrograph. Working with Ferdinand G. Brickwedde and George M. Murphy, he isolated the hydrogen isotope known as heavy hydrogen. A gallon of liquid hydrogen, prepared by the United States Bureau of Standards, was allowed to evaporate away until only a fraction of an ounce was left. The experimenters then examined the residue in a mass spectrograph. On their photographic plates, they found a spot corresponding to a pile of isotopes with atomic weight 2. (The ordinary hydrogen atom has atomic weight 1.) When heavy hydrogen is combined with oxygen, it makes "heavy water," which has proved to be very useful in biological studies.

In the early 1930's, atomic researchers were faced with a vexing problem. It was agreed that the nucleus is made up of positively charged protons and that as many negatively charged electrons whirl around them in different shells, or orbits. Now the weight of the atom is almost entirely in the nucleus, since an electron is about 1/1838 as heavy as a proton. As each proton has the same weight, the atomic weight of a given element should correspond to the number of protons. The nucleus of hydrogen, the lightest element, contains one proton and its atomic weight is approximately 1. If the nucleus consisted entirely of protons, helium, with 2 protons, should have atomic weight 2; iodine, with 53 protons, should have atomic weight 53; uranium, with 92 protons, should have atomic weight 92.

But what do we find? The atomic weight of helium is 4; of iodine, 127; of uranium, 238. Apparently there must be some other particles in the nucleus to account for the proportionately greater atomic weight of the heavier atoms. In 1932 James Chadwick (born in 1891), another pupil of Rutherford, solved the mystery. He demonstrated that the nucleus of the atom contains, in addition to positively charged protons, neutral particles, which have the same mass as the protons but no electric charge. Because of their neutral charge, Chadwick gave these particles the name of "neutrons."

The total number of protons and neutrons in a nucleus gives the approximate atomic weight of the atom. (The mass of each proton and neutron is assumed to be 1.) Ordinary oxygen has 8 protons and 8 neutrons; its atomic weight is 16. The isotopes of a given element have different numbers of neutrons. Thus uranium 235 has 92 protons and 143 neutrons; uranium 238 has 92 protons and 146 neutrons.

We saw that Rutherford had bombarded nitrogen atoms with alpha particles from radium as atomic bullets and that the end products were oxygen and hydrogen. Alpha particles travel at the speed of some 12,500 miles a second; but they are not nearly fast enough or powerful enough to serve as really effective atomic bullets. In 1932, Drs. J. D. Cockcroft and E. T. S. Walton, working in Rutherford's laboratory, pointed the way to bigger and better atomic bombardments: they designed high-voltage electric generators and vacuum tubes that could withstand such high voltages. With apparatus yielding an output of almost 800,000 electron volts, they transformed lithium and hydrogen into helium, releasing energy in the process.

Even bigger atom-smashers were constructed in the United States. Dr. Robert J. Van de Graaff (born in 1901), working

McGill University

Cyclotron at McGill University, Montreal, Quebec.

first at Princeton and later at the Massachusetts Institute of Technology, designed the electrostatic generator that is pictured and described elsewhere (see Index). Van de Graaff's first machine, which was constructed in 1931, had an output of 1,500,000 electron volts. In later models this voltage was increased many fold.

At the University of California Dr. Ernest Orlando Lawrence (1901–58) constructed a new kind of atom-smasher, called a cyclotron. The principal feature of the cyclotron is a huge electromagnet that keeps whirling atomic bullets (electrons, or protons, or neutrons or atomic nuclei) round and round until they gain tremendous speed and voltage. The original model, built in 1931, produced protons with an energy of 80,000 electron volts; later models were more than 2,000 times more powerful. A diagram of the cyclotron is shown elsewhere (see Index).

With such powerful atom-smashers, scientists were able to perform fresh miracles. They broke the atom into its parts; they converted one atom into another. Two young French scientists — the husband-and-wife team of Irène Joliot-Curie (1897–1956) and Frédéric Joliot-Curie (1900–58) — even succeeded in creating artificial isotopes. Mme. Joliot-Curie was the older daughter of Pierre and Marie Curie; M. Joliot-Curie, whose family name was Joliot before his marriage, had been one of Marie's students at the Sorbonne. Ever since their marriage, in 1926, the two young scientists had worked together on the problem of radioactivity.

In 1934, they bombarded aluminum atoms, using alpha particles as atomic bullets. The end product was a hitherto unknown isotope of phosphorus. It gave off gamma rays, thus showing that it was radioactive; then in the brief space of fifteen minutes it disappeared and all that was left in the experimenters' test tube was silicon. The Joliots had created the first artificial radioactive isotope; in 1935 they received the Nobel Prize for their accomplishment. (In later years their unquestioning acceptance of communist doctrines laid them open to justified criticism.) To-

day radioactive isotopes are produced in quantity in nuclear reactors, or atomic piles; they are used extensively in medical and biological research.

The transmutation of elements under bombardment by atomic bullets went merrily on. Another era of alchemy was at hand — the "newer alchemy," as Rutherford called it. (That, incidentally was the title of his last book, published in 1937.) Finally the dream of the old alchemists — the transmutation of a base metal into gold — was achieved. In 1941 Dr. Kenneth T. Bainbridge of Harvard University and an associate, using a late-model atom-smasher, succeeded in converting mercury into gold. This achievement created little excitement, since the amount of gold obtained in this way was exceedingly small and the cost of producing it was great. It would have been more profitable, from a commercial standpoint, to dig for gold in the ground or even to buy it in the open market!

Thus far the transmutation of atoms had taken place only in the laboratories of scientists, and it had had no particular effect upon the world at large. But now certain researchers turned to a new approach, which was destined to have the most momentous results for all mankind. Scientists had known for some time that if a heavy atom like that of uranium (atomic weight 238) could be completely broken up, an enormous amount of energy would be liberated. Even if this atom could be split in two, the amount of energy released would still be fantastically great. Might not this energy provide explosives of terrifying effectiveness?

The lowering clouds of war in the late 1930's and the actual outbreak of World War II in 1939 lent a new and terrible urgency to this question. Soon the scientists of nazi Germany and of certain Allied powers (particularly the United States, Great Britain and Canada) were secretly engaged in a life-and-death race to develop an atomic bomb, which would derive its titanic power from the smashing of heavy atoms. In a later chapter we shall see what came of this race.

Continued on page 21, Volume 9.

Amer. Mus. of Nat. Hist.